The Emperor's New Mathematics

The Emperor's New Mathematics

Western Learning and Imperial Authority During the Kangxi Reign (1662–1722)

Catherine Jami

Centre National de la Recherche Scientifique
& Université Paris Diderot, Sorbonne Paris Cité

UNIVERSITY PRESS

Great Clarendon Street, Oxford OX2 6DP

Oxford University Press is a department of the University of Oxford.
It furthers the University's objective of excellence in research, scholarship,
and education by publishing worldwide in

Oxford New York

Auckland Cape Town Dar es Salaam Hong Kong Karachi
Kuala Lumpur Madrid Melbourne Mexico City Nairobi
New Delhi Shanghai Taipei Toronto

With offices in

Argentina Austria Brazil Chile Czech Republic France Greece
Guatemala Hungary Italy Japan Poland Portugal Singapore
South Korea Switzerland Thailand Turkey Ukraine Vietnam

Oxford is a registered trade mark of Oxford University Press
in the UK and in certain other countries

Published in the United States
by Oxford University Press Inc., New York

© Catherine Jami 2012

The moral rights of the author have been asserted
Database right Oxford University Press (maker)

First published 2012

All rights reserved. No part of this publication may be reproduced,
stored in a retrieval system, or transmitted, in any form or by any means,
without the prior permission in writing of Oxford University Press,
or as expressly permitted by law, or under terms agreed with the appropriate
reprographics rights organization. Enquiries concerning reproduction
outside the scope of the above should be sent to the Rights Department,
Oxford University Press, at the address above

You must not circulate this book in any other binding or cover
and you must impose this same condition on any acquirer

British Library Cataloguing in Publication Data

Data available

Library of Congress Cataloging in Publication Data

Data available

Typeset by SPI Publisher Services, Pondicherry, India
Printed and bound by
CPI Group (UK) Ltd, Croydon, CR0 4YY

ISBN 978–0–19–960140–0

1 3 5 7 9 10 8 6 4 2

PREFACE

In recent years, the need to think of the world as a set of connected entities has increasingly been felt within and beyond academic circles. At the same time, a new awareness of the crucial role of science and technology in the present and future of human societies has fostered their inclusion into historical narratives. While the ambition of this book is to shed light on one episode of the circulation of science within and across cultures, in today's context its subject is relevant to intellectual concerns beyond the circles of specialists in the history of science and of Chinese studies. Accordingly, I hope it will contribute to making China better known to specialists of other regions of the world, and mathematics better perceived as a historical object.

Bearing in mind this possible wider readership, I have striven to make this book readable beyond the colleagues who master the sources that I have used, without denying these colleagues the means to check these sources for themselves. Therefore I have generally chosen to translate technical terms and book titles from Chinese (with the exception of length and weight units), while giving the original in both transcription and Chinese characters at the first occurrence in each chapter, and in the index. I use the *pinyin* transcription of Chinese characters, and follow the East Asian order when mentioning any person from that part of the world: family name followed by given name. If however a modern author has adopted a non-standard representation of his or her name when writing in Western languages, I follow the author's practice. In Chinese sources Westerners are referred to by their Chinese names: in my translations I give their original name, and I refer to Chinese people by their given name (*ming* 名) rather than by their style (*zi* 字). As far as is practical I have added brief explanations and references for those who are unfamiliar with China and its history. I have been highly selective in choosing which mathematical materials to discuss in detail. My aim has only been to provide reasonably accessible evidence and examples for my arguments, rather than to convince the reader by the sheer weight of evidence that China has a rich mathematical tradition. Many publications, a number of which are found in the bibliography at the end of this book, have already established this latter point beyond doubt.

Over the years spent working on this book, I have accumulated many debts of gratitude, which I take pleasure in acknowledging here. My position as a full time researcher at the French CNRS (*Centre national de la recherche scientifique*), and a member of REHSEIS (*Recherches épistémologiques et historiques sur les sciences exactes et les institutions scientifiques*), has enabled me to concentrate on this project for a number of years, and to travel as much as was required to gather the sources used here. The idea of this book first emerged in 2002, during my term as member of the School of Historical Studies of the Institute of Advanced Studies (Princeton, New Jersey). The Needham Research Institute (Cambridge, UK) hosted me during the years it took to complete the task of writing; for two of these years I was a French Government Fellow at Churchill College.

Within and outside these institutions, many colleagues have helped me access sources, provided information and read some passages in draft. They include An Shuangcheng 安雙成 (First Historical Archive, Beijing), Davor Antonucci (University of Rome La Sapienza), Alain Arrault (Ecole Française d'Extrême Orient, Paris), Timothy Barrett (School of Oriental and African Studies, London), Chu Pingyi 祝平一 (Academia Sinica, Taipei), François De Gandt (University of Lille III), Ad Dudink (Catholic University of Leuven), Peter Engelfriet (Amsterdam), Feng Lisheng 馮立昇 (Qinghua University, Beijing), Noël Golvers (Catholic University of Leuven), Han Qi 韓琦 (Institute for the History of Natural Sciences, Beijing), Hashimoto Keizō 橋本敬造 (Kansai University, Osaka), Hu Minghui (University of California, Santa Cruz), Huang Yi-Long 黃一農 (Tsing-Hua University, Hsinchu), Jacques Jami (Paris), Léna Jami (Paris), Jun Yong Hoon 全勇勳 (Kyujanggak Institute for Korean Studies, Seoul National University), Isabelle Landry-Deron (Ecole des Hautes Etudes en Sciences Sociales, Paris), Lee Chia-hua 李佳嬅 (University of Tokyo), Liu Dun 劉鈍 (Institute for the History of Natural Sciences, Beijing), Rui Magone (Max Planck Institut for the History of Science, Berlin), Kenneth Manders (University of Pittsburgh), John Moffett (Needham Research Institute, Cambridge, UK), Michèle Pirazzoli-t'Serstevens (Ecole Pratique des Hautes Etudes, Paris), Qu Anjing 曲安京 (Northwest University, Xi'an), Shi Yunli 石云里 (University of Science and Technology of China, Hefei), Sun Chengsheng 孫承晟 (Institute for the History of Natural Sciences, Beijing), Tegus 特古斯 (Inner Mongolia Normal University, Hohhot), Takeda Tokimasa 武田時昌 (University of Kyoto), Tian Miao 田淼 (Institute for the History of Natural Sciences, Beijing), Watanabe Junsei 度辺純成 (Gakugei University, Tokyo), Yoshida Tadashi 吉田忠 (University of Tohoku, Sendai), and Yuan Min 袁敏 (Northwest University, Xi'an).

Particular thanks are due to Karine Chemla (CNRS, Paris), Annick Horiuchi (University of Paris-Diderot), Antonella Romano (European University Institute, Florence), Simon Schaffer (University of Cambridge), Nicolas Standaert (Catholic University of Leuven) and Pierre-Etienne Will (Collège de France) for their careful reading of and encouraging comments on an earlier version of this book.

But my first and most generous reader was without doubt Christopher Cullen (Needham Research Institute, Cambridge, UK), with whom I discussed most aspects of this book.

Having acknowledged the help of so many colleagues, I should add that I take sole responsibility for any mistakes that the present work may contain.

Last but not least, this book is dedicated to my daughter Chloé, who, at the age of seven, thought that Chapter 6 made a nice bedtime story.

Catherine Jami

Paris, September 2010

CONTENTS

List of Boxes, Tables, Charts and Illustrations — xiii

Introduction — 1

Part I Western learning and the Ming–Qing transition

Chapter 1 The Jesuits and mathematics in China, 1582–1644 — 13
    1.1 Mathematics and literati culture in China c.1600 — 14
    1.2 Mathematics in the Society of Jesus — 22
    1.3 Teaching and translating — 24
    1.4 Jesuit science, 'practical learning' and astronomical reform — 31

Chapter 2 Western learning under the new dynasty (1644–1666) — 35
    2.1 Dynastic transition in Beijing: a new calendar for new rulers — 36
    2.2 Adam Schall, imperial astronomer — 38
    2.3 Jiangnan scholars and the Jesuits — 41
    2.4 *Number and magnitude expanded*: a scholar's mathematics — 44
    2.5 Schall's defeat: the 1664 impeachment — 49

Part II The first two decades of Kangxi's rule

Chapter 3 The emperor and his astronomer (1668–1688) — 57
    3.1 Kangxi's takeover and the rehabilitation of Jesuit astronomy — 57
    3.2 Ferdinand Verbiest, imperial astronomer and tutor — 65
    3.3 Kangxi, student of Chinese and Western learning — 73
    3.4 Philosophy, 'fathoming the principles' and orthodoxy — 78

Chapter 4 A mathematical scholar in Jiangnan: the first half-life of Mei Wending — 82
    4.1 Mei Wending's early career — 83
    4.2 Integrating Chinese and Western mathematics — 86
    4.3 The *Discussion of rectangular arrays*: restoring one of the 'nine reckonings' — 90
    4.4 Writing mathematics: purpose, structure and style of the *Discussion of rectangular arrays* — 93

Chapter 5 The 'King's Mathematicians': a French Jesuit
mission in China 102
 5.1 Setting up a scientific expedition to China 102
 5.2 From Brest to Beijing 108
 5.3 In the capital 112
 5.4 Travels and observations in China 116

Chapter 6 Inspecting the southern sky: Kangxi at the
Nanjing observatory 120
 6.1 The *Imperial Diary* 121
 6.2 Li Guangdi's recollection 127
 6.3 The Jesuits' role 131
 6.4 Imperial investigation of the Old Man Star 133

Part III Mathematics for the emperor

Chapter 7 Teaching 'French science' at the court: Gerbillon
and Bouvet's tutoring 139
 7.1 Chronology of the lessons 141
 7.2 The *Académie*, the Moderns and the way to God 144
 7.3 A typical lesson: 10 April 1690 148
 7.4 The imperial workshop and instruments 151
 7.5 Chinese, Manchu and the control of Western
 science 156

Chapter 8 The imperial road to geometry: new *Elements of
geometry* 160
 8.1 Changing the textbook 160
 8.2 Pardies' *Elemens de geometrie*: the pedagogy of
 geometry in seventeenth-century France 162
 8.3 The double translation 166
 8.4 Diamonds and pearls: ratios in the new *Elements* 169
 8.5 Practical geometry 173
 8.6 The emperor's role in the composition of
 the new *Elements* 176

Chapter 9 Calculation for the emperor: the writings of
a discreet mathematician 180
 9.1 Antoine Thomas (1644–1709) and his *Synopsis
 mathematica* 180
 9.2 The *Outline of the essentials of calculation* 184
 9.3 Practical geometry and tables 191
 9.4 The foundations of calculation: back to Euclid 195
 9.5 Cossic algebra: the *Calculation by borrowed root
 and powers* and its summary 200
 9.6 Symbolic linear algebra: a treatise within the
 treatise 210

Chapter 10 Astronomy in the capital (1689–1693): scholars,
            officials and ruler   214
    10.1 Mei Wending in Beijing   214
    10.2 A patron's commission: the *Doubts concerning
          the study of astronomy*   218
    10.3 Three solar eclipses   222
    10.4 An imperial pronouncement on mathematics
          and classical scholarship   229
    10.5 The officials' responses   233

Part IV Turning to Chinese scholars and Bannermen

Chapter 11 The 1700s: reversal of alliance?   239
    11.1 Locating the Beijing Jesuits *c.*1700   240
    11.2 The mathematical sciences in history   245
    11.3 From favour to distrust: the papal legation   253
    11.4 The Jesuits as imperial cartographers   255
    11.5 Assessing mathematical talent: Chen Houyao's
          interview   257

Chapter 12 The Office of Mathematics: foundation and staff   260
    12.1 The Summer solstice of 1711   260
    12.2 Selecting talented men   262
    12.3 The mathematical staff at the emperor's
          sixtieth birthday   265
    12.4 An editorial project in the mathematical
          sciences   267
    12.5 The imperial princes   273
    12.6 Mathematicians, astronomy and cartography   277
    12.7 The mathematical sciences in examinations   280

Chapter 13 The Jesuits and innovation in imperial science:
            Jean-François Foucquet's treatises   284
    13.1 Foucquet in Beijing: from the *Book of Change* to
          astronomy   284
    13.2 The *Dialogue on astronomical methods*:
          a reform proposal?   287
    13.3 Foucquet's writings on mathematics   294
    13.4 Symbols in the *New method of algebra*   300
    13.5 The Jesuits and the Office of Mathematics   305
    13.6 Tables and the standardisation of
          mathematical sciences   309

Part V Mathematics for the Empire

Chapter 14 The construction of the *Essence of numbers and their principles* — 315
    14.1 Outline — 315
    14.2 Sources — 320
    14.3 The historical narrative of mathematics — 323
    14.4 The structure of imperial mathematics — 327
    14.5 Rephrasing the Jesuits' textbooks — 330
    14.6 Vocabulary and classification — 333

Chapter 15 Methods and material culture in the *Essence of numbers and their principles* — 340
    15.1 Problems and their genealogy: Master Sun and Hieron — 341
    15.2 Inkstones and brushes — 345
    15.3 Weighing up the difficulty of problems — 348
    15.4 Algebra and the problems of the 'Line Section' — 352
    15.5 The remainder problem — 354
    15.6 The construction and use of instruments — 356
    15.7 Time-keeping — 358

Chapter 16 A new mathematical classic? — 364
    16.1 Yongzheng's preface — 364
    16.2 Harmonics and astronomy — 368
    16.3 The contributors — 373
    16.4 Supplements to the *Origins of pitchpipes and the calendar* — 378
    16.5 The study of mathematics in mid-Qing China — 382

Conclusion — 385

Main Units — 393

Bibliography — 395
    1 Mathematical and astronomical manuscripts from the Kangxi court — 395
    2 Editions used for main other Chinese works on mathematics, astronomy and harmonics — 397
    3 Other sources — 397
    4 Secondary literature — 400

Index — 421

LIST OF BOXES, TABLES, CHARTS AND ILLUSTRATIONS

Box 1.1	Calculating devices in imperial China: counting rods and the abacus	14
Table 1.2	The *Five Classics* and *Four Books*	16
Table 1.3	The 'nine chapters' and Cheng Dawei's *Unified lineage of mathematical methods*	19
Box 1.4	Layout for the inkstones and brushes problem in the *Unified lineage of mathematical methods*	21
Box 1.5	Layout for the inkstones and brushes problem in the *Instructions for calculations in common script*	28
Chart 2.1	The Astronomical Bureau in 1649	39
Table 2.2	Structure of *Number and magnitude expanded*	46
Box 2.3	Layout for the brushes and inkstones problem in *Number and magnitude expanded*	48
Box 4.1	Layout of basic operations in *Brush calculation*	89
Box 4.2	Mei Wending's classification of the nine reckonings ('nine chapters')	91
Box 4.3	First layout for the inkstones and brushes problem in the *Discussion of rectangular arrays*	98
Box 4.4	Correspondence between phrasings of the inkstones and brushes problem and layouts in the *Discussion of rectangular arrays*	100
Illustration 5.1	Frontispiece of Tachard's *Voyage de Siam*	108
Illustration 5.2	The King of Siam and the French Jesuits observe the lunar eclipse of 11 December 1685	109
Illustration 5.3	The Beijing Observatory	114
Table 8.1	Table of contents of Pardies' *Elemens de géométrie*	165
Illustration 8.2	Constructing a circle, *Elemens de géométrie*	166
Illustration 8.3	Constructing the midpoint of a line	175
Illustration 8.4	Preface to the *Elements of geometry* manuscript	178
Table 9.1	Contents of the *Outline of the essentials of calculation* compared with those of its possible sources	186
Illustration 9.2	Opening paragraph of the *Elements of calculation*	197
Box 9.3	Cross-reference among the four main treatises	203
Table 9.4	Contents of the *Calculation by borrowed root and powers*	204
Table 9.5	Table of power numbers (transcription and translation), *Calculation by borrowed root and powers*	207

Table 9.6	Abbreviations for the names of powers in Clavius' *Algebra*	208
Table 9.7	Terminology of cossic algebra in Latin and in Thomas' Chinese treatise	208
Box 9.8	Equation and transcription	209
Box 9.9	Layout summarising the solution of a problem	210
Box 9.10	Layout of the linear problem in seven unknowns using symbolic notation	212
Map 11.1	The Jesuits in Beijing *c.*1700	242
Illustration 12.1	Use of a level as illustrated in the *Practice of instruments for measuring heights and distances*	267
Table 12.2	Manchu and Han staff serving in mathematics	268
Box 13.1	Foucquet's solution of the problem set by the emperor	295
Box 13.2	Numerical and literal multiplication in the *New method of algebra*	301
Illustration 13.3	Equations, *New method of algebra*	303
Table 14.1	Contents of the *Essence of numbers and their principles*	317
Illustration 14.2	River Diagram and Luo Writing, *Essence of numbers and their principles*	323
Box 14.3	Diagram summarising the solution of the problem of the two men walking towards each other	337
Box 14.4	Cross-reference in the *Essence of numbers and their principles*	339
Box 15.1	First layout for the 'chickens and rabbits in the same cage' problem in the *Essence of numbers and their principles*	342
Box 15.2	Layout using borrowed root and powers for the 'chickens and rabbits in the same cage' problem in the *Essence of numbers and their principles*	343
Table 15.3	The 'chickens and rabbits in the same cage' problem in various mathematical works	344
Box 15.4	Layout for the inkstones and brushes problem in the *Essence of numbers and their principles*	346
Table 15.5	The inkstones and brushes problem in some Ming and early Qing mathematical works	347
Illustration 15.6	Illustration of the copper scales problem in the *Outline of the essentials of calculation* and in the *Essence of numbers and their principles*	349

Illustration 15.7	Illustration of the wooden stick scale problem, *Essence of numbers and their principles*	351
Illustration 15.8	Proportional compass in the *Outline of the essentials of calculation* and in the *Essence of numbers and their principles*	358
Illustration 15.9	Measuring the flow of a river, *Explanation of the proportional compass*	361
Table 16.1	The contributors to the *Origins of pitchpipes and the calendar*	373
Illustration 16.2	List of contributors to the *Origins of pitchpipes and the calendar*	374

Introduction

In 1669, at the age of fifteen, the Kangxi emperor (b. 1654, r. 1662–1722) overthrew the regent Oboi 鰲拜 and inaugurated his personal rule. This Manchu youth thus took charge of a large and still unstable multi-ethnic empire, of which the former Ming empire conquered in 1644 and mainly inhabited by Han 漢 Chinese was only one part.[1] Despite his lack of experience, Kangxi soon showed himself capable of effective and decisive action, particularly in dealing with the major military threats to his rule from those areas of China not yet under effective control of the Qing dynasty (1644–1911). Here he followed in the tradition of his Manchu forebears, for whom prowess in battle, riding and hunting were essential accomplishments for a ruler. It may therefore seem a little surprising that one of the young emperor's first executive decisions concerned the functioning of the Astronomical Bureau (*Qintianjian* 欽天監).

There his personal intervention resulted in a return to the 'Western methods' (*xifa* 西法) which had been banned for the past five years. This close involvement with mathematical and astronomical matters was to be a striking characteristic of his reign. One of his last projects, completed after his death, was the compilation of a major book that was intended to set imperial standards in mathematics, astronomy and musical theory, the *Origins of pitchpipes and the calendar imperially composed* (*Yuzhi lüli yuanyuan* 御製律曆淵源, 1723). As this title suggests, Kangxi had closely supervised its compilation. Between these two landmarks, he remained an eager student of the mathematical sciences for more than half a century. History has few examples of rulers of major states with such interests, pursued with such energy and determination.[2] Kangxi's contemporaries where quite aware of how unusual all this was: but while European elites were enthusiastic in their reaction to the descriptions they read, the upper ranks of Qing

---

1 The term 'Han' is still used today to refer to the majority ethnic group of the People's Republic of China, whose territory includes most of that of the former Qing empire. In this book I follow the general practice of historians writing in European languages in referring to the Han, as opposed to their Manchu rulers, as 'Chinese', using 'Han' when greater precision is appropriate.
2 Two contemporaries of Kangxi had comparable interests: Peter I of Russia (Gouzévitch & Gouzévitch 2008) and Jai Singh (Sharma 1995).

society were more ambivalent. Both groups were influenced by the same fact, though it affected them in opposite ways: the emperor's tutors in the sciences were Jesuit missionaries.

Why would the ruler of the most populous empire in the world devote so much time and energy to an area of study then regarded by most Chinese literati as merely technical, and strive to make it one of the central concerns of his educated subjects? The answer given to this question by some of Kangxi's Jesuit tutors is that he had accepted the truth and superiority of the European sciences they taught—which for them was a major step towards accepting the truth of the Catholic religion. The Jesuit account of China and of the importance of Jesuit services to the Qing dynasty has shaped Western as well as Chinese historiography. Thus, in the same way that the Jesuits saw it as a failure that the emperor did not convert to Christianity, the fact that he did not enforce the adoption of Western science more radically in the empire has long been regarded by historians as a missed opportunity and a regrettable failure of cross-cultural transmission.[3]

However, recent developments in the various fields to which this book pertains have made it possible to look at this topic anew. For one thing, the main terms involved in defining the subject matter of the present book, including 'China', 'Europe', 'science' and 'cross-cultural transmission', have come under close scrutiny over the past decades, either one by one or in relation to one another.

Whereas China has long been represented as an entity that remained immutable over the millennia spanned by its history, in recent decades the Qing period, among others, has been envisaged anew. Until then only a few specialists in the Manchu language, who published mainly in German and in Japanese, had emphasised the importance of Manchu sources for Chinese history.[4] Taking their point on board, and relying on their work, more and more historians of late imperial China have turned to the Manchu documents in the archives. Combining these with sources in Chinese, they have proposed a new vision of the Qing empire, pointing out that what used to be Ming China only occupied one third of its territory, and that it conquered and absorbed several major central Asian entities. They have argued that Qing rulers aptly constructed different images of themselves to best match the cultures of the various people they controlled, and that only a particular facet of their multiform rulership is visible through Chinese sources.[5] History of science can and should be integrated into this new understanding of

---

3 See e.g. Xi 2000.
4 For a bibliography, see Stary 1990 & 2003b.
5 On 'new Qing history', see Waley-Cohen 2004; on the imperial expansion of the Qing, see Perdue 2005.

the Qing dynasty.⁶ And at the same time, the Qing empire needs to be fully integrated into the narrative of science and empires in the early modern age—a narrative so far mostly concerned with the overseas empires established by European powers. The present work is a contribution to the study of science in the only empire constructed in that age which has defined a geopolitical entity that has changed very little to the present day, albeit under a different name—the People's Republic of China.

On the other hand, 'Europe' has been historicised in the context of its contacts with East Asia. 'Christian', 'Aristotelian' and 'scientific', taken together or separately, no longer seem to describe the actors of European overseas expansion in a satisfactory way. In particular recent research on the Society of Jesus gives us essential insight in understanding the culture of the missionaries who worked in China in the seventeenth and eighteenth centuries.⁷ Furthermore, the great diversity of their itineraries within Europe, on the way to Asia, and finally within China, forces us to a very close scrutiny of the various elements that made up Western learning (*xixue* 西學), as their teachings came to be called in the early and mid-Qing period; all the more so as the intellectual cultures of Europe themselves underwent radical transformations during the two centuries spanned by the Jesuit mission of China.

Joseph Needham, whose monumental work has made Western academia aware of the existence of a wealthy tradition of science in pre-modern China, has argued that, while the Jesuits' transmission of astronomy to China was 'imperfect', the Chinese understood that this knowledge was 'new' rather than 'Western' or 'Jesuit', and that Chinese astronomy then began its integration into modern science, which alone is universal.⁸ Today, we approach the same story by trying to account for what did happen, rather than what did not. We no longer assume that science circulates on the sole strength of its universal truth, but on the contrary that it is the very circulation of knowledge that makes it universal. Neither should we assume that all human beings have striven towards one single modernity with more or less success and that civilisations and states can meaningfully be assessed comparatively, or even competitively, according to their success in this enterprise.⁹ Whereas we can identify 'new', 'modern', 'Western' or 'Jesuit' as categories used by the actors that we study, we are cautious about

---

6 Hanson 2006, Jami 2010.
7 O'Malley *et al.* 1999 & 2006 give a glimpse of the abundance and variety of 'Jesuit studies'.
8 Needham 1954–, 3: 442–458.
9 Benjamin Elman still retains such a framework of analysis (Elman 2005, xxxviii); see Jami 2006b, 406–407.

appropriating these adjectives, fraught with value-judgement as they are, to characterise the knowledge that circulated.

The circulation of science is but one of many aspects of contacts between cultures. In his reflections on the study of cross-cultural contacts, Nicolas Standaert has distinguished four frameworks of analysis that characterise the historiography: 'transmission', which focuses on the message transmitted; 'reception', which centres on the receiver; 'invention', which assumes that the sources give us access only to the transmitter's discourse on the receiver, and finally 'interaction and communication', which posits that both transmitter and receiver are changed in the process of cross-cultural contact. This last framework, which has emerged more recently, bears strong resemblance to the notion of circulation as it has recently become widespread in the recent historiography of science: it is understood as a complex if not always symmetrical process that shapes both the actors and the message.[10]

In short, this book relies on and hopes to further two major changes in the historiography. First, science is no longer seen as an immutable body of ideas that wins assent through its obvious truth (as missionaries once believed that the Christian religion ought to do). Instead, one has to account for the circumstances that gave rise to the circulation of knowledge and for the ways in which knowledge was shaped by this very circulation process. Secondly, this process cannot simply be reduced to a dichotomy opposing China to Europe. Rather, it is necessary to locate the various actors of 'Western learning', both Chinese and European, in order to understand its diverse content and the different stakes behind it.

These actors, whether they were missionaries, Chinese literati, Manchu conquerors or French academicians, worked and established their position within various networks.

Several of the networks involving Western learning centred on Kangxi, whether as an active agent or as a centre of attention. One aim of this book is therefore to ask what Kangxi himself sought to achieve, not only by using Western learning as a tool for statecraft, but also by co-opting its mathematical content into state-sponsored scholarship. Before going any further, however, it is necessary to specify what is meant by 'mathematics' in this historical context: the category itself, far from being immutable, was the object of several reconstructions in the period that we shall consider. The ancient Chinese category *suan* 算, which seems to have referred originally to reckoning using counting rods, was sufficiently well defined to be a subject of teaching in imperial institutions during the Tang dynasty (618–907); in the middle of the seventh century, some pre-existing works were canonised as

---

10   Standaert 2002.

the *Ten mathematical classics* (*Suanjing shishu* 算經十書). Up to 1600, it seems reasonable to equate the subject matter that historians of mathematics typically study with the techniques of *suan*. However, the Jesuits' introduction of the duality between number and magnitude reconfigured the field, so that to some scholars *suan* came to refer only to the processing of numbers (*shu* 數), while the study of magnitudes (*du* 度) was reserved for Euclidean geometry; one of the purposes of this book is to shed light on further reconfigurations of mathematics proposed during the Kangxi reign. In the European scholastic tradition, mathematics, as codified in the *quadrivium*, encompassed not only arithmetic and geometry, but also astronomy and music.[11] On the other hand, there was a long tradition of associating these last two disciplines in China, as evidenced for example by the inclusion of a 'Monograph on pitchpipes and the calendar' (*Lü li zhi* 律曆志) in the *Book of the Han* (*Hanshu* 漢書, the dynastic history of the Western Han dynasty, (206 BCE–24 CE, compiled at the end of the first century CE); both disciplines were of close concern to the imperial state, as they were associated with ritual and cosmological order. The association of harmonics and astronomy is a point of convergence of the Chinese and European traditions. The addition of mathematics to them in the *Origins of pitchpipes and the calendar*, however, was unprecedented. Although it was not uncommon for mathematics and astronomy to be practised by the same persons (some of whom were official astronomers), when Kangxi commissioned his three-fold compendium, it had been more than a millennium since the imperial state had endorsed a body of mathematical writings. And never before had this state found it necessary to issue a definition of mathematics by producing a treatise on the subject.

The treatise in question, the *Essence of numbers and their principles imperially composed* (*Yuzhi shuli jingyun* 御製數理精蘊), appears to have been the largest mathematical work ever printed in imperial China. It is also one of the least studied by historians of mathematics. For it meets neither of the two criteria that have long presided over the historiography of this field: it is not ancient, and it does not seem to include 'anything new', i.e. anything that is not found elsewhere in the world mathematical literature of its time. The present book stems from questions about this lengthy and seemingly unremarkable treatise: what were the circumstances of its compilation? What was the mathematics that it constructed? What underlay the claim that it was 'imperially composed'? It is hardly surprising that answering these questions

---

11 The list of disciplines included in 'the mathematical sciences' practised and taught by the Jesuits who worked for Kangxi is much longer; see pp. 71–72.

involves investigating the politics of learning as well as studying mathematical problems and their solutions.

Starting from questions usually thought to belong in the history of science, the research for this book has led me to venture into the fields of cultural, political and social history, and to call upon the sources and methods of 'Chinese studies'. I have striven to combine these tools in order to write the story of how mathematics, a discipline often regarded as the touchstone of both modernity and universality in the sciences, came to be appropriated as part of the construction of rulership by the Qing dynasty, a dynasty to which historiography has long denied participation in either modernity or universality.

In order to achieve this aim, the present book follows the story of Kangxi's engagement with mathematics; it is divided into five parts of two to four chapters each. The first part provides the necessary historical background. Chapter 1 outlines the beginnings of Western learning in China during the years 1582 to 1644, the last six decades of the Ming dynasty; it discusses the Jesuits' teaching of mathematics in China during that period, and the translations that resulted from their work. The most famous of these is the *Jihe yuanben* 幾何原本 (1607), a rendering into Chinese of the first six books of Euclid's *Elements of geometry*. One of the reasons for the success of the Jesuits' teaching was the perceived relevance of their mathematical knowledge to statecraft. In 1629, some of them were employed to work on a calendar reform, the need for which had been felt for almost half a century. Chapter 2 covers the first two decades of Qing rule (1644–1664): by presenting the result of the Jesuits' work on the calendar to the Manchus, Johann Adam Schall von Bell provided the latter with an important token of legitimacy, while making a niche for himself in the Chinese civil service. Schall's directorship of the Astronomical Bureau, however, gave rise to multiple conflicts; these eventually led to his impeachment, to his condemnation and to that of some of his collaborators, and to the proscription of almost all missionaries.

In the second part of the book, we turn to the first two decades of Kangxi's personal rule, from 1669 to 1689. Chapter 3 opens with his acts on the assumption of power, including the reversal of the verdict against the Jesuits and their astronomy. Ferdinand Verbiest then became an official astronomer, but also an imperial tutor as Kangxi began the study of Western science. This study was integrated in the traditional pattern of imperial study of the Classics, and adjusted to the emperor's special needs. Meanwhile, outside the capital, some Chinese literati also pursued the study of mathematics and astronomy. One of them, Mei Wending 梅文鼎, slowly rose to prominence as a scholar specialising in these fields; Chapter 4 recounts his early career and analyses some aspects of his mathematical work. Court mathematics, however, was to be shaped from much more remote sources: in 1685, six French Jesuits sent by King Louis XIV as 'his mathematicians' set out for China. Chapter 5 discusses their enterprise, which was both a scientific expedition and an evangelisation mission. Five of them reached Beijing in 1688, bringing with them a scientific culture closer to that of the Paris *Académie royale des sciences* than to the tradition of Jesuit colleges, on which their China confreres had previously relied. Two of the Frenchmen entered the emperor's service in the Great Interior. The following year, Kangxi visited the Nanjing Observatory during his Southern tour. This visit, discussed in Chapter 6, gives a striking illustration of his skill in using what was effectively a monopoly on the Jesuits' astronomical knowledge to assert his authority vis-à-vis the most distinguished Chinese scholars who served him as high officials.

Part III is devoted to the tutoring in mathematics that Kangxi received in the 1690s, when his Jesuit teachers composed a number of textbooks for his use. Chapter 7 reconstructs the study plan worked out by the French Jesuits, and the course of the daily lessons, in which mathematical instruments and their use played an important role. Chapter 8 focuses on these French Jesuits' work as translators of a recent geometry textbook, which was substituted for the 1607 translation of Euclid in teaching Kangxi. Significantly, this book was first translated from French into Manchu and then from Manchu into Chinese. Chapter 9 discusses a series of lecture notes, most of which can be ascribed to Antoine Thomas, on calculation, algebra and other topics. Thomas, who had published a mathematical textbook in Europe, seems to have been by far the most prolific Jesuit mathematician at court. As was the case with classical learning, the lecture notes produced by the emperor's tutors were treated as imperial writings. As shown in Chapter 10, this enabled Kangxi to present himself as a teacher of the mathematical sciences vis-à-vis his high officials, while continuing to dismiss Chinese scholars' knowledge of these disciplines.

Part IV shows how the emperor gradually revised this assessment, and eventually recruited some Chinese scholars as editors of treatises on mathematics, music, and astronomy. Chapter 11 recounts how, as the atmosphere of the court grew tenser while the issue of his succession loomed, Kangxi began to distance himself ostensibly from the Jesuits who worked for him. He gave audience to Mei Wending in 1705, and started to look for mathematical talents among the literati. Chapter 12 discusses the creation, in 1713, of an Office of Mathematics (*Suanxue guan* 算學館) where a group of selected young scholars, both Chinese and Bannermen,[12] started to work on the treatises commissioned by the emperor, under the supervision of Kangxi's

---

12 The Manchu, Mongol and 'Chinese military' (*hanjun* 漢軍) members of the Eight Banners (*baqi* 八旗), the units into which the conquerors of China and their allies had been organised since the early seventeenth century.

third son, Prince Yinzhi 胤祉, who in turn reported to his father. In the meantime, however, the emperor continued to employ Jesuits as astronomers and cartographers, and commissioned them to write a treatise on astronomy similar to those on mathematics produced in the 1690s. Chapter 13 focuses on Jean-François Foucquet, the French Jesuit in charge of drafting the new astronomical treatise, and analyses the conflicts that arose among his confreres following his proposal to revise the astronomical constants in use at the Astronomical Bureau. These conflicts prompted Yinzhi and the staff of the Office of Mathematics to further assert their autonomy vis-à-vis the Jesuits.

The final part focuses on the *Essence of numbers and their principles*. Chapter 14 shows how the treatise's introductory chapter places it within a tradition of learning going back to the mythical founders of the Chinese civilisation. While most of the treatise's content was derived from the lecture notes produced by the Jesuits in the 1690s, these were rewritten within the structure of a unified field of mathematics. Chapter 15 analyses some of the problems found in the treatise, bringing out the material culture of imperial mathematics, and showing how the erasure of the history of individual problems allowed the construction of a body of universal knowledge. Finally, Chapter 16 discusses the circumstances that surrounded the completion of the *Origins of pitchpipes and the calendar imperially composed* and outlines the subsequent fate of each of its three components. The publication of the *Essence of numbers and of their principles* was the final outcome of Kangxi's lifelong pursuit of mathematics; it also marked the end of a unique story of imperial engagement with this discipline.

While tracking the various actors of this complex story as closely as the historical material available allows, I have had to make some choices as to the samples of the mathematical sciences to be proposed to the reader's attention. First, the present book focuses in many places on aspects of calculation (*suan* 算), rather than on geometry, a topic which has so far been something of an obsession with both Chinese and Western historians of mathematics in China during this period. That obsession undoubtedly reflects the Jesuits' conviction that the Euclidean geometry they introduced to China was central to their mathematical projects, as well as being radically novel. Qing intellectuals and their ruler formed a somewhat different view, as we shall see. As far as calculation is concerned, we have to deal with a field that is best understood as an area of synthesis rather than as a straightforward reflection of the introduction of European methods into China. That is not to say that only calculation is relevant to our story; I simply propose to cast light on its hitherto obscure place in the landscape of mathematics and of Western learning in early Qing China. Secondly, within the field of calculation, the focus is often narrowed down to what are nowadays described as linear problems in several unknowns.

This choice is dictated by the fact that the mathematical manuscripts written by the Jesuits for Kangxi, which are the most substantial new historical material uncovered during the research for this book, contain successive proposals for solving such problems. Such proposals were effectively in competition with those put forward by Chinese mathematicians, so that calculation was a field that was both common to all actors and highly contested. To highlight this theme of calculation, problems that appear repeatedly during the period under discussion have been traced through their successive occurrences in the literature: in particular, a problem on the trading of brushes against inkstones and another on a cage containing chickens and rabbits will accompany the reader along the way. They serve as reminders that mathematics is a culturally situated practice wherever it is practised, but also that it can be a pleasurable occupation worth a detailed scrutiny rather than a subject of awe as well as dislike, to be contemplated only from a safe distance.

Admittedly, Kangxi never mentioned pleasure as a reason for studying mathematics—or for doing anything else for that matter; but had he not enjoyed it, would he have spent so much time studying it? His lifelong engagement with mathematics and his endeavour to integrate it into scholarship, both abundantly documented, allow some insight into the place of that discipline in Qing China. In this way the book aims at contributing both to a history of science that spans all human civilisations and to a better understanding of the society and culture of late imperial China.

PART I

Western learning and the Ming–Qing transition

CHAPTER I

The Jesuits and mathematics in China, 1582–1644

The story of Western learning (*xixue* 西學) in China begins where Joseph Needham's account of the mathematical sciences in *Science and Civilisation in China* closes, that is, when Jesuit missionaries entered China at the end of the sixteenth century.[1] For the whole duration of their presence (1582–1773), they put their science in the service of evangelisation: the knowledge and know-how that they displayed enhanced the prestige of their religion and served to attract the patronage of officials, as well as that of the imperial state.

The Jesuits' emphasis on science as a tool for proselytisation seems to be unique both among the missionary orders present in China in the seventeenth and eighteenth centuries,[2] and indeed among Jesuit missions around the world at the time.[3] In fact it could be argued that the Jesuits' science had a much more pervasive influence on China than their religion. Whereas Christianity remained a minority, and even marginal religion,[4] Western learning was known to all Chinese scholars interested in the mathematical sciences by the late seventeenth and eighteenth century, whatever their attitude towards it might have been. On the other hand, most Jesuit missionaries devoted their time and effort solely to evangelisation,[5] and only a few 'specialists' among them taught and practised the sciences.

Two factors contributed to shaping Jesuit science in China: on the one hand, the importance that the Society of Jesus gave to mathematics (in the usual sense of this term in early modern Europe)—what we might call the supply and, on the other hand, a renewal of interest in

---

1 '... for our present plan the year 1600 is the turning point, after which time there ceases to be any essential distinction between world science and specifically Chinese science...' (Needham 1954–, 3: 437).
2 The other orders present in China up to 1800 included the Dominicans, the Franciscans, the Augustinians and the Lazarists; see Standaert 2001, 309–354.
3 See Romano 2002 and the other articles in the same volume.
4 For the period under discussion, it is estimated that there were up to about 200,000 Chinese Christians. This maximum figure was reached around 1700, when the Chinese population is assessed at about 150 million; Standaert 2001, 380–386.
5 Brockey 2007 gives a vivid description of the itinerary and activities of the China Jesuits.

'practical learning' (*shixue* 實學) among late Ming scholars—what we might call the demand. Accordingly, this chapter briefly describes and situates mathematics in China around 1600 and then outlines some characteristics of science in Jesuit education; it goes on to discuss the Chinese translations of works on the mathematical sciences during the first decades of the mission and how they were integrated into scholarship by their Chinese translators. Lastly it gives an account of the astronomical reform on which the Jesuits worked from 1629 until the fall of the Ming dynasty in 1644.

1.1 Mathematics and literati culture in China *c.*1600

It is widely admitted that by 1600, some of the most significant achievements of the Chinese mathematical tradition had fallen into oblivion. The *Nine chapters on mathematical procedures* (*Jiuzhang suanshu* 九章算術, first century CE),[6] regarded by many as the founding work of the Chinese mathematical tradition and included in the *Ten mathematical classics* (*Suanjing shishu* 算經十書, 656), had effectively

---

Box 1.1 Calculating devices in imperial China: counting rods and the abacus

1	2	3	4	5	6	7	8	9
|	||	|||	||||	|||||	丅	丆	丌	𠀎
—	=	☰	☰	☰	⊥	⊥	⊥	⊥

Counting rods use a decimal place-value notation: successive numbers are represented alternatively by horizontal and vertical rods. An empty place is left where we would write a zero in our notation. From 1 to 5, one uses the corresponding number of rods. From 6 to 9, a rod laid out in the opposite direction on top of the others represents 5.

The number displayed above can be read as 1,234,567,890.

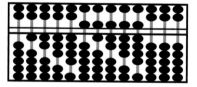

The representation of numbers on the abacus is also based on a decimal place value notation. Here digits from 1 to 9 are represented, so that the number formed is 1,234,567,890 (if the third column from the right is chosen as the unit column).

---

6 Chemla & Guo 2004.

been lost. Furthermore the sophisticated 'celestial element' (*tianyuan* 天元) place-value algebra developed in the thirteenth century had been forgotten. The calculating device on which both were based, the counting rods, had fallen into disuse; the abacus had become the universally used calculating device.[7]

By contrast with this picture of decline in mathematics, some historians have described the sixteenth century as a 'second Chinese Renaissance'.[8] The last decades of the sixteenth century witnessed a strong renewal of interest in technical learning and statecraft.[9] The advocates of 'practical learning' emphasised the social role of literati, underlining that scholarship was of value only if it contributed to welfare and social harmony, while being grounded in verifiable evidence. The lowering of the cost of printing at the time allowed a significant broadening of the book market, which facilitated the circulation of knowledge. The renewal in many fields of scholarship is exemplified by such major works as Zhu Zaiyu's 朱載堉 *New explanation of the study of the [standard] pitchpipes* (*Lüxue xinshuo* 律學新說, 1584), Li Shizhen's 李時珍 *Systematic materia medica* (*Bencao gangmu* 本草綱目, 1593), and Song Yingxing's 宋應星 *Exploitation of the works of nature* (*Tiangong kaiwu* 天工開物, 1637). Cheng Dawei's 程大位 *Unified lineage of mathematical methods* (*Suanfa tongzong* 算法統宗, 1592) can be regarded as belonging to this trend and is representative of the state of mathematics in China by 1600. It was to remain a bestseller to the end of the imperial age (1911).[10]

Mathematics was a form of specialised knowledge, but there is no evidence that it defined a profession throughout the imperial era. When the *Ten mathematical classics* had been compiled, they formed the syllabus for the training of specialists at a College of Mathematics (*Suanxue* 算學); but this training, which concerned only a small group of individuals, does not seem to have outlived the Tang dynasty (618–907).[11] Since the Song dynasty (960–1279), literati culture was centred on the works that formed the syllabus of civil examinations: the *Five Classics*, traditionally regarded as compiled by Confucius, and the *Four Books*, which formed the core of Confucian teachings.[12] The selection of the former collection (among what were previously thirteen classics) and the creation of the latter collection was mainly the work of Zhu Xi 朱熹 (1130–1200), who came to be regarded as the leading figure of the Study of principle (*lixue* 理學), and as the

---

7   Discussions will be found in general surveys of Chinese mathematics in Western languages, including Li & Du 1987, Martzloff 1997 and Yabuuti 2000.
8   Gernet 1972, 370.
9   Cheng 1997, 496–530.
10 In what follows I use the facsimile reprint of the 1716 edition in Guo 1993, 2: 1217–1421; see also Li & Mei 1990; Guo 2000.
11 Siu & Volkov 1999, 88–90.
12 On the *Four Books* see Gardner 2007.

Table 1.2 The *Five Classics* and *Four Books*

The *Five Classics* (*Wujing* 五經)	The *Four Books* (*Sishu* 四書)
Book of change (*Yijing* 易經)	*Great learning* (*Daxue* 大學)*
Book of odes (*Shijing* 詩經)	*Analects* (*Lunyu* 論語)
Book of documents (*Shangshu* 尚書)	*Mencius* (*Mengzi* 孟子)
Book of rites (*Liji* 禮記)	*Doctrine of the mean* (*Zhongyong* 中庸)*
Spring and autumn annals (*Chunqiu* 春秋)	

* Extracts from the *Book of rites*

founder of what Chinese scholars have called the 'Cheng-Zhu school', after Cheng Yi 程頤 (1033–1107) and Zhu Xi, commonly referred to in English as 'Neo-confucianism'. The notion of principle (*li* 理) as developed by Cheng Yi is intrinsic to things rather than transcendental;[13] Zhu Xi recapitulated it as follows:

When it comes to things under heaven, for each of them there must be a reason by which it is so and a rule according to which it should be so. This is what is called principle.[14]

The duality between 'what should be so' (*suodangran* 所當然) and 'why it is so' (*suoyiran* 所以然) would later provide a rationale for discussing the mathematical sciences. The Cheng-Zhu school redefined the investigation of things (*gewu* 格物), advocated in the *Great learning*—then attributed to Confucius—as a means of self-cultivation, in terms of 'fathoming the principles' (*qiongli* 窮理) that underlay these things; again these broad notions were to be applied in particular to fields that in modern terms pertain to the sciences.[15] By the fifteenth century, the Cheng-Zhu school and its interpretation of the Confucian teachings had gained the status of state orthodoxy, a status it retained throughout the period under discussion in the present book.

During the Song dynasty, numbers also came to play an increasing role in cosmology: Shao Yong 邵雍 (1011–1077), a contemporary of Cheng Yi, was the most influential representative of this trend. His approach to and use of numbers was inspired by divination techniques, which also influenced mathematics at the time.[16] For him however, numbers were categories rather than means of quantification; they

---

13 Cheng 1997, 447–452. 'Principle' is the received English translation of *li* 理; therefore I will use it in what follows, although 'pattern' is both more literal and appropriate: *li* does not precede things, but rather resides in them.
14 至於天下之物，則必各有所以然之故，与其所当然之則，所謂理也。*Queries on the Four Books* (*Sishu huowen* 四書或問), in *SKQS* 197: 222; Kim 2000, 19.
15 Gardner 2007, 4–6.
16 On mathematics and divination during the Song dynasty, see Hou 2006.

encapsulated the regularities of change.[17] Shao Yong gave new impetus to 'figures and numbers' (*xiangshu* 象數), a very broad tradition of investigation into the cosmic regularities in which numbers were thought to play a part; it included the study of the relationship between hexagrams, a topic relevant both to divination and to scholarly interpretations of the *Book of change*, but could also encompass subjects today regarded as pertaining to the sciences. For example, in his *Dream pool essays* (*Mengxi bitan* 夢溪筆談, 1088), the famous Song scholar Shen Gua 沈括 (1031–1095) discusses topics such as calendrical astronomy and astronomical instruments, mathematics, harmonics, the hexagrams and divination, under the heading 'figures and numbers'.[18] Thus the *Book of change* provided a kind of link between classical studies and disciplines that are now regarded as belonging to the sciences.

Mastery of the *Five Classics* and *Four Books* and the moral virtues they were supposed to reflect, was the key to success in examinations, and to recognition as a member of the literati class. Such recognition was easier to secure than a good position in the civil service: the examination system was organised in three stages, and it could take the best part of a man's life to reach the top. Passing the first examination, which took place every year at the local level, would secure the title of District Graduate (*shengyuan* 生員), which literati referred to informally as 'cultivated talent' (*xiucai* 秀才); it brought tax benefits with it. The next stage was an examination which took place every three years in provincial capitals: success in this made one a Provincial Graduate (*juren* 舉人, lit. 'recommendee'). Finally, a triennial examination was set in the capital; for the happy few who passed it, there followed a Palace examination that would determine their rank. They finally attained the title of Metropolitan Graduate (*jinshi* 進士, lit. 'presented scholar').[19]

Mathematics played no particular role in literati's social promotion. Nonetheless, throughout the last centuries of the imperial era, some cultivated it, so that the tradition was not entirely discontinued: although Cheng Dawei was never able to read the *Nine chapters on mathematical procedures*, he had access to all the problems it contained through the *Great classified survey of the Nine chapters on mathematical methods* (*Jiuzhang suanfa bilei daquan* 九章算法比類大全, 1450), which took them up. This tradition mattered to Cheng: far from claiming to innovate, he aimed at providing a compilation of earlier treatises that he had spent decades collecting. The *Unified lineage of mathematical methods* is regarded as the final representative of a tradition of 'popular mathematics' which emerged as part of merchant culture; it was based

---

17   Arrault 2002.
18   Shen 1975, 74–96.
19   Elman 2000, Magone 2001.

on abacus calculation, which is traced back to Yang Hui 楊輝 (fl. 1261), in the Southern Song dynasty.[20] Cheng's historical awareness is reflected in his book's title: *suan* 算 was the usual term to refer to mathematics, whereas *fa* 法 refers to the methods by which each problem was solved; *suanfa* occurred in the title of most works on the subject known to him. The phrase 'unified lineage' (*tongzong* 統宗) suggests that he saw himself as the heir of a lineage of scholars versed in mathematics.[21] In his view the lineage could be traced back much further than the Song, as he stated in the opening paragraph of the work:

How did numbers begin? They began from the [River] Diagram ([*He*]*tu* [河]圖) and the [Luo] Writing ([*Luo*]*shu* [洛]書)! Fuxi 伏羲 obtained them and by this means drew the trigrams [of the *Book of change*]. Yu 禹 the Great obtained them and by this means ordered the sections [of the *Great Plan* (*Hongfan* 洪範)].[22] All the Sages obtained them and by this means operated things so as to succeed in affairs. Heavenly officials and earthly clerks deal with harmonics and calendar, military affairs and taxes in great detail, with minute quantities. All involve numbers, so all are grounded in the [*Book of*] *change* and the [*Great*] *Plan*. Here clarifications are aimed at mathematical methods; one brings to light the River Diagram and the Luo Writing at first, so that it can be seen that numbers have an origin (*yuanben* 原本).[23]

Since the Song dynasty, the two charts mentioned here had been represented as layouts of numbers from one to ten and from one to nine respectively,[24] which became the object of mathematical investigation.[25] Thus Cheng also followed a tradition that went back to the Song in his attribution of the creation of mathematics to the Sages who founded Chinese civilisation.[26]

Like most of his predecessors known to him, Cheng referred to the canonical nine-fold classification of mathematics, although the book from which it was drawn was evidently unavailable to him. During the late Ming and early Qing, the phrase 'nine chapters' (*jiuzhang* 九章) mostly referred to that classification rather than to the classic work of that name. But again like most if not all authors before and after him, Cheng Dawei failed to fit all the mathematical knowledge at his command into the headings of the 'nine chapters': his work is divided into

---

20 Here 'popular' should be understood as opposed to what pertained to literati culture; by the Ming dynasty the border between the two was at best quite fuzzy; Yabuuti 2000, 103–121; on Yang Hui see Lam 1977. Bréard 2010 provides a glimpse of some aspects of this 'popular mathematics'.
21 *Tongzong* has often been translated as 'systematic treatise'; see e.g. Li & Du 1987, 185.
22 Both these attributions were the traditional mythical ones. The *Great Plan* is a chapter of the *Book of documents*.
23 Guo 1993, 2: 1227.
24 See Illustration 14.2, p. 323.
25 Smith *et al.* 1990, 120–122; Hou 2006.
26 See e.g. the opening of the *Precious mirror of mathematics* (*Suanxue baojian* 算學寶鑑), preface dated 1513); Guo 1993, 2: 347–348.

Table 1.3 The 'nine chapters' and Cheng Dawei's *Unified lineage of mathematical methods*

'Nine chapters'[1]	*Unified lineage of mathematical methods*[2]
1. Rectangular fields (*Fangtian* 方田) [Areas of fields of various shapes; manipulation of fractions]	1. Basic notions and terms; calculation rhymes
	2. The abacus; associated calculation rhymes
	3. Rectangular fields (*Fangtian* 方田)
2. Millet and rice (*Sumi* 粟米) [Exchange of commodities at different rates; pricing]	4. Millet and cloth (*Subu* 粟布)[3]
3. Graded distribution (*Cuifen* 衰分) [Distribution of commodities and money at proportional rates]	5. Graded distribution (*Cuifen* 衰分)
	6. The Lesser breadth (*Shaoguang* 少廣)
4. The lesser breadth (*Shaoguang* 少廣) [Division by mixed numbers; extraction of square and cube roots; dimensions, area and volume of circle and sphere]	7. Division of fields by cutting off areas (*Fentian jieji* 分田截積) [Problems involving plane figures from which are removed portions of a given area or linear dimension]
5. Consultations on works (*Shanggong* 商功) [Volumes of solids of various shapes]	8. Consultations on works (*Shanggong* 商功)
6. Equitable transport (*Junshu* 均輸) [More advanced problems on proportion]	9. Equitable transport (*Junshu* 均輸)
7. Excess and deficit (*Ying buzu* 盈不足) [Linear problems solved usinig the 'rule of false position]	10. Excess and deficit (*Yingnü* 盈朒)[3]
8. Rectangular arrays (*Fangcheng* 方程) [Linear problems with several unknowns]	11. Rectangular arrays (*Fangcheng* 方程)
9. Base and altitude (*Gougu* 句股) [Problems involving the relation between the sides of a right triangle]	12. Base and altitude (*Gougu* 句股)
	13. Difficult problems (*Nanti* 難題) (rectangular fields; millet and cloth)
	14. Difficult problems (*Nanti* 難題) (graded distribution)
	15. Difficult problems (*Nanti* 難題) (lesser breadth)
	16. Difficult problems (*Nanti* 難題) (excess and deficit; rectangular arrays; base and altitude)
	17. Difficult problems (*Nanti* 難題) (various methods)

1 Translations and description of contents based on Cullen 2007, 28; extensive discussions of the methods that give their titles to the chapters are found in Chemla & Guo 2004.
2 Guo 1993, 2: 1217–1421.
3 Variant title.

seventeen chapters. It opens with the discussion of the River Diagram and Luo Writing quoted above. Chapter 1 contains some general prescriptions for the study of mathematics, a list of the 'nine chapters', short definitions of more than seventy terms used thereafter, lists of powers of tens and units, tables of addition, subtraction, multiplication and division for the abacus, and brief explanations of some terms referring to common operations such as the simplification of fractions or the extraction of cubic roots. Chapter 2 focuses on the abacus, starting with an illustration. Chapters 3 to 17 contain 595 problems presented in the traditional form: question, answer and resolution method. Chapters 3 to 6 and 8 to 12 take up the traditional nine chapters, whereas Chapter 7 introduces a particular type of problems: 'Cutting off the area of a straight field' (*Zhitian jieji* 直田截積). Chapters 13 to 16 contain 'Difficult problems' (*Nanti* 難題), again ordered according to the 'nine chapters'; Chapter 17 gives various methods and number diagrams such as magic squares. Some of the 'difficult problems' as stated as poems:

Moonlight on the Western River[27]

A borrows seven inkstones from the B family, and returns him three fine-haired (*maozhui* 毛錐) handles, compensating in coins four full hundreds and eighty; it's exactly even, and done.
Yet C borrows nine brushes from B, and returns him three items from Duanxi 端溪; one hundred and eighty are compensated to B, it's even; how much should the prices of these two kinds (*se* 色) be?[28]
Answer: The price of a brush is 50 pieces (*wen* 文) the price of an inkstone is 90 pieces.
Method: Lay out the numbers of the problem:[29]

[For reasons of formatting, the layout given here in the original text is reproduced in Box 1.4.]

First, taking the positive 7 inkstones of the right column as factor, multiply all the numbers obtained in the left column (middle and lower).

Yet taking the positive 3 inkstones of the left column as a factor, in return multiply all the right column. The middle negative 3 brushes give 9. Subtract from the negative 63 brushes of the left column, the remainder is negative 56 brushes as divisor.

The price positive 480 gives positive 1440, add the opposite of the price of the left column negative 1260, together one obtains 2700 as dividend. Divide it by the divisor, one obtains the price of a brush: 50 pieces.

Add the left column price, positive 480, to the opposite of the price 150 of negative 3 brushes, together one obtains 630. Divide it by 7 inkstones, one obtains the price of an inkstone: 90 pieces. This matches the question.[30]

---

27 *Xijiang yue* 西江月 is the title of a tune to which the poem could be sung. The form is that of a *ci* 詞 of fifty characters, divided into two lines of twenty-five characters each. If one translated this problem into modern notations each line would yield one of the two linear equations.
28 甲借乙家七硯, 還他三管毛錐, 貼錢四百整八十, 恰好齊同了畢。丙卻借乙九筆, 還他三箇端溪, 百八十貼乙齊, 二色價該各幾。
29 Guo 1993, 2: 1402.
30 Guo 1993, 2: 1402–1403.

Box 1.4 Layout for the inkstones and brushes problem in the *Unified lineage of mathematical methods* (Guo 1993, 2: 1402)

(left)	(right)	左	右
inkstones positive 3	inkstones positive 7 as factor	硯 三 正	硯 七 正 為 法
(middle) brushes negative 9 one obtains negative 63	(middle) brushes negative 3	中 筆 負 九 得 十 負 三 六	中 筆 負 三
(lower) price negative 180 one obtains negative 1260	(lower) price positive 480	下 價 負 八 十 百 得 百 負 六 一 十 千 二	下 價 正 八 四 十 百

The method used here is that of rectangular arrays (*fangcheng* 方程), which gave its name to the eighth of the *Nine chapters on mathematical procedures*. Following the tradition stemming from this work, Cheng uses positive (*zheng* 正) and negative (*fu* 負) numbers in rectangular array problems and exclusively there. In his chapter on this subject, he glosses these terms: 'Positive is a positive quantity; negative is an owed quantity.'[31] There he also gives the algorithm followed in the problem quoted above in the form of a 'rhyme for rectangular arrays of two kinds' (*er se fangcheng ge* 二色方程歌).[32]

---

31 正者正數。負者欠數。 Guo 1993, 2: 1359.
32 Guo 1993, 2: 1359.

The difficulty of the inkstones and brushes problem lies in the literary style in which it is stated: it took an educated reader to know that 'fine-haired' is a type of brush, and that Duanxi (Guangdong province) was famous for its fine inkstones. The answer and the solution, on the other hand, are stated in the plain prose used for all problems in the work. The fact that this problem already appeared in the *Great classified survey of the Nine chapters on mathematical methods* exemplifies the claim apparent in the title of Cheng Dawei's work: as his predecessors had done before him, he is taking up problems from various sources, while providing his own version of the methods to solve them.[33]

The last chapter of the *Unified lineage of mathematical methods* gives 'Miscellaneous methods' (*Zafa* 雜法) that include diagrams such as magic squares and finger calculation mnemonics; the chapter closes with a bibliography of earlier mathematical works, from the Song edition of the *Ten mathematical classics* to works published in Cheng Dawei's lifetime, spanning five centuries. It does not seem that Cheng Dawei actually saw any of the *Ten mathematical classics*. Neither is there any discussion of the use of counting rods in his work.

1.2 Mathematics in the Society of Jesus

Such, then, was the mathematical culture prevalent in China when the Jesuits first arrived. Founded in 1540, the Society of Jesus had soon started setting up colleges across Europe. Many sons of the elites of Catholic countries were educated in them, as were most members of the Society. The latter often trained to be teachers, and for some of them this remained their main occupation. The content and structure of the education provided by the Society were crucial in shaping Jesuit culture, in Europe as well as in China. Having previously studied the *trivium* (grammar, logic and rhetoric), students entering a Jesuit college would typically begin with further training in rhetoric. This was followed by three years devoted to logic, philosophy and metaphysics. Early in the order's history, natural philosophy (or physics) and mathematics were both grouped under philosophy. According to the Aristotelian classification, physics and mathematics addressed two of the ten categories, quality and quantity, respectively. Physics provided a qualitative explanation of natural phenomena; it was based on the four elements theory, according to which all matter was composed of earth, air, fire and water. In the scholastic tradition, mathematics consisted of the four disciplines of the *quadrivium*, namely arithmetic, music, geometry and astronomy.

Mathematics was somewhat redefined in the Jesuit curriculum compared with common contemporary usage. The Roman College,

---

33 Guo 1993, 2: 266.

founded in 1551, set the standards for the Society's educational network. The *Ratio Studiorum* (*Plan of studies*, final version 1599), which defined the Jesuit educational system, gave a new importance to mathematics.[34] Christoph Clavius (1538–1612) was instrumental in establishing it as a subject independent from philosophy. The architect of the Gregorian Calendar Reform of 1582, he taught mathematics at the Roman College from 1565, and was the first to hold the chair of mathematics there, and to assert its status as a science.[35] According to him, 'It is so ordained by nature that eminence in any subject, even of the least importance, causes the eyes of everyone to converge on oneself.'[36] Whereas excelling in learning in general had been a concern of the Jesuits since the foundation of the society, it was the reappraisal of the 'mathematical arts' in sixteenth-century Italy that prompted them to include those arts into the subjects in which they should strive to be eminent.[37]

While establishing mathematics as a subject in the Jesuit curriculum, Clavius defined its structure and produced textbooks for its teaching. Following Proclus, the fifth century CE philosopher who wrote an influential commentary on Euclid's *Elements of geometry*, Clavius divided mathematics into 'pure' and 'mixed', the former consisting of arithmetic and geometry, the latter of six major branches (which were further divided into more disciplines): natural astrology (astronomy), perspective, geodesy, music, practical arithmetic and mechanics. This structure, while evocative of that of the *quadrivium*, broadened the scope of mathematics and extended its fields of application. This is in keeping with the broader conclusion of several historians that 'Jesuit science was concerned mainly with the promotion of areas related to "applied mathematics"'.[38] The works authored by Clavius, first and foremost his editions of and commentaries on Euclid's *Elements* and Sacrobosco's *Sphere*, as well as his textbooks on arithmetic and algebra, formed the basis of mathematical education as he defined it for the Society.[39]

In natural philosophy, the authoritative reference stemmed from commentators working at the Portuguese College of Coimbra. The *Conimbricenses*, as they and their writings are often called, consisting of five volumes of editions of and commentaries on Aristotle's work, including *Physica*, *De Cælo*, *Meteorologica*, *Parva naturalia*, *Ethica Nichomachæ*, *De Generatione et corruptione*, and *De anima*, were published between 1592 and 1602; they were reprinted and used by teachers in Jesuit colleges throughout Europe.

---

34 Pralon-Julia *et al.* 1997.
35 Baldini 1992, Romano 1999, Rommevaux 2005, esp. 24–25.
36 Quoted by Gorman 1999, 172.
37 Gorman 1999, 172.
38 Feldhay 1999, 113.
39 Feldhay 1999, 109–113; see also Engelfriet 1998, 30–32.

Jesuit education in Europe was not entirely uniform: there were local variants in the mathematics taught, and, as with any school curriculum, a number of updates occurred.[40] Thus, since the 1620s, the Ptolemaic system defended and taught by Clavius, in which the Earth lay motionless at the centre of concentric crystalline spheres, was gradually replaced by the Tychonic system, in which the Sun, while revolving around the Earth, was the centre of the orbits of the planets. By and large, the tradition Clavius had established was continued in the sense that many later teachers produced textbooks modelled largely on his, though departing from Clavius in their pedagogical approach.[41] The number of textbooks entitled *Elements of geometry* produced in the seventeenth century, within and without the Society, was such that the phrase, and even the name of Euclid came to refer to a genre—that of geometry textbooks—rather than merely to editions and commentaries of the Greek classic. Including innovations that originated outside the Society was also part of Jesuit policy. Thus whereas Clavius' *Algebra* was one of the last representatives of the medieval tradition of cossic algebra, in which the unknown and its powers are denoted by abbreviations of their names, Vieta's new notations were introduced into teaching in the 1620s.[42]

Clavius, author of the Gregorian calendar, played a considerable role on the Roman, and indeed on the European, 'scientific scene'. However, during the seventeenth century, the authority and prestige of the Society, which he seemed to personify in this field, decreased significantly.[43] In the 1680s, while the Royal astronomers assessed Jean de Fontaney (1643–1710), the Professor of mathematics at the Jesuit College in Paris as a 'good observer', he had taken no part in the definition of the standards according to which he was assessed.[44] Matteo Ricci (1552–1610), the first Jesuit to enter China, had studied with Clavius at the Roman College and brought with him the latter's mathematics; some of his successors in the China mission would present mathematics as it evolved in Jesuit colleges over the next century.

1.3 Teaching and translating

The China mission was part of the Portuguese Assistancy of the Society: following the Treatise of Tordesillas (1494), all Asian missions were under the patronage (*padroado*) of the Portuguese crown.[45] The port of Macao, founded by the Portuguese in 1557, served as their Eastern base. While their Japanese mission was flourishing in the late

---

40 On Portugal, see Leitão 2002; on France see Romano 1999, 183–354; 2006.
41 Feldhay 1999, 111 & 114; Baldini 2000, 77.
42 Reich 1994; Feldhay 1999, 116–126.
43 Gorman 1999.
44 Hsia 1999, 38–42.
45 Alden 1996.

sixteenth century, the Jesuits' efforts to settle in China were unsuccessful until 1582, when Michele Ruggieri (1543–1607), after three years of study of the Chinese language, culture and customs, obtained permission to reside in China. The next year, he established the first Jesuit residence in China in Zhaoqing 肇慶 (Guangdong province), together with Matteo Ricci.[46] The latter has come down in history as the 'founding father' of the Jesuit mission in China.[47] He is usually credited with two related features of the Jesuit strategy there: evangelisation 'from the top down', and the use of science in the propagation of the faith. Recent research has shown, however, that Ricci's itinerary from Zhaoqing to Beijing (1583–1601) was mainly determined by the necessity to get protection from the central authority in order to establish permanent residences in China. Similarly, it was only in order to explain to the somewhat incredulous scholars who visited him where he came from that he first drew a Chinese version of his world map; this was the first translation of a non-religious work. In other words the political organisation and culture of the society they met was no less instrumental in shaping the Jesuits' strategy than their own background and aims.[48]

Ricci very soon started to make use of his master Clavius' textbooks. According to his own account, in 1589 he taught first arithmetic, then the first book of the *Elements* and the *Sphere* to Qu Rukui 瞿汝夔 (1549–1611), a literatus who became one of his sympathisers and advisers, and eventually converted.[49] After Ricci settled in Beijing in 1601, he taught mathematics to Xu Guangqi 徐光啓 (1562–1633) and Li Zhizao 李之藻 (1565–1630); both were high officials who converted and became active protectors of the Jesuit mission. In collaboration with Ricci, they produced works based on some of Clavius' textbooks:[50]

- The *Elements*: the first six books of Clavius' 1574 edition, translated as *Elements of geometry* (*Jihe yuanben* 幾何原本), 1607, Ricci and Xu Guangqi.[51]
- The *Astrolabium*: translated as *Illustrated explanation of the sphere and the astrolabe* (*Hungai tongxian tushuo* 渾蓋通憲圖說), Ricci and Li Zhizao, 1607.[52]
- The *Sphere*: translation of Clavius' commentary of Sacrobosco's *Tractatus de Sphaera* as *The meaning of Heavenly and Earthly forms*

---

46 Brockey 2002, 19–31.
47 The literature on Matteo Ricci is too abundant to list here. Spence 1984 is both inspiring and reliable.
48 Standaert 1999a, 358–360.
49 Jami 2002a, 161–162.
50 Martzloff 1995.
51 Clavius, *Euclidis Elementorum Libri XV...*, Rome, 1574; on the translation, see Engelfriet 1998.
52 Clavius, *Astrolabium*, Rome, 1593; see Ahn 2007, 209–256.

(*Qiankun tiyi* 乾坤體義), 1608, by Ricci and Li Zhizao; essay on isoperimetric figures inserted in his commentary translated as *The meaning of compared [figures] inscribed in a circle* (*Yuanrong jiaoyi* 圜容較義), 1614, Ricci and Li Zhizao.[53]

- The *Arithmetic*: a compilation with several earlier Chinese mathematical treatises, *Instructions for calculation in common script* (*Tongwen suanzhi* 同文算指), 1614, Ricci and Li Zhizao.[54]

To this list one should add *The meaning of measurement methods* (*Celiang fayi* 測量法義, 1608), a brief treatise on surveying completed by Ricci and Xu Guangqi at the same time as the *Elements of geometry*. It was most likely based on Ricci's lecture notes, as it is different from Clavius' *Geometria practica*, published in Rome in 1604.[55]

The structure of mathematics as a whole according to the *quadrivium*, and the Aristotelian duality between number and magnitude as the two instances of quantity that underlay this structure, were presented by Ricci in his preface to the *Elements*:

> The school of quantity (*ji he jia* 幾何家) consists of those who concentrate on examining the parts (*fen* 分) and boundaries (*xian* 限) of things. As for the parts, if [things] are cut so that there are a number (*shu* 數) [of them], then they clarify how many (*ji he zhong* 幾何眾) the things are; if [things] are whole so as to have a measure (*du* 度), then they point out how large (*ji he da* 幾何大) the things are. These number and measure may be discussed (*lun* 論) in the abstract, casting off material objects. Then those who [deal with] number form the school of calculators (*suan fa jia* 算法家); those who [deal with] measure form the school of mensurators (*liang fa jia* 量法家). Both [number and measure] may also be opined on with reference to objects. Then those who opine on number, as in the case of harmony produced by sounds properly matched, form the school of pitchpipes and music (*lü lü yue jia* 律呂樂家); those who opine on measure, in the case of celestial motions and alternate rotations producing time, form the school of astronomers (*tian wen li jia* 天文曆家).[56]

Rather than describing mathematics as the *quadrivium*—with which most Chinese readers would have been wholly unacquainted—this passage is actually proposing to subsume four different disciplines, all of which corresponded to known technical fields in late Ming China, under the broader albeit hitherto unknown field of the 'study of quantity'; here *jihe* renders the Latin *quantitas*. In this light, the title chosen by Ricci and Xu for their translation must have intended to refer not only to

---

53 Clavius, *In sphæram Joannis de Sacro Bosco commentarius*, Rome, 1570; Ahn 2007, 160–162.
54 Clavius, *Epitome arithmeticæ practicæ*, Rome, 1583; see Takeda 1954; Jami 1992; Pan 2006.
55 Engelfriet 1998, 297; Ahn 2007, 146–149.
56 Guo 1993, 5: 1151; comp. Engelfriet 1998, 139; Hashimoto & Jami 2001, 269–270.

geometry, but more broadly to the whole of the *quadrivium*. The claim here is also that the *Elements* provides foundations for a discipline that includes the Chinese tradition of *suanfa* 算法 ('mathematical methods' as in the title of Cheng Dawei's work) as one of its parts. On the other hand, *jihe* 幾何 means 'how much' in classical Chinese. It occurred in every single ancient mathematical text, as many times as there were problems. In the *Unified lineage of mathematical methods*, however, *ruogan* 若干 (a synonym) is the word used, whereas *jihe* appears in the list of terms defined at the beginning: it is glossed by 'same as *ruogan*'.[57] The distinction between the two instances of quantity, rendered by *shu* 數 (number) and *du* 度 (magnitude) respectively, would have been entirely new to late Ming Chinese readers: for them *shu* was more evocative of numerology and the study of the *Book of change* than of procedures of *suanfa* to which Ricci and Xu wanted their translation to relate.

That the above rationale might have seemed somewhat unfamiliar to Chinese readers should not, however, obscure two major points. First, the translations based on Clavius listed above aroused interest; indeed they were often done in response to the perceived curiosity of Chinese scholars about the fields they covered. Secondly, bringing together mathematics, surveying, astronomy and harmonics was not foreign to their tradition: surveying was one of the main themes of mathematical problems, and astronomy and harmonics were discussed in the same section of quite a few dynastic histories. Also, one finds many examples of scholars known both as mathematicians and astronomers. Thus Li Chunfeng 李淳風 (602–670) was the head of the imperial observatory when he compiled the *Ten mathematical classics*; he was also the author of the 'Chimera virtue' astronomical system (*Linde li* 麟德曆), in use from 665 to 728.

Although Euclidean geometry is the best-known branch of mathematics taught by the Jesuits, not everything presented in their mathematical works was unfamiliar to Chinese readers. In the *Instructions for calculation in common script* Ricci and Li Zhizao took up problems from earlier Chinese works and showed how these could be solved using written calculation. Thus rectangular array problems appear in a section entitled 'Methods with miscellaneous sums, differences and multiplications' (*Za hejiao cheng fa* 雜和較乘法). A note in small characters refers to the Chinese method, equating it to one of the methods translated from Clavius' *Epitome arithmeticae*:

---

57 Guo 1993, 2: 1230.

What used to be called rectangular array is also the same as repeated borrowing for mutual comparison (*diejie huzheng* 疊借互徵).[58] Many use the latter as it is more convenient.[59]

The third problem in this section reads:

Question: 3 brushes are exchanged against 7 inkstones, with a contribution to the inkstones of 480 pieces. Apart from this, 3 inkstones are exchanged against 9 brushes, with a contribution to the brushes of 180 pieces. What are the prices of a brush and of an inkstone?[60]

Box 1.5 Layout for the inkstones and brushes problem in the *Instructions for calculations in common script* (Guo 1993; 4: 187)

inkstones positive 3	inkstones positive 7	硯正三十正一二	硯正七
positive 21			
brushes negative 9	brushes negative 3	筆負九十負三六	筆負三負九
negative 63	negative 9		
price negative 180 pieces	price positive 480 pieces	價負一百八十文二負百一六千十	價正四百八十文百正四一十千四
negative 1260	positive 1440		

The solution to this problem is the same as in the *Unified lineage of mathematical methods* and it uses the same terminology, but each step is explained in more detail. This would have suggested to readers that procedures hitherto carried out using the abacus could still be used if

---

58 The double false position method.
59 Guo 1993, 4: 186.
60 Guo 1993, 4: 187.

one adopted written calculation. The presence of this and many other problems in the *Instructions for calculation in common script* result in a 'dissolution' of Chinese procedures into Western learning: rectangular arrays no longer forms a category of its own, but is demoted to one of several 'methods with miscellaneous sums, differences and mutliplications'. It has, in some sense, been disqualified, losing both the specificity and the generality that it had in earlier Chinese treatises.

The translations mentioned above were part of the Jesuits' larger enterprise of 'apostolate through books' in China. This was rendered possible by the flourishing of printing and publishing.[61] The Jesuits' teachings were thus presented as a coherent whole in a compendium edited by Li Zhizao in 1626, the *First collection of heavenly learning* (*Tianxue chuhan* 天學初函).[62] It was divided into two parts: principles (*li* 理, nine works) and concrete things (*qi* 器, ten works). The first part opens with a description of the European educational system, entitled *Outline of Western learning* (*Xixue fan* 西學凡, 1621). Like Ricci, its author, Giulio Aleni (1582–1649), had been a student of Clavius at the Roman College. The work presents the structure of disciplines that was then most common, mathematics consisting of the *quadrivium* and being one subdivision of philosophy.[63] The next six works of the collection discuss mainly ethics and religion. The last work of the first part is an introduction to world geography. Illustrated by several maps, including an elliptical world map, the *Areas outside the concern of the Imperial Geographer* (*Zhifang waiji* 職方外紀, 1623, by Giulio Aleni) describes the Earth as part of the universe created by God, and Europe as the ideal realm where Christianity has brought long-lasting peace.[64]

The second part of the *First collection of heavenly learning* includes five of the six works based on Clavius' textbooks and teaching mentioned above (the *Meaning of Heavenly and Earthly forms* was not included). It also includes three works by another former student of Clavius, Sabatino de Ursis (1575–1620), dealing respectively with hydraulics, the altazimuth quadrant and the gnomon. A short treatise written by Xu Guangqi after he had completed the translation of the *Elements* with Ricci is also included.[65] Only one work pertaining to 'concrete objects' does not stem from the student lineage of Clavius. The author of the *Summary of questions about the heavens* (*Tianwen lüe* 天問略, 1615), Manuel Dias Jr (1574–1659) does not seem to have studied outside

---

61 Standaert 2001, 600–631.
62 On this compendium, see Ahn 2007; for the date of *Tianxue chuhan*, I follow Standaert 2001, 141.
63 Standaert 2001, 606.
64 Aspects of the Christian worldview were also introduced in the Jesuits' presentation of some aspects of European medicine; Standaert 1999b.
65 Engelfriet 1998, 301–313, Engelfriet & Siu 2001, 294–303.

the Portuguese Assistancy: he had completed his studies in Goa and taught theology at Macao before entering China. His *Summary of questions about the heavens* was mainly an account of Aristotelian-Ptolemaic cosmography, which Clavius had defended and taught in Rome; internal evidence suggests that in writing his work, Dias may have relied on the fourth edition of Clavius' commentary on Sacrobosco's *Sphere*.[66] An appendix at the end of the work reported Galileo's invention of the telescope and the observation that he had made with it.[67] This was in keeping with the Society's policy of including innovations in its teaching in Europe. On the whole, the works on 'concrete objects' in the *First collection of heavenly learning* all pertained to the mathematical sciences construed and constructed by Clavius for Jesuit colleges.

No less important in the Jesuits' enterprise was Aristotelian philosophy. Here the main source was the *Conimbricenses*; a number of works published in the late Ming drew on them.[68] They were far from encountering the same success as mathematical works, and their influence on Chinese philosophy remained marginal: they were much more obviously in conflict with Chinese cosmology. By the end of the seventeenth century the term 'heavenly learning' (*tianxue* 天學) had long been supplanted by 'Western learning' (*xixue* 西學) to refer to the Jesuits' teachings, and the latter expression covered only the mathematical and technical subjects. Rather than a split between science and religion—a dichotomy alien to the actors—this can be seen as resulting from a selection from the Jesuits' teachings of what was compatible with Confucian orthodoxy and best served imperial interests. The story of Western learning under the Kangxi Emperor shows how this selection was brought into play.

Education and mission were the two main domains invested in by the Society of Jesus, both of them on a worldwide scale. In China the Jesuits were never in a position to set up colleges as they had done in Europe, Goa and Macao. However, most China missionaries had had some teaching experience before arriving in China. Once there they reinterpreted that role of teacher in a Chinese context, establishing master to disciple relationships with the scholars interested in their teaching.[69] In that sense the mathematical sciences were instrumental in their successful construction of an identity in Chinese literati circles.

---

66   Leitão 2008; Magone 2008.
67   Standaert 2001, 712–713.
68   See Standaert 2001, 606–608; Peterson 1973 focuses on natural philosophy; Wardy 2000 discusses the translation of Aristotle's categories from the point of view of the philosophy of language.
69   Jami 2002a.

1.4 Jesuit science, 'practical learning' and astronomical reform

As mentioned above, the first Chinese translators of mathematical and technical works were also the most eminent converts.[70] To them, 'principles' and 'concrete things' formed a coherent whole: the latter would serve to improve the material life of the people, whereas the former would serve to improve their morality. In their view, heavenly learning could thus provide a response to the concerns of scholars of 'practical learning'. On the other hand, their understanding of heavenly learning was qualitatively different from that of most scholars who were acquainted with it only through books: they had actually studied with Jesuit masters. Thus, Li Zhizao, who, in addition to the *Instructions for calculation in common script* and other works that he included in the *First collection of heavenly learning*, worked on the translation of various parts of the *Conimbricenses*, including Aristotle's 'Categories', would have had quite a clear picture of the landscape in which the 'study of quantity' fitted.[71] It is worth noting this point, as most of the works translated by the Jesuits were originally textbooks rather than writings intended for individual reading without the guidance of a teacher.

Xu Guangqi's scholarship was by no means limited to his translations. Indeed his most famous work, a major outcome of the late Ming intellectual trend of 'practical learning', is a treatise on agronomy, the *Complete treatise on agricultural administration* (*Nongzheng quanshu* 農政全書, 1639).[72] An extensive survey taking into account major previous works on the topic, it owes very little to Western learning. Xu also wrote on military defence, tackling such varied topics as the economic organisation of the army, the layout of troops and the use of Western artillery. In mathematics, he went further than translating, seeking to interpret some Chinese mathematical texts in the light of what he had studied with Ricci.[73]

Li Zhizao's interests encompassed geography and, as mentioned above, Aristotelian philosophy; they were not limited to heavenly learning either. He was also the author of the *Memorial on ritual and music at local schools* (*Pangong liyue shu* 頖宮禮樂疏), a treatise on the history of ritual used in sacrificial ceremonies to Confucius.[74] This suggests how heavenly learning related to the purposes of Confucianism, even in its ritual dimension: music was a crucial element in the proper performance of the rites. In the *quadrivium* it was linked

---

70 On these converts, see Standaert 2001, 404–420.
71 Kurtz 2008.
72 Bray & Métailié 2001.
73 Engelfriet 1998, 297–316; Engelfriet & Siu 2001.
74 Standaert 2005, 90–96; the work is reproduced in *SKQS* 651: 1–415.

to arithmetic—precisely the topic on which Li Zhizao worked with Ricci.

While putting their learning into the service of statecraft, those few officials who had converted also engaged in politics. Thus, implementing Western artillery was part of Xu Guangqi's failed attempt to reform the army in order to counter the progress of the Manchu invasion, around 1630.[75]

Astronomy too was in need of reform. The calendar had always been of utmost political and symbolical importance in China. Issued in the emperor's name, it ensured that human activity followed the cycles of the cosmos. At the beginning of the dynasty, the Ming (1368–1644) had taken up their predecessors' astronomical system, and by the end of the sixteenth century the need for astronomical reform was felt acutely amongst officials; several proposals were put forward.[76] In 1613, Li Zhizao presented a memorial to the throne, recommending that three Jesuits should assist the work for a reform that had been proposed three years earlier. His proposal seems to have been among the causes of what is known as 'the Nanjing persecution' (1616–1617): beside hostility to Christianity from officials, the Jesuits' description of the heavens as consisting of a number of concentric crystalline spheres was regarded by some of them as highly subversive. In Chinese cosmology, correspondences between Heaven and Earth played a major role, so that 'dividing the heavens' could be understood as an allusion to the division of the empire.[77] During the Tianqi reign (1621–1627) the eunuchs had the upper hand, and it was not until 1629 that Xu Guangqi, then a Vice-Minister of Rites (*Libu zuoshilang* 禮部左侍郎), was in a position to successfully request an astronomical reform.

Late Ming Christians were aware that the 'foreignness' of heavenly learning was an obstacle to its adoption. Taking up the idea put forward by the Song philosopher Lu Jiuyuan 陸九淵 (1139–1193), they argued that all men have 'the same heart and the same principles' (*xin tong li tong* 心同理同).[78] However when it came to the adoption of European astronomy by the imperial state, this universalistic argument did not suffice. When he submitted his proposal for astronomical reform, Xu Guangqi, who had long emphasised the similarities between the European and Chinese traditions of scholarship, had to resort to another rationale. First, he argued that the Jesuits' teachings were only a tool that Chinese scholars could use to retrieve the lost learning of Antiquity.[79] Secondly, he pointed out that China had a long

---

75   Standaert 2001, 695; Huang 2001.
76   Peterson 1968; Wang Miao 2004.
77   Dudink 2001.
78   Cheng 1997, 483–486.
79   Hashimoto & Jami, 2001, 270–276.

tradition of resorting to foreign specialists to calculate its calendar. Just like the Muslims, whose astronomy had been introduced at the beginning of the dynasty, the Europeans were to be regarded as a mere foreign tribe: they were no threat to Chinese civilisation. Borrowing from them was a means to restore order, at a time of severe social, political and military crisis. In astronomy, he proposed to 'melt their material and substance to cast them into the mould of the Great concordance' (*rong bifang zhi caizhi ru Datong zhi xingmo* 鎔彼方之材質入大統之型模)—Great concordance being the name of the Ming astronomical system. Xu's rationale for the adoption of heavenly learning was thus phrased in terms of the categories and concerns that he shared with all the Chinese scholars of his time.

Xu Guangqi's proposal was approved, and he was commissioned to set up and supervise the Calendar Department (*Liju* 曆局), where two Jesuits were to be employed.[80] During the following years, this Bureau produced a number of works on mathematical astronomy that were presented to the emperor as they were completed, between 1631 and 1635. These formed the *Books on calendrical astronomy of the Chongzhen reign* (*Chongzhen lishu* 崇禎曆書). The two Jesuits appointed to work on the astronomical reform were Johann Schreck (1576–1630) and Johann Adam Schall von Bell (1592–1666). The latter had studied mathematics at the Roman College under Christoph Grienberger (1561–1636), who had succeeded Clavius at the chair of mathematics. The former, by contrast, had not followed the standard Jesuit curriculum. A medical doctor and a renowned scholar, he was a member of the famous Accademia dei Lincei when he decided to enter the Society of Jesus in 1611. He was the author of the first Chinese treatise introducing European anatomy, as well as a two-volume Latin work on Chinese natural history entitled *Plinius Indicus*.[81] When Schreck died in 1630, Giacomo Rho (1692–1638) took up his position; like Schall, Rho had been a student of Grienberger at the Roman College.

Following changes in the Jesuit curriculum as well as in response to the hostility previously aroused on the part of some officials, Schreck, Schall and Rho used the Tychonic system at the Calendar Department. This was simply called 'new' (*xin* 新) as opposed to the 'old' (*gu* 古) Ptolemaic system.[82] Schall had first discussed the 'new' system in 1626 in an *Explanation of the telescope* (*Yuanjing shuo* 遠鏡說). This work was included in the *Books on calendrical astronomy of the Chongzhen reign*, as

---

80 Hashimoto 1988, 34–46; Hashimoto & Jami 2001, 271–276.
81 Standaert 2001, 787–788, 804; the Latin manuscript is no longer extant. For a biography of Schreck, see Iannaccone 1998.
82 Dudink 2001; Hashimoto 1988, 74–163.

were other works completed before the creation of the new Calendar Department. However, Clavius remained one of the sources used. Thus the *Complete meaning of measurement* (*Celiang quanyi* 測量全義), completed by Rho in 1631, relied partly on Clavius' *Geometria practica*, while its last chapter, devoted to astronomical instruments, was based on Tycho's *Astronomiæ instauratæ mechanica* (1602).[83] In some cases, adaptations of earlier Jesuit works were made: Euclidean geometry was later added to the *Books on calendrical astronomy* in the form of a work entitled *The main methods of geometry* (*Jihe yaofa* 幾何要法, completed in 1623, printed in 1631), which Giulio Aleni had derived from the *Elements* by retaining some the constructions proposed by Clavius that Ricci and Xu Guangqi had translated into Chinese.[84] Mathematics, it should be noted, did not appear as a separate discipline in the *Books on calendrical astronomy*.[85]

Although the *Books on calendrical astronomy of the Chongzhen reign* seem to have been regarded as completed by 1635, the new astronomical system it proposed was never implemented by the Ming dynasty. After Xu Guangqi's death in 1633, it seems that the Jesuits' other protectors amongst high officials were not in a position to carry through the reform according to his initial proposal. The project of astronomical reform, however, had a deep impact on the circulation of Jesuit science in China. Until 1629, their publications were done privately, thanks to the finances of the mission and to the support of some Chinese scholars. Once the Jesuits produced mathematical and astronomical works in the service of the emperor, these works' contents could no longer be presented as evidence in favour of their religious beliefs which they were trying to promote. On the whole, they did more than merely combine the two 'specialisations' of their order, namely mission and education. The personal influence of Clavius as a role model seems to have been overwhelming in their activity. Not content with transmitting his teaching, they took up in Beijing the role he had played on the Roman scene: providing the highest power—spiritual in Rome, temporal in Beijing—with the mathematical expertise needed in order to reassert its standardising and unifying authority. By modelling their action on that of Clavius both within and outside the Society of Jesus, they succeeded in drawing attention to themselves by excellence in the mathematical sciences; the tactics he had devised for Europe turned out to bear fruit in China.

---

83 Standaert 2001, 714–715.
84 Jami 1997.
85 On Xu Guangqi's classification of the books that form the *Books on calendrical astronomy of the Chongzhen reign*, see Hashimoto & Jami 2001, 273–274.

CHAPTER 2

Western learning under the new dynasty (1644–1666)

The dynastic change opened a new era for the Jesuits in China. With Johann Adam Schall von Bell at the head of the Astronomical Bureau (*Qintianjian* 欽天監), the mission had direct protection from within imperial institutions. On the other hand, the Jesuit presence in that position was strongly contested from various sides. Nanjing, the former secondary capital of the Ming, continued to be a major gathering place for scholars until the end of the imperial age; during the first decades of Qing rule, it appears to have been a centre of study of Western learning. After the death of the Shunzhi 順治 Emperor (r. 1644–1661), charges brought against Schall resulted in a trial that put in danger his life and that of some Chinese astronomers working with him, and indeed threatened the whole Jesuit missionary enterprise in China.

The formation of the Manchu dynasty dates back to the end of the sixteenth century, when Nurhaci 努爾哈赤 (1559–1626) unified the Jušen tribes. By the time his son Hong Taiji 皇太極 (1592–1643)[1] succeeded him, his troops, allied with some of the Mongol tribes and organised into the Eight Banners (*baqi* 八旗), administrative divisions that provided the framework for their military organisation,[2] had conquered Liaodong 遼東. It was Hong Taiji who substituted the name Jušen with 'Manchu' in 1635, and the following year chose Qing 清 as a new name for his dynasty.[3] Under his reign the Manchu state became an empire. Korea and what is now Inner Mongolia fell under his control, and his troops made three incursions into Ming territory.[4] In 1643, after Hong Taiji's death, his brother Dorgon 多爾袞 (1602–1650) took over as military chief and assumed the role of regent, as the heir to the throne was only five years old. On 25 April 1644, troops led by the rebel Li Zicheng

---

1 I follow Elliott for the spelling of the personal name of the first emperor of the Qing dynasty (r. 1636–1643); for a summary of discussions of his name, see Elliott 2001, 396–397, n. 71; Hong Taiji was mistakenly known in older Western historiography as Abahai (e.g. in Hummel 1943, 1–3).
2 On the Eight Banners system and on its evolution during the Qing dynasty, see Elliott 2001.
3 Roth Li 2002, 62–63.
4 Roth Li 2002, 56.

李自成 (1605?–1645) took Beijing; the Ming Chongzhen Emperor committed suicide. Li's six weeks' rule over the capital was brutal and bloody enough to make the Manchus appear as liberators and avengers of the Ming when they entered Beijing on 5 June.[5]

2.1 Dynastic transition in Beijing: a new calendar for new rulers

Among the Jesuits who had worked on calendar reform under the Ming, only Schall was in Beijing when it fell to Li Zicheng: he had chosen not to flee, in order to protect the Jesuits' residence and their parishioners. A few days after the Manchus' takeover, he was among the crowd who presented petitions to the new government: his was a plea that his house and book collection be spared, so that he could go on improving the calendar. The Jesuit residence was located in the northern part of the city, which the Manchus had ordered to be vacated and reserved for themselves. Schall eventually succeeded not only in retaining his residence and increasing the size of the attending land, but also in claiming other Jesuit properties in Beijing.[6]

The following month, he petitioned again, offering his calendar to the new rulers, and presenting his prediction for the solar eclipse that was to occur on the 1st September as a sample result of the 'New Western method' (*xiyang xinfa* 西洋新法).[7] He claimed that this prediction would be more accurate than those made following either of the two systems hitherto used at the Astronomical Bureau: the Great concordance (*Datong* 大統) system and the Muslim (*Huihui* 回回) system. Without waiting for the eclipse to occur, Dorgon decided that 'the new method was to be used to correct the calendar' from the following year, which was to be the second of the Shunzhi reign. The new astronomical system, he added, 'should be called the "Timely modelling system" (*Shixian li* 時憲曆), in order to state our sincerest intention of modelling upon the heavens to regulate the people'.[8] This reference to the *Book of documents* (*Shangshu* 尚書),[9] one of the *Five Classics*, and the promulgation by the new dynasty of a calendar originally prepared for the Ming, were a token of the continuity of institutions through dynastic change. Dorgon also ordered an official to observe the eclipse with Schall, in order to ascertain the accuracy of the latter's prediction. The new calendar was duly issued in the tenth

---

5 Dennerline 2002, 79–81. See Wakeman 1985, 225–318.
6 Witek 1998, 111–113; on Beijing's division into Chinese and Manchu cities, see Elliott 2001, 98–105; see Map 11.1, p. 242.
7 Zhang Peiyu 1990, 1042; this was the 1st day of the 8th month of the 1st year of the Shunzhi reign.
8 Zhao 1977, 10019–10020.
9 Legge 1960, 3: 255.

month of that year; the following month Schall was put in charge of the Astronomical Bureau (*zhang Qintian jian shi* 掌欽天監事).[10] The new calendar's title page read: 'Timely modelling calendar printed by the Astronomical Bureau relying on the new Western method and promulgated throughout the Empire' (*Qintianjian yi xiyang xinfa yinzao Shixian liri banxing tianxia* 欽天監依西洋新法印造時憲曆時憲曆日頒行天下). The following year, the *Books on calendrical astronomy of the Chongzhen reign* (*Chongzhen lishu* 崇禎曆書), renamed *Books on calendrical astronomy according to the new Western method* (*Xiyang xinfa lishu* 西洋新法曆書), was presented to the new dynasty.[11]

It took decades for the Manchus to control the whole territory of the Ming empire, and during that time there was more than one centre of power in that territory. By the time Schall became an official astronomer, three Jesuits, Francesco Sambiasi (1582–1649), Michael Boym (1612–1659) and Andreas Xavier Koffler (1612–1652), were serving the Southern Ming court in Nanjing.[12] Two others, Lodovico Buglio (1606–1682) and Gabriel de Magalhães (1610–1677), were, somewhat against their will, held and given the title of 'National masters of heavenly learning' (*Tianxue guoshi* 天學國師) by Zhang Xianzhong 張獻忠 (1601–1647), the brutal and destructive warlord who controlled Sichuan. Astronomy as an imperial monopoly effectively functioned as a link between the Jesuits and whoever claimed to have the heavenly mandate. When the Manchus defeated Zhang Xianzhong, they spared Buglio and Magalhães because of the connection they could claim with Schall.[13] This shows how effective the latter's position was for the protection of his associates in Qing territory.

The transaction concluded between the missionary and the new rulers granted safety to the former and his confreres, and legitimacy to the latter. It took place within Chinese culture and institutions, to which both parts were foreign. This foreignness was to remain an issue on both sides, albeit on completely different scales. The Manchus took up many of the institutions of the Ming: they updated and reformed, sometimes invented, but rarely abolished. Their choice to abide by Chinese cultural imperatives is usually regarded as one of the reasons for the longevity of their rule over China (1644–1911, more than 250 years). The calendar was only one case among others: it had always symbolised the emperor's role as an intermediary between Heaven and Earth, or in other words between the cosmos and the human realm. Dorgon was aware that the task of promulgating a calendar for the new dynasty fell to the emperor.

---

10 Zhao 1977, 10020; *Da Qing huidian shili* 1104: 1a.
11 Zhao 1977, 1658–1659 ; Chu 1994, 72–75; Chu 2007a, 164–168.
12 Pfister 1932–34, 140–141, 266–268, 270–271.
13 Zürcher 2002.

2.2 Adam Schall, imperial astronomer

Schall was the first of a long lineage of Catholic missionaries to hold the office of imperial astronomer.[14] Many Qing institutions had two heads: one Han 漢 Chinese and one Bannerman. However, for the first two decades of the dynasty, Schall presided alone over the Astronomical Bureau.[15]

At the beginning of the dynasty, as under the Ming, the Bureau was an agency under the Ministry of Rites (*Libu* 禮部), which, in addition to state ceremonial and court ritual, was also in charge of managing the civil examination system, and of visits by foreign dignitaries.[16] This meant that all memorials and reports stemming from the Bureau were submitted through the Ministry. In 1658 the Bureau became autonomous, which gave Schall more room to manoeuvre; however, in 1663 it was put back under the Ministry's control. As to the internal structure of the bureau, in the 1640s it was divided into four sections: Section of the calendar (*Like* 曆科 or *Shixian ke* 時憲科), Section of heavenly signs (*Tianwen ke* 天文科), Section of water clocks (*Louke ke* 漏刻科), and Muslim Section (*Huihui ke* 回回科). In these sections, different methods were sometimes used for the same purpose. Thus, both the first one and the last one calculated eclipses; in principle this allowed two different methods to be compared (see Chart 2.1).

Schall's position as imperial astronomer aroused conflict both within the Bureau and amongst the China Jesuits. A thick file devoted to him in the Roman Archive of the Society bears witness to the latter. It contains documents dated between 1634 and 1666, and includes repeated accusations against him, in particular—but not only—for his acceptance of the post of imperial astronomer. One of the most adamant critics of the Jesuit presence at the Astronomical Bureau was Gabriel de Magalhães, who as mentioned above owed his life to Schall's very position.[17] After he returned from Sichuan to Beijing, Magalhães became the first Jesuit to work as a clockmaker for the emperor; but he had no official post in this capacity.[18] It was not the service of the dynasty that he criticised, but what he claimed was the superstitious character of the use made of the calendar that Schall was producing:

For when the fathers do not have the administration of this prefecture [the Astronomical Bureau] and occupy themselves with mathematics, they occupy themselves with a thing that is in itself neutral and indifferent, nay good, if it tends towards a good end, as the growth of Christianity. But when Fr. Adam has

---

14 Standaert 2001, 721.
15 Qu 1997, 48–49.
16 Hucker 1985, no. 3631.
17 ARSI, Jap. Sin. 142 & 143. Pih 1979, 61–101; Romano 2004.
18 Standaert 2001, 844.

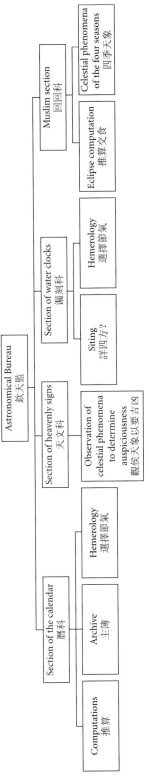

Chart 2.1 The Astronomical Bureau in 1649; based on ARSI, Jap. Sin. 142, f. 70r. (1649); see Romano 2004, esp. 746.

the charge of mathematics with a prefectural office, then he treats of this thing with a bad end.[19]

Magalhães' argument relies on his detailed description of the Astronomical Bureau, including an organisation chart that both locates the institution within the hierarchy of the Chinese civil service and describes its structure.[20] Whereas most Jesuits referred to the Bureau as the 'Tribunal of mathematics', he chooses to render *Qintianjian* 欽天監 as 'College of the veneration of heaven'. This literal translation, unlike the one used by his confreres, 'Tribunal of mathematics', and the term mostly used by historians of science who write in English, 'Astronomical Bureau',[21] emphasises the 'superstitious' functions of the institution that Magalhães reproved. These included the production of almanacs, hemerology and the interpretation of celestial and meteorological phenomena in political terms.

What is nowadays called astrology formed part of Schall's everyday work.[22] It was also one of the focal points of conflict with his Chinese subordinates at the Bureau, and more generally with Chinese scholars: he proposed to translate some European astrological treatises and to substitute their methods to those then in use.[23] However, in the manifold conflicts that put him in opposition to some of the astronomers of the Bureau, and to some Chinese scholars, these matters were not separate from others that we now think of as technical, or even 'scientific'.

Amongst the groups that were hostile to the Jesuit presence at the Astronomical Bureau was that of Muslim astronomers. They formed one of the four sections of the Bureau, the Muslim Section. This section dated back to the Yuan dynasty (1279–1368), when the rulers had summoned to China scholars from other parts of the immense territory under Mongol control. Under the Ming, they specialised in the calculation of eclipses and conjunctions. In 1644 the Muslim Section employed five officials, lead by Wu Mingxuan 吳明炫,[24] a descendant of one of the astronomers brought in by the Mongols; he was the Head of the Autumn Office (*qiuguanzheng* 秋官正). Their methods had been deemed inferior to the Western ones since the beginning of the Shunzhi reign; this threatened their position within the Bureau.

---

19 Romano 2004, 748, 755–756 for the Latin text.
20 Romano 2004, 746–747.
21 Comp. Hucker 1985, no. 1185, 'Directorate of astronomy'; the original term is used in secondary literature in Chinese and Japanese; to my knowledge the literal meaning of the original Chinese term has not been discussed in this literature.
22 Some examples are given in Lippiello 1998.
23 Huang 1993a, esp. 99.
24 The third character of Wu's name is changed to *xuan* 烜 in a number of sources, because the right part of 炫, being part of Kangxi's given name Xuanye 玄燁, became taboo; Huang 1992, 146.

In 1657, after Wu Mingxuan made what turned out to be unfounded accusations that Schall had made faulty predictions, the Muslim Section of the Bureau was closed down.[25]

Schall's success against the Muslim astronomers coincided with the assumption of personal rule by Shunzhi, which increased the Jesuits' political influence. The young emperor showed favour to Schall, but also to Lamaist and Chan Buddhist monks. It seems that, like Chinese literati, they provided him with some room to manoeuvre against the Manchu aristocracy who had overseen the conquest of China and governed it during his minority.[26] This did not preclude conflict amongst the various groups favoured by the emperor. In spite of the Jesuits' antipathy towards the 'Mahometans', the conflict with Muslim astronomers appears to have been mainly a rivalry within the bureaucracy. In the case of Lamaist Buddhism, on the other hand, the Jesuits competed with another religion, one that was favoured by the emperor. On the occasion of the Dalai Lama's visit to Beijing in 1652, Shunzhi, following the advice of Manchu officials, announced his intention to go to meet him. The Astronomical Bureau then reported that the conditions for this imperial trip were highly inauspicious, which caused Shunzhi to cancel it. Moreover, Schall memorialised to warn of various signs that he interpreted as negative effects of the Dalai Lama's visit on the emperor's power and prestige.[27]

By reaching the heart of imperial institutions, the scientific engagement of the Jesuits had changed nature: it was now put in the service of a new, foreign, dynasty, which still had to unify the empire and consolidate its rule. Chinese scholars who had an interest in Western learning were no longer close to imperial power; indeed many of them were Ming loyalists.

2.3 Jiangnan scholars and the Jesuits

Long established as China's culturally and intellectually dominant area, the Lower Yangzi area known as the Jiangnan 江南 opposed strong resistance to the Manchu conquest in 1645. The ensuing massacres remained imprinted in the literati's collective memory.[28] Although those who joined the Southern Ming were relatively few in number, the Jiangnan elite was not easily reconciled to the new rulers. Nanjing had been the Ming capital until 1403, and remained a secondary capital until the end of the dynasty. It was a major intellectual centre, towards

---

25 Zhao 1977, 3324; Hummel 1943, 890; see Huang Yi-Long 1991a.
26 Dennerline 2002, 113.
27 Huang 1991c, 8–11.
28 Dennerline 2002, 86–89.

which scholars converged on the occasion of triennial examinations. During the first decades of the Qing dynasty the concentration of literati in Nanjing was reinforced when many elite families fleeing the conquest moved there.[29]

In the decades around the middle of the seventeenth century, a number of scholars pursued an interest in what Li Zhizao 李之藻 called 'concrete things' (*qi* 器) in the *First collection of heavenly learning* (*Tianxue chuhan* 天學初函). Many of them spent time in Nanjing in the 1650s and 1660s.[30] None of them converted and in most cases their knowledge of Western learning came from reading books rather than from study with Jesuit missionaries. Among them one can mention Xiong Mingyu 熊明遇 (1579–1649), a high-ranking official acquainted with the Jesuits before 1629. Having read the *Books on calendrical astronomy of the Chongzhen reign*, he attempted a synthesis between Western astronomy and Chinese cosmology in the *Draft on the investigation of things* (*Gezhi cao* 格致草, 1648).[31] This work was not widely read, but nonetheless appears to have influenced some later scholars. Xiong Mingyu was a friend of Fang Kongzhao's 方孔炤 (1591–1655), another official of the last decades of the Ming, whose work on the *Book of change* (*Yijing* 易經), the *Compilation of current opinions on the Change of Zhou* (*Zhouyi shilun hebian* 周易時論合編, printed in 1660) included a chapter on Jesuit astronomy entitled *Summary of the Books on calendrical astronomy of the Chongzhen Reign* (*Chongzhen lishu yue* 崇禎曆書約, 1646?).[32] Fang Kongzhao was the father of Fang Yizhi 方以智 (1611–1671), one of the major scholars of the Ming–Qing transition, who also had an interest in Western learning.[33] Another famous scholar of the period, Huang Zongxi 黃宗羲 (1610–1695), was in charge of the calendar at the Southern Ming court for a short time. He was among the first to include a historical dimension to his interest in mathematics and astronomy.[34] Like all specialised scholarship, Western learning was often a family interest. Fang and Huang each had a son who wrote on mathematics: Fang Zhongtong 方中通 (1634–1698) and Huang Baijia 黃百家 (1643–1709).[35] Among the three literati of the period most famous for their expertise in astronomy, Xue Fengzuo 薛鳳祚 (1600–1680), Wang Xichan 王錫闡 (1628–1682) and Mei Wending 梅文鼎 (1633–1721), the first and the last of them spent time in Nanjing that would prove decisive for their respective astronomical works.

---

29 Wakeman 1985, 88–156.
30 Engelfriet 1998a.
31 Zhang 1994, 5–48; Fung 1997, 178–189; Engelfriet 1998, 353–356; Hsu 2008.
32 Fung 1997, 190–198.
33 Peterson 1976.
34 On Huang Zongxi's engagement in astronomy, see Yang 1997.
35 On Huang Baijia see Fung 1997.

When the Jesuits started to work on the calendar reform in 1629, they ceased to publish books on the mathematical sciences privately, in collaboration with Chinese scholars. However there was one exception during the Shunzhi reign: Nikolaus Smogulecki (1610–1656), who worked in Jiangnan in 1646–1647, and returned to Nanjing from 1651 to 1653. He was appreciated in local circles of scholars interested in mathematics and astronomy, and has come down in later sources as the perfect scholar who 'took pleasure in discussing methods of calculation with people, but did not attract people into entering the Society of Jesus'.[36] Both Xue Fengzuo and Fang Zhongtong were among his students. The former came to Smogulecki with quite a deep knowledge of astronomy gained through study of Chinese texts. Together, Xue and his Jesuit master produced a series of small treatises entitled *True source of the pacing of heavens* (*Tianbu zhenyuan* 天步真原, 1646). These were published as part of Xue's work entitled *Integration of heavenly learning* (*Tianxue huitong* 天學會通) or *Integration of learning in calendrical astronomy* (*Lixue huitong* 曆學會通, 1670). Some of these treatises were based on astronomical tables calculated on the basis of the heliocentric model by the Copernican astronomer Philippe van Lansberge (1561–1632); however they did not give any explicit explanation of the model itself.[37] On the other hand, the rest of the *True source of the pacing of heavens* consisted of several astrological writings, dealing with different branches of astrology as then practised in Europe.[38] The *True source of the pacing of heavens* thus appears to be the first Chinese work of European origin discussing the heavens that departed from the late Ming Jesuit writings appropriated by the Manchu dynasty. It was prefaced by Fang Yizhi, under a pseudonym that alluded to his loyalty towards the Ming dynasty.[39] Beside Xue's attempt to integrate the no less than five astronomical systems available to him, the *True source of the pacing of heavens* contains the first applications of trigonometry and logarithms to the computation of an ephemeris.

Like Xue Fengzuo, with whom he studied under Smogulecki,[40] Fang Zhongtong was already versed in mathematics when he came to Nanjing to study with the Jesuits in 1653. During the decade that followed, Fang Zhongtong also met Adam Schall, as he recalled in a poem entitled 'A discussion on astronomy with Master Tang Daowei 湯道未':[41]

---

36 Standaert 2001a, 51.
37 Shi 2000; Shi 2007.
38 Standaert 2001a, 55–65.
39 Shi 2007, 112–116.
40 As recalled by Fang in a poem, Zhang 1987, 248.
41 Shi 2007, 93 dates the meeting 1653–1663.

For a thousand years we have been in the Wu era.
There are a hundred ways to exhaust civilisation.
Among Han methods, Pingzi's 平子 were praised.
Among Tang monks, Yixing 一行 was prominent.
Where writings are concerned, what difference do borders make?
Those who love learning always share the same feelings.
Because I was impressed by the Master's ideas,
Deep down in my heart, day and night I admire him.[42]

Writing poems on the occasion of a meeting with another scholar was common practice. The title of the poem refers to Adam Schall by his style (*zi* 字), Daowei (his Chinese name being Tang Ruowang 湯若望), as one should refer to a fellow literatus: evidently Fang regarded Schall as a member of the same scholarly elite to which he himself belonged. The first verse refers to Shao Yong's division of time into cycles,[43] thus introducing the theme of time-measurement. The evocation of Zhang Heng 張衡 (78–139)[44] and Yixing (683–727)[45] situates Schall as an heir of the lineage of great Chinese astronomers, thereby conferring great prestige on him while appropriating his astronomy as part of the Chinese heritage. Yixing may well have also been mentioned because he was a Buddhist monk; this suggests a parallel between Tang dynasty (618–907) astronomy, influenced by Indian astronomy transmitted with Buddhism, and the imperial astronomy of Fang's own time, influenced by European astronomy, taught by priests preaching another foreign religion. This poem is one in a number of accounts—usually more factual—by Chinese scholars of their encounters with Schall.[46]

2.4 *Number and magnitude expanded*: a scholar's mathematics

Contacts between the Jesuits and some literati and the latter's study of all the works on the mathematical sciences available reshaped their conception of these disciplines. As an example, let us briefly discuss Fang Zhongtong's mathematical work, *Number and magnitude expanded* (*Shudu yan* 數度衍, completed 1661, printed 1687).[47]

---

42 Zhang 1987, 249; Fung 1995, 129.
43 In to Shao Yong's cosmology, time is divided into cycles of cycles. The Wu era (*Wuhui* 午會) is the seventh in a series of twelve cycles of 10,800 years each; according to Shao's calculations, the Wu era had started shortly before his own lifetime; Arrault 2002, 256–260.
44 Zhang Heng is referred to by his style, Pingzi, in the poem; for a biography, see Du 1993, 72–95.
45 See Du 1993, 360–372.
46 For other such accounts, see Chan 1998.
47 Zhang 1987, 250; Fung 1995, 119–121.

This work reflects what a Jiangnan scholar versed in mathematics around 1660 knew. According to the foreword (*fanli* 凡例), Smogulecki had taught written arithmetic as presented in the *Instructions for calculation in common script* (*Tongwen suanzhi* 同文算指, 1614) to the author. The work does not seem to contain any other subject that Fang might have learned from the Jesuits.[48] *Number and magnitude expanded* is an anthology of problems and methods of Chinese and Western origins, arranged according to the 'nine chapters' classification, with commentary by the author.[49] It claims to encompass all the mathematics known at the time, apart from trigonometry. After three prefaces contributed by his father, his younger brother and a friend, Fang Zhongtong explains his viewpoint and purpose in the foreword:

Western learning is refined; indeed in China [the heritage] has been lost. Here, through Western learning one returns to the nine chapters, through the nine chapters one returns to the *Gnomon of Zhou* (*Zhoubi* 周髀). The *Gnomon of Zhou* only discusses base and altitude, whereas the nine chapters are entirely derived from the base and altitude (*gougu* 句股). Therefore I have placed *Base and altitude* at the beginning, then *Lesser breadth* (*shaoguang* 少廣), then *Rectangular fields* (*fangtian* 方田), then *Consultations on works* (*shanggong* 商功), then *Differential distribution* (*chafen* 差分), then *Equitable transport* (*junshu* 均輸), then *Excess and deficit* (*yingnü* 盈朒), then *Rectangular arrays* (*fangcheng* 方程), then *Millet and cloth* (*subu* 粟布).[50]

The nine chapters cannot do without the four operations (*sifa* 四法) of addition, subtraction, multiplication and division. The four operations are combined in the four calculations (*sisuan* 四算). Therefore the methods of beads, brush, rods and ruler (*zhu bi chou chi* 珠筆籌尺) are expanded before the nine chapters.

Numbers are exhausted by the nine chapters. However some things cannot belong under one of the chapters. Therefore I have called them external methods (*waifa* 外法), and expanded them after the nine chapters. The nine chapters are also reliable in their applications.[51]

For late Ming converts, Western learning had mainly been a tool for 'practical learning' and statecraft, while also providing the means to retrieve China's ancient tradition of mathematics. For Fang Zhongtong, it was this latter role that was the most important. To seek the origins of mathematics, Fang went back to the *Gnomon of Zhou* (*Zhoubi* 周髀), a work on mathematics and astronomy now dated to the first century CE (that is, about the same period as the *Nine chapters on mathematical procedures*), which was then thought to date back to the Zhou dynasty, a thousand years earlier. The work is regarded as

---

48 Jing 1994, 2: 2558.
49 Engelfriet 1998, 362–371.
50 The nine chapters are listed here in the following order: 9, 4, 1, 5, 3, 6, 7, 8, 2; see Table 1.3, p. 18.
51 Jing 1994, 2: 2557–8.

Table 2.2 Structure of *Number and magnitude expanded* (Jing 1994, 2: 2555–2872)

Chapter number	Chapter title and 'nine chapters' classification[1]
I–III	[Opening chapters 首卷][2]
1–5	[Four calculations 四算]
6–8	Base and altitude 句股 (9) (8 items)
9–14	Lesser breadth 少廣 (4) (12 items)
15	Rectangular fields 方田 (1) (2 items)
16	Consultations on works 商功 (5) (2 items)
17–18	Differential distribution 差分 (3) (7 items)
19	Equitable transport 均輸 (6)
20	Excess and deficit 盈朒 (7) (2 items)
21	Rectangular arrays 方程 (8)
22	Millet and cloth 粟布 (2) (3 items)
23	[External methods]

1 Headings between square brackets do not pertain to the nine chapters; some of the nine chapters are explicitly subdivided in different items in the text; the number of items is then indicated between brackets. The order of the 'nine chapters' is indicated between brackets after each corresponding chapter title.
2 The version included in the *Complete library of the four treasuries* (*Siku quanshu* 四庫全書, 1782) has only two preliminary chapters. The *Summary of geometry* is at the end, as chapter 24; SKQS 802: 233–592.

containing the earliest statement of the relationship between the sides of a right-angled triangle in Chinese sources. There, as in all later texts prior to the translation of Euclid's *Elements* in 1607, the right-angled triangle is referred to by its perpendicular sides, called respectively base (*gou* 句) and altitude (*gu* 股), rather than described as a subcategory of triangle characterised by one of its angles. The *Gnomon of Zhou* was included in the Tang collection *Ten mathematical classics* (*Suanjing shishu* 算經十書, 656), and became known thereafter as the *Mathematical classic on the Gnomon of Zhou* (*Zhoubi suanjing* 周髀算經).[52] As mentioned earlier, it

---

52 For a study and translation of the *Gnomon of Zhou*, see Cullen 1996, esp. 81–92 on the 'Pythagorean relation'.

was Xu Guangqi who initiated the discussion of the base and altitude in terms of Euclidean geometry; thereafter these two branches of mathematics were often equated. Whereas there was no claim of a hierarchy amongst the chapters in the 'nine chapters' tradition, Fang Zhongtong structured mathematics according to what he knew of its historical development, its foundation being, in his view, its most ancient part. On the other hand, his reordering of the nine chapters follows the links that he believes exist between them. These links are not stated as strictly deductive, but rather in terms of 'producing' (*chu* 出 and *sheng* 生).[53] This is more evocative of relations between the Five Phases (*wuxing* 五行), a concept central to Chinese cosmology,[54] than of deduction as practised in Euclidean geometry. However, unlike the Five Phases, the system constructed by Fang is not circular: the idea that each of the nine chapters could in some way be derived from the previous one(s) is evocative of the structure of Euclid's *Elements*. Fang's project entailed a reorganisation of mathematical knowledge taking into account the Jesuits' teachings.

In so doing, he did not fit everything he had collected into the ninefold classification: the mathematics in his work expanded (*yan* 衍) to include many preliminaries, and a few external methods (*waifa* 外法) recorded after the main part, which of course consisted of nine chapters. The external methods were not methods unknown to the 'nine chapters' tradition, but rather methods pertaining to it—like the reduction of fractions to the same denominator—that in Fang's view did not belong under any of the chapter headings. And indeed the title of his work points to the fact that mathematics as he constructs it expands beyond the traditional nine-fold classification.

In Fang's reordering of the nine chapters, rectangular arrays come last but one. The corresponding chapter is divided into two parts: 'Methods with miscellaneous sums, differences and multiplications' (*Za hejiao cheng fa* 雜和較乘法), the heading under which rectangular array problems appear in the *Instructions for calculation in common script*, and 'Methods for establishing the positive and negative' (*Li zhengfu fa* 立正負法).[55] Within each of these parts the problems are classified according to the number of kinds (*se* 色) that they involve (up to five kinds). The inkstones and brushes problem is the only one that involves only two kinds in the second part of the chapter:

3 brushes are exchanged against 7 inkstones, with a contribution of 480 pieces (*wen* 文) towards the price of the inkstones. Again, 9 brushes are exchanged against 3 inkstones, with a contribution of 180 pieces towards the price of the brushes. Question: how much is each price?[56]

---

53 Jing 1994, 2: 2563; *SKQS* 802: 236.
54 Cheng 1997, 243–247.
55 Jing 1994, 2: 2825.
56 Jing 1994, 2: 2850.

Box 2.3 Layout for the brushes and inkstones problem in *Number and magnitude expanded* (Jing 1994, 2: 2850)

left	right		
inkstones positive 3	inkstones positive 7	硯正三	硯正七
brushes negative 9 63	brushes negative 3 9	筆負九 三六十	筆負三 九
price negative 180 pieces 1260	price positive 480 pieces 1440	價負一百八十文 一千六百十二	價正四百八十文 一千四百十四

The solution follows the procedure given in Cheng Dawei's *Unified lineage of mathematical methods* (*Suanfa tongzong* 算法統宗, 1592), while giving more hints as to the derivation of the layout from the question, especially emphasising how to determine whether a number should be positive (*zheng* 正) or negative (*fu* 負). On the whole, compared with his predecessors, Fang both clarifies important steps of the method of rectangular arrays and restores it to the status that the *Instructions for calculation in common script* had denied it.

Euclidean geometry is discussed in the 'Summary of geometry' *Jihe yue* 幾何約), the final chapter of Fang's work:[57] as this title indicates, the author intended to provide an outline of the *Elements of geometry*. In his introduction to *Number and magnitude expanded*, Fang pointed to the many geometrical figures he had included:

When there is a method without a diagram many people do not understand (*jie* 解) it. Here for each method I have been obliged to add a diagram. Pointing

---

[57] In some editions, including the one reproduced in Jing 1994, the 'Summary of geometry' forms the third chapter of *Number and magnitude expanded*, thus preceding the nine chapters; see Fung 1995, 129–138.

to the diagram while discussing the method makes people understand the method easily. When the explanation is easy I did not do a diagram.[58]

This concise albeit precise description of the use of geometric diagrams is revealing of the widely perceived difficulty of the 1607 translation of Euclid's *Elements*. The 'Summary of geometry' opens with most of the definitions of the first six books of the *Elements of geometry*. These are followed by the axioms, and then by the propositions. The proofs are most of the time shortened or omitted. The diagrams do play a major role: they are sometimes substitutes for a verbal definition. On the whole, Fang puts greater emphasis on the geometrical objects and their properties than on the deductive structure of the work, an attitude that seems to have been common to his contemporaries who studied the *Elements*.[59] Nevertheless the overall structure of *Number and magnitude expanded* indicates that Fang did conceive derivation as a relevant tool to reconstruct mathematics. This reconstruction is an important feature of his appropriation of Western learning.

2.5 Schall's defeat: the 1664 impeachment

While in Nanjing some scholars pursued their interest in Western learning; in Beijing Schall's position and his conflicts with other groups continued to make him enemies. The one who turned out to be the most dangerous, Yang Guangxian 楊光先 (1597–1669), was an outsider to the Astronomical Bureau. The eldest son of a family of minor hereditary military officials, Yang had given up his right to a position in favour of his younger brother, and was therefore a mere commoner. In the last decade of the Ming, he had already made himself known in the capital by his memorials attacking high officials, which he presented at the risk of his life; for this he was banished to Liaodong in 1637.[60] After 1644, he seems to have spent some time in Nanjing. Back in the capital, he attacked the Jesuits on both religious and astronomical grounds: to him they were a threat to the Confucian orthodoxy that he had always championed.[61] Yang's accusations mainly concerned the heterodox character of the Jesuits' teaching and their setting up of what was effectively a religious sect in China.[62] His first attack that had some bearing on the Astronomical Bureau followed the death of one of the emperor's sons: in his *Deliberations on hemerology* (*Xuanze yi* 選擇議, 1659), he claimed that the Bureau had mistakenly selected an inauspicious time for the burial of Prince Rong 榮, a son of the Shunzhi

---

58 Jing 1994, 2: 2558.
59 Martzloff 1980; Engelfriet 1998, 362–439.
60 Hummel 1943, 89–890; Moortgat 1996, 261.
61 Chu 1994, 17–20.
62 Fu 1966, 35–36; Kessler 1976, 58–60.

emperor who had died of smallpox in infancy. This, Yang insinuated, could have unfortunate consequences for the imperial family.[63] The matter was investigated, but had no further consequence. In some of his other tracts against the Jesuits, Yang took up Wu Mingxuan's criticisms against their astronomy. Among the erroneous innovations (*xin* 新) enforced by Schall at the Bureau, he pointed out that closing down the Muslim section of the Bureau had brought to an end the practice of comparing the results obtained using different methods. He singled out the phrase 'according to the new Western method' (*yi xiyang xinfa* 依西洋新法) printed on every calendar as a symbolic surrender to the West that threatened the very legitimacy of the new dynasty, just as did the Westerners' seditious teachings. Similarly, Yang pointed out that Schall had improperly presented a calendar for 'only two hundred years', which suggested that the Qing dynasty would not be long-lived.[64] Another change that he disapproved of related to time units: whereas the day used to be divided into 100 *ke* 刻 for the purpose of time-measurement, the Jesuits had changed this unit so that a day would only contain 96 *ke*. This was of course consistent with the European system of time-measurement, this new *ke* corresponding to 15 minutes. However it also fitted in with the Chinese division of the day into twelve 'double-hours', each of these containing 8 new *ke*. Finally, Yang denounced alleged mistakes in the 1661 calendar.[65]

Schall's favour with the emperor made these attacks vain. However, while the Jesuit astronomer maintained good relations with a number of Chinese scholars, he had fewer friends among the Manchu elite. When Shunzhi died in 1661, his son Xuanye 玄燁 (1654–1722), for whom the reign name Kangxi 康熙 was chosen, was only seven years old: four Manchu regents were nominated. According to most historians, they reversed Shunzhi's policy towards Chinese officials, and chose to 'rule from horseback'.[66] They were suspicious of the Jesuits as well as of the Buddhists who had been close to the late emperor.[67] This, combined with the hostility felt by a number of Chinese scholars and officials against the Jesuits' religion and against the position given to Schall in the civil service, may suffice to explain why in 1664 Yang Guangxian's accusations against the Jesuits finally found a favourable ear.[68]

---

63 Yang 2000, 41–42.
64 Yang 2000, 47.
65 Yang 2000, 43–47; Chu 1997, 11–14.
66 Oxnam 1975, 4; Struve 2004 argues that this view needs some qualification.
67 Oxnam 1975, 148.
68 The secondary literature is too abundant to be listed here exhaustively; the missionaries' accounts are taken up in Väth 1991, 295–319; Huang 1991a–d, 1993a–b has discussed the various aspects of Schall's practice as imperial astronomer on which Yang Guangxian's accusations rested; Moortgat 1996 interprets Yang's attacks against Schall's astronomy in the light of his belief that astronomy in its most technical aspect should be consistent with the world-view to which he adhered; Chu 1997 analyses the political and

CHAPTER 2 | Western learning under the new dynasty (1644–1666)

What may have prompted Yang to present a further memorial at that time was the publication of the *Survey of the transmission of heavenly learning* (*Tianxue chuangai* 天學傳概, 1664), where, for the first time, it was stated in print—among other heterodox views—that the Chinese were descendants of Adam and Eve, and that during the Zhou dynasty the Christian God had been worshipped in China. The author, Li Zubai 李祖白 (d. 1665), was a Christian and an official astronomer; a Hanlin Academician, Xu Zhijian 許之漸 (1613–1700), wrote a preface to the work.[69] Yang's *I cannot do otherwise* (*Budeyi* 不得已), published 1665, which gathered all his writings against the Jesuits, duly denounced Li and all those who had contributed to the publication of the *Survey of the transmission of heavenly learning*.[70]

Yang had allied with Wu Mingxuan against the Jesuits' astronomy; however the detailed criticisms put forward in *I cannot do otherwise* revealed his own limited understanding of the discipline. It would have been easy for the Jesuits to respond to him on that ground;[71] but those who were to decide on the validity of Yang's charges knew no more than him about the technicalities of astronomy. Yang, however, was entirely consistent in this respect, readily admitting his own incompetence in astronomical methods (*fa* 法) and claiming to be concerned only with the underlying principles (*li* 理), as a scholar following Zhu Xi's philosophy should be.[72] In his view, the principles that underlay the cosmos were more important than the methods, and no one who misunderstood or challenged these principles should be entrusted with the tasks performed at the Astronomical Bureau, as these tasks had such essential bearing on whether human life conformed to the cosmos. What turned out to be his most serious accusation was in keeping with this point of view: the death of Prince Rong's mother and then that of the emperor himself had followed the miscalculation of the time for the burial of the infant prince within two years. Both the parents died of smallpox like their infant son. It was well known at the time that the Manchus were particularly vulnerable to this disease, and very strict measures were taken to prevent its propagation among them.[73] It was against a background of predominance of the traditional Chinese worldview, rather than in reference to specifically Manchu beliefs, that the link between the mistaken

---

cultural stakes of the trial; see also Hashimoto 1970a & 2007, Zurndorfer 1993, Elman 2005, 133–144.

69   Standaert 2001, 429–430, 513–514.
70   Yang 2000, 10–11; on the *Budeyi*, see a.o. Moortgat 1996, Menegon 1998.
71   As Verbiest later did in the *Lifa bude yi bian* 曆法不得已辨 (Refutation of the 'I cannot do otherwise' in astronomy, 1669); Yang 2000, 139–193.
72   Moortgat 1996, 271.
73   Chang 2002.

calculation and the emperor's death implied by Yang took all its significance.

In September 1664, his accusations were formally received at the Ministry of Rites.[74] Two months later, the Beijing Jesuits were arrested, as well as some official astronomers.[75] At the time four Jesuits resided in Beijing: Schall, Buglio, Magalhães, and Ferdinand Verbiest (1623–1688), who had been Schall's assistant in astronomy since 1660. The missionaries working elsewhere in China were sent to the capital to await judgement. A couple of months later, before the sentence was issued, the Ministry of Rites decided that 'the phrase 'according to the new Western method' on the cover of the calendars published every year should be replaced by the phrase 'approved after submission to the emperor' (*zouzhun* 奏准)'.[76]

The case was not concluded until the following spring. In the meantime a spectacular celestial phenomenon was seen in the whole empire: this was the famous 1664 comet, which was widely observed around the world.[77] Between November 1664 and April 1665, there are ten mentions of the comet in the *Veritable records* (*Shilu* 實錄) of the dynasty, to say nothing of many in local gazetteers (*difangzhi* 地方志).[78] In China this would be understood as a sign that serious reconsideration of policies might be needed. Thus, some officials took this occasion to call for a restoration of earlier practice in imperial examinations.[79] The parallel with Yang's call for a return to the methods used under the Ming at the Astronomical Bureau is striking; however there seems to be no mention of the comet in the materials that refer to the astronomers' trial.

In April 1665, a judgment was submitted to the regents for approval: although astronomical matters were deemed difficult to decide, it was apparent that some of Yang's accusations were founded; in particular, it was agreed that the wrong astrology book had been used to choose Prince Rong's burial time. Consequently a death sentence by dismembering was proposed for eight officials of the Astronomical Bureau, including Schall himself; some of their relatives were to be decapitated.[80] According to Jesuit and Chinese Christian sources, this sentence was issued on 15 April.[81] In the *Veritable Records* however, it only appears on 30 April, with the regents' response, which called for

---

74 Väth 1991, 299.
75 Väth 1991, 303.
76 Zhao 1977, 1664.
77 See e.g. McGuire & Tamny 1985 on Newton's observations of the comet; in England it was associated with the outbreak of the Great Plague of London; for an Annamese record of the comet, see Ho 1964, 134.
78 Beijing tianwentai 1988, 466 (*SL* 13–14, *passim*); for mentions of the comet in local gazetteers see Beijing tianwentai 1988, 466–469.
79 Struve 2004, 20–21.
80 *SL* 14: 27a–28a; see Fu 1966, 37.
81 Väth 1991, 312; Brockey 2002, 132, n. 14; Han & Wu 2006, 302–303.

reconsideration, on the grounds that Schall himself was not a specialist in hemerology, that he had presided successfully over the choice of a number of graves for the imperial family, and that, having served for many years, he was now elderly.[82] Dates are important here: on 16 April, a strong earthquake was felt in the capital. Three days later, the regents issued a statement acknowledging that the comet and the earthquakes were warnings that 'government affairs were not run in a wholly appropriate manner'.[83] Therefore, it is very plausible that Schall owed the regents' leniency to the earthquake, as suggested by Christian sources and historiography: what the rulers of China had to take as a warning from the Heavens appeared as a sign of divine providence to the Christian convicts.[84] It should be noted, however, that it was not uncommon for harsh sentences against civil servants to be commuted by rulers. A second sentence, which appears in the *Veritable Records* dated 17 May, proposed that Schall and three of the Chinese astronomers only be flogged and banished, while the others would still be executed; again the regents' response softened it, as Schall and the three for whom the death penalty had been raised were pardoned; so were their relatives.[85] According to an often quoted nineteenth century Chinese compilation, Grand Empress Dowager Xiaozhuang 孝莊 (1613–1688), the widow of Hong Taiji, and the mother of the late Shunzhi emperor, persuaded the regents that Schall should be released and allowed to join his fellow Jesuits in house detention.[86]

Five of the astronomers under trial, who were Christians, still were to be executed. They were beheaded instead of dismembered, which was another gesture of leniency. Among them was Li Zubai, the author of the *Survey of the transmission of heavenly learning*. So perhaps the main motive for the condemnation was the threat posed by Christianity to Chinese beliefs concerning the origins and foundations of civilisation, rather than Schall's practice as an imperial astronomer and the mistake in the time of the prince's funeral. It is also interesting to note that the arguments used to justify the sentence point out departures from traditional practices of the Astronomical Bureau as inherited from the Ming, rather than the violation of any new rules that the Manchus were striving to enforce. Thus, the astronomers' condemnation did not necessarily result from the regents' hostility against Christianity. It appears rather, as an attenuated endorsement on their part of the hostility felt by some Chinese literati who claimed to defend the

---

82 *SL* 14: 28a–28b; see Fu 1966, 38.
83 Kangxi 4/3/2 for the earthquake; Kangxi 4/3/4 for the edict; *SL* 14: 17a–17b.
84 See e.g. Väth 1991, 313.
85 Kangxi 4/4/3; *SL* 15: 1b–2a.
86 *Zhengjiao fengbao* 正教奉褒; Han & Wu 2006, 302–303; for an assessment of the *Zhengjiao fengbao*, see Standaert 2001, 133–134.

foundations of a social order that the Jesuits' teachings and their inadequate running of the Astronomical Bureau was threatening.[87]

For the Jesuit mission, the trial was an unmitigated disaster: Buglio, Magalhães and Verbiest remained under house arrest in Beijing with Schall. All the other missionaries were banished to Canton.[88] This put a halt to the whole enterprise that had by then been continued for more than eighty years. Schall died the following year. The very means by which he had, for twenty years, ensured the continuity and the prosperity of that enterprise had caused its collapse.

During the two first decades of the Qing dynasty, Western learning was appropriated in two separate settings. Schall, acting as an official, turned Jesuit astronomy into the basis of the Manchu dynasty's calendar. Thereafter the mission no longer needed to rely on the patronage of Chinese officials. However, this very closeness to the centre of power evoked strong criticism and enmity; the risk that it entailed eventually materialised with disastrous consequences. Outside the capital, the pattern of collaboration from which translations had resulted during the last decades of the Ming dynasty had been replaced by one in which Chinese scholars studied these translations, mostly on their own but sometimes under a Jesuit teacher, and incorporated Western learning as an element of their own scholarship. Another point of contrast with the late Ming was that there was a divorce between the scholars versed in Western learning and the state. Concomitantly these scholars laid more emphasis on the links between mathematics and philosophy—especially studies on the *Book of change*—than on those between mathematics and statecraft. At the same time, another rupture was finalised: that between the mathematical sciences and the rest of the Jesuits' teachings. Not only were there no converts among the scholars versed in Western learning, but these scholars showed less and less interest in what pertained to the Jesuits' natural philosophy.[89] While the condemnation of Schall and his colleagues was a telling blow to the mission as well as to Jesuit astronomy, there is little evidence regarding the extent to and ways in which it affected Jiangnan scholars' interest in and studies of Western methods.

---

87 This is suggested in Struve 2004, 21, n. 75; comp. Chu 1997, 14–16.
88 Brockey 2007, 129–130.
89 Peterson 2002.

PART II

The first two decades of Kangxi's rule

CHAPTER 3

The emperor and his astronomer (1668–1688)

3.1 Kangxi's takeover and the rehabilitation of Jesuit astronomy

The four regents who ruled in Kangxi's name during his minority appear to have been generally distrustful of Chinese literati and hostile to Chinese culture. Nonetheless, the condemnation of Schall and of his Chinese colleagues under the regents' rule is usually presented as a victory of conservative literati over what they saw as a threat against Confucian orthodoxy. This apparent contradiction hints at the complexity of factions and alliances in court politics during the regency.

Schall had been one of Shunzhi's favourites, whereas Yang Guangxian 楊光先 seems to have had the patronage of Suksaha 蘇克薩哈 (d.1667), one of the four regents.[1] After the 1665 verdict, Yang Guangxian was appointed Director of the Astronomical Bureau, despite his reluctance to take up a post for which, he insisted, he had no technical competence. Wu Mingxuan 吳明炫 was to supervise the calendar, as a Vice-Director (*jianfu* 監副) of the Bureau; its Muslim Section was revived.[2] Also, for the first time, a Manchu Director was appointed for the Bureau in parallel with Yang Guangxian, according to the rule that applied elsewhere in the administration;[3] Mahu 馬祜, the first Manchu Director, does not seem to have known much about astronomy. In 1666, the number of Chinese Students in Astronomy (*tianwensheng* 天文生), which had been settled at sixty-six in 1644, was increased to ninety-four. Thus new recruits, unfamiliar with Schall's astronomy, began their training under Wu Mingxuan.

Meanwhile, the four Beijing Jesuits remained under house arrest at the Eastern Church (Dongtang 東堂), a residence that had been given to Buglio and Magalhães by the Shunzhi emperor.[4] Schall died in 1666; the other three did not discontinue activities that could serve their relations with officials, Manchu and Chinese. Verbiest improved his

---
1   Fu 1966, II-452, n. 67; Pih 1979, 204.
2   *Da Qing huidian shili* 1103, 2b; Golvers 1993, 73.
3   Qu 1997, 49.
4   Standaert 2001, 583.

skills in astronomy; Buglio went on painting using perspective, with the help of at least one Chinese assistant who was a convert; Magalhães also continued to make and repair clocks and other instruments.[5] In 1666 Suksaha, who lived close to the Eastern Church, used the services of Magalhães to repair some of his clocks, and asked the Jesuits to build a machine that could take water out of a well: despite his support of their enemy, he had recourse to their technical skills.[6] In March 1668, when a white light (*baiguang* 白光) was seen in the southwestern part of the Beijing sky,[7] some official astronomers consulted Verbiest on this novel phenomenon. He provided them with a short essay, in which he included a detailed description of what he had observed, with illustrations, estimates of the size and position of the object in space, and what he himself described as a 'prophecy' (*vaticinium*). He took this phenomenon to be related to the unusually mild weather of that winter, during which various epidemics had plagued the capital.[8] Evidently, Jesuit contacts with the Manchu rulers and with officialdom at various levels were not severed. The Jesuits were quite well informed on the developments of court politics, which they reported in their correspondence.[9]

Kangxi's official rule began on 25 August 1667; however it took him another two years actually to seize power from Oboi 鼇拜 (d. 1669), who then held effective power: of the three other regents, Soni 索尼 died of illness in 1667, Oboi had Suksaha executed in that same year; Ebilun 遏必隆 (d. 1674) appears to have been supportive of or submissive to Oboi.[10] The young emperor's takeover was a complex process that entailed a shift of balance between Manchu factions. His grandmother, Empress Dowager Xiaozhuang 孝莊, was a long-term ally of his; it is not known to what extent she guided him towards victory.[11]

The calling into question of the 1664 impeachment is usually regarded as an episode in Kangxi's struggle against Oboi that led to the latter's downfall. Indeed according to the Jesuits, the Regents supported Yang Guangxian and Wu Mingxuan. But it would be simplistic to assume that this was the sole motivation for the emperor's concern with astronomical matters. What seems to be the first expression of this concern could well have been triggered by the celestial phenomenon about which Verbiest was consulted in March 1668; it would not have escaped the emperor's attention that it came after a winter of plagues in

---

5   Neither did they stop the production of texts related to proselitisation. For example, Buglio translated the *Missale Rituale Romano*; Standaert 2008, 106–110.
6   Pih 1979, 204; Golvers 1993, 63, 171–172, n. 7, 299, n. 4.
7   Zhao 1977 1468, 1483; other records of this phenomenon are given in Beijing tianwentai 1988, 471–473, in the section on comets.
8   Josson & Willaert 1938, 122–129, esp. 125.
9   Pih 1979, 204–205, 215; Golvers 1993, 19–20, 172.
10 Oxnam 1975, 167–181, 185, 189–191.
11 Spence 2002, 128.

the capital. On 28 March an edict was sent to the Ministry of Rites, which at the time oversaw the Astronomical Bureau:

Celestial phenomena are of major consequence. It is necessary to get people who are well versed in them and skilled; then predictions will be verified without mistake. Let the Governors and Governor generals of Zhili 直隸 and all provinces inform all the districts under their jurisdiction of this edict. If there are men versed in astronomy there, once they have been sent to the capital and examined, they will be employed at the Astronomical Bureau. They will be promoted and transferred like all officials of the Ministries and Offices.[12]

Thus the young monarch's first measure in relation to astronomy was to recruit new staff, selected for their competence, and who had no connection with Wu Mingxuan. In the following months, Kangxi had evidence that the Astronomical Bureau did not work as it should: on 26 June, after an earthquake hit Beijing, he rebuked the Bureau for failing to submit astrological interpretations of the phenomena on which they had earlier reported.[13] Two months later, Wu Mingxuan, in his capacity as Vice-Director, reported that the 'old method' (*gufa* 古法, i.e. that of the Great concordance (*Datong* 大統) system in use until 1644) was faulty, and that there were discrepancies between the calendar calculated according to it and the one produced by the Muslim section. This suggests that the condemnation of Christian astronomers had not put an end to dissensions within the Astronomical Bureau. The Ministry of Rites, ordered to investigate the issue, sided with Wu, and it was decided accordingly that from the following year Wu Mingxuan should calculate and submit calendars.[14] At this stage the emperor seems to have been unaware of the Jesuits' presence in the capital. The first motivation of his attitude towards the Bureau was indeed, as he stated in his first edict, to ensure that it should produce accurate calculations as well as complete reports of observations.

Every year, on the first day of the tenth month, the calendar for the following year was formally presented to the emperor, and then circulated in the whole empire and sent to vassal states.[15] The first calendar submitted by Wu Mingxuan aroused controversy. According to Kangxi's later reminiscence:

In the seventh year of Kangxi (1668), after adding an intercalary month and promulgating the calendar,[16] the Astronomical Bureau memorialised again,

---

12 *SL* 25: 14b–15a; this edict was later reproduced in the *Da Qing huidian shili*, omitting the phrase 'then predictions will be verified without mistake' (*nai ke zhan yan wu wu* 乃可占驗無誤); Jami 1994a, 237.
13 *SL* 26: 4a–4b.
14 *SL* 26: 16b–17a, 25a–26b.
15 Golvers 1993, 75–78.
16 康熙七年，閏月頒曆之後，欽天監再題，欲加十二月又閏. The translation of the first part of this sentence is controversial; Han Qi, Chu Pingyi and Wang Yangzong

wishing to add another intercalary month after the twelfth month. Many people talked about this, rejecting the proposal on the grounds that never, since the calendars of Antiquity, had one heard of a year with two intercalary months.[17]

There is no trace of such an incongruous proposal in the *Veritable (Shilu* 實錄*)*[18] or in the other sources that I have consulted; but it would indeed explain why the calendar became the object of wide discussion. It is unclear who at the Bureau submitted the memorial to which Kangxi refers: was it Wu himself or the proponents of the 'old method' who first cast doubt on the published calendar? In any case, it seems to be as a consequence of this memorial that Kangxi heard that some Westerners versed in astronomy were still in the capital; he sought their opinion on Wu Mingxuan's calendar.[19]

On Christmas day some officials visited their residence for that purpose.[20] Verbiest immediately replied that 'the ephemerides were indeed teeming with mistakes'; in particular, he stated that Wu had wrongly inserted an intercalary month after the twelfth month.[21]

Let us briefly explain the issue: the Chinese calendar was luni-solar. This means that the first day of each month should contain the instant of a new moon, i.e. a conjunction of the Sun and Moon. As 235 lunar months take up almost exactly the same time interval as nineteen tropical years,[22] an intercalary month needed to be added in seven out of nineteen years in order to keep the calendar in step with the seasons, which means that there would be thirteen months instead of twelve in the year approximately once every three years. In order to

---

have all proposed to put the first punctuation mark after 閏月, so that the sentence would read 'During the intercalary month of the 7th year of Kangxi, after the calendar was issued...' Han 1997, 14; Chu 2005, 75; Wang 2006a, 118. However, there was no intercalary month in that year (Xue & Ouyang 1956, 334); in any case, in dates intercalary months are usually numbered according to the month that they follow. The punctuation given here follows the layout of the original print-out of this text (dated to 1704), which is bilingual Chinese and Manchu (3a). According to Prof. Watanabe Junsei (personal communication), the Manchu version of the text is consistent with the translation proposed here.

17 *Discussion of triangles and computation imperially composed* (*Yuzhi sanjiaoxing tuisuanfa lun* 御製三角形推算法論, 1704); IMARL, no. 2335; a reproduction of another copy is available at http://archive.wul.waseda.ac.jp/kosho/ni05/ni05_00410/ni05_00410.html. The Chinese version of the text is reproduced in the collection of Kangxi's writings first printed in 1732, *Shengzu renhuangdi yuzhi wenji* 聖祖仁皇帝御製文集; see SKQS 1299: 156–157; on this text see pp. 248–251.

18 The *Veritable Records* of that period tend to contain fewer materials than those for later years of the reign.

19 The Jesuits did not know who first mentioned them to the emperor, and to my knowledge there is no evidence on this point; Pih 1979, 214.

20 Golvers 1993, 174 n. 13.

21 Golvers 1993, 58.

22 This period is conventionally called the Metonic cycle; a tropical year is the interval between two winter solstices; it is the interval at which the seasons renew.

determine when such a month should be inserted, the tropical year was divided by twenty-four equally spaced instants: twelve so-called central *qi* (*zhongqi* 中氣) alternated with twelve nodal *qi* (*jieqi* 節氣).[23] The central *qi* included the solstices and equinoxes. In this system, the average time that it takes for the Sun to move from one central *qi* to the next one is approximately $365.25 \div 12 \approx 30.44$ days, while the duration of a lunar month is only 29.53 days. Therefore, at some stage in the cycle a lunar month would not contain a central *qi*: such a month was designated as intercalary (*run* 閏); it was numbered identically to the previous one. The official astronomers' request that an intercalary month should be added after the twelfth month of the published calendar must have relied on calculations that predicted that the Sun would not reach *yushui* 雨水 (which should be the first central *qi* of a year) during the month that followed the twelfth month of the published calendar. But suggesting that two intercalary months were needed in the same year amounted to acknowledging that there had been a mistake in a previous calendar: the calendar should never be more than one month out of step with the seasons. Verbiest's claim that there should be no intercalary month at all in that year was in itself easier to accept, but its implications were no less serious: if the Jesuit turned out to be correct, the whole empire would have to be informed that the Bureau had made a mistake, and the copies of the calendar already circulated would have to be replaced. Therefore the emperor's reaction was prompt: the next day he sent a joint order to imperial astronomers and to the three Jesuits:

Heavenly signs are very profound; calendrical methods are linked to essential affairs of the State. All of you should not think of your long-standing enmity, in which each has held fast to his own view, taking the other party to be wrong, and fighting. What is correct should be followed as appropriate; what is incorrect should be revised further. Your task is to correct astronomy, so as to perfect the very best method.[24]

Calling for cooperation between the parties could hardly suffice to solve the problem: it was necessary to decide who was right as regards the intercalary month. For that purpose, Kangxi gave audience to all those involved. He asked Verbiest to inform him whether there was 'any apparent sign by which it could be proven before [his] eyes whether the calendar calculation does or does not correspond with the Heavens', and then accepted the Jesuit's suggestion that the calculation of the length of the shadow of a gnomon at a given time was such

---

23 In everyday usage, *jieqi* has come to refer to the solar periods between two of the 24 points, as well as the time when the Sun reaches them. Imperial astronomers were expected to determine this time by daily measurements of the Sun's shadow rather than simply to keep count of the days; *SL* 26: 26a.
24 Han & Wu 2006, 73.

a test.[25] Thus the emperor was provided with a means to arbitrate a controversy between specialists in a field of which he knew nothing, on the basis of a visual criterion. More generally, this implied that officials could control astronomers even though the former were not trained in the computational aspects of the latter's profession. Indeed on the following day, a number of officials accompanied Yang Guangxian, Wu Mingxuan and Verbiest to the Observatory. Wu declined to carry out a calculation; the Jesuit successfully predicted where the shadow of the gnomon would reach by noon. This was reported to Kangxi, who ordered that the test should be repeated on the following two days, once inside the Forbidden City, and once more at the Observatory; these repeats were further successes for Verbiest.[26] The latter, who was aware of the errors that were unavoidable in such measurements, due in particular to the mounting of instruments, attributed this success to 'God's most clement providence'.[27]

This was only a beginning. The Jesuit then wrote a detailed report on Wu Mingxuan's calendar: in addition to a wrong intercalary twelfth month, it contained, among other conspicuous mistakes, two spring equinoxes and two autumn equinoxes.[28] Kangxi then ordered 'the Princes of the Deliberative Council (*yizheng wang* 議政王), the Princes of the third rank (*beile* 貝勒), the Grand Ministers (*dachen* 大臣) and the Nine Ministers (*jiujing* 九卿)' to make a decision about this case: clearly he treated the matter as an affair of state. According to Verbiest, referring the correction of the calendar to the Deliberative Council was unprecedented.[29] Thereupon, according to the emperor's reminiscence, 'the Princes, the Nine Ministers and others examined the issue again and again; but there was no one at Court who understood the calendar.'[30] An *ad hoc* committee of twenty officials was nominated to investigate the matter further.[31] Together with Mahu, the Manchu Director of the Astronomical Bureau, they took part in further observations that were to decide between Verbiest and Wu Mingxuan. The former had proposed as tests some points for which he found severe discrepancies in Wu's calculations. These were the moments when

---

25 Golvers, 1993, 60.
26 Golvers 1993, 62–65; Han & Wu 2006, 304.
27 Golvers 1993, 64–65; on the issue of error, see also Greslon 1672, 7.
28 *SL* 27, 24a. To my knowledge, Wu Mingxuan's calendar is no longer extant. Therefore one can only speculate that it may have given both the equinoxes calculated using the Great concordance system and those calculated using the Muslim system; this would be consistent with Verbiest's statement that Wu 'mixed Chinese with Arabic, and Arabic with Chinese so that the Calendar could be called 'Arabic–Chinese' (Golvers 1993, 65).
29 Golvers 1993, 66.
30 *SKQS* 1299: 156.
31 *SL* 27: 24a–24b.

the Sun would reach two particular *qi*,³² and the positions of the Moon, of Mars and of Jupiter at certain given times. In all cases Verbiest's calculations turned out to be more accurate than those of Wu; in particular, observations of the Sun showed Wu Mingxuan's calendar to have misplaced the time when the Sun would reach *yushui* by more than a day.³³ Therefore the committee recommended that the Jesuit should henceforth be put in charge of the calendar.³⁴ The emperor's response to this suggests that his purpose was not merely to displace Yang Guangxian and his allies and to promote the Jesuits:

When Yang Guangxian first accused Schall and the Assembly of the Princes of the Deliberative council and the Grand Ministers deliberated, on what grounds did they find Yang to be right? When their advice was followed, on what grounds did they find Schall to be wrong? Because thereupon you call off (*tingzhi* 停止) their advice, and replace the advice of that day by the advice of today, it cannot be appropriate for you [to do so] without finding out the details from Mahu, Yang Guangxian, Wu Mingxuan and Verbiest and reporting minutely. It is ordered that you should deliberate again in a more reliable fashion.³⁵

This order appears to be an exercise in authority. Clearly, for the young emperor, the process of decision-making mattered as much as the decision itself; if the decision taken four years earlier was to be reversed, this should be done with full knowledge of the facts, in an incontrovertible way, so as to close the issue for good. It has been suggested that he hoped to obtain not only the reversal of a decision taken under the regents, but also their direct implication in the case.³⁶ Ten days later, the Princes and Ministers submitted a new memorial, in which they reported that Mahu and three deputy heads of the Bureau, who had been promoted under Yang's directorship,³⁷ all believed that Verbiest's calendar was 'in accordance with celestial phenomena' (*he tian xiang* 合天象), and that moreover, 'although the division of the day into 100 *ke* had a long history, Verbiest's method for calculating [with] 96 *ke* was entirely in accordance with celestial phenomena'.³⁸ Therefore this division could be enforced from the following year on. After recommending the adoption of some corrections that Verbiest had proposed to the calendar of the current year, the committee concluded by advising that both Wu Mingxuan and Yang Guangxian should be

---

32 Namely *lichun* 立春, the nodal *qi* closest to the Chinese New Year, and *yushui* 雨水, the first central *qi*.
33 Here the *SL* and Verbiest's reminiscence converge; Golvers 1993, 66–69; see also Greslon 1672, 3–4.
34 On the deliberations that resulted in this advice, see Golvers 1993, 202, n. 2.
35 *SL* 28: 6b; Kangxi 8/1/26 (26 February 1669).
36 Kessler 1976, 63.
37 These were Yitala 宜塔喇, Hu Zhenyue 胡振鉞 and Li Guangxian 李光顯; *SL* 28: 8b; Qu 1997, 49.
38 See p. 50.

dismissed and that the former should be handed over to the Ministry of Justice to be judged, whereas the latter should be spared.[39] Verbiest has left an account of the deliberations that resulted in this final advice. According to him, the two regents still in power at the time, Oboi and Ebilun 'who were [his] adversaries and the initial instigators of all quarrels',[40] attended the debates, and most of the Chinese officials present were hostile to him as well. The proceedings went on for four days. It was over the issue of the division of the day into 100 or 96 *ke* that Yang Guangxian sealed his own fate, by exclaiming: 'If the Manchus, by repealing four quarters of the natural day, should side with Ferdinand and the European astronomy, then the Manchu empire in China will not last long!'[41] For a Chinese to evoke the possibility that the Qing dynasty might be short-lived in front of such an assembly was an irreparable gaffe.

In April, Verbiest was appointed to the Astronomical Bureau, and the calendar for the current year was corrected according to his calculations: the contentious intercalary month would be inserted after the second month of the following year.[42] However, Jesuit astronomy did not fully regain the status it had had under Shunzhi: the calendar's title did not revert to boasting a foreign origin. Moreover, unlike Schall, Verbiest was never actually made the head of the Bureau. His title was eventually changed to a new one, which seems to have been created especially for him: he was Administrator of the Calendar (*zhili lifa* 治理曆法), 'acting as Chinese Director' (*chong Han jianzheng* 充漢監正);[43] the Jesuits' standing within the civil service was not entirely regained.

Two months later, in June 1669, Kangxi had Oboi and his foremost supporters arrested and tried.[44] In September, following an accusation from Verbiest and some of his colleague astronomers, Yang Guangxian was tried for, among other charges, being a follower of Oboi. He was sentenced to death and pardoned, while the astronomers condemned in 1665 were rehabilitated; Schall was then granted the funeral that befitted his rank.[45] However missionaries who had worked outside the capital were not immediately allowed to return to their residences; it was only in March 1671 that Verbiest's personal favour with Kangxi brought this permission.[46] The former summarised this success in the

---

39 *SL* 28 8b–9b; Kangxi 8/2/8 (7 March 1669).
40 Golvers 1993, 69.
41 Golvers 1993, 70.
42 *SL* 28: 15b–16a, Kangxi 8/3/17 (17 April 1669).
43 Qu 1997, 49; *Da Qing huidian shili* 1103: 1a.
44 Kessler 1976, 71–73.
45 *SL* 31: 4a–5a; Kangxi 8/8/11 (5 September 1669); *SL* 31: 7b; on Schall's funeral see Standaert 2008, 190–192.
46 Brockey 2007, 136.

preface of his triumphant account of 'European astronomy called back from the Dark into the Light'[47] published in Dillingen in 1687:

> The rehabilitation of our astronomy has been the only cause—after God!—of the rehabilitation of our Religion, and [...] our astronomy remains even now the most important root for the propagation of that Religion all over China.[48]

Although there continued to be dissenting voices, such as that of Magalhães,[49] the rationale that put forward astronomy as the fundamental support and protection of the mission remained dominant among Jesuits throughout the Kangxi reign.

3.2 Ferdinand Verbiest, imperial astronomer and tutor

The Bureau where Verbiest took office had only three sections: the Muslim Section was finally closed down in 1669.[50] Two years later, the Bureau regained its independence vis-à-vis the Ministry of Rites: thereafter the former institution was to report to the latter only once a year, for the agrarian rites performed on the day of the Beginning of Spring (*lichun* 立春).[51] According to his description, the three sections of the Bureau were located in three different parts of Beijing. The Section of heavenly signs, specialising in observation, had its seat in the Eastern part of the city, near the Observatory. The Section of water clocks which seems to have been the largest, was located centrally, near the Outer Palace.[52] The Section of the calendar was adjacent to the first Jesuit residence, in the Western part of the Inner City, near the Xuanwu 宣武 Gate. This was in the former location of the Calendar Office created in 1629 and first supervised by Xu Guangqi, where the calendar reform had been elaborated before the fall of the Ming dynasty. Verbiest's title suggests that formally he was only in charge of the Section of the calendar. However, both his account and Chinese sources reveal that his responsibilities entailed giving interpretations of the configurations of the sky that he predicted:

> Each half quarter of a year, therefore eight times a year, I [Verbiest] am to draw up the figure of the firmament. I also have to forecast very meticulously the future disposition of the Heavens at a given half-quarter, and the change in the atmosphere, including its consequences like plague or other diseases, food shortage, etc., indicating also the days on which wind, lightning, rain, snow

---

47 *Astronomia Europæa ex umbra in lucem revocata*... Golvers 1993, 51.
48 Golvers 1993, 55, 349.
49 See pp. 38–40; Pfister 1932–34, 254.
50 Golvers 1993, 73.
51 *Da Qing huidian shili* 1103 2b. The Beginning of Spring is one of the central *qi*; it falls on 4 February of the Gregorian calendar.
52 Golvers 1993, 216 n. 7; see Map 11.1, p. 242.

and other similar phenomena will appear. All this must be presented to the Emperor by means of a petition.[53]

A number of memorials reporting observations made by the Section of heavenly signs submitted by the Astronomical Bureau during the period when Verbiest was Administrator of the Calendar have been preserved. Most of them were signed by the Manchu Director of the Astronomical Bureau rather than by Verbiest. These memorials all have the same structure: they first describe what has been observed, for example the direction of the wind when the Sun reaches a particular *qi*, solar and lunar eclipses, planetary conjunctions, meteorological phenomena and earthquakes.[54] They go on to quote the orthodox divinatory interpretation of these phenomena as given in the official divination book (*zhanshu* 占書). It seems that Verbiest was somewhat reluctant, or at least wary, about such interpretations.[55]

Another task that fell on Verbiest was to teach the Bureau's Students in Astronomy (*tianwensheng* 天文生). Beside the Chinese students mentioned above, Kangxi was also concerned that talents should be spotted among Bannermen: as elsewhere in officialdom, it was crucial to have representatives of the conquest elite among astronomers. In 1670, when Verbiest asked for an increase in the number of staff of the Bureau,[56] the emperor's response was directly aimed towards Bannermen:

From each Banner select six Manchu official students and four official Chinese Military students. They are to be detached to the sections [of the Astronomical Bureau] to study. The accomplished ones are to serve as Erudites (*boshi* 博士).[57]

These students (eighty of them altogether) could also enter the Imperial College when there were vacancies there by submitting essays. In other words, for some Bannermen astronomical skill might allow a shortcut on the path of examinations; this must have been expected to increase the attraction of study at the Astronomical Bureau for them.[58] The Bannermen thus selected learned astronomy with the Chinese students previously recruited at the Section of the calendar. The figures given by the successive edicts are consistent with Verbiest's assessment of the number of students he had: according to him, Kangxi's second

---

53   Golvers 1993, 79.
54   Cui & Zhang 1997, 3–11, 134, 141–142, 161–169, 172–174, 184–185.
55   Huang 1993a, 100.
56   Golvers 1993, 213 n. 15.
57   *Da Qing huidian shili* 1103, 1a; see Jami 1994a, 237. According to Golvers 1993, 213 n. 15, this edict is dated Kangxi 9/9/30 (12 November 1670).
58   This edict is also quoted, in slightly different terms, in the *Comprehensive gazetteer of the Eight Banners, imperially commissioned* (*Qinding baqi tongzhi* 欽定八旗通志), under the section entitled 'Mathematics' (*suanxue* 算學); *SKQS* 665: 744. On Bannermen's education and their recruitment into the civil service, see Crossley 1994.

decree resulted in an increase in this number from about 100 to almost 300. A few years later, in 1676, Verbiest gave an estimate of 160 students, which probably reflected the longer-term effect of Kangxi's measures.[59] One would assume that by that time the newly recruited students fully recognised Verbiest's authority. However, the 'old method', that is, the Great concordance system, remained in practice; it continued to be used along with the 'new method' for predicting eclipses, and he continued to argue in favour of the latter. Thus, after a solar eclipse occurred on 11 June 1676, Verbiest reported that the prediction of the magnitude of the eclipse according to the new method was much closer to what had been observed than the prediction according to the old method; he went on to explain that the former prediction was not entirely exact because of refraction.[60] Three months later, the emperor publicly reminded the Manchu Director of the Bureau that:[61]

At the offices of the Astronomical Bureau which administer heavenly signs and the calendar, those who are in charge must undertake to study until they are highly proficient. Formerly, there were controversies as to whether the new method and the old method were right or wrong. Since it is now fully known that the new method is correct, the Manchu officials who study heavenly signs and the calendar at your Bureau must be ordered to study it devotedly and industriously. From now on those who become highly proficient will be promoted. Those who have not studied will not be promoted.[62]

This wording reveals the continuation of tensions between supporters of the 'old' and the 'new' method; it also suggests that Bannermen who had entered the Astronomical Bureau since 1670 treated it more as a shortcut to better positions in the civil service than as a desirable career in itself.

Verbiest's position also entailed updating the instruments used at the Bureau. Between 1669 and 1673, he had a number of them made for the Observatory. These instruments are still kept there today; made of bronze, they were modelled on Tycho Brahe's instruments.[63] Concomitantly, he wrote an illustrated treatise describing them and explaining their uses, the *Compendium on the newly constructed instruments of the Observatory* (*Xinzhi lingtai yixiang zhi* 新製靈臺儀象志, 1674).[64]

---

59 Golvers 1993, 213 n. 15.
60 *SL* 61: 1a, 4b–5a.
61 Yitala, who had replaced Mahu in 1669, was Director until 1679 (Bo 1978, 88; Qu 1997, 49–50).
62 Dated Kangxi 15/8/10 (17 September 1676); Zhongguo diyi lishi dang'an guan 1984, 268.
63 The old instruments were removed from the platform and put away in one of the rooms of the Observatory; in 1688 the French Jesuits got a glimpse of them, 'buried away in dust and in forgetfulness' (Lecomte, 1990, 98).
64 Facsimile reprint in Bo 1993, 7: 7–457; Halsberghe 1994; see also Chapman 1984.

The illustrations of this treatise were published in Europe; in particular the first one, which shows the imperial observatory with the newly constructed instruments, was reproduced in many works and hence widely circulated.[65]

Beside the tasks related to his position, Verbiest was called upon for other matters in which he was known or supposed to have some competence. Thus, in September 1674, he was ordered to supervise the casting of cannon to be used in the war against the Three Feudatories, Chinese surrendered generals from the former Ming dynasty who had been put in charge of the Southern provinces during the Shunzhi reign and now threatened the dynasty.[66] As had been the case for the construction of the observatory's instruments, the work was carried out under the supervision of the Ministry of Works (*Gongbu* 工部). In 1682, after the final suppression of the rebellion, he was promoted in reward for the efficiency of these cannon.[67] Such a use of the skills of civil servants outside the domain covered by their position was not uncommon.

As had been the case for Schall, Verbiest's position in the civil service enabled him to operate as a member of the political elite, and in particular to build up a network of connections with other officials; he exchanged visits with them. Thus, in 1675, Lu Longqi 陸隴其 (1630–1693), a scholar-official posthumously canonised as an upholder of Neo-Confucian orthodoxy, paid several visits to Verbiest and Buglio, while waiting to be sent to his next post in Jiangsu province. Lu recorded in his diary that they gave him several of their works in Chinese, including a copy of the *Refutation of 'I cannot do otherwise'* (*Budeyi bian* 不得已辨, 1669), which they had written in response to Yang Guangxian's 1665 pamphlet against them.[68] Lu's general assessment of the Jesuits' teachings was reserved:

The Western methods are not easy to assess: they contain untrustworthy things, especially the stories about Adam, Eve, and the incarnation of Jesus.[69]

Lu Longqi discussed astronomy with Verbiest, and felt he had gained further understanding of it from these conversations. The Jesuit also gave him a copy the *Tables of the orbits of the Sun* (*Richan biao* 日躔表) included in the *Book on calendrical astronomy according to the new method* (*Xinfa lishu* 新法曆書)—as the compendium

---

65 For one such reproduction, see Illustration 5.3, p. 114.
66 On the Rebellion of the Three Feudatories (*San fan zhi luan* 三藩治亂), see Spence 2002, 136–146.
67 Fu 1966, 48, 58; Golvers 1993, 105–109; see Stary 1994, Shu 1994.
68 Xu 2000, 228–235; see p. 51.
69 Wu *et al.* 1993, 235.

compiled during the Chongzhen reign was now called.[70] Three years later, Lu Longqi, who was back in Beijing, had further exchanges with Verbiest.[71]

Another scholar who visited Verbiest was Li Guangdi 李光地 (1642–1718), who passed the metropolitan examination in 1670. He reported a conversation that they had while he was studying Manchu as a Hanlin 翰林 academician—a post granted to the most talented Metropolitan Graduates, who, among other things, worked on drafting imperial pronouncements:[72]

Some month during the 11th year of Kangxi,[73] I met the Western scholar Ferdinand Verbiest. Ferdinand deeply denigrated the theory of the square earth and the circular heavens, and the mistaken view that China consists of nine provinces.[74] His words were:

'The heavens surround the earth like the yolk is inside an egg;[75] there has never been a square yolk inside a round egg. How can people say that the earth is flat and square relying on what they see around them? Heaven and earth being round, what one calls the centre of the earth is the centre of the heavens. The mere fact that two minutes around the equator is the zone where the gnomon has no shadow at noon [shows that] this is correct.'[76]

When Verbiest and his colleagues came, they themselves passed through this zone; that is what they call the centre of the earth. I humbly replied to this:

'Don't heaven and earth separate into square and circle, and into motion and rest? For the set-up (ji 機) of motion must be round, the set-up of rest must be square. Then, although the heavens are not round, nothing prevents them from being round; although the earth is not square, nothing prevents it from being square. And when one speaks about the Middle Kingdom, one means that [it is the place where] rites, music, government and teachings have reached the correct principles of heaven and earth. How should these be located within form? To use an analogy, the heart is in the middle of a man in a different sense than the navel; and finally one must take the heart as the centre of a man. How could this be [a matter of] shape?'

Reading Wu Caolu's 吳草廬[77] theory of the centre of the earth,[78] I happen to remember this and record it here.[79]

---

70 Wu *et al.* 1993, 235; Hummel 1943, 547.
71 Xu 2000, 228–235.
72 Hummel 1943, 473.
73 30 January 1672 to 6 February 1673.
74 This division was thought to date back to the legendary Yu the Great.
75 These words echo Zhang Heng's description of the universe; see Needham 1954–, 3: 217.
76 On Verbiest's proof of the sphericity of the Earth, see Moortgat 1993.
77 Wu Cheng 吳澄 (1249–1333), styled Caolu, was a follower of Zhu Xi's philosophy who served the Mongol Yuan dynasty; the parallel with Li Guangdi's own biography would have been obvious to his readers: a follower of Zhu Xi, he served the Manchu.
78 Here *tu* 土, one of the Five Phases (*wuxing* 五行), is used instead of *di* 地 which denotes the Earth in astronomical context: Li Guangdi is reading a traditional cosmological discussion.
79 *Rongcunji* 榕村集. *SKQS* 1324: 809.

This is a later recollection; it does not inform us of the details of Li Guangdi's exchange with Verbiest. Li's purpose in recording it seems to display how he himself defended the compatibility of classical learning with what the Jesuits reported, the latter being treated as reliable experience, whereas the former had to be understood in a metaphysical rather than physical way. Judging from his recollection, his interest in Western learning at the time appears to have been limited to the issue of its compatibility with Chinese classical cosmology.

Verbiest recounted contacts such as these to his European readers, using characteristic allegories:

> Thus the Holy Religion makes her official entry as a very beautiful queen, leaning on the arm of Astronomy, and she easily attracts the looks of all the heathens. What is more, often dressed in a starry robe, she easily obtains access to the rulers and prefects of the provinces, and is received with exceptional kindness, so that she can safely protect her churches and priests thanks to their sympathy. Hence, when showing the rehabilitation of our astronomy in China in this compendium, I am showing the rehabilitation of our Religion at the same time.[80]

As suggested here, Verbiest paid as well as received visits: this was part of normal sociability among officials in Beijing. When an official was sent to a post in one of the provinces, the Jesuit astronomer went to present his congratulations to the newly appointed magistrate before he left the capital, in order to recommend to him the priests and Christians to be under his administration. Indeed astronomy protected religion, but in an indirect way: it was not Verbiest's expertise in itself, but the recognition from bureaucratic institutions and the emperor's favour which this expertise had won him, that enabled astronomy to protect religion. This was a new feature in the articulation of the way science served religion by Jesuit missionaries in China. For Ricci, the mathematical sciences prepared one's mind for the understanding of the higher truths of religion: the very success of the policy he had promoted combined with the dynastic change resulted in two major changes. First, instead of the scholastic category of mathematics (the *quadrivium*) the Chinese category *lifa* 曆法 (astronomical methods) came to define the specialised skills required from Jesuits in China. Secondly, the link between science and religion, originally an intellectual one in Jesuit culture, became first and foremost a tactical one, matching the structure and needs of the Qing imperial order. Verbiest remained aware that this tactic entailed a risk: the triumph of European

---

80 Golvers 1993, 56–57; on the Chinese depiction of the muses of the sciences, see Golvers 1995.

astronomy could never be regarded as final, due to the very nature of this discipline:

> When I consider more closely the fact that often a great discrepancy has been found between the tables and calculations of even the most prominent astronomers in Europe and the observed heavenly movements, then I do not doubt that it was due to God's extraordinary Providence that, in the course of so many years that [the Chinese] have been comparing our astronomy and our calculations with the heavenly movements, not one conspicuous deviation has been found! This is, I insist, because the Divine Clemency has concealed the errors, if there were any: by carelessness of the observers, by clouds, or by some similar favour from the indulgent Heaven, and because it leads all things to the good of our Religion.[81]

The Bureau's interest as an imperial institution, whether under the Jesuits or before their time, was that there should be as few discrepancies as possible between predictions and observations.[82] Conflict within the Bureau, however, remained possible. As mentioned above, Manchu students may have lacked zeal in their study of astronomy. The Chinese staff, for their part, were by no means unanimously convinced of the superiority of the Jesuits' astronomy, and within the Bureau some of them continued to express preference for the Great concordance system.[83]

The third change in the ways the Jesuits engaged in the sciences resulted directly from Kangxi's personal involvement that began with the reversal of the verdict of the 1664 impeachment: the emperor was now their main target when it came to introducing sciences from Europe. Verbiest's account again indicates this in the same allegorical style:

> After Astronomy, marching like a venerable queen between the Mathematical Sciences and rising above all of them, had made her entry among the Chinese and had ever since been received by the Emperor with such an amiable face, all the Mathematical Sciences also gradually entered the imperial Court as her most beautiful companions. They followed astronomy, adorning themselves with all the extraordinary or beautiful things they carry with them as if they were gold and precious stones, to find more favour in the eyes of such a great majesty. This happened with Geometry, Geodesy, Gnomonics, Perspective, Statics, Hydraulics, Music, and all the Mechanical Sciences, each of them dressed in such precious and skilfully woven clothes that it conveyed the impression that they were competing with each other in beauty. However, the aim of their fervent desire to please was not to keep the Emperor's eyes only upon themselves, but to direct them fully towards the Christian Religion,

---

81   Golvers 1993, 75.
82   Needham 1954–, 3: 191–192, gives evidence of this in the Song dynasty (960–1279).
83   There is evidence of this to the end of the years when Verbiest was Administrator of the Calendar; Widmaier 1990, 9.

whose beauty they all professed to worship, in the same way as smaller stars worship the sun and the moon.[84]

Here Verbiest conveyed to his European audience that he was working towards the emperor's conversion, a goal he sought to achieve by means of the sciences: in other words he claimed to be continuing the 'Ricci strategy'. He went on to list in successive chapters what each science had presented to Kangxi, which usually took the form of instruments or of engineering work. Although ballistics was not mentioned in the list above, it did appear in one of these chapters, where Verbiest recounted how he had cast cannon.[85] In other chapters he did not fail to acknowledge the contributions of several of his colleagues.[86] On the whole, apart from cannon and the calendar, the gifts provided by Verbiest's sciences were courtly presents rather than tools for statecraft.

As Verbiest himself noted in his reports to Europe, he did not provide all these on his own: his colleagues all developed and applied a variety of skills as they were required. It was Magalhães who constructed the gnomon thanks to which Verbiest succeeded in outdoing Wu Mingxuan in predictions at the crucial moment. The Portuguese Jesuit's role as clockmaker was taken over by Tomás Pereira (1646–1708) who arrived in Beijing in 1673. The latter also applied wide-ranging musical skills: he made and played musical instruments, taught Kangxi to play the harpsichord, eventually composing for him a treatise on harmonics the *Essentials of the [standard] pitchpipes* (*Lülü zuanyao* 律呂纂要).[87] Two years before Pereira's arrival, the Beijing Jesuits had been joined by Carlo Filippo Grimaldi (1638–1712), who made 'several famous new inventions, on the basis of the principles of mathematics which he had brought with him to Beijing'; he also mastered perspective and anamorphosis techniques in painting.[88] In 1685, Antoine Thomas (1644–1709), another confrere, brought fresh mathematical expertise to the Beijing residence.[89] Thus, whether self-trained on the spot or bringing with them skills acquired as part of their education and earlier career, all these companions of Verbiest partook in a technical culture in fields related to the mathematical sciences, a culture that seems to have given more importance to the production of artefacts than to that of texts.[90]

---

84 Golvers 1993, 101.
85 Golvers 1993, 104–109; see Stary 1994.
86 See Pih 1979, 222.
87 Wang 2002, 2003; Jami 2008b; what could be an earlier version of this treatise was drafted by Verbiest himself in 1685; Josson & Willaert, 1938, 491; Golvers 2010, 297.
88 Josson & Willaert 1938, 341; Golvers 1993, 115; Golvers 2010, 283–285.
89 See pp. 181–183.
90 Golvers 2010.

3.3 Kangxi, student of Chinese and Western learning

The Kangxi emperor's identity as a Manchu was not a simple one. His paternal grandmother, Empress Dowager Xiaozhuang, who deeply influenced him, was a Mongol princess; there were also Chinese amongst his ancestors.[91] His identity as a ruler was carefully constructed: he had to be acknowledged as a sovereign in control of the Eight Banners and of the territory they had conquered, but also as an emperor according to Chinese cultural standards. It is no surprise that the latter identity comes out as predominant in Chinese sources. His native language was Manchu, and it seems that in his childhood he first learned Chinese secretly with some eunuchs, possibly against the regents' will.[92] After his takeover, he established tutorials so as to complete his own education in the Confucian tradition.

The study of the Classics had long been part of emperors' education. During the Ming dynasty, ceremonial Lectures on the Classics (*jingyan* 經筵), during which the emperor discussed them with eminent officials, were held twice a year (once in the spring and once in the autumn). Shunzhi took these up in 1657, after establishing Daily Tutoring (*rijiang* 日講) in 1655. This tutoring took place in two sessions during the year, each session starting after a Lecture on the Classics and lasting for about three months. This allowed for actual instruction: the daily sessions enabled the emperor to get real knowledge about what was discussed at the formal Lectures. Kangxi re-established both the Lectures and the Tutoring in 1670. Three years later he decided that the tutoring should continue all year round.[93] Further institutionalisation took place in 1677, when Kangxi set up the Southern Study (*Nanshufang* 南書房), located near his sleeping quarters in the Great Interior (*Danei* 大內). There, a team of scholars selected from among Chinese Hanlin academicians were in charge of Daily Tutoring to the emperor as well as of teaching the imperial princes on the Classics and Histories, and more broadly on all 'literary' matters. Their duties included drafting edicts, setting questions for the metropolitan examinations, but also writing poetry and calligraphy for the emperor and keeping daily records of his actions in the *Imperial Diary* (*Qiju zhu* 起居注).[94] They were the only Chinese in the Great Interior, the part of the Forbidden City otherwise staffed by Manchus.[95]

Some of the lecture notes prepared by Kangxi's tutors were eventually revised under his supervision and printed: seven 'Explanations of

---

91 Kessler 1976, 53–54.
92 Spence 2002, 131.
93 Kessler 1976, 138–144; Smith 1991, 113–114, describes some tutorials on the *Book of change* (*Yijing* 易經).
94 On the role of calligraphy, see Hay 2005: 315.
95 Wu 1968; Kessler 1976, 143; Bartlett 1991, 30, 47.

the meaning [of the *Four Books* and each of the *Five Classics*] during the daily lectures' (*Rijiang…jieyi* 日講……解義) were thus published between 1676 and 1749.⁹⁶ According to the Confucian model of the sage ruler, Kangxi's study resulted not only in his own education, but also in that of the whole empire. Manchu or bilingual editions of several works were also prepared: this was part of an effort to create a corpus of literature in Manchu, and to direct imperial instruction towards Manchu officials as well as their Chinese counterparts. It was in keeping with the philosophical stands of the main imperial tutors that these works perpetuated the status of imperial orthodoxy conferred on the Cheng-Zhu school under the Ming dynasty.⁹⁷

A major editorial project set up by Kangxi in the first decades of his reign was the compilation of the official *Ming History* (*Mingshi* 明史), which was completed in 1739, under his grandson. Compiling the *Veritable Records* of the previous dynasty into a History (*shi* 史) had long been a task taken up by new rulers. Kangxi took this opportunity to seek further reconciliation with some of the most brilliant scholars of the transition period, many of whom were Ming loyalists. Early in 1678 he issued an edict, asking officials to nominate candidates for a special examination that would be held in the Palace in order to select, through personal examination by the emperor, those worthy of the title of 'Profound scholars of vast learning' (*boxue hongru* 博學宏儒). A number of the most eminent scholars recommended declined to take part in the examination, while some of their disciples attended it: from that point of view the imperial initiative was only a mitigated success. Nonetheless the examination, which was effectively a shortcut on the civil service examination path, was held in the spring of 1679. Fifty successful candidates were thereafter made Hanlin academicians, and the first compilers of the *Ming History* were thus appointed.⁹⁸

At the time when he reinstated daily tutoring on the Classics, Kangxi also started studying mathematics and astronomy with Ferdinand Verbiest. This was quite a novelty: there are no earlier records of a Chinese emperor setting out systematically to learn about such technical subjects. They were far less prestigious topics than those discussed above, which were the key to an official career. Kangxi later explained that his resolution to undertake this study stemmed from the review of the verdict in the 1664 affair:

---

96 Zhao 1977, 2483, 4221, 4226, 4236, 4244, 4248. Kangxi followed in his father's footsteps: in 1656, Shunzhi had commissioned the *Expanded meaning of the Classic of filial piety* (*Xiaojing yanyi* 孝經衍義; Zhao 1977, 4325); the work was completed in 1682 and prefaced in 1690; Standaert 2008, 271, n. 52.
97 Chan 1975; Cheng 1997, 447–452.
98 Wilhelm 1951, Kessler 1971, 191–196; Kessler 1976, 157–166.

Seeing with Our own eyes [that no one at Court understood the calendar], We felt sick at heart. During the little leisure time left to Us by the many affairs [of the State], We have devoted Ourselves to astronomy for more than twenty years, so that We have taken a view of its broad outlines and will not come to be confused about it.[99]

This is characteristic of the emperor's attitude in government as well as cultural affairs: he displayed great eagerness to acquire enough competence to be personally in control. It is worth noting that Kangxi names astronomy (*tianwen lifa* 天文曆法) as the field he studied, and does not qualify it as 'Western'.

Verbiest, who was Kangxi's first tutor in this field, used words that echoed those of his student when describing the latter's interest in the newly made astronomical instruments: '[... he] has concentrated himself with great passion on their use and understanding, whenever he managed to find some leisure in his public duties.'[100] The Jesuit further recounted how he taught the emperor:

[... T]he Emperor, over a continuous period of more than five months, used to summon me daily to the Inner Court and even to his *Museum*.[101] Almost entire days he kept me there, for no other reason than to examine during his spare time matters related to mathematics, and especially astronomy. The first day, he brought forward all the astronomical and other mathematical books, which our fathers had once written in Chinese—they easily reach the number of 120—and he wanted them to be explained one by one.[102]

The books in question must have consisted mainly in the *Books on calendrical astronomy according to the new method*. Verbiest himself is credited with an impressive number of Chinese writings to do with the sciences;[103] a number of them were adapted from publications of his late Ming predecessors. The continuity with their teaching also appears in the fundamental role that he ascribed to Euclidean geometry:

When the emperor heard from me that the books of Euclid made up the prime elements of the whole mathematical science, he immediately wanted the first six books of Euclid, which had once been translated into Chinese by Father M. Ricci, to be explained to him. He asked for the meaning of each proposition from A to Z, with the persistence of his obstinate or (to use the word) stubborn mind. Although he knew Chinese very well, and although he could paint the Chinese characters with a most facile hand, he nevertheless wanted the Chinese Euclid to be translated into Manchu, to gain some further help from this.[104]

---

99 *SKQS* 1299: 156.
100 Golvers 1993, 98.
101 The Great Interior (Rawski 1998, 31–34) and the emperor's private study, next to the Southern Study (Golvers 1993, 263, n. 92).
102 Golvers 1993, 98.
103 Bernard-Maître 1945, 369–382, lists more than thirty works.
104 Golvers 1993, 99.

The Manchu translation of *Elements of geometry* (*Jihe yuanben* 幾何原本) is no longer extant; we do not know whether it was completed.[105] It seems that it remained in the form of a manuscript, which some French Jesuits had in their hands in 1690. Verbiest was helped by at least one Manchu scholar for this translation.[106] The fact that Kangxi asked Verbiest to do this translation suggests that at this stage he might have felt more comfortable in his native language than in Chinese; but it could also have formed part of his endeavour to create a corpus of literature in Manchu. The implication here is that he found the *Elements* worthy of appearing among this corpus.[107]

The parallel between Kangxi's study of the Classics and of the mathematical sciences is revealing. Verbiest, who had a position in the civil service, seems to have played in the sciences a role similar to that of Daily Tutors and scholars appointed to the Southern Study in literary matters. The lack of formality in the exchanges with the sovereign, which he and his successors interpreted as signs of imperial benevolence towards them and their religion, also characterises Kangxi's relationship with the scholars of the Southern Study. It placed the tutors, Chinese or Jesuit, outside court ritual, rather than above it. That Verbiest was made to learn Manchu especially in order to translate the *Elements*[108] seems to reinforce the parallel with other officials who were close collaborators of Kangxi: the mastery of the conquerors' language by Chinese officials seems to have made quite a difference in imperial favour, and therefore in their career prospects.

Verbiest's teaching was not limited to geometry and astronomy; it extended to what the Jesuits termed 'mixed mathematics', which both master and student seem to have enjoyed more:

> After the Emperor had mastered the elements of Euclid, and in order to proceed properly and gradually, he wanted the mathematical analysis of the triangles to be explained, not only of the rectilinear but also of the spherical ones. After he had braved all the traps and thorns of mathematics, he turned to more pleasant problems with even greater joy, viz. to applied geometry, geodesy, chorography, and other attractive sciences in the field of mathematics in which he took highest pleasure.[109]

The sessions of intensive study described by Verbiest took place in 1675, during the first years of the Rebellion of the Three Feudatories, when it most threatened the survival of the dynasty.[110] This timing may seem

---

105 Golvers 1993, 266 no. 100; Engelfriet 1998, 136–137.
106 Landry-Deron 1995, II, 38. See p. 160.
107 Jami 2010; see Crossley and Rawski 1993, 91–95.
108 Golvers 1993, 99.
109 Golvers 1993, 99–100.
110 Spence 2002, 137–146.

surprising at first: it was hardly a period when Kangxi had much leisure, energy or funds to devote to scholarship. Looking at it more closely, however, it appears that there was a correlation between the various instruments that Verbiest made and the technical treatises he wrote for the emperor. The latter was keen to understand how to handle them. Thus in 1671 Verbiest wrote a short *Explanation of the examination of qi* (*Yanqi shuo* 驗氣說),[111] which gave details of how to use the 'thermoscope' that he had presented to Kangxi.[112] More significantly, the same seems to have applied to ballistics: the *Compendium on the newly constructed instruments of the Observatory* as presented to the emperor in 1674 included ballistic tables. These were probably deemed too 'sensitive' to be included in the printed version.[113] It was in the same year that Verbiest was commissioned to make cannons; the lessons in geometry and applied mathematics started a few months after this commission. When Kangxi later inspected the finished pieces, he displayed remarkable competence as a gunner:

> Before these cannon were sent to the provinces, they were first transported outside the city walls of Peking, to the Western Hills, and tested one by one in the presence of the most prominent generals. Mostly even the Emperor in person was present and occasionally he himself aimed (those pieces) at the bull's eye, which they also pierced with an accurate shot.[114]

This suggests that Kangxi was seeking more than pleasure when he obstinately explored 'the traps and thorns of mathematics'. As he could read in Matteo Ricci's preface to the *Elements of geometry*—the only part of the book for which he would have needed no explanation from a Jesuit tutor:

> Of all specialisations that borrow their methods from mathematics, it is above all the art of warfare—in which security in the major affairs of the state is rooted—that requires excellence in this knowledge. Therefore, a wise and courageous general must give priority to the study of quantity. To he who does otherwise, wisdom and courage will be useless. How could a good general trust such things as his perception and auspicious dates? A good general's concern is in the first place to plan food for soldiers and horses, distances to be covered, terrains' configuration, difficulties, dangers and possible losses; in the second place to plan the proper battle order: in a circle to make his troop look few; in a horn to make them look numerous; in a moon crescent to encircle the enemy, or in a pointed shape to penetrate and disperse them. Next, in devising offensive and defensive weapons he has to give much thought to what is advantageous, to manage to make them efficient, and to always use

---

111 Bernard-Maître 1945, no. 447; the title is often translated as 'Explanation of the thermometer'.
112 Golvers 1993, 129.
113 Golvers 1993, 104.
114 Golvers 1993, 108–109.

what is newest. Looking at what the Histories of various countries have recorded, does one read of anyone who, having developed a new skilful weapon, wasn't victorious in battle or firm in defence?[115]

To sum up, it seems that Kangxi took mathematics seriously as a foundation of warfare, and it could well be that he decided to study it as part of the preparation for his military campaigns against the Three Feudatories. If such was the case, his study of Euclidean geometry had motivations parallel to those of his study of astronomy: he was eager to be personally in control of technical matters that were crucial for the survival of the dynasty.

3.4 Philosophy, 'fathoming the principles' and orthodoxy

In Verbiest's mind, his tutoring of Kangxi was a means to 'direct the Emperor's eyes towards religion', in particular by making apparent the links between the latter and astronomy. In the Jesuit curriculum, philosophy was a prerequisite to the study of theology. Therefore Verbiest prepared a compendium of works on philosophy, the *Study of the fathoming of principles* (*Qiongli xue* 窮理學), which he presented in 1683, asking that it be printed and promulgated.[116] This was an attempt to substitute the Chinese Classics with the Jesuit version of scholastic education, just like Schall had done with the corpus of Western astronomical methods in 1644. *Qiongli* 窮理 was the Chinese term that Verbiest used, as his predecessors had done, to render *philosophia*.

Only part of the *Study of the fathoming of principles* is still extant; divided into three parts, the work consisted of a number of treatises on various subjects, and contained an expurgated version of the partial translation of the *Conimbricenses* (commentaries on Aristotle written by the Coimbra Jesuits at the end of the sixteenth century) by Francisco Furtado (1589?–1653) and Li Zhizao 李之藻, entitled *Investigation into the principles of names* (*Mingli tan* 名理探, 1631), from which Verbiest had deleted all references to God and to divine law.[117] The first part opens with an adaptation of the introduction of the Coimbra course, followed by an explanation of Aristotle's dialectics derived from *Isagoge Porphyrii* (*Porphyry's Introduction* to Aristotle's categories). The second part consists of a translation of *Analytica Priora* (Prior Analytics). The third part deals with Aristotelian physics. It includes a number of technical works by Verbiest himself and by his predecessors of the late Ming period.[118]

---

115 Engelfriet 1998, 456–457; Guo 1993, 5: 1152–1153.
116 See Dudink & Standaert 1999; Golvers 1999.
117 On the *Investigation into the principles of names*, see Kurtz 2008, who renders *li* 理 as 'patterns' in this title; see also Wardy 2000.
118 This overview of the *Study of the fathoming of principles* relies on Dudink & Standaert 1999, 20–29.

While generally leaving out explicit references to divinity, the *Study of the fathoming of principles* followed the pattern of the *First collection of heavenly learning* (*Tianxue chuhan* 天學初函) compiled by Li Zhizao in 1626. As Li had done, Verbiest constructed the *Study of the fathoming of principles* so that it presented the technical works as a sequel to the Aristotelian treatises (on logic, dialectics and physics). The whole thus formed a corpus of 'Western fathoming of principles', which included what we might call science, technology and philosophy. The work emphasised the derivation of principles (*tuili* 推理), which Verbiest argued was a prerequisite for the understanding of astronomy.[119] This was a way to draw attention to what in his view characterised the 'European sciences' in contrast to Chinese learning: a systematic method for ordering knowledge.[120] What he submitted to Kangxi can indeed be characterised as a Jesuit version of the fathoming of principles.[121]

The *Study of the fathoming of principles* was presented to the emperor on 16 October 1683, together with a memorial that expounded Verbiest's motivations for writing it, and what made it essential in his eyes:

> I am presenting works on the study of the fathoming of principles, in order to clarify the principles of the calendar, in order to widely open the door to all kinds of studies, to be left to generations to come ever after. I humbly represent that when administering the calendar and clarifying the seasons, the Emperor is fulfilling his foremost duty. Now Your Majesty is administering the calendar and clarifying the seasons in a way that surpasses all previous generations, as the sun outshines all the stars. However, in calendar methods there are numbers pertaining to the methods and there are principles to establish the methods. Having only the numbers pertaining to the methods, (*fashu* 法數) without having the principles of the methods (*fali* 法理) is like having a body without having a soul; it is as if in the Heavens there were only fixed places, and no oversight of motions.[122]

Verbiest's first metaphor is the same one that he used in the *Astronomia Europaea*, albeit adapted to those for whom the memorial was written. It is now the emperor, instead of the Christian religion, who is compared to the Sun. Such a comparison would not seem strange to Kangxi and to Chinese literati, although the former carefully avoided boasting that he outshone all previous emperors of China. The next sentence of the memorial also proposes a double metaphor, in a style quite familiar to them. The terms of the comparison, however, probably left them unmoved: the body and soul dichotomy, the necessary supervision of planetary motions, would hardly seem evocative to them.

---

119 Standaert 1994, 408.
120 Golvers 2004a, 5.
121 On the Jesuit interpretation of *gewu qiongli*, see Standaert 1994.
122 Xu Zongze 1958, 191; some parts of the memorial are translated in Standaert 1994, 408–409.

After pointing to the lack of study of the principles as the cause of discrepancies in the calendars of the previous dynasties, Verbiest went on:

> It can be said that the Canon of the calendar (*lidian* 曆典) is of utmost brightness. I am begging, not in order to enhance the inner brightness of the calendar's principles, but rather in order to enhance their outer brightness, to open the principles recorded in that Canon to the fathoming of the principles, so as to clarify them. Then those versed in the calendar will know not only its numbers, but also its principles. Then the brightness [of these principles] will show outwardly. At present those versed in the calendar understand its numbers, but not its principles; the reason why they do not understand its principles is that they do not understand the method of derivation of principles. For one sees the astronomical principles expounded in all calendar treatises; but not knowing the method to derive them is like having a gold vein in the earth and not knowing the access for opening the mine.[123]

In this attempt to extend imperial patronage of Western learning beyond the Astronomical Bureau, Verbiest's room to manoeuvre was quite narrow: on the one hand, he was the last person who could suggest that calendar methods could or should be improved. Questioning their perfection would have amounted to casting doubt upon his own competence. On the other hand, in order to promote philosophy as he understood it, he had to argue that the *Study of the fathoming of principles* would in some way benefit astronomy. The latter was his acknowledged field of competence, whereas he was hardly regarded as an expert in the study of the principles (*lixue* 理學) as the emperor and the literati understood it. Their perusal of the *Study of the fathoming of principles* did not prompt them to think otherwise. Two months after the work had been submitted, a discussion on it took place between Kangxi and Hanlin academicians:

> His Majesty said : 'The contents of these books are rebellious, erroneous and obtuse.'
> Mingju 明珠[124] and others submitted : 'All he says about knowledge and memory pertaining to the brain and so on indeed strays away from principles.'
> His Majesty said : 'Let the Ministry reply that there is no need to promote Verbiest; the books he has written should be returned to him.'[125]

Since the *Study of the fathoming of principles* infringed on the territory of classical learning, the elite of officials were called upon to assess it. However it was apparently the emperor who set the tone; they only pointed to a particular impropriety to strengthen his position. Rather than recording a debate, this dialogue displays a pre-established consensus. It reveals that Kangxi was eager to show his officials that he kept

---

123 Xu Zongze 1958, 192.
124 Mingju (1635–1708) was one of the Manchu officials that Kangxi promoted when taking over at the end of the Regency. In 1683 he was a Grand Secretary; he was demoted in 1688 following an accusation of corruption; Hummel 1943, 577.
125 Zhongguo diyi lishi dang'anguan 1984, 2: 1104. See Dudink & Standaert 1999, 17.

the Jesuits under control, and would only promote aspects of their work that were compatible with the Chinese worldview. The imperial rejection of Western philosophy was a way of reminding Verbiest of his position. He was a specialist in charge of some technical matters, not a scholar. Kangxi had better things to do than turn his eyes towards Verbiest's fathoming of the principles, let alone towards his religious teachings. Despite the latter's flowers of rhetoric, what he saw as part of his missionary work was not deemed by the emperor as acceptable in his service. Just like Yang Guangxian, who understood only the principles, was not fit to serve as an astronomer, Verbiest, who understood only the methods, had nothing to contribute to philosophical debates. As we shall see, such debates on the relationship between principles (*li* 理) and numbers (*shu* 數) took place among scholars. A main contributor to these debates and a scholar well versed in the mathematical sciences, who played a major role in their development during the Kangxi reign, will be introduced in the next chapter.

CHAPTER 4

A mathematical scholar in Jiangnan: the first half-life of Mei Wending

As Kangxi remarked, there was little or no knowledge of astronomy among the officials and Bannermen who surrounded him at the Court. This was a consequence of the fact that this science was of little use to a literatus' career. But then, one might ask, what prospects, if any, did specialisation in the mathematical sciences open to a member of this class? Following the itinerary of Mei Wending 梅文鼎 (1633–1721), who has come down in history as the greatest specialist in these sciences of the early Qing period, will shed light on this issue.

Mei was certainly a prolific writer: a bibliography that he himself established in 1702 listed 62 titles on astronomy, 17 of which had then been published, and 26 titles on mathematics, 16 of which had then been published.[1] His works are known mainly through two posthumous editions. The *Complete writings on astronomy and mathematics* (*Lisuan quanshu* 曆算全書, 1723), which included 28 titles (16 in astronomy and 12 in mathematics), eventually found its way into the *Complete library of the four treasuries* (*Siku quanshu* 四庫全書, 1782), a huge collection commissioned by the Qianlong emperor (r. 1736–1795); it included all the works that were deemed of importance.[2] The *Selected essentials of Mr Mei's collection* (*Meishi congshu jiyao* 梅氏叢書輯要, 1761), which included 21 titles (13 in mathematics, 8 in astronomy), was edited by Mei's grandson Mei Juecheng 梅瑴成 (1681–1763). This wealth of material has been partly studied by historians of science, but much remains to be done.[3] In what follows, Mei's career until his mid-fifties and some of his early mathematical work will be discussed, as an—admittedly outstanding rather than average—example of the knowledge, interests and opinions of an early Qing scholar versed in the mathematical sciences.

---

1 *SKQS* 795: 961–992; Mei 1939; see Hummel 1943, 570.
2 *SKQS* 794 & 795: 1–818. On the compilation of the *Complete library of the four treasuries*, see Guy 1987.
3 See, among others, Hashimoto 1970b & 1973, Martzloff 1981a & 1981b, Liu 1986, Jami 1994c, Li 2006, Tian & Zhang 2006.

4.1 Mei Wending's early career

Mei Wending was born in 1633 in the Xuancheng 宣城 district of the Anhui province, in the Jiangnan region. His family had a very ancient origin and a couple of illustrious ancestors in the Song dynasty. He was the eldest of four sons by the same mother (sisters, if there were any, are not mentioned). In addition to the usual education in the Classics and Histories, they learned from their father and grandfather about the Song dynasty commentaries on the *Book of change*, and especially about 'figures and numbers' (*xiangshu* 象數).[4] Astronomy was part of the Mei family's traditional study: Wending later recalled that he had observed the stars since his childhood. Two of his brothers, Wennai 文鼐 (1637–1671) and Wenmi 文鼏 (1642–1716), are known to have shared his scholarly interests.[5] As early as 1647 he passed the district examination, thus becoming a 'cultivated talent' (*xiucai* 秀才); like many scholars of his generation, however, he did not go further along the path of civil examinations. In his late twenties, he studied astronomy under the guidance of a scholar from the same district, Ni Guanhu 倪觀湖 (1616–after 1695), investigating eclipses in the Ming dynasty Great concordance (*Datong* 大統) system.[6] This resulted in his first work, the *Superfluous learning on calendrical astronomy* (*Lixue pianzhi* 曆學駢枝), for which he wrote a preface in 1662. Two years later, he completed a work on surveying. In 1671, his younger brother Wenmai, with whom he had been working, passed away; the next year, Wending lost his wife. He never remarried, which was an unconventional choice.[7] In 1672, the completion of the first draft of the *Discussion of rectangular arrays* (*Fangcheng lun* 方程論), a study of the traditional method for solving systems of linear equations in several unknowns, crowned the twenty years that Mei had said he had spent studying the subject. The following year, Mei was requested by Shi Runzhang 施閏章 (1618–1683), an official and a poet also from Xuancheng, to draft the section on field allocation (*fenye* 分野), for the local gazetteer of the district that the latter was compiling.[8] Field allocation consists in establishing correlations between the twelve divisions of the zodiac and the subdivisions of a territory; these correlations form the basis for the interpretation of celestial phenomena as omens for the various parts of the territory thus mapped.[9] This first work commissioned to Mei Wending is an indication of his local reputation as a scholar versed in

---

4   See pp. 16–17; there was a similar interest in 'figures and numbers' in Fang Zhongtong's family; Fung 1995, 146–147.
5   Li & Guo 1988, 11–15, 17, 19.
6   Li & Guo 1988, 22.
7   Li & Guo 1988, 22–23.
8   On Shi Runzhang, see Hummel 1943, 651.
9   On field allocation, see Hsu 2009, esp. 4–17.

astronomy. Shi also knew Mei's nephew, Mei Geng 梅庚 (1640–c.1722), whose poems he admired.[10]

In 1675 Mei went to Nanjing, which remained the cultural capital of Jiangnan, to take the provincial examination, which he failed. Three years later he was still there and he retook the examination, together with Mei Geng. An anecdote recounted by the latter suggests that the main motivation of Wending's presence in the provincial capital may not have been eagerness to become an official. Concerned that his uncle was spending too much time seeking and studying books on Western astronomy, Mei Geng hid the book that he was currently reading, which aroused Wending's fury.[11] Indeed the single-minded study of mathematics and astronomy was no way to meet his family's expectations regarding his career. Whereas Wending never became a Provincial Graduate, Mei Geng passed the examination in 1681.[12]

During the five years that Mei Wending spent in Nanjing, he made the acquaintance of a number of scholars, including Fang Zhongtong, whose *Number and magnitude expanded* (*Shudu yan* 數度衍) was still in manuscript form,[13] Pan Lei 潘耒 (1646–1708) and Huang Yuji 黃虞稷 (1629–1691), both of whom would later contribute to the compilation of the official *Ming History*. Thanks to these new acquaintances, he was able to read mathematical and astronomical works that were not available for purchase. His work had hitherto relied mainly on Chinese sources written prior to the Jesuits' arrival. Kangxi's reversal of the verdict against Schall, however, had aroused further interest in Western astronomy among scholars.[14] Thus Mei read a manuscript copy of the *Books on calendrical astronomy according to the new method* (*Xinfa lishu* 新法曆書), the Jesuits' astronomical compendium. The *Explanation of the proportional compass* (*Bili gui jie* 比例規解) was missing from that copy but he eventually had access to it as well.[15] Mei also read the *Integration of learning in calendrical astronomy* (*Lixue huitong* 曆學會通) by Xue Fengzuo 薛鳳祚, the scholar who had worked with Nikolaus Smogulecki;[16] he thereafter exchanged correspondence with Xue, who then lived in Shandong. After Xue's death in 1680, Mei wrote a series of four poems as a tribute, in which he hinted at an issue that had obviously concerned him:

> When, very late, I had access to your works,
> I was enlightened as a beginner.

---

10 Zhao 1977, 13359.
11 Liu 1986, 55.
12 Zhao 1977, 13359.
13 See p. 44.
14 Hummel 1943, 570.
15 *SKQS* 795: 984.
16 See p. 43.

> Although you never served Jesus,
> Yet you were able to fathom their techniques.[17]

Thus one of the things that Mei had drawn from reading Xue's works was that Christians did not have the monopoly of Western learning. Mei Wending investigated it systematically. Through Ma Decheng 馬德稱, a Muslim scholar from Nanjing who shared his interest, he also had access to translations of 'Muslim' astronomical texts dating back to the fourteenth century.[18] Moreover, thanks to Huang Yuji, who was a bibliophile, Mei was able to read the first chapter of the mathematical classic *Nine chapters on mathematical procedures* (*Jiuzhang suanshu* 九章算術) in a Song edition.[19] In 1684, the first five chapters of the classic were reprinted, also from a Song edition.[20] There is no evidence, however, that Mei Wending ever saw this edition; it seems unlikely that he would not have mentioned it if he had.

In 1680 Mei assembled nine mathematical treatises he had written under a programmatic title, *Integration of Chinese and Western mathematics* (*Zhongxi suanxue tong* 中西算學通). These works never appeared as a collection. Only six of them were ever printed; the first one, a treatise on Napier's *Calculating rods* (*Chousuan* 籌算), was printed in that same year, thanks to the generosity of a certain Cai Rui 蔡璿, from Nanjing, who had sought Mei's advice on matters related to the mathematical sciences; a fascicule entitled *First collection of the integration of Chinese and Western mathematics* (*Zhongxi suanxue tong chuji* 中西算學通初集), intended as the front material of the collection, was also printed under Cai's auspices. He wrote a preface for it, as did Fang Zhongtong and, of course, Mei himself.[21]

Mei Wending returned home in the same year his first work was printed. His grandson Mei Juecheng, who was to continue his grandfather's interest and to play an important role in his posthumous fame, was born in 1681. In the years that followed, Mei Wending seems to have shared his time between his hometown, Nanjing and Hangzhou. Hangzhou then had an important Christian community, and there he met Prospero Intorcetta (1625–1696), with whom he seems to have discussed astronomy, although the Jesuit was not regarded as a 'specialist' among missionaries.[22] While continuing to write new works, Mei was looking for funding to publish his mathematical collection. He eventually succeeded in finding a patron; but still,

---

17 Mei Wending 1995, 239.
18 Li & Guo 1988, 26.
19 *SKQS* 795: 987.
20 Chemla & Guo 2004, 72.
21 A copy of this fascicule (the only one extant to the best of my knowledge) is kept at the Library of Tsing-hua University (Beijing); Tong & Feng 2007.
22 Li 1998: 528–529.

only the work mentioned above, *Calculating rods*, had been printed by 1689.[23] It seems that it was during these years that Mei became acquainted with the works of Wang Xichan, 王錫闡 whom he esteemed on a par with Xue Fengzuo if not more so:

Among those who study the Western method while continuing to respect Chinese principles, in the North there is Xue, in the South there is Wang.[24]

To the present day, Chinese historians of science refer to 'Wang in the South and Xue in the North' (*nan Wang bei Xue* 南王北薛) when discussing seventeenth century astronomy.[25] This bears witness to Mei Wending's overwhelming posthumous influence. He wrote 'Revisions and commentaries' (*Dingzhu* 訂注) of the works of both Xue and Wang.[26] In 1689 Mei Wending went to Beijing for the first time; then aged fifty-six, he was already known to a wide circle of scholars. His stay in the capital, which will be discussed further,[27] opened a new era in his career.

4.2 Integrating Chinese and Western mathematics

According to its front matter, the *Integration of Chinese and Western mathematics* was to consist of the following nine titles: *[Napier's] Calculating rods* (*Chousuan* 籌算), *Brush calculation* (*Bisuan* 筆算), *Calculation by measure* (*Dusuan* 度算), *Calculation of proportions* (*Bili suan* 比例算), *Summary of geometry* (*Jihe zhaiyao* 幾何摘要), *Trigonometry* (*Sanjiao fa* 三角法), *Discussion of rectangular arrays* (*Fangcheng lun* 方程論), *Measurement of base and altitude* (*Gougu celiang* 句股測量), and *The nine reckonings preserved from antiquity* (*Jiushu cungu* 九數存古). In 1701, about twenty years later, Wang Shizhen 王士禎 (1634–1711), a high official regarded as one of the greatest poets of his time,[28] listed these nine works, adding to these some of Mei's astronomical works and concluded: 'He is one of the extraordinary men of our times.'[29] Wang, who does not seem to have been versed in mathematics, had probably seen the fascicule of front matters; it functioned as a pamphlet advertising Mei's work. The collection itself does not seem to have circulated much as a manuscript. In the bibliography of his works, Mei mentioned eight of the nine works that formed the *Integration of Chinese*

---

23 *SKQS* 795: 69.
24 Mei 1995, 28.
25 Li & Guo 1988, 27.
26 *SKQS* 795: 981–982.
27 See pp. 214–222.
28 Hummel 1943, 831–833.
29 Wang Shizhen, *Juyi lu* 居易錄. *SKQS* 869: 425.

and Western mathematics as the 'first collection' (*chubian* 初編) of his mathematical works, albeit in a slightly different order.³⁰

Unlike Fang Zhongtong and other scholars of his time, Mei did not attempt to bring together all the mathematics he knew in one single work. Instead, he wrote a number of independent treatises on subjects that he thought needed clarification or new foundations. Only at a later stage did he bring these treatises together; there is no evidence that he had planned an overall coverage of the field from the onset. The fact that there were nine treatises in the collection can hardly be read as a reference to the 'nine chapters' tradition: it is not claimed by him or by anyone else to be of significance. According to Wang, the last treatise mentioned in both lists above, which seems to be no longer extant, consisted of nine chapters that took up the traditional list. Mei's note in his bibliography does not further specify its content: he described the book as consisting of ten chapters: the first nine are indeed those found in the first century CE classic; the tenth is entitled *Pangyao* 旁要.³¹ That this work was never published probably indicates that Mei Wending himself did not regard leaving to posterity yet another work structured according to the 'nine chapters' classification as a priority.

Among the nine titles intended for inclusion in the *Integration of Chinese and Western mathematics*, each one seems to point to the origin of its content: as all Mei's contemporaries who were interested in mathematics would have known, Napier's rods, written calculation, proportions, trigonometry and geometry were 'Western', whereas rectangular arrays, base and altitude, and the 'nine reckonings' referred to the Chinese tradition. However, Mei challenged the view that everything that the Jesuits had imported was new. Thus he pointed to the presence of *gelosia* multiplication in Ming mathematical texts as evidence that Napier's rods were of Chinese origin.³² Moreover he repeatedly argued for the unity of mathematics. In the preface he wrote for the collection he mentioned the issue in passing: 'How could ancient and modern, Chinese and foreign, not be viewed as one?'³³ More than ten years later, when the *Brush calculation* was at last published, he wrote a preface for it. There he insisted at much greater length on this unity, giving his argument a more philosophical turn:

---

30   *SKQS* 795: 983–987.
31   The meaning of this term is unclear; it is sometimes found as a substitute for *gougu* 句股 (base and altitude) for the name of the last of the 'nine chapters' (e.g. in the *Ritual of Zhou - Zhouli* 周禮); Chemla & Guo 2004, 51.
32   *SKQS* 795: 983.
33   Mei 1995, 52.

> Those who study principles (*li* 理) resort to principles. Those who study numbers (*shu* 數) judge according to numbers. Numbers and principles are united. In China and in the West this does not differ.³⁴

Here the unity of mathematics is subordinate to a more fundamental one: that of numbers and principles. The latter is reasserted in many passages, where Mei articulated the relations between the two more fully:

> Outside numbers there are no principles; outside principles there are no numbers. Numbers are the delimitation and ordering of principles. Numbers cannot be conjectured without grounds; neither can discussions of principles be based on shadows.³⁵

Confident in the primacy of mathematics as well as in its unity, Mei did not limit himself to bringing together elements stemming from two different origins. He reworked them so as to create his own system, which had to form a whole, consistent both in itself and with the Confucian culture within which it was to operate. Thus, in *Brush calculation* he transposed the operations of written arithmetic that he took up from the Jesuits so that numbers written using place value notation would be laid out vertically rather than horizontally:

> Horizontal rows are Western countries' writing. In Arabia, characters are from right to left, whereas in Europe characters are from left to right, all laid out horizontally in rows. All their characters are thus. Their characters all being horizontal, written calculation is also horizontal to make its use convenient, and not in order to seek difference from us. Our characters all being vertical, written calculation should be vertical, to make its use convenient likewise, and by no means in order to boast outdoing them.³⁶

In addition to this refusal to compete with Westerners, Mei asserted an important feature of the mathematics that he aimed to construct: its practice was to be part of scholars' privilege, as it was to be done solely in writing. In this respect he made explicit the implication of the title that Ricci and Li Zhizao 李之藻 had given to their translation of Clavius' textbook of practical arithmetic: *Instruction for calculation in common script* (1614). For Mei, mathematics had to be practised using writing, and in each civilisation the notations should therefore harmonise with the writing. This points to a double meaning of 'integration': it was not only into China as opposed to other civilisations, but also and above all into scholarly learning that he wanted to integrate mathematics, by rooting its practice in the act of writing. In this light the treatise on *Brush calculation* can be seen as the basis of Mei's entire mathematics. This is probably the reason why this treatise is placed first in the

---

34 *Meishi congshu jiyao* 1761, 1: 1b.
35 Mei 1995, 34; Engelfriet 1998, 430.
36 *Meishi congshu jiyao* 1761, 1: 1b–2a.

Box 4.1 Layout of basic operations in *Brush calculation*

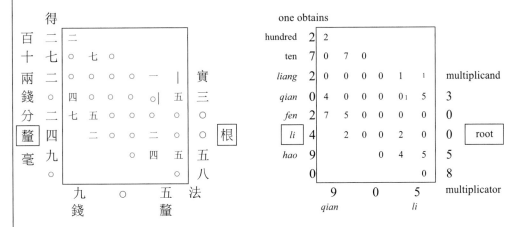

Layout of 892 + 1088 + 350 = 2,330

Layout of 300.58 × 905 = 272,024.90

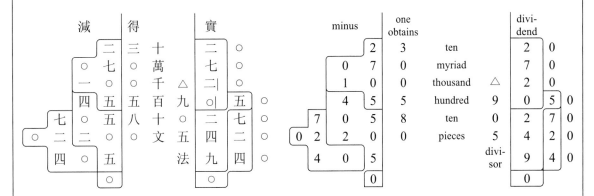

Layout of 27,202,490 ÷ 905 = 30,058

Source: *Meishi congshu jiyao* 1761, *Bisuan* 筆算, 1: 9b, 2: 4a, 14b.
Addition (and subtraction) follow the layout common nowadays, rotated by 90°. Multiplication follows the 'gelosia' layout (*pudijin* 舖地錦), found in Ming mathematical works prior to the Jesuits' arrival. Division follows the 'galley method' used by Clavius and presented in the *Instructions for calculation in common script*, rotated by 90°.

1761 edition of his works. That he indeed intended to teach scholars how to substitute written calculation for the abacus is further indicated by the lengthy commentaries to the titles of each chapter in a 1706 edition of the *Brush calculation*: there readers could see at a glance what each chapter referred to in terms of the equivalent operation performed using the abacus.[37] Mei applied the same 90° rotation to Napier's rods, still for the sake of consistency with writing.

4.3 The *Discussion of rectangular arrays*: restoring one of the 'nine reckonings'

Mei's double agenda of clarification and reconstruction is especially apparent in his earliest mathematical work, the *Discussion of rectangular arrays*.[38] By 1674, two years after its completion, the first draft of the treatise had circulated among some scholars known to him. Their positive reactions seem to have prompted him to write a preface that spelt out the work's ambition at the onset:

Rectangular arrays is one of the nine reckonings (*jiu shu* 九數). What is so unique about it? This treatise states that rectangular arrays—together with the base and altitude—is one of the two acmes of mathematics. Thus both consolidate the nine [reckonings] and make mathematics what it is. Quite often one only deeply reflects on the base and altitude and neglects rectangular arrays. Moreover, many mistakes are reproduced fortuitously. The reckonings are nine; if one is erroneous, how could one but discuss this?[39]

Mei thus put the eighth and ninth reckonings (or chapters) on a par, as the two most important ones. In some editions of the *Discussion of rectangular arrays*, the preface is followed by 'Other remarks' (*Yulun* 餘論)[40] in which Mei Wending made his vision of mathematics, and of his treatise's contribution, more explicit:

Mathematics is ninefold; reducing it to the essentials, it has two branches: calculation and measurement. Measurement consists in: using length and remoteness to seek distances—among the Western methods, this is called measurement of lines (*cexian* 測線); mutually seeking square and circle, arc and sagitta, area and diameter—among the Western methods, this is called measurement of areas (*cemian* 測面); seeking the volume of spheres and piles —among Western methods, this is called measurement of solids (*ceti* 測體). Among the ancient nine chapters, these were Rectangular fields, Lesser breadth, Consultations on works, Base and altitude. Calculation consists in using

---

37 Mei 1706; copy consulted: New York Public Library, call no. *OVQ 09–285.
38 On the *Discussion of rectangular arrays*, see Hashimoto 1973, 253–264; Martzloff 1981a, 161–234. It should be noted that Mei Wending interpreted *fangcheng* as meaning "comparison of arrangements" ("方者比方也。程者法程也。", *SKQS* 795: 67; see Martzloff 1980, 166–168); however a more standard translation is used consistently throughout the present work.
39 *SKQS* 795: 64.
40 This includes the *Lisuan quanshu*; *SKQS* 795: 65–66.

> Box 4.2 Mei Wending's classification of the nine reckonings ('nine chapters')
>
> | Measurement 量法 | Calculation 算術 |
> |---|---|
> | Rectangular fields 方田 (1)* | Millet and cloth 粟布 (2) |
> | Lesser breadth 少廣 (4) | Graded distribution 衰分 (3) |
> | Consultations on works 商功 (5) | Equitable taxation 均輸 (6) |
> | Base and altitude 句股 (9) | Excess and deficit 盈朒 (7) |
> | | Rectangular arrays 方程 (8) |
>
> *The numbers between brackets give the traditional order of the 'nine chapters'

fluctuations, excess and deficit, multiplication and division, and changes to differentiate quantities, examining the comings in order to fathom the goings —among the Western methods, this is called proportions; homogenising the numerators and denominators of parts, arranging and making uniform; naming what cannot be exhausted with the divisor—among the Western methods, this is called decimal parts; as to hidden intricacies and interlocked difficulties that are solved by examining the apparent and exploring the hidden in all its depth—there is nothing worth comparing, therefore among the Western methods, the rule of false position is established separately for use, although it is also proportions. Among the ancient nine chapters, these are Millet and cloth, Graded distribution, Equitable taxation, Excess and deficit and Rectangular arrays.[41]

Here Mei Wending reorganises the nine reckonings, that is, mathematics, into two branches. These are clearly taken up from the Aristotelian duality of magnitude (*du* 度) and number (*shu* 數) introduced by the Jesuits.[42] However in this 'remark' he adopts a different terminology, which also departs from the received Chinese one used in the title *Integration of Chinese and Western mathematics*, in which *suan* 算 referred to the whole of mathematics. In contrast, in the passage above he uses *shu* 數 to refer to the whole of mathematics, while he calls its two branches calculation (*suanshu* 算術) and measurement (*liangfa* 量法). This paves the way for two major changes: one regarding the status of mathematics, and the other regarding its structure and the respective roles of ancient learning and Western learning within it. First, mathematics is identified with numbers, a major concept in the philosophical tradition of the Cheng-Zhu school; this gives a very specific meaning to the claim that Mei often repeated—as he did in the preface of *Brush calculation* quoted above—that numbers and principles belong together,[43] and that the former are a key to accessing the latter. More specifically, the quotations above can be interpreted in this light, as

---

41 *SKQS* 795: 65.
42 Hashimoto 1973, 238–241.
43 See also *SKQS* 795: 64.

meaning that mathematics is the key for analysing the principles that underlie the universe. Secondly, *suan* 算, the traditional term for mathematics, now refers only to one of its two branches, the one in which ancient methods have no counterpart among Western ones.

These two [branches] are mutually necessary; one cannot apply oneself to one and neglect the other. However, calculation can cross the limits of measurement, while measurement cannot exhaust the transformations of calculation. Why is this? What can be measured is visible. In the universe, many things are invisible. If it was not for calculation, how could they be managed? Therefore measurement has limits and calculation has no limits. When one measures and obtains *lü* 率,[44] one measures one item and wishes to know one hundred. The Western use of proportions also amounts to assisting measurement with calculation. However, while comparing ratios is not measurement, it is ruled by the principles of measurement. My friend Fang Zhongtong from Tongcheng says that the nine chapters proceed from the base and altitude, from which they are constructed. However, my view is that the use of rectangular arrays, with its addition and subtraction of similar or opposite positive and negative, cannot be managed by means of the base and altitude, even though the latter can generate proportions. My humble opinion is therefore that calculation cannot be neglected.[45]

Whereas Fang Zhongtong sought to reconstruct the 'nine chapters' so that they would all be derived from the single premise of the base and altitude,[46] Mei Wending, while allowing that measurement was one of the two branches of mathematics, argued that calculation was more powerful.[47] This new vision had some bearing on the way that he went on to assess Westerners and their methods:

In mathematics too there are two schools: that of the ancient methods and that of the Far West. The teachings of the Far West are clear and meticulous; the Ancients' methods are quick and simple. They can shed light on each other. But in ancient works only calculations have been kept, whereas they are sketchy as regards measurement. The Far West are detailed about measurement and somewhat omit calculation. In my view what the school of the Far West says concerning the quadrant, triangles, the eight [trigonometric] lines, the division of the circle, and in the *Elements of geometry* is complete. One can say that in their excellent use of the base and altitude, there may be new ideas that do not proceed from the ancient *lü*—though this has not yet occurred. As to what is translated in the *Instructions for calculation in common script*, it probably uses the three *lü*[48] to modify ancient methods. As to Excess and deficit and Rectangular arrays, their procedures cannot go that far. Therefore they take up the

---

44 In their Chinese writings on proportions, the Jesuits used *lü*, a notion familiar to Chinese mathematicians, to refer to the terms of a proportion. *Lü* are numbers that are mutually related to one another, as in the rates of exchange of various types of cereals, or in the two numbers associated respectively with the diameter of a circle and to its circumference.
45 *SKQS* 795: 65–66.
46 See pp. 45–47.
47 Chu 1994, 164–166.
48 The rule of three.

Ancients' methods and pass them on: these [procedures] have not been passed on by Ricci. Among the wonders of calculations, none is comparable to Excess and deficit and to Rectangular arrays, and the Far West have neither: among the nine chapters, they lack two. Yet they say that they surpass the ancient methods! Moreover the Far West school want to change the Empire with their teachings. Therefore they have to write ingratiating memorials to achieve their aim and to win over scholars' confidence. As to Excess and deficit and Rectangular arrays, they are absolutely unable to discuss them. Because they are unable to discuss Excess and deficit, they have established the method of borrowing grades (*jie cui* 借衰) [false position] as a substitute. They say that it is prodigious and that one can give up the ancient method; but in the end one cannot give up Excess and deficit. On the chapter of Rectangular arrays, they are not only unable to discuss it, but also to use it. They do not go beyond taking up what has remained from the Ancients; they are incapable of establishing another method to replace that one. They have reproduced all the errors of other works and cannot check on them. It is visible that, not knowing [the method], they cannot use it; when they can discuss it, they are not thorough.[49]

This severe attack on 'the Far West' is two-fold. On the one hand, while he acknowledges the value of the *Elements* and of other methods pertaining to measurement taught by the Jesuits, Mei Wending points to their limitations in calculation, the branch of mathematics for the primacy of which he has previously argued: he professes a very poor opinion of the *Instructions for calculation in common script*, although he derived his own *Brush calculation* from the layout of operations presented in this work. On the other hand, he blames them vehemently for their claim of superiority over the ancients, a claim that is groundless in his view; his strongest criticism is moral rather than mathematical. Beside his lucid gaze on the Jesuits' mathematics, Mei appears to be well informed of their situation at court and of their politics. While this text was written before Mei read the *Books on calendrical astronomy according to the new method*, and thus became closely acquainted with the Jesuits' astronomy, this passage was still included in one of the posthumous editions of his works, the *Complete writings on astronomy and mathematics*: it seems that for all his interest in the Jesuits' astronomy, Mei did not change his mind about the shortcomings of their calculation methods—nor about their arrogance.

4.4 Writing mathematics: purpose, structure and style of the *Discussion of rectangular arrays*

Having situated his book in the landscape of the mathematics of his time, and thus justified why he had written it, Mei Wending went on to explain how he had written it. This was indeed necessary: just as he had exclusively focused on one of the 'nine chapters', the way he organised

---

49 SKQS 795: 66.

his lengthy discussion of the topic was unusual. The 'Introduction' (*Fafan* 發凡) that followed the 'Remarks', divided into eight points, explained and justified these original features. It should be noted that he did not boast of these as novelties. The use of a term such as *xin* 新 (new)—which at the time, it should be remembered, characterised the methods used by the Jesuits at the Astronomical Bureau—would have contradicted the whole rhetoric of the preface and 'Remarks', in which the *Discussion of rectangular arrays* was presented as a reconstruction of a lost mathematical method. Mei's claim is that he is reconstructing the method as the ancients (*guren* 古人) had established it. He opposes the latter to his recent predecessors—mostly Cheng Dawei—that he regards as merely 'old' (*jiu* 舊). The first four points of the 'Introduction' discuss the history and meaning of *fang-cheng*. There the narrative is in keeping with the late Ming and early Qing perception of Ming scholarship as somewhat decadent: while the method dated back to antiquity, it fell into oblivion after the Yuan dynasty (1279–1368), when mathematics in general was no longer cultivated by scholars. This oblivion has continued up to Mei Wending's time because of the fashion of geometry associated with Western learning.

In the next three points, Mei expounds 'the reasons why the clauses of rectangular arrays are different from that of the old ones', 'the reasons why the chapters are named according to the discussion', and how 'the examples, detailed or abridged', 'cast light on one another'.[50] He proposes a four-fold typology of rectangular arrays problems, according to the operations involved in stating and solving them; he goes on to explain how his text is structured, positing a distinction between examples and discussion:

When mathematical books give examples and no discussion, then one does not know the foundations of the methods: moreover in transmission there are many mistakes. For this reason, the present book wishes to clarify the principles of calculation (*ming suan li* 明算理). Therefore it gives more discussions (*lun* 論) than examples (*li* 例). Each chapter opens with a general discussion to give an outline. Thereafter one takes examples to substantiate explanations (*shi qi shuo* 實其說). (These are [introduced by] 'suppose' (*jiaru* 假如).)[51] Moreover when an example includes some dubious points, there is always some explanation to cast light on all its aspects, so that the reader can clarify things without the slightest obstacle. In general there are seven tenths of discussion and three tenths of examples. The chapters are named according to the discussions, bringing out their substance.[52]

---

50 *SKQS* 795: 68; *Meishi congshu jiyao* 1761, 5b–6b.
51 In fact the term *jiaru* is also used within the explanations.
52 *SKQS* 795: 68; *Meishi congshu jiyao* 1761, 6a.

Discussion is deemed important enough to be mentioned in the book's title. The term had been used in a mathematical context in the translation of Euclid's *Elements*, where *lun* was used to render 'proof'. In the absence of teachers or treatises explaining this entirely new usage of *lun*, Mei Wending seems to have taken up the term to refer to the parts of the text that are devoted to clarifying the matter under scrutiny. This is fully consistent with the aim he states above, that is, to help the reader understand the principles of calculation. It contrasts sharply with Ricci's view of the *Elements* and of their reception in China: 'It was manifest that Euclid showed his geometrical propositions so clearly that even the most obstinate was convinced.'[53] For Ricci, clarity was conducive to winning over reluctant readers. Despite the controversial dimension of the opening matters of his *Discussion of rectangular arrays*, Mei Wending's approach to writing mathematics seems to have been that of a teacher, taking readers as potential disciples, rather than as possible adversaries in a controversy, or as heathen who are to be convinced of the truth of Heavenly learning.

The *Discussion of rectangular arrays*, which altogether contains 90 problems,[54] consists of six chapters, four of which have titles more evocative of literati culture than of mathematical procedures. The first chapter, in which Mei Wending explains his classification of problems, is thus entitled 'Correcting names' (*zhengming* 正名), and opens with the famous quote from the *Analects*: 'When names are not correct, what is said will not sound reasonable.'[55] In this case the names to be corrected are those involved in the classification of rectangular arrays problems. Mei's criticism is addressed to the classification according to the number of kinds (*se* 色) involved found in the *Unified lineage of mathematical methods* (*Suanfa tongzong* 算法統宗, 1592) (Mei does not name the work, but his readers would have been familiar with it), where, in his view, the distinction between sum and difference is unclear. Instead he sorts problems into four different categories: 'numbers in sums' (*heshu* 和數), that only involve added kinds (and therefore do not require the differentiation between positive and negative); 'numbers in differences' (*jiaoshu* 較數) that involve differences between the kinds; 'sums and differences mix' (*hejiaoza* 和較雜), in which the problem involves both numbers in sums and numbers in differences; and 'sums and differences alternation' (*hejiao jiaobian* 和較交變), involving at least three kinds and in which numbers in sums can be transformed into numbers in differences during the procedure.

The titles of Chapters two and three are both borrowed from the 'Appended statements' (*Xici* 繫辭) section of the *Book of change*.

---

53 Quoted in Hashimoto & Jami 2001, 265.
54 Martzloff 1981a, 169.
55 Lau 1979, 118.

'Extreme numbers' (*Jishu* 極數) discusses problems in which fractions occur; 'Practice' (*Zhiyong* 致用) gives various devices for simplifying calculations. Chapter four, 'Correcting mistakes' (*Kanwu* 刊誤), lists six different types of errors found in previous authors; sixteen problems are discussed there. As might be expected, the *Unified lineage of mathematical methods* and the *Instructions for calculation in common script* are the targets of criticism here as well. Chapter five is entitled 'Measurement' (*Celiang* 測量): in keeping with Mei's claim that calculation can be a substitute for measurement; the chapter deals with astronomy. Rectangular arrays are used as an interpolation method in two situations: when the sky is clouded—and the stars are therefore invisible—and when no time-keeping device is available. The last chapter, 'Managing various methods by means of rectangular arrays' (*Yi fangcheng yu zafa* 以方程御雜法), shows how problems that belong elsewhere in those of the 'nine chapters' that Mei has classified as pertaining to calculation can be solved using rectangular arrays. These include Differential distribution (*Chafen* 差分), Equitable taxation (*Junshu* 均輸), and Excess and deficit (*Yingnü* 盈朒).[56] The subsuming of several chapters under rectangular arrays could well have been inspired by the *Instructions for calculation in common script*, where problems pertaining to several of the 'nine chapters' were brought together with rectangular arrays problems.[57] This is one of the many ways in which Western learning could have induced Chinese scholars to reconsider the structure of mathematics.

This short overview of the content of the *Discussion of rectangular arrays* gives a glimpse of the ways in which the various claims in the work's front matter relate to Mei's actual practice as a mathematician. His reorganisation of material that he found in previous works leads him to reconstruct the field at different levels. First, he classifies rectangular arrays according to the type of operations involved in the procedure. Secondly, by using rectangular arrays to solve problems that belong elsewhere in the nine chapters classification, he gives a very substantial meaning to his claim that this method constitutes the acme of calculation: it unifies the chapters that constitute this branch of mathematics. Last, but not least, he points to the constant use of calculation in what he (like all scholars of his time) regarded as the most important instance of measurement, that is, astronomical observation. In short, his claim of the primacy of calculation, and his singling out of rectangular arrays as the acme of calculation were substantiated in ways that must have seemed quite convincing to the readers who made their way through the book.

---

56 Chapters 3, 6 and 7.
57 The order given here is that of the *Complete writings on astronomy and mathematics*. In the *Selected essentials of Mr Mei's collection* Chapters 5 and 6 are reversed.

To illustrate Mei's practice of discussion, let us examine his treatment of the inkstones and brushes problem found both in the *Unified lineage of mathematical methods* and in the *Instruction for calculation in common script*, and also in *Numbers and magnitude expanded*.[58] It is the first problem given in the category of 'numbers in differences':

Suppose 7 inkstones are exchanged against 3 brushes; the price of the inkstones is 480 pieces (*wen* 文) more. If 9 brushes are exchanged against 3 inkstones, the price of the brushes is 180 pieces more.

Question: what are the prices of brushes and inkstones?[59]

Mei explains in great detail how the numbers of the problem are to be laid out for calculation in the order in which they are given in the problem, and how to decide which 'denomination' (*ming* 名) they are to be given for the calculation (see Table 1). As in earlier texts starting with the *Nine chapters on mathematical procedures*, this denomination can be either positive (*zheng* 正) or negative (*fu* 負), and functions only within the rectangular arrays method.[60] Mei Wending's originality here is that he illustrates how to determine denomination by alternative statements of the problems (four altogether), with one of the four coefficients stated first in each case. The rule he proposes to follow is that this first coefficient should always be placed in the top right place and taken to be positive: this rule is visualised by four corresponding layouts, and an algorithm (*fa* 法, lit. 'method') followed by an explanatory discussion (*lun* 論) is given for each of them.[61] Thus the first one reads:

Lay out each in its place:

[For reasons of formatting, the layout given here in the original text is reproduced in Box 4.3.]

First, multiply all the numbers obtained in the right column by the left column inkstones' negative 3. (The first places have different denominations; one must change one column so that they comply with one another. Therefore the inkstones' positive is changed into negative, the brushes' negative is changed into positive, the price's positive is changed into negative; all the numbers obtained are changed.)

Next multiply all the numbers obtained in the left column by the right column inkstones' positive 7. (Once the right column has been changed, then the left column should not be changed once more; therefore the inkstones' negative, brushes' positive and price's positive all remain as before.)

Thereupon subtract the similarly denominated upper inkstones' negative 21 for each; they exhaust.

---

58   See pp. 20–21, 28, 47–48. Martzloff 1981a, 191–196.
59   Guo 1993, 4: 330.
60   Chemla & Guo 2004, 624–629.
61   Guo 1993, 4: 330–332.

97

Box 4.3 First layout for the inkstones and brushes problem in the *Discussion of rectangular arrays*

	left		right	
upper	inkstones		inkstones	
	negative 3		positive 7	
	one obtains negative 21	subtract exhausted	one obtains negative 21	
middle	brushes		brushes	
	positive 9		negative 3	
	one obtains positive 63	subtract remainder 54	one obtains positive 9	
lower	price		price	
	positive 180		positive 480	
	one obtains positive 1260		one obtains negative 1440	
		add one obtains 2700		

上	右研		左研
	正七 得 負二十一	減盡	負三 得 負二十一
中	筆 負三 得 正九	減餘五十四	筆 正九 得 正六十三
下	價 正四百八十 得 負一千四百四十	併得二千七百	價 正一百八十 得 正一千二百六十

98

Next subtract both the similarly denominated middle brushes' positives; the remainder 54 is the divisor.

Once more add the differently denominated lower prices, left positive, right negative; one obtains 2700; it is the dividend. Divide the dividend by the divisor; one obtains 50 pieces; it is the price of a brush.

Multiply the price of a brush by the left column brushes' 9; one obtains 450. Subtract from this the similarly denominated price positive 180; the remainder is 270. Divide this by the left inkstones' negative; one obtains 90; it is the price of an inkstone. Alternatively, add the differently denominated total price of the right brushes' negative 3, 150, to the price's positive 480; the total is 630. Divide this by the right inkstones' 7 on the right; one also obtains the price of an inkstone, 90.[62]

The method of elimination is followed: it consists in multiplying each column by the number in the top row of the other one, and then subtracting the resulting columns from each other: the kind in the top row (in this case brushes) is thus eliminated, and the price of a brush is then obtained by dividing the result of the subtraction in the third row by that of the result of the subtraction in the second row.

The discussion that follows explains what the numbers in the layout represent and how they are derived; the lavish details given suggest that the author must have expected his reader to study his book on their own rather than with a teacher. This being said, in Mei's view, the correspondence between the problem and the layout functions both ways: thus, he argues, the layout given by Cheng Dawei implies a statement of the problem different from the one he gave. Similarly, Fang Zhongtong, who also gave the problem in his *Number and magnitude expanded*, had not justified his choice of making the inkstones positive and the brushes negative: he just took up the layout he found in previous works, just like Ricci and Li Zhizao had done before him. By contrast, Mei Wending uses this example to bring out a one to one relation between problem and layout, and to show his reader how to deduce the latter from the former (see Box 4.4).

Mei is less innovative as regards the treatment of numbers. In the layouts, numbers are not written in place value notation, but in the same way as in the text. In fact written calculation is not a prerequisite for understanding rectangular arrays; Mei's *Brush calculation* was written later. Moreover here, as in earlier works, including the *Instructions for calculation in common script*, positive and negative denominations function only in the context of the rectangular arrays method: this is why *ming* 名 has been rendered as 'denomination' rather than as 'sign' in the translation above. One can remark in passing

---

62 Guo 1993, 4: 330.

Box 4.4 Correspondence between phrasings of the inkstones and brushes problem and layouts in the *Discussion of rectangular arrays* (Guo 1993, 4: 330–332)

Text	Layout	
	left	right
Suppose 7 inkstones are exchanged against 3 brushes; the price of the inkstones is 480 pieces more. If 9 brushes are exchanged against 3 inkstones, the price of the brushes is 180 pieces more. Question: what are the prices of brushes and inkstones?	inkstones negative 3 brushes positive 9 price positive 180	inkstones positive 7 brushes negative 3 price positive 480
	left	right
Alternatively the question says: 3 brushes are exchanged against 7 inkstones; the price [of the brushes] is 480 pieces less. Again there are 3 inkstones exchanged against 9 brushes, the price [of the inkstones] is 180 pieces less.	brushes negative 9 inkstones positive 3 price negative 180	brushes positive 3 inkstones negative 7 price negative 480
	left	right
If, as in the layout in "Difficult problems",[1] one takes the inkstones as positive, the brushes as negative, the question ought to say: 7 inkstones are exchanged against 3 brushes; the price of the inkstones is 480 pieces more. 3 inkstones are exchanged against 9 brushes; the price of the inkstones is 180 pieces less. Then the price is positive on the right and negative on the left. (In "Difficult problems", the denomination is written together.)	inkstones positive 3 brushes negative 9 price negative 180	inkstones positive 7 brushes negative 3 price positive 480
	left	right
If the brushes are positive and the inkstones are negative, then the prices are: right negative and left positive.	brushes positive 9 inkstones negative 3 price positive 180	brushes positive 3 inkstones positive 7 price negative 480

1 This refers to Cheng Dawei's treatment of the problem; see pp. 20–22.

that the technical sense of *zhengming* 正名, rendered as 'Correction of names' in the title of Chapter one, is 'positive denomination'.

Last but not least, the very length of Mei Wending's treatment of the problem of inkstones and brushes, and of the rectangular arrays method in general, is quite a radical departure from the style of mathematical works that have dealt with the subject before him. His choice of thoroughness over concision may well have been an adjustment to the practice of reading mathematical texts among scholars of his milieu: solitary reading of works seems to have predominated over their use as textbooks under the guidance of a teacher. But his criticism of the Far West school hints at another interpretation:

> Westerners can be said to discuss base and altitude, trigonometry, the division of the circle and the *Elements of geometry* in detail. As to rectangular arrays, they add in all kinds of *lü*, are careless and do not fathom causes (*wei qiong qi gu* 未窮其故).[63]

In the light of this assessment one could say that in the *Discussion of rectangular arrays* Mei uses the Westerners' minute style of explanation as he has seen it at work in the *Elements* and in other fields of measurement. He applies this style to calculation, the branch of mathematics that they (and Mei's other contemporaries) have in his view neglected, and of which the *Discussion of rectangular arrays* constitutes an apology. This is one of the many ways in which Mei Wending's mathematical work can be characterised as syncretistic. As one of the scholars best versed in the mathematical sciences in his time, he provided his contemporaries with treatises that should enable them to acquire at least some of his expertise. The limited readership that he seems to have had up to 1689 brings us to a question that he addressed himself: why should a scholar study mathematics? On the whole his contemporaries regarded it as a rhetorical question—with the notable exception of one man: the emperor. But, much to Mei Wending's dismay, Kangxi recruited teachers that had travelled to Beijing from much further than Jiangnan. By the time Mei set off for the capital, a new group of them was settling down there; they had been sent by none less than the King of France.

---

63 *SKQS* 795: 75.

CHAPTER 5

The 'King's Mathematicians': a French Jesuit mission in China

While Western learning drew the attention of both the Beijing court and Jiangnan scholars, interest for China continued to develop in Europe, with new actors stepping in. First, within the Catholic missionary enterprise, Franciscans and Dominicans were now present in China, as was the *Société des missions étrangères de Paris*; all challenged the Jesuit mission, and in particular its interpretation of Confucian rites as compatible with Christianity.[1] Secondly, as the decline of the Portuguese maritime power was becoming more evident, other European nations became rivals for the Asia trade: among these, France was the most significant Catholic competitor. Thirdly, a number of newly founded institutions whose aim was the advancement of the sciences sought to include China in their fields of enquiries. The circulation of knowledge between Europe and China, and therefore Western learning itself, were shaped by all these new actors.

5.1 Setting up a scientific expedition to China

In France, during the reign of Louis XIV (1643–1715), interest in Asia was multi-sided; it involved two royal institutions. Established in 1664 by Colbert (1619–1683), the *Compagnie française des Indes orientales* (French East India Company), unlike its Dutch and English counterparts, was a state enterprise. The Indian Ocean was its main target;[2] its agenda—trade expansion—was indissociable from diplomatic action and the royal patronage of Catholic missions. Two years after the *Compagnie*, following Colbert's advice, Louis XIV founded the Paris *Académie royale des sciences*; its first members were nominated by Colbert, and it had no formal statutes until 1699. In 1669, one of its most eminent members, Gian-Domenico Cassini (1625–1712), was put in charge of the newly built Paris Observatory. Cartography and the determination of longitudes at sea were among the first commissions of the *Académie*. These tasks were of course

---

1 Standaert 2001, 680–781.
2 Ames 1996.

directly relevant to maritime trade, as well as to the survey of the French territory. Scientific expeditions were organised to make observations in various parts of the world. In the early 1680s, the *Académie* began to train Jesuits before they departed for missions: it was a more economical way of collecting observations than to set up expeditions at its own expense. Such collaboration became all the more important after Colbert's death, as his successor Louvois (1641–1692) was by no means as generous to the *Académie* as he had been.³

The Jesuits were also associated with Colbert's effort to improve navigation in another way: they were appointed to teach hydrography in various French cities. A number of royal chairs were founded at colleges run either by them or by the Oratorians.⁴ These Jesuit professors cultivated relations with the *Académie*. Thus Ignace Gaston Pardies (1636–1673), who taught mathematics at the Collège de Clermont, the Jesuit establishment in Paris, dedicated his geometry treatise to the academicians.⁵ Published six years after the *Académie*'s foundation, this dedication was an expression of acceptance of its authority:

> It is true that we do not have in France this kind of judicature one finds in China, where a Court of learned Mathematicians judges in the last resort all that regards mathematics, which is one of the most important State matters in that country. If the laws of our Kingdom have not granted you this jurisdiction, you have it, Gentlemen, by your own merit. [...] What can one say, when seeing this great edifice standing so magnificently, save that it is a Palace being built for a new Tribunal, and that the King, who surpasses the Chinese Emperors in the structure of this building, wants maybe to imitate their policy in the erection of this new Company? You know, Gentlemen, that the Chinese Tribunal of Mathematics is held ordinarily in two Observatories, both located close to the two imperial cities.⁶

By comparing the Academicians to Chinese official astronomers, Pardies was paying them an elaborate compliment. This bears witness to the image of China in seventeenth-century France: an empire to be emulated, nay surpassed, regarding the state's magnificence and support of the sciences. Such a reference to the institutions of astronomy of imperial China was more relevant in absolutist France than it would have been in parliamentary England, where the Royal Society never was as dependent on royal power as its French counterpart.

Plans for a Jesuit expedition to China were discussed in 1681. During the *Académie*'s session on 29 November of that year, Cassini read a *Project for geographical observations*, which proposed among other things 'to send skilled mathematicians to China as missionaries'. He named

---

3 Stroup 1987, 44–45; Tits-Dieuaide 1996, 4–6; the beginnings of the *Académie* are also discussed in Stroup 1990, 3–10.
4 Hsia 1999, 10–21.
5 Hsia 1999, 24; on Pardies see Ziggelaar 1971.
6 'Epitre', in Pardies (1671) 1673, aii v.– aiii r.

the man who, in his view, was best qualified for such an expedition: 'the Reverend Father de Fontaney, Professor of Mathematics at the College de Clermont, who has long shared observations with the *Académie*, could particularly serve there.'[7]

Before teaching at the College de Clermont,[8] Jean de Fontaney (1643–1710) had held the chair of mathematics at the Jesuit college in La Flèche, then the chair of hydrography in Nantes. Unlike his predecessors at the Paris chair, he did not produce any textbooks: astronomical observation was his main engagement with science. His name appears together with those of Cassini, Picard, La Hire and Roemer, as one of the observers of several astronomical events between 1678 and 1684.[9] Thus when Cassini mentioned de Fontaney's name at the *Académie*, the former's assessment of the latter was based on collaboration in astronomical observations: from this he derived the confidence that de Fontaney was a 'good observer', in other words that he could produce accounts of astronomical events that met the standards established by the *Académie*. In particular, following its new method, he would have recourse to the occultation of Jupiter's satellites to establish longitudes. He could therefore be sent safely on an astronomical expedition.[10]

While evidence suggests that the choice of Jean de Fontaney was made by Cassini, the plan to send Jesuit observers to China had motivations other than scientific. Together with trade and diplomacy, mission was a way of extending French influence worldwide. Although Portugal still retained a monopoly over Asian missions, its declining power as well as the growing number of Christians in China meant that the Jesuit mission was short of personnel. Verbiest was keenly aware of this; in 1678, when he was Vice-provincial of China in the Society of Jesus, he sent out a letter to his Jesuit confreres in Europe, describing the situation of the China mission, and calling for more missionaries to be sent there.[11] The letter was widely circulated and eventually published in Paris. In 1681, Louis XIV's Jesuit confessor, François de la Chaize (1624–1709) duly drew the king's attention to it. As well as taking up Portugal's role as patron of missions, sending a royally sponsored mission to China was also in keeping with Louis XIV's 'Gallican' religious policy that asserted the autonomy of the French church vis-à-vis Rome: in this regard 'sending to China Jesuit mathematicians was but an expedient of the King's government in order to have in China missionaries who would not only be

---

7 AdS, Procès verbaux, 29 November 1681, 9bis, f. 125; see Landry-Deron 2001, 430.
8 The College was renamed 'Louis-le-Grand' in 1683.
9 Landry-Deron 2001, 455.
10 Hsia 1999, 27–38.
11 See Golvers 1993–94 for a French translation of this letter.

representatives of the pope.'¹² Thus the project combined the interests of diplomacy, trade, as well as religion and its politics, with those of the sciences.¹³

In a letter he later wrote from China, de Fontaney recalled how Colbert had first presented the plan of a Jesuit expedition to China:

> The Sciences, Father, do not deserve that you should take the trouble to cross the seas and to reduce yourselves to live in another world, far from your homeland and friends. But, as the desire to convert infidels and to win souls over to Jesus Christ often prompts your Fathers to embark upon such travels, I should wish that they took the opportunity, when they are not so busy preaching the Gospel, to gather on the spot a great many observations that we lack for the perfection of the sciences and the arts.¹⁴

It is not surprising that this account, published in the *Lettres édifiantes et curieuses*, showed Colbert giving priority to mission over science for the *Académie*. However, European learned circles did not fail to establish a link between the Jesuits' scientific expertise and their efficacy as missionaries. Some scientific instruments were deemed especially appropriate as visual evidence of this expertise. This was the case of the two machines constructed by Ole Roemer (1644–1710) for the *Académie* in 1680–1681. One of them represented the motions of the Planets, the other the solar and lunar eclipses.¹⁵ In January 1682, the *Journal des sçavans* published a report on these machines:

> They had been found to be so accurate, so beautiful and so extraordinary that the Jesuit Fathers did not believe that they could bring to the Indies anything that would surprise the Indians more, give them more esteem for our nation, or make the preaching of the Gospel to them easier.¹⁶

The links between the wonders of academic science and Christianity were further articulated by Leibniz in a letter to Colbert:

> A King of Persia will cry out in admiration for the telescope's effect, and a Chinese Mandarin will be delighted and amazed when he understands the infallibility of a Geometer Missionary. What will these people say, when they see this wonderful Machine that you have had built, which really represents the state of the heavens at any given time? I think they will have to acknowledge that human nature has something divine, and that this divinity is communicated especially to Christians. The secret of the heavens, the magnitude of the Earth and the measurement of time are all of that nature.¹⁷

---

12 Pinot 1932a, 40.
13 Landry-Deron 2001, 425–429.
14 De Fontaney, letter to de La Chaize, 15 February 1703. *LEC* 7: 65–66.
15 On the latter machine see Frémontier 1996.
16 Quoted in Hsia 1999, 40 n. 121.
17 Klopp 1864–84, 3: 211.

The 'wonderful machine' mentioned here is evocative of Roemer's machine for planetary motion; however the letter was probably written in late 1678 or early 1679.[18] Thus astronomical instruments seem to have been widely regarded as aids to evangelisation. A century after Ricci had entered China, the idea that scientific knowledge could pave the way for the Christian faith was thus widespread in Europe. This complements rather than contradicts the high prestige of China in France at the time as expressed by Pardies in his *Elemens de geometrie*. The great and admirable scientific institutions of China were a token that 'these people' were civilised enough to appreciate Christian superiority from such sophisticated evidence as astronomical instruments and the world-view that underlay them. Leibniz's phrasing nicely articulates the way in which teaching and practising science in China formed an integral part of the missionary agenda.

At the time of Colbert's death, in 1683, there seemed to be no immediate prospect of carrying out Cassini's plan. During the following year, interest in things Asian was further aroused in France by the arrival of an embassy sent by the king of Siam in September 1684.[19] The same month, Louis XIV also received in Versailles Philippe Couplet (1622–1693), a Jesuit whom Verbiest had sent back to Europe to seek support for the mission, and who was touring European capitals; he was introduced to the king by his confessor de La Chaize.[20] The coincidence of the two visits seems to have resulted in the decision to send the Jesuit expedition for China on the ship that was to convey an embassy to Siam. The Jesuits would go to Siam with the ambassador, and then find a way to continue to China from there.

During Couplet's stay in Paris, the *Académie*, at Louvois' request, prepared a list of 'Questions to be asked to the R.F. Couplet on the Kingdom of China'. These opened with queries on Chinese chronology and history, and went on as follows:

Whether the Reverend Jesuit Fathers have made any observations of longitudes and latitudes of China.
About the sciences of the Chinese, and about the perfection and defects of their mathematics, astrology, philosophy, music, medicine, and pulse-taking.
About tea, rhubarb, and their other drugs and curious plants and whether China produces some kind of spices. Whether the Chinese use tobacco.

---

18 Klopp (1864–84, 3: 211) dated this letter 1675. However in it Leibniz refers to the phosphorus that he has sent to the *Académie*; this would suggest that the letter was probably written not long after September 1678; Aiton 1985, 78.
19 On the relations between France and Siam, see Van der Cruysse 1991.
20 Foss 1990.

Other questions concerned weapons, fortifications, ships, soldiers, as well as various aspects of geography, technology and customs.[21] Much of the literature on China published by the Jesuits in the eighteenth century, while constituting a defence of their mission, provided replies to these questions.[22] In the shorter term, they provided an agenda for the investigations of the expedition that was set up during the last months of 1684.[23]

As planned three years earlier by Cassini, Jean de Fontaney was to lead the expedition. He in turn selected his travel companions. As time was pressing he chose all of them at the Jesuit College of Paris: Guy Tachard (1648–1712), Joachim Bouvet (1656–1730) and Claude de Visdelou (1656–1737) were selected first. None of them had de Fontaney's experience in astronomy: Tachard had just returned from South America; Bouvet and Visdelou were in their third year of study at the College. In December 1684, the four of them visited the *Académie*, where they learned from Denis Dodart (1634–1707) 'what matters of natural history the *Académie* would like them to correspond about'.[24] They also took part in the observation of a lunar eclipse at the Observatory, together with Cassini.[25] The latter gave de Fontaney a copy of his tables of Jupiter's satellites, so that the Jesuit could use them in his later observations, while another academician, also personally acquainted with him, Jacques Borelly (d. 1689), a chemist and a lens maker, gave him several large lenses.[26] When it was known that funding allowed for two more members to join the expedition, Louis Le Comte (1655–1728) and Jean-François Gerbillon (1654–1707) were nominated as well. The latter had taught mathematics and grammar at various colleges: however he did not belong to any scholarly circle. In short, only the leader of the group can be regarded as having been fully trained for the task set to the expedition by the *Académie* regarding astronomy.[27]

Before they set sail from Brest on 3 March 1685, aboard the ship *Oiseau*, the four Jesuits who had visited the *Académie* and the Observatory received from the Ambassador letters patent by which each of them was appointed by the king 'our Mathematician'.[28] Rather than an acknowledgement of their skills in astronomy, this was a device to avoid the necessity for them to swear allegiance to the king of Portugal, as Louis XIV intended them to work under his sole authority. On the

---

21 Pinot 1932b, 7–9.
22 Pinot 1932a, 141–185; Landry-Deron 2002.
23 Hsia 1999, 47, n. 139.
24 Quoted in Stroup 1990, 212.
25 Tachard 1686, 13–15, Landry-Deron 2001, 432.
26 Gatty 1963, 15; Chabbert 1970.
27 Gatty 1963, 14; Hsia 1999, 6–8; Landry-Deron 2001, 431–433.
28 Quoted by Landry-Deron 2001, 435.

other hand, it was in this capacity that they benefited from royal funding: beside pensions, the mathematicians were provided with books and scientific instruments, and most generously so: the amount spent on these instruments is greater that the total allotted to astronomical instruments for the *Académie* during the years it was supervised by Louvois.[29] This may be explained in part by the fact that these instruments were intended not only for the Jesuits' own use, but also as diplomatic presents, as was the case with the two machines of Roemer included among them.

5.2 From Brest to Beijing

Given that the Jesuit mathematicians had boarded a royal ship and travelled with an embassy, it is not surprising that there are several accounts of their crossing from Brest to Siam.[31] From these it appears that beside taking turns in preaching, the Jesuits practised astronomy. For this they used an astronomical map by Pardies that de Fontaney had edited and published after the former's death.[32] They also studied Portuguese, the *lingua franca* of Europeans in the East Indies in general and of the Jesuit mission in China in particular;[33] on the other hand, they do not seem to have done any study of the Chinese language. In the absence of textbooks, let alone teachers, outside China itself, Jesuit missionaries under the *Padroado* themselves started learning the language only once they had made it to Macao or to the China mission itself.[34] The French were even less in a position to study Chinese during their journey.

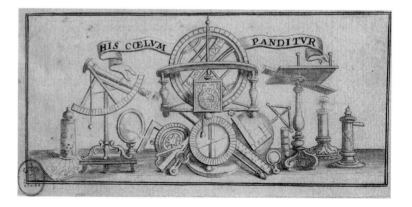

Illustration 5.1 Frontispiece of Tachard's *Voyage de Siam*, 1686 (BnF, Paris).[30]

---

29 Tachard 1686, 9–11; Stroup 1990, 246–247.
30 'By means of these things, heaven is revealed' (*His cælum panditur*); see Hsia 2009, 72–75.
31 These are listed in Choisy 1995, 439–441.
32 Choisy 1995, 49, n. 52.
33 Choisy 1995, 49.
34 Brockey 2002, 313–374.

After a crossing that seems to have been remarkably easy, the embassy reached Siam in September 1685. Tachard, who returned to France with the embassy, published an account of his trip to Siam in 1686. As stated in the title, it contained the six Jesuits' 'astronomical observations and their remarks on physics, geography, hydrography and history'.[35] Thus the first report of the expedition was published within two years of its departure, but not under the auspices of the *Académie*. It described how the Jesuits had followed Cassini's instructions in the observations they had made at the Cape of Good Hope,[36] and at Lopburi ('Louvo'), the secondary capital of Siam, where they had kept King Narai (1629–1688) company during a lunar eclipse, on 11 December 1685. According to Tachard, the King, enthusiastic about the Jesuits' skills in astronomy, had sent him back to France to recruit more Jesuit mathematicians to come and work in an observatory that he intended to have built.[37]

The rest of the trip to China was not as easy for the five remaining Jesuits. No provision had been made in France for taking them further than Siam. Their first attempt to sail on a ship chartered by the Siamese having failed, it was only in July 1687 that they reached Ningbo 寧波, a port in Zhejiang. They had been advised not to attempt to land at Macao, as the Portuguese would not welcome their presence there.[38]

Entering China by this route was illegal, and they were detained while a report of their presence was sent to the Ministry of Rites, which was in charge of foreign affairs. In turn the Ministry reported its decision concerning the five foreigners to the emperor:

Illustration 5.2 The King of Siam and the French Jesuits observe the lunar eclipse of 11 December 1685 (Tachard 1686, 286; BnF, Paris).

---

35 Tachard 1686, 326–355.
36 Choisy 1995, 99.
37 Tachard 1686; see Hsia 1999, 49–53.
38 Landry-Deron 2001, 441–444; Choisy 1995, 276–277.

> The Ministry of Rites memorialises:
> Your subjects have received a report from Jin Hong 金鋐, Governor of Zhejiang, stating that five Westerners, de Fontaney and others, have arrived in Zhejiang from Siam, aboard the ship of Wang Huashi 王華士, a merchant from Guangdong. According to what they say, they want to go to the churches of Suzhou and Hangzhou to visit their co-religionists. If allowed, they would like to remain inland durably.[39] Having checked, it turns out that they have no permit. Should one allow them to reside inland and visit their co-religionists, or order to send them back to their country? Following the Board's decision, henceforth travelling merchants are not allowed to take it upon themselves to bring foreigners inland, in order to prevent such trouble with stowaways etc. Checking regulations, foreign merchants should not be allowed to stay inland. It is inappropriate that the five, de Fontaney and others, should reside inland. They are to be handed over to that Governor, expelled and ordered to return to their country. Henceforth Chinese travelling merchants are strictly forbidden to bring in foreigners. Those who will bring in foreigners in violation of this will be severely punished.
>
> This order can be enforced on the day it is received by local magistrates and maritime customs officers.[40]

This routine report was submitted for approval on 25 September 1687. The Governor of Zhejiang's readiness to deport the newcomers, and the response of the Board of Rites seem to have been characteristic of officials' attitudes towards missionaries: no hostility against their religion was expressed. Rather, they were reacting against an attempt at illegal entry on Chinese territory. The imperial response to the memorial came on 11 October of the same year:

> It has not been possible yet to determine whether among the five Westerners, de Fontaney and others, there might be someone versed in astronomy. Send them to the capital and let them wait to be employed. Those who are not to be employed can reside where they please.[41]

Between the memorial and the emperor's reply, Verbiest, informed of the five French Jesuits' presence, had pleaded with the emperor that they should be allowed to stay in China.[42] This is typical of the role of court Jesuits in the mission: their favour with the emperor was essential for other missionaries to obtain permission to reside in China.

---

39 *Xichao ding'an* 熙朝定案 BnF Chinois 1330, 1a–1b. Another version of *Xichao ding'an*, BnF Chinois 1331, reproduced in Wu 1966, 3: 1725, omits the following passage.
40 *Xichao ding'an* BnF Chinois 1330, 1a–1b. The passage omitted in BnF Chinois 1331 ends here. This second version of the *Xichao ding'an* generally tended to omit passages unfavourable to Christianity in the official documents it reproduced; Standaert 2001, 133–134.
41 *Xichao ding'an* BnF Chinois 1330, 1b–2a. Comp. translation by Fu 1966, 93. This decree was mentioned by de Fontaney in a letter dated 2 November 1687; Landry-Deron 2001, 444.
42 Lecomte 1990, 51.

The French Jesuits' presence was also known to Prospero Intorcetta, who was then the Vice-provincial of China and resided in Hangzhou, where Mei Wending met him. Somewhat worried that their illegal entry into China might endanger the mission, he asked them what motivations had brought them: in response they alleged it was the scientific mission that Louis XIV had sent them to fulfil. Kangxi's positive reaction to their presence, however, reassured Intorcetta.[43]

It seems to have been while they were in Ningbo that the five Jesuits decided a division of labour amongst themselves, as de Fontaney explained in a letter to the *Académie* dated 8 November 1687:

We have sorted the matters on which we will work henceforth, and have divided them in five parts:
The 1st contains the history of Chinese astronomy and geography, and the daily observations of the sky, to respond to those that are done in Paris at the Observatory.
The 2nd the universal history of China, ancient as well as new; the origin of characters; what regards those same characters and the Chinese language.
The 3rd the natural history of plants and animals, and Chinese medicine.
The 4th the history of all arts, liberal as well as mechanical.
The 5th the present state of China, of the police, the government, and the customs of the country; and the other parts of physics that are not included in the 3rd article.
I have taken charge of the 1st part, F. Visdelou of the 2nd, F. Bouvet of the 3rd, F. Le Comte of the 4th and F. Gerbillon of the 5th.[44]

Thus the responsibility for astronomical observations fell to de Fontaney alone, while the ambition of the group remained to cover the various fields in which the *Académie* expected them to provide information about China.[45] Before leaving Ningbo, de Fontaney also wrote to Louvois asking for an increase in their pension, attaching to his letter a memorial in which he listed the instruments and books that each of the five Jesuits further required in order to carry out their specific tasks.[46] De Fontaney also addressed letters to Siam and to his confreres in Paris: the range of his correspondents and the contents of his letters bear witness to the group's continued assumption of the double identity of missionaries and royal mathematicians.[47]

---

43 ARSI, Jap. Sin. 127, 136v–137r. Brockey 2002, 164.
44 AMEP, V. 479, 32; see Bernard-Maître 1942, 279–280; Hsia 2009, 112–113.
45 Hsia 1999, 132–133.
46 AMEP, V. 479, 37–39.
47 AMEP, V. 479, 32 ff.

5.3 In the capital

Starting from Ningbo on 26 November 1687, the five Jesuits still took more than two months to reach their final destination. On 7 February 1688, when they arrived in the capital at last, they found it in a state of public mourning following the death of Kangxi's grandmother, Grand Empress Dowager Xiaozhuang, 孝莊 whose loss greatly affected the emperor.[48] Less visible in the city's life, but of more direct consequence for the King's Mathematicians was that Verbiest had passed away on 28 January, one day after the empress. He had been replaced by Tomás Pereira as the superior of the Beijing mission. This certainly was a change for the worse for the French Jesuits: whereas Verbiest had actively sought support for the mission from all European courts, Pereira, himself a Portuguese, was first and foremost eager to uphold the *Padroado*, in effect the exclusive patronage of the mission by the Portuguese crown.[49] In his view the newcomers disrupted the order's discipline and were competitors rather than reinforcements. Kangxi had summoned him from Macao to Beijing in 1673, after Verbiest recommended him as a talented musician. It was Pereira who introduced the five French Jesuits to the emperor; they were as yet unable to speak either Chinese or Manchu. According to them, he deliberately sabotaged their enterprise by minimising their competence. This is corroborated by a further report from the Ministry of Rites, following up on the one quoted above:

We communicate respectfully: After we transmitted [the decree] to the governor, we have now allowed de Fontaney, Le Comte, Visdelou, Bouvet and Gerbillon, and the armillary sphere and other instruments that they bring, in a total of more than 30 trunks, small, medium and large, to reach the Ministry. Once they arrived at the Ministry, we made it our duty to put de Fontaney and the others in relation with the Astronomical Bureau for information and clarification. It may well be that they are not versed in astronomy.[50]

This negative assessment could well have stemmed from Pereira. Meanwhile, the Frenchmen were staying at the Jesuit residence, where they did not feel at ease: some orders were issued (most likely by Pereira) that directly affected their scientific mission:

That no one leaving the college or travelling could take with him any instrument of mathematics under any pretext.

That during travels one would not take the latitude of places or any other observation that might induce suspicion in the Princes in whose territories we are to work only to the salvation of souls.[51]

---

48 Spence 1974, 104–105; Standaert 2008, 181.
49 Sebes 1961, 137–138; see also Josson & Willaert 1938, 543; Jami 2008b.
50 Wu 1966, 1727.
51 ARSI, Jap. Sin. 127, ff. 145v–146r.

These bans, which made it impossible for the French to carry out the tasks for which Louis XIV had so generously endowed them, reveal a double concern. On the one hand, Pereira very likely wanted the Beijing mission to keep all the instruments brought by the French: as the local superior, he assumed that he should have control over all the money and goods in the residence. On the other hand, the concern that the Qing state might regard the collection of cartographic information by foreigners as undesirable seems quite reasonable. In short, he defended both discipline and the interest of the mission, which in his view were indissociable. De Fontaney, on the other hand, held that he and his companions owed obedience first and foremost to the king of France, to whom '[their Jesuit] superiors had given [them]':[52] in his view the scientific expedition seems to have had priority over the evangelising mission at that stage. This had consequences beyond a Franco–Portuguese rivalry: in their reluctance to take orders from anyone but their king, the Frenchmen sought to be exempted from taking the oath of fidelity to the Pope that was then demanded of all missionaries.[53] The interests of the mission and those of French academic science turned out to be contradictory, rather than complementary as had been assumed in Paris.

The French Jesuits, while waiting to be received by the emperor, visited the capital. In the work he published in Paris in 1696, Le Comte gave a general description of the city, with some details about the Observatory and engravings showing the building and the main instruments; these illustrations were taken up from Verbiest's works.[54] He quoted Pardies' earlier enthusiastic account of the Observatory in the front matter of his *Elements of geometry* (1671), which was concluded as follows:

In one word, it seems that China slighted all other nations, as though, with all their science and all their riches, they could not produce anything similar.[55]

This praise of the Beijing observatory was part of Pardies' rhetoric in his eulogy of the *Académie* and of the Paris Observatory. Le Comte, however, rejected it with some vehemence:

Truth to tell, if China slights us by the magnificence of her observatory, she is well advised to do it from six thousand leagues away, for she would be ashamed to compare herself to us more closely. We went there all prejudiced with these great ideas [...][56]

---

52 AMEP, V. 427, 635; quoted in Landry-Deron 2001, 445.
53 Landry-Deron 2001, 429, 445.
54 Lecomte 1990, 97–114.
55 Pardies (1671) 1673, 'Epître', iii–iii v.; Lecomte 1990, 97.
56 Lecomte 1990, 97–98.

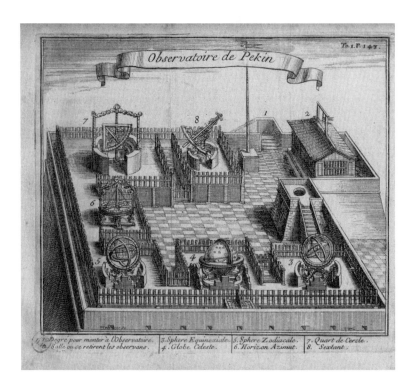

Illustration 5.3 The Beijing Observatory, Le Comte 1696 (BnF, Paris).

His first disappointment was with the size of the observatory. The Beijing observatory was by then more than two centuries old,[57] but Le Comte, far from being impressed by that fact—of which he may well have been ignorant—was quite severe about what he described as the Chinese taste for what is ancient: 'Antiquity, although defective, has for them charms that the most perfect novelty could not diminish; in this they differ from Europeans who only appreciate what is new.'[58] In the context of late seventeenth-century France, this can be read as a way to side with the Moderns against the Ancients in a controversy entirely alien to China. Le Comte's assessment of the new instruments made under Verbiest's supervision, however, was hardly more positive:

They are large, well cast, decorated all over by figures of dragons, laid out very well for the use to be made of them; and if the fineness of divisions matched the rest of the work, and that instead of pinnules, one applied telescopes, following the *Académie Royale*'s method, we would have nothing in this matter that could be compared to them. But whatever care this father took to have the circles divided exactly, the Chinese craftsman has been either very negligent, or unable to follow exactly what he had been asked; so that I would rely more on a one-

---

57   Its construction had started in 1442; Deane 1989, 242.
58   Lecomte 1990, 103.

foot quadrant made by our good Parisian craftsmen than I would on the six-foot one at the tower.[59]

Le Comte nonetheless went on to give a detailed description of the six instruments of which he also provided engravings. De Fontaney, who had probably pointed out the shortcomings of Verbiest's instruments to his companions, assessed them in a similar way in a letter to de La Chaize written in 1703:

[The instruments] are beautiful and worthy of the emperor's magnificence. But I do not know if they are as accurate as would be needed to make exact observations, because they have pinnules, their divisions are visibly unequal, and their transversal lines do not join in several places.[60]

In designing the instruments, Verbiest had made use of Tycho Brahe's *Astronomiæ instauratæ mechanicæ*, published in 1598.[61] In the three decades that elapsed between Verbiest's departure from Europe in 1656 and that of the French Jesuits, however, the size and accuracy of astronomical instruments had changed drastically.[62] It is therefore little surprise that his instruments failed to meet the standards of the *Académie*, and thus that they were inadequate in de Fontaney's eyes: contrary to what Le Comte assumed, the responsibility for this did not lie solely with the Chinese craftsmen. The French Jesuits' visit to the Observatory must have taken place during the day: they did not make any observations from there.

On 21 March 1688, they were given audience by the emperor. Two days later, he informed them that he had chosen to keep two of them in his service: Bouvet and Gerbillon. De Fontaney understood this to be the result of Pereira's own choice as to whom he was least disinclined to keep in Beijing, rather than a well-informed decision on the emperor's part:

It is certain that I was marked first on the Portuguese's papers [...]. F. Le Comte and F. Visdelou were those who had most practised mathematical observations, and I had indicated it. F. Gerbillon was more accommodating than us all to F. Pereyra. F. Bouvet thought of Tartary, the emperor learnt about this and kept him to this end.[63]

In Pereira's eyes, de Fontaney's skills in the sciences were a danger rather than an asset for the mission. He probably understood his own loyalty to Portugal as one and the same as his obedience to the Jesuit hierarchy. He was also eager to keep internal dissensions hidden from Kangxi, which may have explained his wish to move de Fontaney away

---

59   Lecomte 1990, 99.
60   Vissière & Vissière 1979, 122.
61   Chapman 1984; Halsberghe 1994.
62   Chapman 1984, 439–440.
63   AMEP, V. 479, 76; quoted in Landry-Deron 1995, 1: 39.

from the capital: as the superior of the French, the latter was the main challenge to Pereira's authority. This was all the more crucial as at the time the emperor was due to nominate Verbiest's successor at the Astronomical Bureau. Beside the status that this successor would have as an official, he would also make observations from the imperial observatory, which the French Jesuits were unable to do. Kangxi, however, would have been unlikely to appoint a newcomer who knew neither Chinese nor Manchu: past experience must have made him aware that only someone familiar with China's customs and institutions could aptly fulfil an official position. The two Frenchmen who remained in the capital were therefore hardly competitors for the post. On 9 April, the emperor announced that he had chosen Claudio Filippo Grimaldi to take up the position of Administrator of the Calendar. The choice was quite logical: Grimaldi had been Verbiest's assistant at the Astronomical Bureau since 1671, and he had a good command of the Manchu language.[64] As he had gone back to Europe at the time, the emperor appointed Antoine Thomas, who had been Verbiest's assistant since 1685, to fulfil the new Administrator's task until his return, together with Pereira. It seems that the emperor had first nominated Pereira as Verbiest's successor, but the latter declined the honour and put forward both Grimaldi's name and the proposal that Thomas and himself should jointly deputise for him in the interim.[65] In his correspondence, Gerbillon, while acknowledging Pereira's musical talent and his skill for repairing clocks and musical instruments, judged him too ignorant in astronomy for this appointment. In fact Pereira acted as an interpreter while Thomas did the technical work; the latter's language skills were allegedly too limited to enable him to communicate fluently with the emperor.[66]

5.4 Travels and observations in China

Throughout their stay in China, the French Jesuits continued to send reports and observations to the French *Académie*. De Fontaney, Le Comte and Visdelou, who left Beijing on 31 March 1688, were in a better position to do this than their two confreres who stayed in Beijing, for two reasons: first, observations serving to improve cartography can obviously best be accumulated when travelling; secondly, by leaving the capital they escaped Pereira's authority:

> The three fathers, who did not have to show so much consideration for the Portuguese, had not felt obliged to leave the mathematical instruments that the King had given them for his service. Father de Fontaney took the best part of

---

64 Golvers 1993, 267 n. 102.
65 Rodrigues 1990, 17.
66 De Thomaz de Bossierre 1994, 41.

them to Nanjing, firmly resolved to make there all the observations for which the Emperor had with great pleasure given them all freedom, far from being induced to the suspicions that the Portuguese were talking about.[67]

The three of them thereafter did make use of their instruments as they travelled: altogether they sent to Paris observations made in twelve Chinese cities.[68] De Fontaney's ultimate ambition though, was compatible with missionary work. He intended to settle in Nanjing, where he hoped he would be able 'to set up an observatory that will be in correspondence with those of Beijing and Paris, it being expedient that observations should be made in different places'.[69] Nanjing, as the former secondary capital of the empire, also had an imperial observatory. It is unclear whether de Fontaney hoped to use it, or whether he intended to have another observatory built. His plan of a trilateral correspondence, however, suggests that he hoped that Bouvet and Gerbillon would have access to the Beijing observatory, which did not turn out to be the case. De Fontaney's plan was not approved by the Society's Visitor Francesco Saviero Filippucci (1632–1692), who ordered him, Le Comte and Visdelou, to go and work in the provinces where the mission was short of staff, and forbade them to make observations; clearly Pereira's attitude to the French did not stem from personal hostility, but from a commitment to discipline hierarchically enforced within the Portuguese Assistancy of the Society of Jesus.[70]

As mentioned above, in 1686 Tachard had published some observations made by the group on its way to and in Siam before he returned to Europe with the French ambassador. This, however, was an individual and general account of the journey meant for a wide audience, quite different from what the Paris *Académie* was expecting from the Jesuits. Two years later, in 1688, the results of their work during this first part of their journey to China, including some observations made in Siam, were published in a format more in keeping with the practice of the Paris Academicians, that is, as a book devoted solely to observations, the title of which mentioned both their 'quality' of the 'King's Mathematicians' and the *Académie royale des sciences*.[71] The *Observations* were edited by Thomas Goüye (1650–1725), who succeeded de Fontaney at the mathematics chair of the Collège Louis Le Grand. The volume opened with an account of the dissection of three

---

67 ARSI, Jap. Sin. 127, 157r–157v.
68 Landry-Deron 2001, 446.
69 ARSI, Jap. Sin. 127, 165r.
70 Brockey 2007, 160; Hsia, 2009, 122.
71 The title of the work can be rendered as: *Physical and mathematical observations to be of use to natural history and to the perfection of astronomy and geography; sent from Siam to the Royal Academy of Sciences in Paris by the French Jesuit Fathers who are going to China in their capacity as the King's mathematicians; with reflections by Messrs of the Academy and some notes by F. Goüye of the Society of Jesus.*

crocodiles and a description of some other animals, but its main focus was on astronomy; a good part of the observations published within it had been done by Antoine Thomas rather than by the French Jesuits.[72] Goüye made several corrections to the records of astronomical observations so that they would match the *Académie*'s precepts more closely; and indeed the work was published with the *Académie*'s approval.[73]

Four years later, a second volume of their observations was published in the same format; it included observations made in China. The French Jesuits' correspondence bears witness to their continuous and diversified work for the *Académie*. Thus in 1691, Bouvet sent to Le Comte, who was then in Fuzhou, various observations to be forwarded to the *Académie*:

All I can send you is 1) an anatomical description of a tiger with the figure, 2) a translation of the Chinese description of this animal, from their natural history, 3) several remarks about the elephants of Siam ... 4) descriptions of some singular fishes that we saw in the Gulf of Siam [...]. I am sending you the ephemerides of the daily changes that occurred in the air since 12 November 1690 until around the end of October 1691, which I have observed in Beijing this year.[74] These observations are of four kinds: serenity of the weather, wind, heat and heaviness of the air. I have made them everyday at three different times, i.e. around four in the morning, around eight in the evening and at noon, the latter being equally distant from the two former times. I missed a good part of the noon observations, that is, when I had to go to the palace. The observations of the heaviness of the air have been made on a double barometer filled partly with quicksilver and partly with *acqua fortis*. The observations of the heat of the air have been made on a thermometer 18 inches long, which gives the degrees of heat and cold by rarefaction and condensation of the liquor or spirit it contains. The two instruments are of the making of Mr Hubin.[75]

Bouvet's observations pertained in part to natural history, of which he had been put in charge when the group arrived in China; part of the information he sent relied on Chinese sources. With his quantitative observations he gave details of the timing and instruments used: this was part of the information the *Académie* would expect together with the data itself. Hubin was an enameller and instrument maker who worked for the Paris *Académie*; his meteorological instruments, regarded as excellent by the academicians, were quite expensive. That Bouvet used some of these instruments is an indication of the generosity with which the King's mathematicians had been endowed before their departure.[76]

---

72 Hsia 2009, 107–108.
73 Goüye 1688; see Hsia 1999, 54–59; Brian & Demeulenaere 1996, 110.
74 These observations are extant: BnF, Ms. Fr. 17240, ff. 37r–42v; see Landry-Deron 1995, 2: 143.
75 ARSI, Jap. Sin. 165, f. 100v, 102r; the letter is dated 20 October 1691.
76 Stroup 1990, 193.

Contrary to what Bouvet had hoped, it was Gerbillon who travelled north to Tartary. Between 1688 and 1698, he took eight trips there, the first two as a member of a diplomatic mission, the others in the emperor's retinue. His most famous contribution to the knowledge of China in Europe was his account of these trips. It was published independently from the *Académie*, appearing in 1735, in Du Halde's famous *Description de la Chine*.[77] Although it was a descriptive narration rather than a series of quantitative observations, this account supplemented the observations sent by his confreres, who did not go to that part of the empire. The *Description* had a very wide readership in Europe, and this may explain that, of the five French Jesuits sent to China by Louis XIV, it was Gerbillon's name rather than that of de Fontaney that was to remain inscribed in the French collective memory: in 1867 a street was named after the former in Paris.[78]

Although the 'King's Mathematicians' did not succeed in reproducing in China the model of the Paris Observatory, nor in having access to the imperial observatories there, they were a significant source of data for the *Académie royale des sciences*, and more broadly of knowledge about China in Europe. The two of them who entered Kangxi's service, Bouvet and Gerbillon, found themselves in a position to propose French academic science as a model for the construction of imperial science. In the years that followed their arrival in Beijing, they strove to lay out this proposal to the emperor.

---

77 On the *Description*, see Landry-Deron 2002.
78 Hilairet 1963, 1: 586.

CHAPTER 6

Inspecting the southern sky: Kangxi at the Nanjing observatory

1689 was a turning point in the story of the mathematical sciences during the Kangxi reign. During that year all the actors who were to contribute to the shaping of these sciences gathered in Beijing. Mei Wending arrived there, and began to work on a draft of the astronomical chapters of the *Ming History* (*Mingshi* 明史). Meanwhile, the emperor decided to take up the study of Western science on his own account, with Jesuits of both the French and Portuguese missions as tutors. Their continued favour with him was not uncontroversial: astronomy and the related sciences seemed to be the one field of learning where he did not take Chinese sources and the scholars that tutored him as sole authorities. This gave rise to tensions between him and some of his high officials. The present chapter presents a significant instance of such tensions: it focuses on an imperial visit to the Nanjing Observatory that took place on the 27th day of the 2nd month of the 28th year of the Kangxi reign, that is, on 18 March 1689. Different eye witnesses recorded how the emperor interacted with his officials on matters of astronomy: it is worth lending an ear to all of them. Other materials shed further light on the stakes of the confrontation that took place.

The city of Nanjing (lit. 'Capital of the South') had been renamed Jiangning 江寧 after the fall of the Ming, which signified that the city had lost its status of secondary capital. The emperor's visit to its observatory took place during his second Southern tour (*nanxun* 南巡), in early 1689. Altogether, during his reign, Kangxi made six such inspection tours to the 'South', that is, to Jiangnan, the lower Yangzi region. The main official purpose of these tours was to inspect water conservancy works, a task that Kangxi took to heart: this was part of his style of personal government. Jiangnan had resisted the Manchu conquest and had consequently suffered very heavily from it; it was also the main breeding-ground of Chinese scholars, many of whom had been Ming loyalists. The emperor's repeated personal appearances in this region seem to have had a double aim. On the one hand he wished to display interest in the living conditions of its inhabitants. On the other hand, as a complement to this, he wished to impress his Chinese subjects with the martial discipline of the Manchu court.¹

An impressive retinue usually followed imperial travels. The 1689 tour seems to have been quite modest in this respect, with only about three hundred persons following the emperor. Beside his eldest son Yinti 胤禔 (1672–1734), this retinue included high officials like Zhang Yushu 張玉書 (1642–1711), Minister of Rites (*Libu shangshu* 禮部尚書),² Gao Shiqi 高士奇 (1645–1703), who had been among the first to work at the emperor's Southern Study,³ Li Guangdi, then Chancellor of the Hanlin Academy (*zhang yuan xueshi* 掌院學士), Zhang Ying 張英 (1638–1703), Right Vice-Minister of Rites (*Libu you shilang* 禮部右侍郎) and Tuna 圖納 (d. 1697), Minister of Justice (*Xingbu shangshu* 刑部尚書). Each of these officials either left an account of the visit to the Observatory, or is mentioned in one of them.

6.1 The *Imperial Diary*

The institution of the *Imperial Diary* (*Qiju zhu* 起居注) goes back as far as the Han dynasty. Records of the activities and pronouncements of the emperor served, directly or indirectly, as sources to compile official histories. Kangxi established an Imperial Diary Office (*Qiju zhu guan* 起居注館) staffed by Hanlin academicians.⁴ The Diary of the Kangxi reign has only been preserved in part;⁵ the *Veritable Records* (*Shilu* 實錄) of that reign that were compiled, as was customary, at the beginning of the Yongzheng reign (1723–1735) were mainly based on the Diary, which consisted of accounts by eye-witnesses. The record of the emperor's visit to the Observatory given in the *Veritable Records* simply repeats the one in the Diary, omitting a few sentences and characters.⁶ In the translation below, the omitted words will be given in italics and between square brackets.

The record of the day *yichou* 乙丑,⁷ the 27th of the 2nd month, offers a spectacular contrast between the events of the day and those of the

---

1 Spence 1966, 124–134; Chang 2007, esp. 75–86.
2 Zhang Yushu passed the metropolitan examination in 1661. In 1679 he was one of the two scholars appointed to direct the *Ming History* project. He was first a Minister of Justice, then a Minister of Rites; Hummel 1943, 65–66.
3 See p. 73; Hummel 1943, 413–414.
4 Hucker 1985, n. 617 & 618.
5 Zhongguo diyi lishi dang'anguan 1984; other parts are kept at the National Palace Museum in Taipei.
6 The *Veritable Records* version is discussed in Hashimoto 1973, 69–70; Chu 1994, 177–180; Hu 2002, 8–12. Xu 2000, 237–241 uses the *Imperial Diary*; Han 1997 relies mostly on Li Guangdi's account.
7 In the *Veritable Records* as in many other sources, the date is indicated using the hexadecimal cycle of heavenly stems and earthy branches (*tiangan dizhi* 天干地支) that also served to count the years.

evening. In the morning, the emperor treated local military officers to a banquet. On this occasion, he allowed their commander to return home to mourn his mother; the latter shed tears of gratitude.[8] This gives a glimpse of the emperor's attitude to the military, and to the subordinates that he met during the inspection tours.[9] One is duly impressed by his benevolence, but also struck by how well informed he was. What is recorded is in fact a well-prepared staging of the main Confucian virtues. To his subject's filial piety and loyalty, Kangxi did not fail to respond with great humaneness, skilfully handling the cultural values that prevail in the empire that his ancestors had conquered: being a virtuous monarch entailed more than studying the Classics.

By contrast, the account of that evening's events displays a less perfect harmony between the emperor and his interlocutors, mainly the officials of his retinue:

[*At the hour You ,*][10] His Majesty went to the Observatory. Calling all the officials of the Ministries and the Academy [*to the fore*], His Majesty asked: 'Is there any one among Han officials who knows astronomy?' All submitted: 'Your subjects do not understand it.' His Majesty again asked the Chancellor of the Academy Li Guangdi: 'How many lodges do you know?' Guangdi submitted: 'Your subject is unable to know all the twenty-eight lodges.'

The twenty-eight lodges (*xiu* 宿) were groups of stars that defined a division of the celestial sphere into as many zones of right ascension of unequal widths; a reference star (*juxing* 距星) defined the beginning of each lodge.[11] Li Guangdi, as Chancellor of the Hanlin Academy—an institution staffed by the most outstanding among the successful candidates of the metropolitan examinations—was expected to be the very model of an erudite; however he does not seem to have been in favour with the emperor at that particular time.[12] This could well be the reason why Kangxi singled him out as his main interlocutor, and went on testing his knowledge of the lodges:

His Majesty ordered him to point to those he knew, and asked: 'In the old astronomical system [the sequence was] Zui 嘴, Shen 參; today it is Shen, Zui; what is the reason for this?' Guangdi submitted: 'Your subject has not yet [*been able to*] penetrate this reason.' His Majesty said: '[*This must have been a mistake regarding the reference stars.*] If one measures them with the instruments of this Observatory, the lodge Shen definitely reaches the middle of the sky [i.e. crosses the meridian] before the lodge Zui. Observing this suffices to know that today's system is not wrong.'

---

8 Filial piety prescribed a period of three years of mourning for each parent; this was granted to officials as leave in their service.
9 Dai 2000.
10 6 p.m.
11 Cullen 1996, 17.
12 Han 1997; Ng 2001, 61–62.

Here Kangxi puts the conversation on technical ground. In the traditional sequence, Zui ('Beak') was the 20th lodge and Shen ('Triaster') the 21st;[13] the reversal of this order had been one of the issues in the conflict that had opposed Schall and Yang Guangxian.[14] Since the Zui lodge defined a very narrow portion of the sky, the Celestial Pole's motion over the centuries—that is, the phenomenon of the precession of equinoxes, which corresponds to what Chinese astronomers called annual difference (*suicha* 歲差)[15]—resulted in an inversion in the sequence. Kangxi treated this as a mistake of Chinese astronomers, rather than as an adjustment rendered necessary by the changes that had occurred in the heavens since the time of ancient records. This gave him an opportunity to commend the Jesuits' astronomical system. It is noteworthy that he did not mention them explicitly: keeping to the official wording, he talked only about 'new methods'. His next questions suggest, however, that he knew about precession:

He asked again: 'By how many degrees[16] have the centred stars of the *Canon of Yao* moved by now?'[17] Guangdi submitted: 'According to scholars of the past, the difference is already more than 50 degrees.' His Majesty asked again: 'Do the stars[18] move?' Guangdi submitted: 'Your subject would not know. However, the new system talks about the stellar heaven also moving, but its motion is minute.' His Majesty said: 'The reason why Guo Shoujing's instruments could not be used nowadays is that he did not know about the movement of the stellar heaven.'

Here it is becoming apparent that Li Guangdi was not entirely ignorant of astronomy. His answers, however, referred to texts, and he makes no claim of having gained any knowledge from observation. Moreover, his phrasing suggests that whereas he fully trusted texts authored by 'scholars of the past', he was not so positive about the status of the 'new system'. His attitude contrasted with that of Kangxi: the latter directly attacked Guo Shoujing 郭守敬 (1231–1316), the main author of the calendar still in use under the Ming, and the maker of the instruments that he had had replaced by those designed by Verbiest.

---

13   Cullen 1996, 18.
14   See pp. 49–54. Huang 1991d.
15   The year-difference is the difference between the *tropical* year and the *sidereal* year.
16   Kangxi, having studied astronomy with Verbiest, used the term *du* 度 with the meaning of 1/360th of a circle introduced by the Jesuits, rather than with the previous meaning, which corresponded to a division of the circle into 365 ¼—probably the one that Li referred to in his answer; Li (1826) 1999, 487.
17   The stars that 'centred' (i.e. crossed the meridian) and dawn or dusk were used to check on the progress of the seasons. The *Canon of Yao* (*Yaodian* 堯典) is the first chapter of the *Book of documents*, consisting of records of the legendary ruler Yao 堯; it contains what seem to be the earliest references to the Chinese calendar in the scribally transmitted canon; Cullen 1996, 19.
18   The Chinese term used is *hengxing* 恆星 (lit. 'permanent stars'), which refers to the stars as opposed to planets and transient phenomena.

Whether out of ignorance or of deference, Li Guangdi made no attempt to question this allegation, which, if it was indeed uttered, reveals the limits of Kangxi's knowledge of the Chinese astronomical tradition: by Guo Shoujing's time, Chinese astronomers had been aware of the phenomenon of precession for a millennium.[19] It is not impossible, however, that the Imperial Diarist, somewhat overwhelmed by the technicalities in the conversation, did not transcribe them faithfully.

[*His Majesty asked again:* '*Astronomers talk about the Five Planets being like strung pearls;*[20] *is this reliable?*' *Guangdi submitted:* '*The Five Planets have longitudes and latitudes, the idea of strung pearls seems extravagant.*' *His Majesty said, laughing:*] 'In all times, much of the astronomical methods in the chapters of the Histories has been unreliable. Assessing them by means of principles (*li* 理), they are all the same kind of vapid words without substance.'

The likening of the Five Planets to strung pearls was the standard way of referring to conjunctions in the Dynastic Histories; these were regarded as highly auspicious omens; imperial astronomers therefore tended to report such conjunctions more often than they actually occurred. Kangxi, probably amused by the fact that Li Guangdi comments on the literal meaning of the phrase rather than on the actual phenomenon, goes on to cast wider doubt upon what is recorded in the Histories as regards astronomy. He is in fact targeting the use of astronomical observations as omens. It is probably not incidental that the *Veritable Records* compiled after Kangxi's death left out the account of the ruler's sarcasm about texts that were essential to Chinese learning. In this expurgated version, then, the emperor's assessment of the calendar monographs as 'vapid words without substance' seems to apply specifically to Guo Shoujing and his instruments.

[*His Majesty, again turning to all the officials, said:*] 'Take, for example, the story of Mars going back a lodge; that the heavens send warnings is indeed true in principle. [*But*] if [Mars] [*had*] really gone back one lodge, how [*indeed*] would those who did astronomical computations thereafter have been able to accumulate calculations?'[21]

The story of 'Mars going back one lodge' (*Yinghuo tuishe* 熒惑退舍) is found in the *History of the Three Kingdoms*,[22] one of the Dynastic Histories, where it is reported as an ancient event. Kangxi's ground for disbelieving it seems to be that such a sudden change of position of a planet is incompatible with the continuity in the records and

---

19 Needham, 1954–, 3: 356.
20 The expression *wuxing ru lianzhu* 五星如聯珠 used here is found in the Dynastic Histories (where it is mostly written *wuxing ru lianzhu* 五星如連珠) to refer to conjunctions of the Five Planets. The phrase is first found in the *Book of documents*.
21 Here Kangxi is contrasting real and apparent motion.
22 Chen 1982, 1467.

predictions of astronomers ever since. Having further cast doubt on the reliability of the Dynastic Histories, and made it plain that, in astronomy, his authority is greater than that of the Hanlin Academy, the emperor continues to assume the role of a teacher:

> His Majesty, pointing to the Three Enclosures, again questioned Guangdi who was unable to cite all their names.[23] His Majesty pointed his finger and gave repeated explanations to the officials of his retinue. The Minister Zhang Yushu, Tuna and others submitted: 'Your Imperial Majesty clearly understands the breadth of the heavens; You have seen its pattern and examined its principles. Truly, Your foolish subjects could not have looked up to this.'

What etiquette required from the officials at this stage was evidently an expression of admiration rather than a display of any knowledge they might have had on the subject of astronomy. The lecture went on:

> His Majesty again, unfolding a small star map, and according to its orientation, pointed to a large star close to the horizon to the South, and, speaking to all His officials, said: 'This is the Old Man Star.'[24] Guangdi submitted: 'According the Histories and Commentaries, the Old Man Star's visibility is evidence of humaneness and longevity in the Empire.' His Majesty said: 'It can be derived from the altitude of the Celestial North Pole that this star is visible in Jiangning. How could its visibility be an issue?' [*Having said these words, He headed back to the Palace. This day, His Majesty dwelt in the Jiangning Prefecture.*][25]

In the mention of the Old Man Star, Li Guangdi saw an opportunity to combine a compliment to the emperor with a display of his mastery of classical learning; the latter was certainly not expected to take exception to the commonplace astrological interpretation of this star.[26] It is possible that Li was also attempting to uphold the traditional interpretation of heavenly phenomena against what Kangxi had just said. The emperor, however, ignored the compliment and rebuked him for failing to abide by his previous point: what could be predicted by imperial astronomy was no longer open to the interpretations based on literati culture. This was not a refutation of 'astrology' in the name of a 'scientific approach' to natural phenomena: there is ample evidence that the emperor treated meteorological reports quite seriously as possible indicators of bad administration.[27] Rather, he was drawing a boundary that restricted literati's authority on astronomical matters, over which he claimed to exert sole control. Just as he had, in the morning, exercised his skill in manipulating Confucian virtues, in the

---

23 The Three Enclosures (*San yuan* 三垣) are three groups of stars close to the Pole Star: Taiwei 太微, Ziwei 紫微 and Tianshi 天市.
24 Canopus, a very bright star in the southern celestial hemisphere.
25 Zhongguo diyi lishi dang'anguan 1984, 1843–1844.
26 Ho 2003, 148.
27 Wu 1970, 34–35; Elvin 1998, 219–220.

evening Kangxi displayed his command of the idiom of Chinese cosmology.

Further in this cosmological perspective, it is significant that this scene took place during an imperial Southern tour. While inspecting the Jiangnan area, displaying concern with earthly affairs such as water conservancy, and human affairs such as the bereavement of an officer, Kangxi did not fail to inspect the Southern sky from the observatory built by the Ming, thereby asserting his control on Heaven, Earth, and Man, the 'three powers' (*sancai* 三才) that appear in the *Book of change* and that were thereafter crucial to Chinese worldviews.

The significance of this scene in the construction of the Kangxi Emperor as a teacher is further highlighted by the fact that an excerpt of it appeared in his *Sagely instructions*, which were published posthumously by the Yongzheng Emperor in 1731.[28] It is made up of cuttings from the *Imperial Diary* or from the *Veritable Records*, selected so as to form a monologue:

> Day *yichou*.
> His Majesty stopped in Jiangning and went to the Observatory.
> He informed Chancellor Li Guangdi: 'The reason why Guo Shoujing's instruments could not be used nowadays is that he didn't know about the movement of the stellar heaven. Since Antiquity[29] much of the astronomical methods in the chapters of the Histories has been unreliable. Assessing them by means of principles, they are all the same kind of vapid words without substance. Take, for example, the story of Mars going back a lodge; that the heavens send warnings is indeed true in principle. If [Mars] really went back one lodge, how would those who did astronomical computations thereafter have been able to accumulate calculations?'[30]

The stakes apparent in the long dialogue with Li Guangdi and other scholars are no longer visible in this summary. Instead it reads like a straightforward criticism of the Chinese tradition of official astronomy: the latter is characterised by unusable instruments, unreliable methods and forged observations. In the *Sagely Instructions*, this criticism functions as a legitimisation of emperor Yongzheng's (r. 1723–1735) continuation of his father's policy of employing Jesuits at the Astronomical Bureau. A generation later, Qianlong (r. 1736–1795) evoked his grandfather's visit to the Nanjing observatory in one of his many poems: the event was indeed of dynastic significance.[31]

---

28 *Shengzu renhuangdi shengxun* 聖祖仁皇帝聖訓. *SKQS* 411: 145–808.
29 Here the *Sagely Instructions* read *zigu* 自古, whereas both the *Imperial Diary* and the *Veritable Records* read *zilai* 自來 ('in all times').
30 *SKQS* 411: 205.
31 Qianlong, *Yuzhi shiji* 御製詩集. *SKQS* 1309: 358–359.

6.2 Li Guangdi's recollection

Li Guangdi, the emperor's main interlocutor in the conversation above, was one of the main scholar-officials of the Kangxi reign. After passing the metropolitan examination in 1670, he had been appointed to the Hanlin Academy to study Manchu; knowledge of Manchu was crucial in personal contacts with the emperor. Li's encounter with Verbiest dated back to that time.[32] The scholar played a major role in the articulation of the Qing state orthodoxy: he edited several works by Cheng Yi and by Zhu Xi. Later in his career, he supervised several imperial compilations that contributed to promoting this tradition to the status of imperial orthodoxy.[33] He also appears to have been an able politician: he never lost Kangxi's favour entirely throughout his career, despite his indirect involvement in a case of treason and the fierce rivalry among officials.[34] Factions were a characteristic feature of court life and politics during the Kangxi reign, a feature about which the emperor repeatedly expressed exasperation. Nonetheless, his own use of factions to gather and verify information and to implement his policy form an integral part of his style of rulership; it certainly contributed to their survival, if not to their coming into existence.

Li Guangdi also wrote down his recollection of the emperor's visit to the Nanjing Observatory in his account of 'the present affairs of this dynasty' (*Benchao shishi* 本朝時事), reminiscences that were not intended for publication.[35] He interpreted the scene as an episode in attempts by his enemies—first and foremost Xu Qianxue 徐乾學 (1631–1694), another important scholar-official of the period, who had been among the staff of the Southern Study when it was founded[36]—to denigrate him to the emperor. The twenty-eighth year of the Kangxi reign was not a very auspicious one for Li: several officials whom he had previously recommended had been found guilty of corruption. He was nevertheless among the emperor's retinue during the Southern tour. On the day of the visit to the Observatory, Kangxi met Xiong Cilü 熊賜履 (1635–1709), also a former member of the team who drafted memorials for him. In Li's account, a private conversation between the emperor and his former secretary preludes the visit to the Observatory:

---

32 See pp. 69–70.
33 See pp. 267–269.
34 Hummel 1943, 473–475, Durand 1992, Ng 2001, 51–68.
35 Li 1995, 2: 690–763. Li Guangdi's account is also found in Li (1826) 1999, 511–513; it is discussed in Han 1997: 8–13 and Xu 2000, 237–241.
36 Hummel 1943, 310–311; Struve 1982.

During the Southern tour, I followed the imperial carriage to Nanjing. There I saw Xiaogan 孝感[37] entering around mid-day. His Majesty dismissed those who surrounded Him, and had a conversation with him until dusk. His Majesty asked Xiaogan:

'How is that Li in scholarship?'
'He is illiterate, and plagiarises other people to talk wildly; it always tastes of deception.'
His Majesty said: 'I hear that he knows astronomy and calendrical methods.'
'He knows nothing of it, let Your Majesty try and ask him about the stars in the sky, he won't know one.'

The way Li Guangdi refers to places and persons contrasts with that of the official record: thus under his brush Nanjing keeps the name in use under the Ming, which continued to be used by scholars. Xiong Cilü had been an examiner for the metropolitan examination in 1670, when Li Guangdi had passed it. This placed the latter in the position of a 'student' of the former. However, the two diverged as regards interpretation of the *Book of change*;[38] furthermore, Xiong had sided with Xu Qianxue in court factions.[39] This explains the very harsh judgment on Li Guangdi that, according to the latter, Xiong Cilü gave to the emperor.

No sooner had Xiaogan withdrawn than His Majesty promptly went up to the Observatory. Everyone rushed uphill; [the path was] uneven with scattered rocks, all were covered in sweat. His Majesty summoned [us] in an extremely pressing manner. Jingjiang 京江[40] and I walked up leaning on one another, gasping for breath on the verge of exhaustion. His Majesty's face was red, and He asked me angrily:

'Do you know the stars?'
I submitted: 'No, I don't. I never went further than transcribing a few sentences from calendar books, and I have never reached deep knowledge. As to the patterns of stars, I wouldn't know them at all.'
Pointing to the Shen constellation, His Majesty asked: 'What is this constellation?'
I replied: 'This is the Shen constellation.'
His Majesty said: 'You said you don't know, how come you still know the Shen constellation?'
I submitted: 'Among the canonical stars, there may be a few that everybody knows. But the stars in the sky are so many, I really don't know any other.'

Li Guangdi thus interprets the emperor's questions to him as a verification of what the latter has just been told by Xiong Cilü. This sheds a

---

37 Xiong Cilü is referred to by the name of his native place, as is Zhang Yushu further in the text.
38 Han 1997, 11.
39 Hummel 1943, 308–309.
40 Zhang Yushu.

different light on the scene: the position of courtier could sometimes be very trying. The demeanour of the emperor as depicted here is quite different from the evenness of temper suggested by the *Imperial Diary*. Li emphasises Kangxi's distrust of him, rather than his interest in astronomy. The dialogue went on in the same mode:

His Majesty said again: 'That is the Old Man Star.'

I explained: 'According to books, when one sees the Old Man Star, there is Great Peace in the Empire.'

His Majesty said: 'What's this got to do with it? Nonsense! The Old Man Star is to the south, it is naturally not visible in Beijing, and it is naturally visible when we are here. If one went to your Fujian or to the Two Guangs,[41] one would even see the South Pole Star. The Old Man Star is always in the sky, how can you say that when one sees is there is Great Peace?'

In Li's recollection, the exchange on the Old Man Star comes quite early in the conversation. Its meaning and implications are similar to those apparent from the *Imperial Diary*, albeit less explicit. The context in which Li Guangdi mentions it makes it appear mainly as an opportunity created by Kangxi to rebuke him.[42]

His Majesty's anger had not calmed down; he pressingly summoned someone from the Astronomical Bureau. The man had got drunk at home, and upon being summoned pressingly, he got his horse to come back; arriving on the mountain, he fell off and killed himself. His Majesty was again remonstrating about the man's delay, when someone came to report that he had fallen off his horse. His Majesty said: 'Get him some spirits.' The messenger drew near His Imperial Majesty and whispered in His ear: 'He is dead.' His Imperial Majesty calmed down at once and he lowered his voice.

This unfortunate accident was omitted from the *Imperial Diary*; as above it contributes to a less idealised depiction of the scene. Before returning to the conversation on astronomy, Li Guangdi also recalls an insignificant exchange, which, while not worthy of being recorded in the *Imperial Diary*, shows the familiarity that existed between the ruler and those who worked with him:

He drew out a star map drawn on black glazed paper, allowing [me] to look at it with [Him]. I submitted: 'My vision is blurred, and I didn't bring my spectacles.'[43]

---

41 Fujian was Li Guangdi's native province; the Two Guangs, i.e. Guangxi and Guangdong, are the southernmost provinces of China.
42 This is the interpretation put forward in Han 1997.
43 Li Guangdi wrote a *Rhapsody on spectacles* (*Yanjing fu* 眼鏡賦); *Rongcun ji* 榕村集, *SKQS* 1324: 1073; on lenses in China at the time see McDermott 2001.

'Your vision is blurred already?' He allowed us to teach, asking about the phrase 'stellar heaven'.[44] I wanted to respond, [but] His Majesty said: 'Stop, let Zhang Yushu explain.' Zhang said: 'I don't know.' I started to say: 'This is the ancient theory of annual difference. The Westerners call it "stellar heaven."' His Majesty asked: 'Who is right?' I said: 'It seems that the foreigners' (*yangren* 洋人) theory is right.'

Zhang Yushu's involvement in this part of the conversation was not mentioned in the *Imperial Diary*. The suggestion here is that Kangxi is aware that Li Guangdi knows more than most other officials on astronomy, and wants to test someone else on the subject. The issue of the 'stellar heaven' is not phrased in terms of 'old' and 'new', as the official language used in the *Imperial Diary* would have it; it is more bluntly a matter of acknowledging that the foreigners are right against the Chinese. As above, the emperor's demand that Li should make a choice suggests that he does not see the 'annual difference' and the motion of the 'stellar heaven' as two different models of the same phenomenon, although the fewer details given here make imperial ignorance of Chinese astronomy somewhat less conspicuous. Li's somewhat hesitant choice of 'who is right' echoes an assessment of the Jesuits that he had expressed to the emperor two years before, which seems to have been the object of a consensus at the court: 'It seems that the Westerners' learning is utterly absurd, whereas what they say about astronomy is accurate'.[45]

The closing of the scene brings to the fore the tension between Chinese officials and their Manchu ruler:

His Majesty then headed back. Before leaving he commanded: 'Han officials can't ride horses, let the Manchu staff of all offices support their Han colleague officials on the way back, so that no one should stumble and fall. If there is any problem, you will hear from Us!' I was tightly supported by Kugong 庫公 on the way down, and luckily everyone was safe.[46]

Kangxi's display of superiority in the field of astronomy comes out an exasperated response to the officials on whom he depended for ruling the empire, and who once and for all took for granted the superiority of their learning—and by silent implication, the superiority of the Chinese over the Manchus, while constantly denouncing each other to him.

Li Guangdi's story does not stop there: he goes on to report further enquiries made by Kangxi on his scholarship. For him, the visit to the

---

44 *hengxingtian* 恆星天 is the Chinese rendering of 'sphere of fixed stars'.
45 Li (1826) 1999, 487–488.
46 Li 1995, 2: 741–742.

Nanjing Observatory was but one episode in the attempts made by his colleagues of the adversary faction to discredit him.[47] His recollection of the scene left out most of the technical discussion recorded by the Imperial diarist. On the other hand, it sheds light on some of the more earthly implications of the scene for the actors involved, while revealing the existence of minor figures who always remain in the backstage of the imperial presence.

The unfortunate astronomer who died, on the other hand, did not command Li's attention very much. But another official who witnessed the scene, Zhang Ying, centred his whole recollection around him, mentioning his name no less than six times.[48] According to Zhang, as the emperor set out for the observatory, he summoned 'Mucengge of the Astronomical Bureau' (*Qintian jian* Mucengge 欽天監穆成格); Zhang devoted only a couple of sentences to the discussion on astronomy that Li Guangdi remembered so vividly. He noted, on the other hand, that the emperor gave a hundred taels of silver for Mucengge's funeral.[49] Thus the views of those who took part in the emperor's visit to the observatory differed as to what the most significant event of the evening had been: the imperial peroration on astronomy, their own humiliation, or the death of a low ranking Manchu official.[50]

6.3 The Jesuits' role

The accounts above only show the court and its attendants, that is, Chinese and Manchu actors. However the Jesuits, whom Kangxi and Li Guangdi both referred to in the conversation, played a direct role in its very occurrence, principally Jean de Fontaney, who, after leading the French Jesuit mission to Beijing, had established residence in Nanjing in 1688. During the Southern tour of 1689, de Fontaney and his Italian colleague Giandomenico Gabiani (1623–1694) did not fail to pay respect to the emperor, who reciprocated the courtesy by sending them messengers. Gifts were exchanged: the Jesuits presented him, in particular, with a barometer and a thermometer; in turn they received fine food from his table.[51] De Fontaney not only witnessed Kangxi's interest in the Old Man Star, but may well have prompted the emperor

---

47 Hummel 1943, 474.
48 Zhang Ying wrote *Eight poems on the Southern tour* (*Nanxun shi ba shou* 南巡詩八首); one of them evokes 'The ascent to the South Pole Pavilion' (*Deng Nanji ge* 登南極閣), as the observatory was known; it contains no hint of the tragedy he recounted in prose, but instead refers to 'virtue and longevity' (*renshou* 仁壽), as one should when seeing the Old Man Star; *Qingchao wenying* 清朝文穎, *SKQS* 1450: 463–464.
49 Zhang Ying 1994, 216–217.
50 I have found no other evidence concerning Mucengge; his seems to have been a fairly common name among Manchus.
51 Du Halde 1736, 4: 341.

to visit the Observatory just when he did. De Fontaney's recollection is dated 1703, that is, fourteen years after the event:

> During the Emperor's stay in *Nankin*, we went to the Palace every day, and he also did us the honour of sending to us one or two Gentlemen from his Chamber every day. He had someone ask me whether one saw the *Canopus* in *Nankin*. It is a beautiful Star of the South that the Chinese call *Lao-gin-sing*, the Star of the Elderly, or of the people who live long;[52] and upon my replying that it appeared at the beginning of the night, the Emperor went one evening to the old Observatory, called *Quan-Sing-tay*,[53] just to see it.[54]

This shows how the emperor made use of the Jesuits as informants: his questions were promptly answered. His visit to the Observatory was made at the time of night when de Fontaney had predicted the star would be visible. The visit enabled him to verify the information provided by the French Jesuit, and to lecture about it to his officials. For all his competence in astronomy, on the other hand, de Fontaney apparently did not know that the Chinese name of Canopus did not refer to elderly people in general, but rather to a Daoist immortal, who was a symbol of longevity. This star, it should be noted, was of no less interest to the Paris *Académie* than to the emperor of China: Antoine Thomas' observation of Canopus, carried out in Siam in January 1682, had been published in Paris in the observations edited by Thomas Goüye one year before the imperial visit to the Nanjing Observatory.[55]

A more precise account of how the information was requested and provided is included in a collection of imperial edicts and memorials that bear witness to the imperial favour in which the Jesuits were held:

> The Imperial Guard Zhao came to the Church upon the Emperor's order to ask whether the Old Man Star of the South Pole was visible in Jiangning; if so how high above the horizon it was in Guangdong and in Jiangning. (Gabiani and de Fontaney)[56] explained things one by one to the Imperial Guard Zhao who then hurried back on his horse to deliver the reply. Afterwards, (Gabiani and de Fontaney) feared that their reply, made hastily, might hardly be accurate. When night came, at the beginning of the *xu* 戌 hour[57] they separated to observe celestial phenomena and checked the number of degrees of the rising and setting of the Old Man Star [with respect to] the horizon; after several detailed verifications to clarify the matter, they prepared a sketch that was sent in to the temporary Palace early on the 28th day.[58]

---

52 *Laoren xing* 老人星.
53 *Guanxing tai* 觀星臺.
54 'Lettre du Père de Fontaney au Révérend Père de la Chaize', 15 February 1703; *LEC*, 7: 183.
55 Goüye 1688, 147–150; see pp. 117–118.
56 The two names are in small characters in the text, which is rendered by putting them between brackets in the translation.
57 That is, around 7 p.m.
58 *Xichao ding'an* 熙朝定案, in Wu 1966, 3: 1770; quoted by Xu 2000, 237–241.

The conversation with Zhaochang 趙昌, a Manchu guardsman who regularly served as a go-between with the Jesuits, and whom they regarded as sympathetic to them,[59] took place on the day of the visit to the Observatory, and they must then have observed the Old Man Star at the time of the visit. As to the two Jesuits' respective roles in the conversation with the guardsman, it is possible that Gabiani, who had by then lived in China for decades, acted as interpreter, whereas de Fontaney provided the technical explanations. The guardsman seems to have conveyed the answer back to the emperor orally: the written report of the observations only reached the latter on the morning of the next day. There is a close resemblance between the questions asked by the guardsman and the emperor's words as reported by the Imperial Diarist; for example, Kangxi mentioned the fact that the star would also be visible in Guangdong. This suggests that Zhaochang understood astronomy well enough to repeat what the Jesuits had told him in some detail.

What went on behind the scene of Kangxi's visit to the observatory sheds light on the imperial information resources. On the one hand, it must have been the first time that an astronomer trained by the Paris *Académie* put his skills in the direct service of the emperor. This training is reflected in the recourse to double observation to ensure that the latter's query would receive as accurate a reply as possible. On the other hand, the means through which the emperor was able to obtain this information was certainly not specific to astronomy or to the Jesuits. This is but one glimpse of the efficient information network that enabled him to be extremely well informed on all affairs.

6.4 Imperial investigation of the Old Man Star

Kangxi's interest in the Old Man Star has not merely come down to us through records of his spoken words. He also left a short discussion of it in his *Collection of the investigation of things in leisure time* (*Jixia gewu bian* 幾暇格物編).[60] The complete *Collection* was first published in 1732, ten years after the emperor's death. The compilation of the work started in 1701; but the last section, to which the note on the Old Man Star belongs, seems to have been compiled posthumously. It is therefore difficult to date this note.

The Old Man Star

We happened to read in the chronicle of Muzong 穆宗 in the *Liao History* (*Liaoshi* 遼史),[61] that, in the 2nd month of spring of the 12th year of Yingli 應

---

59 Sebes 1961, 177.
60 *SKQS* 1299: 566–603; Li 1993.
61 Tuotuo *et al.* 1974, 1: 77.

歷, Xiao Siwen 蕭思溫[62] submitted: 'The Old Man Star is visible; we beg that an amnesty should be promulgated.'[63] But, although the stars follow heavenly motion, their visibility and invisibility [can be discussed in terms of] coordinates. The Old Man Star can be seen in the area of today's Yangzhou 揚州 during the second and the third month.[64] If one is in the North, it is not visible. This can only be understood if you point it out on a celestial globe. Thus [the fact that] it is called the Old Man of the South Pole tells us that this star belongs to the South. Zhang Shoujie's 張守節 Commentary on the *Book of heavenly officials* (*Tianguan shu* 天官書) of the *Records of the historian* (*Shiji* 史記)[65] says that 'the Old Man Star is to the south of the Bow;[66] at the autumn equinox it is visible at dawn at *Bing* 丙, at the spring equinox it is visible at dusk at *Ding* 丁.'[67] Both *Bing* and *Ding* are to the south.[68] This is clear evidence (*mingzheng* 明證). The Liao capital in Linhuang 臨潢 prefecture is located to the far North East. How could it make any sense to see the Old Man Star there?[69]

This text does not focus on the observation of the star. Instead it is evocative of the approach of eighteenth-century Chinese scholars to astronomy: rather than study celestial phenomena *per se*, they are often said to have been mostly interested in ancient astronomical records, and in the use of astronomy for the purposes of evidential scholarship (*kaozhengxue* 考證學).[70]

The target of Kangxi's note was again an event recorded in one of the Histories. Knowledge of the position and visibility of the star allowed him to conclude that the record of the *Liao History* was not reliable. He did not further speculate as to whether it was a mistake or fraud on the part of official astronomers, or simply a cosmological dressing up of a political decision. As during the visit to the Observatory, his main concern was to point out an inaccurate record in one of the Histories. However, the evidence he gave regarding the Old Man Star's position and visibility was not his own observation, or a contemporary source. Rather, it was taken from an authoritative commentary on the founding model of all Histories. In other words, Kangxi's investigation of the Old Man Star in this note was purely textual, and perfectly respectful of

---

62 Xiao Siwen (d. 970) was an official of the Khitan Liao dynasty.
63 The quotation is accurate, but the *Liao History* record is dated the 11th year of Yingli; this date corresponds to 21 March 961 (Julian calendar).
64 This would usually correspond to the time around the spring equinox.
65 Zhang Shoujie (618–709) is the author of the *Exact meaning of the Records of the historian* (*Shiji zhengyi* 史記正義) one of the three standard commentaries of the *Records of the historian* (the first of the twenty-five Dynastic Histories, completed *c.* 90 BCE; it set the standard for the genre).
66 *Hu* 弧: a constellation of the southern hemisphere.
67 Sima 1959, 4: 1308.
68 *Bing* and *Ding* are associated with the South in the Five Phases (*wuxing* 五行) correlations. They do not correspond to specific directions. Here they seem to refer to the eastern and the western side of the meridian respectively.
69 *SKQS* 1299: 587; see Li 1993, 43–44 & 99.
70 Elman 1984, 79–85.

the Chinese tradition: he used an older, more venerable, source to refute a more recent one.

Moreover, in quoting Zhang Shoujie's commentary, the emperor omitted the interpretation of the visibility of the Old Man Star as an omen of peace that is mentioned there, as in most if not all occurrences of this star in the Histories. This omission allowed him to avoid the issue of the meaning given to the visibility of the Old Man Star: as he had argued at the Nanjing Observatory, the predictability of when and where the star was visible ruled out any interpretation of it as an omen. Instead of challenging the *Records of the historian* and his commentator, he retained from them only what conformed to his own observation, and used their authority in support of his criticism of the *Liao History*. This again placed the note in accordance with the vision of textual tradition that dominated in eighteenth-century China. Genuine wisdom and knowledge could be found in ancient texts, but had not been understood in later periods: therefore ancient sources were more reliable than more recent ones.

Not surprisingly, reverence for the *Records of the historian* as an authority on astronomy underlies the emperor's note on the Old Man Star. He had learned from his Chinese tutors how to use ancient sources to criticise more recent ones. It is reasonable to suppose, however, that Kangxi's views on the visibility of the star and its lack of divinatory significance were based on his personal observation while in Nanjing and his knowledge of astronomy, rather than on his study of the *Records of the historian*. In this note, as during his visit to the observatory, the emperor deliberately underplayed the role of his Jesuit informants: instead he represented himself as the sole intermediary between Heaven, Earth and Men—as was appropriate.

PART III

Mathematics for the emperor

CHAPTER 7

Teaching 'French science' at the court: Gerbillon and Bouvet's tutoring

In 1690, just a few months after returning from his Southern inspection tour, the Kangxi Emperor set up regular sessions of study of Western science, with four Jesuits as his tutors: Pereira, Thomas, Gerbillon and Bouvet. These sessions were modelled on the Daily Tutoring (*rijiang* 日講) on the Classics that Kangxi had started twenty years earlier, especially regarding their ultimate outcome: in both cases the lecture notes drafted by tutors were revised and eventually published as products of imperial scholarship.[1] Although not all of the Jesuits' teaching delivered to Kangxi was turned into such publications, in mathematics and in harmonics their lecture notes provided a significant source for the major compendium in which these fields were brought together with astronomy, the *Origins of pitchpipes and the calendar imperially composed* (*Yuzhi lüli yuanyuan* 御製律曆淵源), compiled during the last ten years of the Kangxi reign.[2]

At least three series of tutorials in the sciences took place during that reign. The first one took place in 1675, with Verbiest as a tutor.[3] The second one started in early 1690, and probably went on at least until the mid-1690s, although sessions seem to have been less frequent after the first two years.[4] A third series was to follow in the 1710s, partly related to the compilation of the *Origins of pitchpipes and the calendar*.[5] The second series is the most important one for the production of mathematical lecture notes. It is also the best documented: no Chinese or Manchu sources on the actual course of the tutoring seem to be extant, but the diaries kept by the two French tutors, Gerbillon and Bouvet, have both been preserved. Together, they cover the period from January 1690 to November 1691 in

---

1 See pp. 73–76. Kessler 1976, 137–146.
2 See pp. 330–333.
3 See p. 73.
4 Von Collani 2005, 17–20.
5 See pp. 287–299.

great detail.⁶ They inform us of the topics they taught, of the sources on which they relied, as well as of the emperor's demands and of the location and format of the tutoring. Their general narrative reflects the Jesuits' agenda, while placing 'French science', as embodied by the books and artefacts brought to China by the 'King's Mathematicians', at the heart of the imperial study of Western learning. This is the focus of the present chapter. The mathematical subjects taught will be discussed in the next two chapters. For these a number of manuscript lecture notes are extant, allowing for a detailed analysis of the successive stages of the construction of imperial mathematics, and also for an assessment of the actual importance of 'French science' at court.

Some Jesuit letters, and the *Historical Portrait of the Emperor of China* written by Bouvet for Louis XIV, allow some insights into the context of the lessons, as well as into Kangxi's motivations and attitude.⁷ Like the diaries, they represent the Beijing court in terms that their French readers would use to talk about Versailles. Thus de Fontaney, who knew of the first two years of tutoring through his confreres' accounts, evoked it more than a decade after it took place:

> That prince, seeing all his Empire in a profound peace, resolved, in order either to amuse or to occupy himself, to learn the sciences of Europe. He himself chose Arithmetic, Euclid's *Elements*, practical Geometry, and Philosophy. F. Antoine Thomas, F. Gerbillon and F. Bouvet had orders to compose treatises on these matters. The first one had Arithmetic as his lot, and the two others Euclid's *Elements* and Geometry. They composed their proofs in Tartar: those they had been given as masters in this language revised these with them; and if some word seemed obscure or less proper to them, they substituted others. The Fathers presented these proofs and explained them to the Emperor, who, understanding easily all that was taught to him, admired the solidity of our sciences more and more, and applied himself to them more and more.⁸

The depiction of Kangxi turning to the sciences as an amusement once he had solved all military problems is somewhat evocative of the French royal rhetoric: Louis XIV's *bon plaisir* (wish) was a sufficient justification for his taste for dance as well as for his political decisions. Although by all accounts Kangxi enjoyed mathematics, he never presented his study of the sciences as proceeding merely from his wish. Rather, he strove to master technical as well as scholarly learning as part of his endeavour to control the state; studying was an important element in his rulership.⁹ Moreover, although the lessons started a few months after the peace treatise with Russia signed at Nerchinsk, in the summer of 1690 there was another military campaign, against the

---

6 Landry-Deron 1995 is a thorough study of these diaries; in what follows I rely on her transcription of Bouvet's diary, which has been preserved as a manuscript.
7 Bouvet 1697.
8 *LEC*, 7: 283–284.
9 See pp. 73–74.

Ölöds,[10] during which the Jesuit teachers continued to prepare their teaching material. That the emperor should 'understand easily' what they taught him was indeed their main aim, as we shall see below; his 'admiration' for it, which would only follow from that understanding, was by no means the sole criterion that he used in deciding what he would further pursue, and what should be published in Chinese. This being said, the facts mentioned by de Fontaney (who the teachers were, the subjects they taught and the language in which they taught them, the fact that they had to compose lecture notes) all played a part in the shaping of imperial science.

7.1 Chronology of the lessons

By 1690, Tomás Pereira was the most senior in the Beijing Residence, its superior and, it seems, the most fluent in Chinese. He was Kangxi's main Jesuit interlocutor, having taught him music for more than fifteen years.[11] Gerbillon and Bouvet, on the other hand, had no access to the emperor during their first year in Beijing; and as we have seen, Pereira was not keen to facilitate this access. It was only in the autumn of 1689 that things changed, owing to Gerbillon's participation in a diplomatic mission.

After his final victory over the Three Feudatories in 1681, Kangxi had turned his attention to the frontier regions of his empire. Settlers from the Russian empire had moved to the Amur region, where they had founded two bases: Nerchinsk in 1658 and Albazin in 1665. In the mid-1680s, Kangxi decided to regain full control of the area. In an attempt to do so by force, the Qing troops besieged Albazin twice, but failed to put an end to the movements of settlers around the region; a diplomatic solution was then sought. In May 1688, a Qing delegation left Beijing for Selenginsk, a city to the west of Nerchinsk, to meet their Russian counterparts. As planned before Verbiest's death, Pereira was sent to act as an interpreter in peace negotiations: his knowledge of languages would enable him to communicate with both sides. Following the emperor's instructions that he should select another Jesuit to go with him, he chose Jean-François Gerbillon as his companion. Because of unrest in the region, the delegation was called back to Beijing; they set out again a year later. This time, they headed for Nerchinsk, where a treaty was signed on 22 August 1689; this treaty demarcated the frontier, stipulated that Albazin should be destroyed, while allowing trade across the border and making provision for embassies to be exchanged.[12] The mediation of the two Jesuits and of a Latin–Russian interpreter made communication possible between the two parties:

---

10 Spence 2002, 153–154; Perdue 2005, 174, 182.
11 Jami 2008b; see pp. 72, 112.
12 Spence 2002, 150–153.

they helped prepare the multilingual treaty. Both Pereira and Gerbillon left accounts of the negotiations, in which each emphasised his personal role; however, the hierarchies of both the Qing court and the Society of Jesus gave authority to Pereira over his confrere who was still a newcomer, and had apparently little knowledge of Chinese or Manchu at the time. Pereira's account was a report written for the Society's hierarchy,[13] whereas Gerbillon's was intended for a broader readership. Jean-Baptiste Du Halde eventually published it in his *Description de la Chine* (1735): the 1688 false start and the 1689 journey and negotiation form the first two of Gerbillon's eight 'trips to Tartary'.[14] During these trips he would typically record the distance covered every day and the direction in which it had been covered; whenever possible, he measured the Sun's altitude, and noted his observations on the climate, the vegetation, and the customs of the local inhabitants: clearly his diary's purpose was to collect information as systematically as possible.[15] Thus while serving the emperor as an interpreter he continued to fulfil his task as one of the French 'King's Mathematicians'.

The two leaders of the Qing delegation sent to Nerchinsk were both protectors of the Jesuits. Tong Guogang 佟國綱 (d. 1690), a maternal uncle of the emperor, had received baptism in 1672;[16] Songgotu 索額圖 (1636–1703), the son of one of the regents and a minister, had supported Kangxi against Oboi. The leader of one of the major court factions, he remained very influential until his disgrace in 1703. Although he did not convert, he played a major role in the issuance of the 1692 'Edict of Toleration'.[17] It seems that, having taken the French Jesuits in particular under his protection, he praised Gerbillon to Kangxi after the delegation returned to Beijing.[18] In January 1690, the emperor decided to resume the study of Western science that he had started with Verbiest in the 1670s. Gerbillon then offered him his services and those of Bouvet, proposing to teach mathematics and philosophy in Manchu. Kangxi first considered sending them to Manchuria to complete their linguistic training, then set the two of them to study this language intensively in Beijing.[19] In the meantime he took up the study of mathematics in the Chinese language with Thomas and Pereira, the former providing the technical expertise while the latter acted as interpreter:[20] Thomas' Chinese, it was alleged, was not fluent enough for him to communicate directly with Kangxi. Pereira's

---

13 Sebes 1961.
14 Du Halde 1736, 4: 103–259; on this work see Landry-Deron 2002.
15 Sebes 1961, 147–148.
16 Hummel 1943, 794–795; Standaert 2001, 444.
17 Hummel 1943, 663–666; Witek 1988, 90–92; Bouvet 1697, 236–237.
18 Landry-Deron 1995, 1: 60–61.
19 APF, Informazioni vol. 118, ff. 438–438v.
20 Du Halde 1736, 4: 267.

presence was also an expression of his precedence in both Jesuit and court hierarchies. In February Gerbillon and Bouvet showed the emperor their first writing in Manchu, completed with the assistance of their language teachers: it was devoted to digestion. In March they were deemed ready to teach in that language, and from then on until at least the autumn of 1691, lessons took place regularly, sometimes on a daily basis, the four Jesuits usually going to the Palace two by two.

The first subject systematically taught by Gerbillon and Bouvet was Euclidean geometry: by July they had completed the subject, and the emperor decided to go on to practical geometry.[21] This occupied them until January 1991, when he started to revise with them the Manchu treatise on the *Elements of geometry* that they had produced, as well as its Chinese translation.[22] This confirms that in his mind the production of a text was closely associated with the tutoring. About a month later, he ordered the two Frenchmen to start teaching him philosophy. They attempted to begin by logic,[23] but then turned to medicine. Although the diaries are less detailed for the rest of 1691, a letter from Bouvet to Le Comte dated 20 October 1691 shows that by that date they had started teaching physics (in the seventeenth-century sense of natural philosophy), leaving the more difficult subjects of logic, ethics and metaphysics to be treated last.[24] It seems likely that they then included their writing on digestion in a wider treatise on medicine. There is no evidence, however, as to the completion of such a treatise. What became of the intensive tutoring given by Gerbillon and Bouvet after 1691 is not known; as we shall see further, Thomas, at least, went on tutoring Kangxi and writing for him for several years.[25]

Several elements in what the French Jesuits tell us about their tutoring closely resemble what we know about the Daily Tutoring: first, the study of each subject was limited in time; secondly, the texts produced were revised several times by the emperor. Thirdly, he treated his European tutors, who repeatedly marvelled at his generosity, as he treated the Chinese scholars with whom he studied the Classics, providing them with both good food and warm clothing for the winter. Moreover, he behaved towards them with great familiarity, ignoring ritual and etiquette. This was the counterpart for their having to work all day—and sometimes day and night—to prepare for lessons.[26] Finally, apart from the Jesuits and the Chinese tutors, only Bannermen were admitted to the Great Interior (Danei 大內) on a regular basis.[27]

---

21 BnF, Ms. Fr. 17240, f. 279 v.
22 BnF, Ms. Fr. 17240, f. 287 r.-v.; Du Halde 1736, 4: 295.
23 BnF, Ms. Fr. 17240, f. 289 v.
24 ARSI, Jap. Sin. 165, f. 101 r.
25 *LEC*, 17: 285; Landry-Deron 1995, 1: 17; see pp. 184, 200–201.
26 Kessler 1976, 141–142.
27 Rawski 1998, 31–32.

The location of tutoring in the sciences within the imperial palace will be further discussed below.

7.2 The *Académie*, the Moderns and the way to God

Gerbillon and Bouvet had a well-articulated plan for tutoring the emperor. It was French science and, whenever possible, that produced by the Paris *Académie Royale des Sciences* that Gerbillon and Bouvet intended to teach the emperor. For them, this was the means to a greater end: developing exchanges between France and China, and friendship between their respective monarchs. This would in turn pave the way for the Kangxi Emperor's conversion. Bouvet made this quite explicit in his letter to Le Comte of October 1691:

> If these two great monarchs knew each other, the mutual esteem they would have for each other's royal virtues could not but prompt them to tie a close friendship and demonstrate it to each other, if only by an intercourse in matters of science and literature, by a kind of exchange between the two crowns of everything that has been invented until now in the way of arts and sciences in the two most flourishing empires of the Universe. If Heaven graced us with the achievement of this goal, we would feel we had made no small contribution to the good of Religion which, under the auspices and protection of two such powerful princes, could not fail to progress considerably in this empire.[28]

Thus, it was through their position as tutors of the emperor that Bouvet and Gerbillon sought to fulfil their task of missionaries. Gerbillon's initial offer to teach Kangxi philosophy as well as mathematics should be understood in this light. As Bouvet recounted in his *Historical Portrait of the Emperor of China*:

> With some reason, we believed [philosophy] to be of greater consequence than all [our] other [tasks], for there is no better means to dispose minds, especially those of Chinese scholars, to receive the truths of the Gospel, than a well written philosophy. And this is what obliged us to redouble our application. Among all works of philosophy ancient and modern that we then consulted, having found none that seemed more appropriate to our aim than the Ancient and Modern Philosophy by Mr Duhamel, of the *Academie Royale des Sçavans*,[29] because of the solidity, the neatness, and the purity of the doctrine of this excellent Philosopher, it was one of the main sources on which we drew to compose our work.[30]

Jean-Baptiste Du Hamel (1623–1706), an Oratorian priest, was the first Secretary of the Paris *Académie royale des sciences*, which was a stronghold of the Moderns against the Ancients in the controversy that was then raging in France. His *Philosophia vetus et nova* was a textbook

---

28 ARSI, Jap. Sin. 165, f. 101 r.
29 Du Hamel, *Philosophia Vetus et Nova* (1678; 3rd edition 1684).
30 Bouvet 1697, 148–150.

commissioned for colleges, which he revised several times; it was regarded as striking a balance between the Ancients and the Moderns, as its title suggested.³¹ The Manchu treatise that the two French Jesuits started to compose on this basis in 1691 was devoted to logic.³² Bouvet gave some indication as to the style in which they wrote: they 'treated [this topic] in the briefest and clearest way that [they] could, removing all there is of complicated terms and of pure chicanery, following the Moderns' style'.³³ The production of this treatise, written in a new style, was in fact a response to the order that they had received from Kangxi on 20 January 1691:

> The Emperor had us told that he wanted us to start very shortly to explain philosophy to him and to put it into Tartar. And for that we should not attach ourselves to the one that the late F. Ferdinand Verbiest had presented to him in the past, but we should choose and digest matter as we wished.³⁴

In October of that year, Bouvet still believed that the emperor intended to study and circulate their philosophy: in his letter to Le Comte he mentioned 'the philosophy course, to the study of which the emperor wants to apply himself for good, and have it entirely in the language to eventually publish it in his empire and leave it to posterity'.³⁵ According to Bouvet there was a convergence between supply and demand: the subject was to be tackled in a way different from that used by Verbiest a decade earlier in his *Study of the fathoming of principles* (*Qiongli xue* 窮理學), a work which the emperor no doubt remembered having dismissed following the advice of Hanlin academicians.³⁶ The derivation of principles (*tuili* 推理) that Verbiest had advocated in it seems to have been precisely what Bouvet dismissed as the 'pure chicanery' that was left out of the Manchu treatise on logic. In other words, Aristotelian philosophy and its emphasis on modes of reasoning were no longer deemed crucial for the evangelising enterprise. Verbiest had relied upon the irrefutable God-granted truth of the sciences as taught in Jesuit colleges during the first half of the seventeenth century. Instead, the French Jesuits chose to display the power and glory that resulted from the advancement of the sciences under Louis XIV's auspices. Science as approved by the Paris *Académie* replaced Jesuit science as the first step towards religion, whereas the Sun King's power replaced that of God as the guarantor for the validity of that science. This change

---

31 Fontenelle 1709, 161.
32 Bouvet 1697, 151; on the probable date of composition, see Landry-Deron 1995, 1: 76.
33 ARSI, Jap. Sin. 165, f. 101r.
34 BnF, Ms. Fr. 17240, f. 288v.
35 ARSI, Jap. Sin. 165, f. 100 v.
36 See pp. 78–81.

reflected both the evolution of teaching methods in European colleges and the role played by the emperor in the shaping of what the Jesuits taught in China.

The earthly might of another sovereign may well have spoken more to Kangxi than the heavenly omnipotence of the Christian god. In any case, Gerbillon and Bouvet followed similar lines when they replied to Kangxi's questions on matters related to medicine and anatomy, in which he showed some interest.[37] As mentioned above, the first text that the two Frenchmen wrote in Manchu, while they were still training in the language, was 'an explanation of the digestion of food, the way taken by chyle in the animal's body before reaching the heart, the circulation of blood; in one word how nutrition works'.[38] It was presented to Kangxi in February 1690 with illustrations. He first looked at these carefully,

> especially those of the stomach, the heart, the viscera, the veins, and compared them with those of a Chinese book which he had brought to him, and which deals with these matters, and found that there was much relation. He then read [their] writing from beginning to end, and praised its doctrine, which he said was very subtle.[39]

The next month, however, Gerbillon and Bouvet started teaching Euclidean geometry, and they did not continue explanations of the human body until 1691, when it appeared that their teaching of logic would be no more successful than that of Verbiest. Their actual use of Du Hamel's *Philosophia vetus et nova*, for example, turns out to have been very limited. The 1684 edition of the work comprised two volumes: the first one dealt with logic, metaphysics and ethics, the second one with physics. Gerbillon and Bouvet only had the first volume.[40] The latter's diary suggests that Kangxi, who reiterated his injunction that their philosophy should not follow that of Verbiest, may not have been wholeheartedly enthusiastic about the subject, which he found difficult—and perhaps pointless?[41] After the introduction was finally completed to his satisfaction, there is no mention in the diaries that the two tutors wrote any more on the matters contained in the first volume of Du Hamel's work. It was at that point that they decided to turn to physics (in the sense of natural sciences), and to write on anatomy:

> After having completed this introduction to philosophy in Tartar [i.e. Manchu], [...] we started with physics, and knowing that this prince has a very high opinion of European medicine and that he most wishes to know the structure

---

37 On Kangxi and medicine, see Puente Ballesteros 2009, esp. 50–182.
38 BnF, Ms. Fr. 17240, f. 264 r.
39 Du Halde 1736, 4: 271.
40 Landry-Deron 1995, 1: 78.
41 BnF, Ms. Fr. 17240, f. 289 r.-v.

of the human body, we started this part of philosophy by the science of the human body, in which, in addition to a short anatomy, with all the figures and their explanation, and all the beautiful discoveries of the authors, both ancient and modern, we will put, with the help of God, all the curious observations that we have here in the first part of the memoirs of the gentlemen of the *Académie des sciences* on animals, and all those that we will get from these gentlemen on the same subject [...][42]

The emperor was pleased with the beginnings of this work:

No sooner had he seen about fifteen propositions and as many figures, that he displayed extraordinary joy and satisfaction; to the extent that to show his esteem for [the work], he ordered his first painter, who above all excels by the delicacy of his brush, to leave all other works he had in his hands so as to work himself on these figures. Nevertheless, [...] this science, although pleasant and very curious, demands application, and this Prince, still somewhat weak, was not so capable [of it], as he had us informed. Meanwhile he exhorted us to continue working as usual, and to hold matter ready for when he would have leisure.[43]

Kangxi did not take tutorials in European medicine as he did in mathematics. Nevertheless he was very keen to gather information on the subject. He never ceased his effort to have European doctors at his court and it seems that, during the second half of his reign, about twenty Westerners practised medicine there.[44] The outcome of the French Jesuits' efforts in that field seems to have consisted of a number of short treatises on various diseases, written in Manchu on imperial commission.[45]

In his *Historical Portrait*, Bouvet mentioned the authors on whom they drew to write on anatomy: 'the famous Mr du Verney' and 'other savants of the *Académie Royale*, who have distinguished themselves in this matter, as well as in everything else, above all other nations'.[46] Joseph Guichard Duverney (1648–1730), a protégé of Du Hamel,[47] was a Professor of anatomy at the Jardin du Roy, an *Académicien*, and a tutor of Louis XIV's son. In 1687, upon the French Jesuits' arrival in China, when Bouvet was allotted to work on natural history and medicine for the Paris *Académie*, he asked that some of Duverney's writings should be sent to him.[48] Duverney, on the other hand, added 'Reflections' to the account of the dissection of a crocodile carried out by the French Jesuits in Siam, which was published in Paris in 1688.[49] It was indeed

---

42 ARSI, Jap. Sin. 165, f. 102r. See also von Collani 2005, 79.
43 Von Collani 2005, 79.
44 Dong 2004, 75.
45 Bouvet 1697, 155.
46 Bouvet 1697, 152–153; see also von Collani 2005, 79.
47 Fontenelle 1709, 158–159.
48 AMEP, V. 479, 39.
49 Goüye 1688, 1–47; see pp. 117–188.

appropriate that works by one of the Dauphin's tutors should be used in tutoring Louis XIV's alter ego in China on anatomy. Again, when the two Frenchmen made medicines for the emperor, they relied on the pharmacopoeia published by a Royal Apothecary, Moyse Charas (1619–1698), entitled *Pharmacopée royale galenique et chymique* (1672).[50] Thus Gerbillon and Bouvet strove to set up in Beijing a version of imperial science that would be modelled on 'French science', that is, royal science as produced under the auspices of the Paris *Académie*. In so doing, they effectively challenged the authority of their predecessors of the *Padroado* who had introduced into China learning that was now regarded as pertaining to the side of the Ancients. Although the Moderns' style apparently suited Kangxi, what the two Frenchmen taught him under the heading of philosophy turned out to be unfit for 'publishing in the empire and leaving to posterity'. Bouvet's hopes in that respect remained unfulfilled.

7.3 A typical lesson: 10 April 1690

The two diaries that record the tutoring are of different natures: Bouvet's is a manuscript in his own hand, while Gerbillon's was edited by Du Halde and published several decades after it was written. The two sometimes differ by a few days in their dating.[51] Under 8 April 1690, Bouvet wrote:

The Emperor came back to Beijing, and he came to the Yam cim tien [Yangxindian], where we were. First we gave him our explanation of Euclid that he understood very well, then, having had brought [to him] the tables of sines, tangents and secants with their logarithms that F. Thomas had had put into Chinese characters for him, he wanted to see a few uses of them in an observation that was made on the spot with a semi-circle.[52]

Gerbillon's entry for 10 April 1690, on the other hand, reads:

The Emperor came back to Beijing to honour, according to the custom, the Emperors his predecessors; after the ceremony he expedited that day's affairs, & came to the apartment where we were. He stayed with us for more than two hours, having the propositions of geometry that we had prepared explained to him as well as having calculations of triangles done by the logarithmic tables that had just been put in Chinese digits by his order; he took much pleasure in seeing the advantage he drew from what he had already studied of the elements

---

50 Bouvet 1697, 157–158; Charas 1676; the second edition (1682) was in the Jesuit Library of Beijing; Verhaeren 1949, no. 174. The Manchu manuscript entitled *Si yang-ni okto bithe* (*Xiyang yaoshu* 西洋藥書, 'Book on Western medicines') kept at the Beijing Palace Museum Library may well have been written in this context; Li 1999, Walravens 2000, 97, Hanson 2006, 145.
51 Landry-Deron 1995, 1: 6–17, 149–156.
52 BnF, Ms. Fr. 17240, f. 269 r.-v.

of geometry to facilitate his understanding of practices of geometry, the explanation of which he had asked for.[53]

The two entries obviously recount the same events. The *Veritable Records* (*Shilu* 實錄) did not mention the lessons, which were part of the emperor's 'private' life; however they enable us to solve the discrepancy between the two dates mentioned above.[54] They state that 'His Majesty came back from the Garden of pervading spring (Changchunyuan 暢春園) to the Palace' on the 2nd day of the 3rd month of the 29th year of the Kangxi reign, which corresponds to 10 April 1690.[55] While in the capital, he did not always reside in the Imperial Palace: for the previous ten days, the diaries report that the Jesuits had gone to the Garden of pervading spring his villa located in the 'suburb' of Haidian 海澱, to give their lessons. In the Imperial Palace, on the other hand, they always worked on their lecture notes and saw the emperor at the Hall of the Nourishment of the Mind (Yangxindian 養心殿). The lesson of 10 April 1690 combined several activities in a way that is quite typical. It began as usual with the Jesuits explaining what they had prepared in writing beforehand: this was both a tutorial, in which Kangxi played the role of a student, and the editing of a geometry textbook, in which he supervised the revisions. Then other matters were tackled, at the emperor's behest: the same duality between study and compilation is apparent for the tables that Thomas had had translated. The division of subject matters between him and the French concerned what they were to write about, but both sides were expected to be able to explain what the other had prepared. It was also quite typical for a lesson to comprise a practice (*une pratique*): although Kangxi's interest was not restricted to practice, his purpose in studying geometry or other subjects was to master the many uses to which, in his view, mathematics could be put.

Gerbillon's diary quoted above mentions Kangxi's pleasure at acquiring the means to understand practical geometry. In general, both diaries recorded the emotions that the study of geometry aroused in the emperor: pleasure, satisfaction and joy followed from understanding; incomprehension, on the other hand, caused pain (*peine*), in both the senses of difficulty and sadness. These emotions were clearly of importance to the Jesuits as they might trigger increased favour or loss thereof. The emperor also referred to them explicitly:

The 21st [May 1690]. The Emperor, not understanding very clearly the explanation that we gave him of Euclid, gave us to know that this pained him. [...]

---

53 Du Halde 1736, 4: 278. The villa was completed in that same year; Rawski 1990, 23, 35.
54 On the *Veritable Records*, see p. 121.
55 *SL* 145: 1b.

The 22nd. He understood perfectly that day's lesson, which was of importance, and that he had desired to know for a long time, it was the proposition that serves to prove the rule of three. So that on that day he was very satisfied, and told us, as if to apologise for not having seemed so pleased the day before, that he saw that when he did not clearly understand our thoughts, he well saw that this pained us, but that we should not be worried about that, he really wished to learn.[56]

Thus to the Jesuits' desire to please the sovereign, expressed in terms that would not have been out of place at the French court, he responded by assuring them, using the Confucian rhetoric of reverence towards one's teachers, that their position was not put at risk by the emotions of one day, as he intended to carry out the plan of study he had set for himself.

As mentioned above, the main subjects about which the French wrote in 1690–1691 were Euclidean and practical geometry, logic and anatomy. Their diaries mention in passing a number of mathematical subjects for which Thomas produced material: arithmetic, algebra and numerical tables. Beside these, a number of topics were discussed orally, without any written material being produced. Thus one day the emperor asked 'various questions on Europe, its size compared to that of China and how many countries it had, if we used hypocausts similar to the Chinese ones,[57] how and of what material were our houses built [...]'.[58] This was the continuation of an interest in the world outside the Qing empire to which Verbiest had responded by presenting a world map, the *Complete map of the earth* (*Kunyu quantu* 坤輿全圖, 1674) accompanied by an expanded version of Giulio Aleni's *Areas outside the concern of the Imperial Geographer* (*Zhifang waiji* 職方外紀, 1623).[59] On another occasion Kangxi asked 'various questions on rain, thunder, the propagation of sound, the pendulum, the compass and the variation of magnet'.[60] These topics belonged to physics in the Jesuits' curriculum, and to the 'investigation of things' (*gewu* 格物) in the classification of knowledge that Kangxi followed when he later wrote on some of them, using information gathered in these conversations—as he did in the case of the Old Man Star.[61] One day the discussion was on 'a few diseases, such as the stone and small pox and the way to cure them; the composition of theriac and its excellence: on which His Majesty, who had had a few trials of it done on various sick persons, having found that good effects had ensued, asked us

---

56 BnF, Ms. Fr. 17240, f. 275 v.
57 The *kang* 炕—the 'Chinese hypocaust'—is a hollow bed usually made of brick or clay, that can be heated by fire; it was still common in Northern China in the twentieth century.
58 BnF, Ms. Fr. 17240, f. 270 v.
59 Debergh 1993; Standaert 2001, 755.
60 BnF, Ms. Fr. 17240, f. 275 v.
61 Jami 2007a; see pp. 133–135 & 360–361.

whether we still had any [...]'.⁶² These examples show the diversity of Kangxi's interests; they also reveal that in most cases he effectively appropriated what he obtained from the Jesuits, be it a piece of information, the use of an instrument or a medicine.

7.4 The imperial workshop and instruments

As mentioned above, the tutoring took place in the Great Interior; the Hall of the Nourishment of the Mind, located in the Western part of the Forbidden City, was part of the private space that, except for the Chinese scholars of the Southern Study, was staffed by Bannermen. During the Kangxi reign, the Hall of the Nourishment of the Mind accommodated the Imperial Workshops (*Zaobanchu* 造辦處);⁶³ Bouvet called it 'the location of His Majesty's Academy of Arts'.⁶⁴ As this name suggests, the French Jesuit claimed that the workshops were deliberately modelled after Louis XIV's *Académie royale des sciences*: Bouvet thus told Louis XIV that the emperor wanted to have more Jesuits working for him 'in order to form in his Palace, with those who are already there, a kind of Academy subordinate to your Royal Academy'.⁶⁵ The two Jesuits described the French institution to the emperor at some stage of their teaching. But one may well doubt that Kangxi ever intended to set up an institution that would be subordinate to any other that he did not himself control, let alone one that depended on the king of France. It was not a novelty that the Imperial Palace had its own workshops. What Kangxi did was to reorganise earlier imperial institutions to suit his own purpose rather than to try and imitate those set up by Louis XIV.⁶⁶ In any case, it was there that Gerbillon and Bouvet prepared their lecture notes with the help of their Manchu teachers and gave their lessons. The location of the Jesuits' encounters with the emperor reflects their position and function at court: they were in the direct service of the emperor, whose trust enabled them to enter the Great Interior. At the same time, their service was associated with technical objects and matters, as opposed to the literary ones that were dealt with at the Southern Study. The Jesuits were thus situated outside the hierarchy that court rituals served to enact, while Western learning was kept apart from classical scholarship. After his first visit to the Hall of the Nourishment of the Mind in January 1690, Gerbillon described what he had seen there:

---

62 BnF, Ms. Fr. 17240, f. 273 r.
63 From the Yongzheng reign on, the Hall of the Nourishment of the Mind served as the emperor's main residence. Guo 2003; Standaert 2001, 824.
64 BnF, Ms. Fr. 17240, f. 270 r.
65 Bouvet 1697, 197–200, 247.
66 On the general tendency in historiography to see Kangxi's initiatives as imitations of those of Louis XIV's, see Jami 2006a.

We were shown into one of the Emperor's suites, named *Yang sin tien* [Yangxindian], in which some of the most skilled craftsmen, painters, turners, goldsmiths, coppersmiths etc. work. There we were shown the mathematical instruments, which His Majesty had had placed in rather neat boxes or racks made for that purpose. There were no considerable instruments: it all consisted of a few proportional compasses, almost all imperfect; several ordinary dividers large and small of several kinds, a few squares and geometric rules, a divided circle of about half a foot in diameter with its pinnules. All of which made rather crudely, and quite far from the neatness and accuracy with which the instruments we had brought were made, as the Emperor's servants themselves admitted. His Majesty had us told that we should examine well all the uses of these instruments, so as to explain them to him clearly. He added that on the next morning we should bring the other instruments fit for measuring the elevations and distances of places and for taking the distance of stars that we had in our house.[67]

Thus the allocation of quarters to the Jesuits within the Hall of the Nourishment of the Mind was first and foremost pragmatic. Instruments and their practice were central in the sciences that Kangxi strove to master; the Jesuits not only provided but also maintained these instruments. Clocks, which were kept together with instruments, were the most spectacular and have remained the most famous among the Western objects collected by the emperors. Their collection, started by the last Ming emperors, was continued by Kangxi's successors. Concomitantly, clock making continued to be one of the skills required from some of the Jesuits of the China mission throughout the eighteenth century.[68]

Gerbillon's assessment of the imperial instruments, which is no less severe than that of his confreres de Fontaney and Le Comte concerning those they saw at the Beijing Observatory, is best understood in the light of the position he was hoping to secure at court. The French Jesuits' standards, set by the instruments that they had used in Paris with French academicians and those that they had brought to China, were certainly quite high. On the other hand, the fact that the emperor's collection did not meet these standards and that his craftsmen seemed to be no match for the great Parisian instrument makers obviously suited their purpose: displaying the superiority of what France had to offer over what had hitherto been put in the emperor's service. In fact, fixing the instruments was a task commonly entrusted to the Jesuits; teaching their use to Kangxi formed an important part of their tutoring. Instruments consequently played a crucial role as gifts that the Jesuits could present to him; he mostly reciprocated with silk or clothes and food from his table. Thus in May 1690, Giandomenico

---

67 Du Halde 1736, 4: 262–263.
68 Standaert 2001, 843–845; Pagani 2004.

Gabiani, who was then the Jesuit Vice-provincial for China, arrived in Beijing where he spent a year, trying to act as an arbiter between the French and the Portuguese. As he was the most senior of all China Jesuits in their hierarchy, it was essential to his Beijing confreres that he should be given audience by the emperor as soon as he arrived. Entrance into the Palace was denied to him at first, so that his confreres had to intervene:

> F. Thomas went to the Palace to see chao laoye [Zhao laoye 趙老爺, i.e. Zhaochang], to ask him to arrange for F. Gabiani to enter the Palace. Meanwhile F. Gerbillon and I were working on the machines of eclipses and planets to bring them to the Emperor, who had asked for them on the previous day. And F. Gabiani, on his side, had brought with himself the repeater clock that F. de Fontaney had sent to F. Thomas, and that this father handed over to F. Gabiani, so as to secure him a better welcome from His Majesty. He also brought the barometer and the thermometer that F. Fontaney had presented to the Emperor last year. On top of these he still brought a picture of our lady in miniature. The emperor accepted the clock because he said he believed it to be good, he also received the barometer and thermometer and he gave back the holy Picture to the father. When the father had arrived at the palace, His Majesty had him told that he should enter the yam cin tien [Yangxindian], that is to say in the innermost palace, and that he did not treat people of his sort as ordinary people. Shortly thereafter His Majesty admitted him in his presence with F. Pereyra whom he called together with him, then the emperor did him the favour of having Tartar *cha* [tea] sent to him.[69]

Presenting a clock made Gabiani 'the sort of people' that the emperor received at the Hall of the Nourishment of the Mind, and on whom he bestowed some of his own tea in return. In the capital as during imperial travels, the guardsman Zhaochang—to whom the Jesuits referred using the appellation *laoye*, as they did for officials and princes—remained the main intermediary between Kangxi and the Jesuits.[70] He had some understanding of the subjects they taught, and gave them valuable advice as to how they should behave. The presents brought by Gabiani were typical of the kind of objects that they used for establishing and maintaining contact with officials. Both the clock and the meteorological instruments were provided by the French. At the time royal and academic support from Paris provided them with items that their confreres might have difficulties in obtaining through the networks of the *Padroado*; despite the French breach of the Portuguese monopoly, the 'King's Mathematicians' would often cooperate with their confreres in the interest of the mission. Ricci had already presented religious paintings to the Wanli Emperor at the beginning of the century; such paintings usually greatly impressed their recipients,

---

69 BnF, Ms. Fr. 17240, f. 276 v.
70 See pp. 132–133.

especially due to the use of perspective.[71] Kangxi's restitution of the religious picture presented by Gabiani contrasted with his laudatory comments on the clock. Rather than any personal antipathy against the Catholic religion, this was a public gesture, which epitomised his policy towards the Jesuits: he was interested in the skills and learning they could put in his service, to the exclusion of their religious teachings. On the same day, Roemer's machines were brought back to him:

> The Emperor then stopped to consider the 2 machines, of eclipses and of the Planets, and asked us to explain them. We had the pleasure of explaining them to him in Tartar, and to make him understand their beauty, and to give him about them the idea that they deserve. He asked whether one could not accommodate them to the Chinese year so as to be able to use them; one replied that by the means of the çié ki [jieqi 節氣],[72] this could be done, but that, however, if His Majesty wished one could give Him in writing a method for using it according to the Chinese year. The Emperor asked how long ago these machines had been made, whether the person who had made them was still alive and whether he was regarded as skilled. We replied to him that he was still alive, that he was a very skilled man, and that all the Science of Astronomy had been captured in these two machines.[73]

What Kangxi first saw in Roemer's machines is the limit of their universality. In France they had been deemed the most appropriate items that the Jesuits could take to China as evidence of the truth of their science and therefore of that of their religion.[74] However the machines relied on solar cycles for the measurement of time, which corresponded to years in the Gregorian calendar. The Chinese calendar, on the other hand, was luni–solar, and the position of the Sun on either machine would therefore not give any direct indication as to the date in the latter calendar: thus the machines could not be of obvious use to the emperor, who expected the Jesuits to adapt them to the Chinese calendar for his convenience. He eventually had the machines placed on both sides of his throne.[75] In the days that followed the discussion on these machines, he enquired repeatedly about the clock presented by Gabiani. Thus on 26 June:

> [The Emperor] stopped to consider quite at length the repeater clock and all its movements, of which F. Pereyra gave him the explanation. His Majesty considered in particular the pieces that make it repeat the hours. Then His Majesty, having asked who was the inventor of this clock, we told him that it was a

---

71 Standaert 2001, 493; Golvers 1993, 117.
72 The twenty-four points that determine the periods into which the interval between successive winter solstices is divided; see pp. 60–61.
73 BnF, Ms. Fr. 17240, f. 275–275 v.
74 See pp. 105–106.
75 Bouvet 1697, 139.

famous French clockmaker, and that it was the same who had also made the 2 machines for eclipses and planets.[76]

Thus, while it was Pereira who explained how the clock worked, Bouvet and Gerbillon did not fail to emphasise its French origin. It is likely that the skilled clockmaker to whom they refer was Isaac II Thuret (1630–1706), the Paris *Académie's* clockmaker who had built the two machines devised by Roemer.[77] The image that Kangxi formed of French instruments was an important stake in the eyes of both the French and the Portuguese: according to de Fontaney, 'the Portuguese fathers at court, in order to minimise the industry of the French, erase[d] their names from the watches, clocks and pendulum clocks, and put others'.[78] It is not known whether the signature on any of the instruments extant at the Palace Museum nowadays has been thus altered.

Kangxi also showed interest in the thermometer and barometer presented by de Fontaney, albeit more than a year after they were brought to the Palace: in June 1691 he asked Gerbillon for 'the explanation of the uses of a thermometer and a barometer that were there, and that F. de Fontaney had given him in Nanjing.'[79] Two months later he ordered Bouvet to make use of these instruments and to 'mark every [day] on the calendar the degrees of heat and cold, of heaviness and lightness of the air shown on the barometer and thermometer that F. de Fontaney had given' him.[80] At the time Bouvet was making similar observations for the French *Académie*, using instruments that the Jesuits kept at their residence.[81] The French Jesuits could put the same scientific practice simultaneously in the service of both the king of France and the emperor of China.

Of the four Jesuit tutors, it was Pereira, not the French, who was in charge of instruments, be they musical, astronomical or mathematical. When he took up the study of mathematics, the emperor ordered many mathematical instruments such as proportional compasses to be made, many of them modelled on those brought by the French; he gave some of them to his sons.[82] Some instruments, however, were tailored to suit the emperor's needs. Thus, in 1690 Pereira supervised the manufacture of a 'mathematical table' made of camphor

76 BnF, Ms. Fr. 17240, f. 276 v.
77 Frémontier 1996, 320.
78 BnF, Ms. Fr. 25060, f. 238 v.; quoted by Landry-Deron 1995, 2: 208, n. 94.
79 Du Halde 1736, 4: 341.
80 BnF, Ms. Fr. 17240, n.p.; Landry-Deron 1995, 2: 195.
81 ARSI, Jap. Sin. 165, f. 102 r.
82 APF Informazioni 118, ff. 439 v.–440 r.; many of these instruments are kept at the Palace Museum in Beijing; Liu 1998, no. 53–77.

(*nanmu* 楠木), and designed for 'Manchu style' sitting: it is only 32 centimetres high. Inside the table there is room for putting away calculation and drawing instruments; it is inlaid with silver plates on which various lines were engraved, so that 'simply with a pair of dividers, many arithmetical and geometrical operations can be carried out'.[83] The table was, it seems, the only luxury object in Kangxi's study inside the Hall of the Nourishment of the Mind.[84]

Just as instruments were central to the emperor's approach to the sciences, his interest in European medicine went beyond texts and the occasional conversation mentioned above; it entailed an 'experimental' dimension. He had set up a 'kind of laboratory', equipped with silver instruments and utensils, where during three months the two Frenchmen had various medicines made; these were tested and then incorporated into the imperial pharmacy. This was the case with the theriac mentioned above, which was among the drugs used by the missionaries themselves.[85] The most famous case was that of cinchona, which cured Kangxi from a fit of malaria in early 1693.[86] This cure turned out to be as providential for the French as for him: as a reward he gave them a house located within the Imperial City (Huangcheng 皇城), which at last enabled them to live separately from the Portuguese.[87] Beyond this well-known anecdote, Bouvet and Gerbillon depict the imperial workshop as the location of the imperial appropriation of Western learning, achieved through the production of textbooks in Manchu and Chinese, the use and reproduction of instruments, and the replication of French royal apothecaryship.

7.5 Chinese, Manchu and the control of Western science

As mentioned above, Manchu was the language that Gerbillon and Bouvet studied first, because the emperor ordered them to teach him in his native language. It should be noted that he was fluent in both Chinese and Manchu, so that this choice was not due to any difficulty he might have had in understanding Chinese. On the other hand, as we have seen in the case of Li Guangdi, the elite of Chinese officials had to learn Manchu: the knowledge of this language, though not widespread among Chinese literati, was not limited to Manchus.[88] The two Jesuits'

---

83   APF Informazioni 118, f. 440 r.; Liu 1998, no. 90.
84   APF Informazioni 118, ff. 440 v.–441 r.
85   De Thomaz de Bossierre 1977, 59; the silver instruments used by the Jesuits may well be those still kept at the Palace Museum in Beijing; Liu 1998, no. 234.
86   Von Collani 2005, 79–84; Puente Ballesteros 2007.
87   This residence came to be known as the Northern Church (Beitang 北堂) Von Collani 2005, 90–103; Bouvet 1697, 156–162.
88   Crossley & Rawski 1993, 73–74.

CHAPTER 7 | Teaching 'French science' at the court: Gerbillon and Bouvet's tutoring

diaries give some insight into Kangxi's attitude towards the issue of language.

Gerbillon had a chance to practise Manchu during his trip to Nerchinsk, but when he was summoned by Kangxi in January 1690 he was not yet able to discuss technical matters in this language:

His Majesty had us[89] called to the Palace very early: we stayed in his presence for more that two hours, explaining to him various geometrical practices: he always talked to us with great generosity and familiarity: he had repeated to him the use of several instruments that F. Verbiest had had made for him in the past. I always spoke to him in Tartar, but I would not undertake to give mathematical explanations: I apologised to His Majesty that I did not know either the Chinese or the Tartar language well enough to speak to him relevantly, especially regarding the sciences, not knowing the appropriate Chinese or Tartar terms: but I told him that when we would have learnt Tartar well, F. Bouvet and I would be able to give him lessons in mathematics or in philosophy very clearly and neatly, because the Tartar language by far surpasses the Chinese language, in that the latter has no conjugations, no declensions, no particles to link discourses, whereas in the former they are very common. The emperor seemed pleased with this speech, and turning to those around him: this is true, he told them, and this defect makes the Chinese language much more difficult than the Tartar one.[90]

This is how Gerbillon offered his services and those of Bouvet (who had not yet been admitted into the Hall of the Nourishment of the Mind at the time) as tutors. The two Frenchmen thus bypassed the ban that the Jesuit visitor, Francesco Saviero Filippucci, had put on the study of Manchu.[91] In any case, Gerbillon knew that proposing to teach in Manchu rather than in Chinese on the grounds that the former language was more appropriate for the study of mathematics and philosophy could not fail to please the emperor. Be that as it may, Gerbillon and Bouvet both did find Manchu much easier than Chinese. The former's diary records that a few days later, Kangxi, 'unable to understand a geometrical practice that [Thomas and Pereira] had explained to him in Chinese',[92] made provisions for the two Frenchmen to be taught Manchu intensively at the Imperial Household Department (*Neiwufu* 內務府), which was entirely staffed by Bannermen:[93] it was intended from the beginning that they should produce texts in Manchu. The two Frenchmen both claimed that Pereira could

---

89 Thomas and Pereira were with him.
90 Du Halde 1736, 4: 264–265.
91 Landry-Deron 1995, 1: 50, 54.
92 Du Halde 1736, 4: 266.
93 Gerbillon refers to it as the 'Tribunal of Poyamban' (Du Halde 1736, 4: 266); the Manchu term he transcribes as 'poyamban' was transliterated in Chinese as *baoyi angbang* 包衣昂邦, 'Commandant of Bondservants' (Hucker 1985, no. 4483).

speak a little Manchu but could not read it.⁹⁴ He rendered Thomas' explanations in Chinese. Bouvet mentions that Kangxi once addressed Pereira and Thomas on the issue of language, saying 'with a smile: as for you who content yourselves with Chinese, study it well'.⁹⁵ On another occasion, Kangxi asked the two Frenchmen to translate Pereira's treatise on harmonics into Manchu; they declined to do so, on the grounds that they had never studied 'practical music'.⁹⁶ They were reluctant to undertake any work that would entail collaboration with him, and unwilling to help promote his work in any way. The Jesuits all knew that the emperor wished to promote Manchu as the language of Western learning: Zhaochang 'had given [them] to understand that His Majesty intended to put all [their] sciences in his language to make them current in his empire'.⁹⁷ This must have been the reason why the guardsman had prompted the French to learn Manchu rather than Chinese. For this purpose he gave them a book that he had borrowed from the emperor; it was 'a new translation of Confucius, handwritten and very neatly bound' that the emperor 'wanted [them] to read in order to improve [their mastery of the Manchu] language'.⁹⁸ This must have been part or whole of the *Four Books* (*Sishu* 四書), which were translated into Manchu in 1677, but do not seem to have been printed in this language until 1691.⁹⁹

Having Gerbillon and Bouvet translate their geometry into Chinese amounted to 'declassifying' it: as mentioned above, using the Manchu language effectively restricted access to written materials; in particular, it would exclude all Chinese literati outside officialdom. Limiting circulation even within officialdom was certainly a concern for the emperor: as soon as he heard that the two Frenchmen had composed their first text in Manchu, the treatise on nutrition, he took steps to ensure that it should not even circulate within the Imperial Household Department where they had been studying Manchu:

He had us told: not to translate anything of our sciences at the Tribunal where we were, but only in our house; that this advice he gave us was only a precaution, and that we should not fear to have given rise to it by some mistake or indiscretion, since he was entirely pleased with us.¹⁰⁰

The emperor further reminded them of what had happened to Schall: according to him, the Manchus liked the Jesuits, but the Mongols and

---

94 BnF, Ms. Fr. 17240, f. 269 r.; Du Halde 1736, 4: 278.
95 BnF, Ms. Fr. 17240, f. 264 r.
96 Wang 2002; see p. 72; BnF, Ms. Fr. 17240, f. 277 r.
97 BnF, Ms. Fr. 17240, f. 271 r.
98 BnF, Ms. Fr. 17240, 266 v.
99 Rawski 2005, 313–314.
100 Du Halde 1736, 4: 272.

Chinese were hostile to them.[101] Pereira, through whom the French received this warning, blamed it on the subject of their first essay: apparently he thought that it was not a good idea to inform the emperor of what 'French science' had to say on digestion. This left Bouvet somewhat incredulous: 'as though a piece of paper that only contained a short explanation of the manner in which nutrition works and in which blood circulates in the veins could have brought us such serious advice'.[102] It appears, however, that Kangxi, while interested in the European approach to the body, did not want scholar-officials to know that he encouraged the production of writings on the subject: he remembered well—and so did Pereira—that when the Hanlin academicians had rejected Verbiest's *Study of the fathoming of principles* in 1683, they had singled out the location of thought in the brain as especially objectionable.[103] It is therefore not incidental that Bouvet and Gerbillon were instructed to complete their treatise without mentioning it to anyone. It seems that nothing of what the Jesuits wrote in Manchu on the human body was ever translated into Chinese.[104]

Thereafter, when the French taught Euclidean geometry and wrote on it, they went to the palace especially to do their translations, even when the emperor was absent from the capital: the production of mathematical texts in Manchu itself was safely confined within the walls of the Hall of the Nourishment of the Mind.

In summary, the motivations for Kangxi's preference for the Manchu language were two-fold. On the one hand, it was a means of restricting the circulation of the Jesuits' writings; on the other hand, it could contribute to the corpus of Manchu literature that he strove to create. Since it was commonly accepted that all civilisation and knowledge stemmed from the Chinese scholarly tradition, the emperor may well have been intent on importing knowledge into China via the Manchu language. If this were indeed the case, the two motivations mentioned above were but the two sides of the same coin: he placed himself in a position of arbiter as to what Western learning should circulate and who could access it.[105] As a result of this, among the many results of 'French science' that Gerbillon and Bouvet told their French confreres and audience they were presenting to him, only Pardies' geometry was to contribute to the shaping of imperial scholarship.

---

101 Du Halde 1736, 4: 272.
102 BnF, Ms. Fr. 17240, f. 264 r.-v.
103 Chu Pingyi 2007b; see p. 80.
104 On the best known of these writings, the 'Manchu Anatomy', see Walravens 1996, Hanson 2006, Watanabe 2005b; Puente Ballesteros 2009, 290–300, gives a synthesis on the subject.
105 Jami 2010.

CHAPTER 8

The imperial road to geometry: new *Elements of geometry*

[T]hey say that Ptolemy [King of Egypt] once asked [Euclid] if there was in geometry any shorter way than that of the elements, and he answered that there was no royal road to geometry.[1]

8.1 Changing the textbook

When the emperor had studied geometry with Verbiest in the mid-1670s, Ricci and Xu Guangqi's *Elements of geometry* (*Jihe yuanben* 幾何原本, 1607) had been translated into Manchu.[2] In 1690, once he was satisfied with the two Frenchmen's progress in Manchu, the emperor assigned them to teach him geometry. They started with proposition I.1 of the *Elements* on 8 March 1690, preparing notes before each session. As they were still struggling with the language, the guardsman Zhaochang suggested that they use the existing Manchu translation of Ricci and Xu's *Elements of geometry*:

Tchao laoye pointed out to [the emperor] that the first six books of Euclid translated into Chinese with Clavius' explanation by Father Ricci, had also been translated into Tartar a few years earlier by a skilled man that His Majesty had nominated Himself; and although this translation was neither accurate nor easy to understand, it could not but be of great help to us in preparing the explanations of Euclid and in making them more intelligible, especially if one summoned the translator to help us and to write them in Tartar. The emperor very much appreciated this proposal, and ordered that the Tartar translations should be put into our hands and that the translator should be summoned.[3]

Verbiest's translation, carried out with the help of the scholar mentioned here (but unfortunately not named), seems to have remained in the form of a draft. This may explain why some historians have doubted its existence.[4] In any case when Gerbillon and Bouvet started teaching Euclidean geometry in Manchu, there was someone in the Palace who had already had a go at putting the subject in writing in the emperor's

---

1 Proclus, quoted by Heath in Euclid, 1956, 1: 1.
2 See pp. 75–76.
3 Du Halde 1736, 4: 273; Landry-Deron 1995, 2: 48.
4 Golvers 1993, 266, n. 100.

CHAPTER 8 | The imperial road to geometry: new *Elements of geometry*

native language: thus Zhaochang must not have been the only member of the imperial staff versed in Western learning. The emperor, however, found the work quite difficult: according to Bouvet, on 13 March, during their sixth session, his tutors offered to use 'modern ways to explain these sciences, easier, less thorny and shorter, as [they] showed him in the *Elements of Geometry* of F. Pardies',⁵ a geometry textbook in French by one of de Fontaney's predecessors at the mathematics chair of the Jesuit college in Paris. At first, Kangxi asked them to keep to the translation from Chinese into Manchu; however a few days later, he accepted the change of textbook they proposed:

24th [March]. The emperor came to the yam cim tien [Yangxindian] and had us explain the five next propositions of Euclid. He read carefully the explanation that we had prepared in Tartar, finding it clear and most intelligible; he approved above all of a wholly new way of proving the 7th* that we had proved in two ways, and following what we had previously insinuated and that we took this opportunity to repeat to him, he allowed us to take the way we pleased for the explanation of these propositions, leaving us free to follow the Elements of geometry by Father Pardies that we had proposed to him as the most appropriate for his Majesty.

* If two straight lines drawn from the extremities of the same straight line meet at a point, one will not be able to draw from the same extremities to another point, and on the same side, two other straight lines equal to the first two respectively.⁶

Thus the two tutors, while following Euclid in the earlier Manchu translation, proposed a different proof for one of the propositions. This was what convinced the emperor that they should follow the textbook they had proposed to him instead. Gerbillon's account is less specific, but brings out the general features of their proposed way to geometry:

As he showed us his eagerness to know as soon as possible what was most necessary in the *Elements* for understanding practical geometry, we pointed out to him that should he wish it we would choose only the most necessary and most useful among Euclid's propositions, and that, without attaching ourselves to follow the way of proving given in the Chinese translation, we could abbreviate that work considerably, and prove more perfectly the most necessary and the most beautiful [propositions]. His Majesty agreed to this proposal, and we resolved to follow the order of Father Pardies' geometry, endeavouring to make his proofs even easier to understand.⁷

Thus, in geometry as in other disciplines, Gerbillon and Bouvet substituted what their predecessors had taught with a work whose approach they deemed more 'modern', a quality that entailed clarity,

---

5   BnF, Ms. Fr. 17240, f. 266 r.; Landry-Deron 1995, 2: 39–40.
6   BnF, Ms. Fr. 17240, f. 266 v.; Landry-Deron 1995, 2: 41. The text added by Bouvet as a footnote is that of proposition I. 7 of Euclid's *Elements*.
7   Du Halde 1736, 4: 275; Landry-Deron 1995, 2: 50.

ease and speed. According to them this met the emperor's demand: what he wanted to master were the principles of practical geometry. Thereafter the teaching followed Pardies' textbook; the lecture notes closely parallel seven out of the nine books which it comprises.

8.2 Pardies' *Elemens de geometrie*: the pedagogy of geometry in seventeenth-century France

Pardies' work was quite successful and broadly used in mathematical education in Europe in the decades that followed its publication in 1671. It underwent several editions and reprints up to 1724, and was translated into Latin, Dutch and English—as well as Manchu and Chinese. Ignace Gaston Pardies S.J. (1636–1673) held the chair of mathematics at the Collège de Clermont from 1670 until his premature death in 1673. The geometrical treatise, published during his tenure there, was one among a large number of rewritings of Euclid's *Elements* produced in seventeenth-century Europe: between 1620 and 1680, no less than six French Jesuit mathematicians produced such rewritings.[8] Thus Claude-François Milliet Dechales S.J. (1621–1678), Pardies' immediate successor in the Paris chair,[9] also authored a work entitled *Elements of geometry*, published first in Latin in 1660, then in French in 1672, which was even more successful than that of Pardies.[10] Both titles are found in the catalogue of the Beijing Jesuit library, and there is evidence that Kangxi had a copy of the first French edition of Dechales' geometry in his hands at some stage.[11]

Pardies' textbook is a small volume: its first edition, in duodecimo, has 120 pages, plus twelve pages of front matters.[12] Its full title announces the approach chosen by the author: 'Elements of geometry in which by a short and easy method one may learn what must be known of Euclid, Archimedes, Apollonius, and the most beautiful inventions of the ancient and modern geometers'. Two claims are made here: on the one hand, the work neither contains the whole of Euclid's *Elements*, nor limits itself to it: the classic is no longer regarded as the repository of 'what must be known'. On the other hand, against the axiomatic and deductive style that characterised this classic, which implied lengthy developments, Pardies chose shortness and ease. Both claims reflected the priority he gave to pedagogy over rigour. He was not the first to challenge Euclid's *Elements* as a textbook; in fact he followed a tradition that can be traced back to Petrus Ramus

---

8 Karpinski & Kokomoor 1928, 22–31; Kokomoor 1928; Romano, 1999, 628–636.
9 De Dainville 1954, 111.
10 Sommervogel 1890–1912, 2: 1040–1042.
11 Verhaeren 1949, no. 170–172, 548–550, 1259, 1261; esp. 172.
12 The second edition (1673, xviii+168 p.) has been used in the present study. A quick comparison with the 1671, 1678 and 1684 editions has revealed no major differences in content, so that it is difficult to ascertain which editions Gerbillon and Bouvet used.

(1515–1572).[13] In composing his textbook, Pardies seems to have drawn on another book, published in 1667: Antoine Arnauld's (1612–1694) *Nouveaux elemens de geometrie*.[14] This double filiation—Ramus was a protestant, Arnauld the leading figure of Jansenism in his time—indicates that, almost a century after the publication of Clavius' Latin edition of Euclid's *Elements*, Jesuit education by no means developed in isolation; on the contrary, it reflected the changes that occurred in the society from which Jesuit colleges drew their students. In a time when these colleges competed with other teaching institutions to attract the sons of the French elite, it is not surprising that Clavius' edition of Euclid was no longer adapted to an audience that had widened considerably. The fact that Pardies' textbook was in French rather than in Latin should also be understood in this perspective.

In Pardies' long dedication of his *Elemens* 'to the gentlemen of the Royal Academy', he compares them to Chinese astronomers,[15] praises the king's magnificence, predicts that 'the arts and [the] sciences have never reached the point of perfection to which [the Academicians] will bring them', and closes by protesting that he 'holds them in the greatest respect imaginable'.[16] This reveals his acute awareness of the changing landscape of science and scientific authority that surrounds him.[17] This dedication came from someone who had already made a place for himself among the seventeenth-century French scientific community;[18] leaving aside personal claims, it placed Jesuit mathematical education under the auspices of royal science.

The claims made in the title, the choice of the French language and the dedication all place the work on the side of the Moderns. In his preface, Pardies further justified the style he had chosen for his textbook, strongly criticising the Euclidean model:

[...] one of the things that make the reading of Euclid and of ordinary Authors difficult and boring, is that in the rigorous exactness that they have applied in letting nothing that can be proved pass without a proof, easy as it may seem otherwise, it often happens that what would have been clear if one had been content with proposing it to the mind, as it naturally appears, becomes difficult and confused, when one wants to reduce it to a regular proof.[19]

This could have been one of Bouvet's inspirations when he recounted how he and Gerbillon chose to leave out 'pure chicanery' and to follow

---

13 Gillispie 1970–1980, 11: 286–290; Ziggelaar 1971, 50; Engelfriet 1998, 44–45.
14 Arnauld 1667; see Ziggelaar 1971, 63–64.
15 See p. 103.
16 Pardies 1673, 'Epitre', f. a iv v.
17 See pp. 102–105.
18 Ziggelaar 1971 *passim*. Gillispie 1970–1980, 10: 314–315.
19 'Preface' in Pardies (1671) 1673, f. x r. I have translated all quotations of Pardies' *Elemens de géométrie* from French as literally as possible, rather than using an English edition.

the Moderns' style in teaching the emperor.[20] In the 1600s Matteo Ricci had argued that the *Elements* were 'agreeable' to the Chinese precisely because 'Euclid showed his geometrical propositions so clearly that the most obstinate were convinced.'[21] In his mind clarity was linked with irrefutability. To Pardies, Gerbillon and Bouvet, on the other hand, clarity meant ease and concision. The evolution of pedagogy in Jesuit colleges in Europe, which matched the change in their audience, was thus reflected in the two successive teachings of geometry to the emperor. He too favoured the more 'pleasing' version devised for the sons of the French elite over the more demanding one that had contributed to shaping the identity of the Society of Jesus as a teaching religious order.[22]

Descartes' influence on Pardies is perceptible in the latter's preface, where he argued that mathematics leads to God, and in no indirect manner. Infinity constituted the link: it was in 'the measure of asymptotic spaces'[23] that he saw 'an invincible proof of the existence of God'.[24] However un-Aristotelian Pardies' argument might be,[25] articulating the links between mathematics and religion was a continuation of the approach to mathematics in the Jesuit curriculum as defined by Clavius; it was certainly in keeping with the rationale that underlay the teaching of this discipline by Jesuit missionaries in late Ming China. Pardies' work thus doubly suited Gerbillon and Bouvet's purpose: it proposed a 'clear and easy' way to geometry, while reasserting that geometry led to God.

Pardies' preface strongly contrasts with the preface to Arnauld's *New Elements of Geometry*, written by Pierre Nicole (1625–1695), another major figure of Jansenism, who had co-authored with Arnauld the treatise that came to be known as the Port-Royal *Logic* (1662). Nicole's moral judgment on geometry is wholly negative:

> It is no great evil not to be a geometer, but it is a considerable one to believe that geometry is a very estimable thing, and to esteem oneself for having filled one's head with lines, angles, circles, proportions. It is very blameful ignorance not to know that all these sterile speculations contribute nothing to make us happy; that they do not relieve our miseries; that they do not cure our pains; that they can give us no real and solid contentment; that man is not made for that [...][26]

---

20 See p. 145.
21 Ricci & Trigault 1978, 569–570.
22 Jami 1996.
23 Pardies (1671) 1673, 'Preface', f. a vii r.
24 Pardies (1671) 1673, 'Preface', f. a vii r.; see Ziggelaar 1991, 54–60.
25 Ziggelaar 1991, 55.
26 Arnauld 1667, f. a iii v.

CHAPTER 8 | The imperial road to geometry: new *Elements of geometry*

Table 8.1 Table of contents of Pardies'
Elemens de géométrie (1671)

I. Lines and angles
II. Triangles
III. Quadrilaterals and polygons
IV. Circle
V. Solids
VI. Proportions
VII. Incommensurables
VIII. Progressions and logarithms
IX. Problems or practical geometry

Nicole would hardly have sought proof of the existence of God in something as frivolous as asymptotes; he would have disapproved of the satisfaction that geometry brought to the emperor of China. Against this, Pardies' preface, and perhaps his whole work, can also be seen as a Jesuit response to the Jansenist approach to geometry. The structure and style of his work are quite distinct. Unlike Arnauld and most of his other predecessors, Pardies did not organise the text into definitions, axioms, postulates, and propositions with proofs: indeed these words hardly occur in the work.[27] Its nine books are simply divided into numbered paragraphs. Terms appear in italics when they are defined, and, as was common at the time, an index at the end of the work refers the reader to these definitions. Book I opens with an explanation of what geometry is about:

1. By the name of *Quantity* we mean a thing, which, being compared to another of the same nature, can be called greater, or smaller, equal or unequal: as are extension, number, weightiness, time, movement; and all these things, as they can thus be compared according to the more or the less, form the objects of Geometry.
2. One nevertheless stops to consider expanse in particular, as the one that can serve as example and rule to measure all other *Quantities*.[28]

Geometry thus defined addresses physical entities rather than abstract ones; it constitutes the foundation of mathematics because expanse is taken as an example of quantity. One is quite far from the Euclidean model. This is confirmed by the list of titles of the nine books that make up the work (see Table 8.1).

This departure from Euclid is also apparent in the way Pardies defines objects and works on them. For example, in book I he introduces the circle as follows:

---

27   Pardies (1671) 1673, 'Preface', f. ax v.
28   Pardies (1671) 1673, 1.

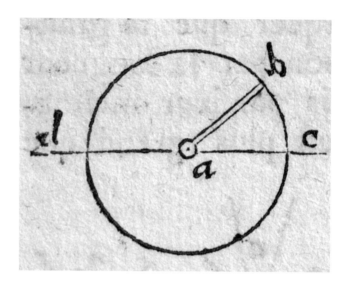

Illustration 8.2 Constructing a circle, *Elemens de géométrie*, 1671, 4 (BnF, Paris).

10. If we imagine a line *ab* tied by its end *a* to the midpoint of the line *dc*, and that moreover we make this line move around point *a*; when it has returned to the point from which it had started to move, the end *b* will have described a curved line called a *Circle*, or rather *Circumference* of a circle: for strictly speaking the *Circle* is the whole space enclosed in this circumference.[29]

Imagination, in Pardies' words, 'the material faculty that we have to imagine by the means of our organs', is deemed insufficient to embrace the whole of mathematics: he contrasts it to 'a wholly spiritual [faculty] for thinking and for reasoning' that proceeds from the soul.[30] While referring to materiality (as opposed to spirituality rather than to abstraction), imagination is frequently invoked in the work, as well as sight, as a means of attaining evidence and constructing objects. It thus plays an important role in the study of geometry.

8.3 The double translation

The Manchu textbook based on Pardies' *Elemens*, entitled *Gi ho yuwan ben bithe*, was completed by July 1690.[31] To my knowledge, three copies of it are still extant.[32] Some features of the copy that belonged to the

---

29 Pardies (1671) 1673, 4.
30 Pardies (1671) 1673, a vij r.
31 BnF, Ms. Fr. 17240, f. 279 r. Landry-Deron 1995, 2: 98.
32 Gugong, IMARL, Manchu 210, St Petersburg, C 210.

imperial library, now kept at the Palace Museum Library in Beijing,[33] give evidence as to the way in which a mathematical terminology was created in the Manchu language. This language is written vertically, from left to right; the Chinese equivalent of each geometric term is added between the columns, next to the corresponding Manchu word. This suggests that the terminology was constructed from Chinese rather than from French and that, rather than inventing a terminology anew, Gerbillon and Bouvet relied on the earlier Manchu translation of the 1607 Chinese text. Terms like 'line', 'angle' or 'straight', were drawn from the Manchu lexicon. For other terms, phonetic transliterations of the Chinese terms were used; it is the case for the title of the work, *Gi ho yuwan ben bithe*.[34] That this title was erroneously transcribed back into Chinese on the cover of one of the extant copies[35] indicates that the work and the subject it dealt with remained little known among those who were literate in Manchu in the following centuries. The cyclical characters, *jia* 甲, *yi* 乙, *bing* 丙, *ding* 丁 ... used in Chinese to refer to points in geometric figures, are also transliterated in the Manchu script, rather than rendered using the usual Manchu names of the cyclical characters, which are different.[36] All this gives a glimpse of a technical language in the making, which had not undergone integration into the Manchu language—a language to a large extent constructed by the first emperors of the dynasty.[37]

In the following year Kangxi had the *Gi ho yuwan ben bithe* translated into Chinese, still under his close supervision.[38] The Chinese version was entitled *Jihe yuanben* 幾何原本, again like the 1607 translation of the first six books of Euclid's *Elements of Geometry*, following the European custom of giving textbooks on geometry the title of the classic.

None of the manuscripts are signed; Bouvet and Gerbillon's diaries designate the two of them as the authors of both the Manchu and Chinese texts, with the help of Manchu and Han Chinese teachers or secretaries, including the translator mentioned above: all these are anonymous, as are most if not all the Chinese and Bannermen who assisted the Jesuits in their service of the emperors. However, on a draft of the Chinese translation, two names are mentioned in marginal

---

33 Huang & Qu 1991, no. 0963 (1).
34 Pang & Stary 2000, 49; *bithe* means 'book'.
35 I am grateful to Dr Tegus 特古斯 for providing information concerning the copy kept in Hohhot, and then for helping me to access it.
36 I am grateful to Prof. An Shuangcheng 安雙成 for his help in studying the Gugong manuscript; Huang & Qu 1991, no. 0963 (1); see also Watanabe 2005a, 199.
37 Crossley & Rawski 1993, 80–87.
38 Du Halde 1736, 4: 295; Landry-Deron 1995, 2: 161; Liu 1995, 8–13; see Martzloff 1993b, 208–211.

notes: Antoine Thomas (An Duo 安多) and Jean-François Gerbillon (Zhang Cheng 張誠).³⁹ It was usual to mention the names of leading officials or persons in administrative correspondence; the presence of these two names simply reflects the fact that the four Jesuit tutors were regarded as a single team, in which Thomas and Gerbillon were in charge of the compilation of texts. It is clear from the two Frenchmen's diaries that Gerbillon had precedence over Bouvet; Thomas, who was senior to both of them, was regarded as the most versed in mathematics and astronomy. It is possible that the division of labour between the two teams was based on language rather than on fields of competence, and that Thomas actually participated in the elaboration of the Chinese version.

The structure, content and style of this new textbook are quite close to that of the French source, although two books have been omitted, and some contents from the others have been left out or supplemented.⁴⁰ The content of book VII, 'On incommensurables', and book VIII, 'On progressions and logarithms', were not altogether excluded from the Jesuits' teaching of the emperor. They were discussed in other lecture notes; by the end of March 1690 Kangxi already had logarithmic tables in Chinese.⁴¹ Thus the omission of two chapters from the Manchu and Chinese translations of Pardies' work reflects a choice regarding the structure of mathematics and the scope of geometry rather than an exclusion of some subjects from imperial tutoring. The discussion below is based on the Chinese version.⁴²

In keeping with its French source, the text is divided into paragraphs called items (*jie* 節), most of which have the same structure. Typically, an item opens with a statement beginning with *fan* 凡, literally 'any', which expresses generality; this term was used in the same way both in the earlier Chinese mathematical tradition, and in the 1607 translation of the *Elements*.⁴³ This is followed by an instantiation introduced by *sheru* 設如, 'supposing', in which the statement is rephrased for a particular instance: the cyclical characters are used to name objects, usually points referring to a figure. In the first chapter of the draft of the Chinese translation from Manchu, the phrase originally used to introduce the instantiation was *she yan zhi* 設言之, which can be rendered as 'suppose this is expressed'; the last two characters have been systematically crossed out and replaced by *ru* 如 between the columns.

---

39   Liu 1995, 11.
40   Liu 1991, 92–93.
41   BnF, Ms. Fr. 17240, f. 268 r.; Du Halde 1736, 4: 276; Landry-Deron 1995, 46, 54.
42   Mostly on Taipei NCL, Rare Books 6398 & 6399.
43   Engelfriet 1998, 154.

In later chapters, *sheru* is used from the onset.[44] In the 1607 translation, the corresponding part in a proposition, namely the setting out and the specification was called *jie* 解 (explanation).[45] The instantiation is followed by a question: *heze* 何則 (why?), which opens the justification; this often closes with *ke zhi yi* 可知矣, lit. 'can be known', which also marks the end of the item. In the 1607 translation, the proof was called *lun* 論; it is not named as such in Gerbillon and Bouvet's Chinese treatise. The division of the items found in the latter into general statement, instantiation and proof corresponds to the structure of the French source: in this respect Pardies followed the tradition, as did the 1607 translation. But, as mentioned above, he did not name these various parts. In the emperor's geometry, these parts were marked by terms that belong to common vocabulary and did not need to be defined in order for the reader to understand what they mean in this particular text. This avoided the problem posed by the meta-mathematical terminology used by Ricci and Xu Guangqi: it was nowhere defined in the Jesuits' Chinese printed works, and was diversely interpreted and appreciated by Chinese scholars throughout the seventeenth century.[46]

8.4 Diamonds and pearls: ratios in the new *Elements*

Following Pardies, proportions are given a prominent place in the new *Elements of geometry*. With 90 paragraphs, chapter 6 is by far the longest in the treatise. It is lengthier than its source (which only comprised 74 paragraphs), as illustrated by the opening definition, which, in the French textbook, reads:

1. When talking of magnitude, and saying that a quantity is large, one always makes some comparison of this quantity with some other of the same nature, with respect to which the former is said to be large. Thus we can say of a mountain that it is small and of a diamond that it is large, because we compare this mountain with other mountains, in comparison to which it is small; and that similarly we compare this diamond to other diamonds, in comparison to which we say that this one is large.[47]

The Chinese version is more explicit:

1. In general (*fan* 凡), when wishing to discuss the size of things, one must borrow a thing of the same kind in order to compare it; only then can one talk about it being large or small. Supposing (*sheru* 設如) there is a mountain, it must be compared to another mountain; only then can one talk about it being large or small. If it is a pearl, it must be compared to another pearl, only then

---

44 Taipei NCL, Rare Books 6398, *passim*.
45 Engelfriet 1998, 151–152.
46 Martzloff 1980, Engelfriet 1998: 147–154.
47 Pardies (1671) 1673, 58.

can one talk about it being large or small. Why (*heze* 何則)? If (*jiaruo* 假若) one compares a mountain to a pearl, then since they are of different kinds, although the mountain is larger than the pearl, one cannot say that it is large; although the pearl is smaller than the mountain, one cannot say that it is small. One must compare mountains to mountains, pearls to pearls; only then can one talk about mountains being large or small, pearls being large or small. Therefore in any case if one wants to discuss the measure of a thing, one must compare it with things of the same kind. Only then can one say that this thing is small or large. Suppose (*jiaru* 假如) one compares the measure of a line to the measure of another line, it will be longer or shorter, its value will be greater or smaller, only then can one talk about it; if (*ru* 如) one compares an area with an area, the measure of the area will be greater or smaller, the value of its surface will be greater or smaller, only then can one talk about it; if (*ru* 如) it is a solid, it must be compared to another solid; the thickness of the solid, the value of its volume, only then can one talk about it. If (*ruo* 若) one carelessly compares a line to an area or a solid, since line, area and solid are not of the same kind, one cannot talk about the length of the line, the size of the area, the thickness of the volume.[48]

The translators read the French as containing two parts: a general statement about quantities, followed by an instantiation that takes mountains and diamonds (pearls in the translation) as the particular case. To this they have added the question 'Why?' and the answer to it. This answer is a very explicit statement that comparing a pearl with a mountain makes no sense as regards the study of size. This illustrates the general fact that the third part of the paragraphs of the new *Elements*, while providing explanations and clarifications of what precedes, by no means corresponds to a modern notion of mathematical proof. The Chinese text goes on to discuss lines, areas and volumes, stating in detail that each can only be compared within its category. Although the discussion of mountains and pearls is not strictly parallel to that of lines, areas and solids, the chains of reasoning followed in both discussions are similar. In other words, the emperor's textbook chose unambiguousness instead of the concision favoured by Pardies. In the passage quoted above, another modification has been made to the French text: diamonds have been replaced by pearls—this is one of the cultural adjustments needed to turn a textbook intended for the sons of the French elite into a Chinese imperial work.

Proportions play an important role in Pardies' geometry: they underlie the definition of similar geometrical figures. The Pythagoras theorem also appears in the chapter devoted to proportions.[49] The Chinese version includes more propositions than its French source; among others, it gives the ratios between the following volumes: a cylinder and the sphere inscribed in it, a semi-sphere and the cone inscribed in it, a sphere and an ellipsoid, and the ratio between the area

---

48 Taipei NCL, Rare Books 6398, *juan* 6, item 1.
49 Pardies (1671) 1673, 86–87: (paragraph VI.61); Taipei NCL, Rare Books 6398, *juan* 6 item 63.

of a sphere and that of one of its meridian circles.[50] To study solids, Kangxi had recourse to models:

> [The emperor] did not limit himself to speculation, he combined it with practice; which made study pleasant to him, and made him understand perfectly what one taught him. When, for example, one taught him proportions of solid bodies, he took a ball, had it weighed exactly, and measured its diameter. He then calculated what weight another ball of the same material, but having a bigger or smaller diameter, should have, or what diameter a ball of a bigger or smaller weight should have. He then had a ball made of these diameters or weights, and he noted whether practice fitted speculation. He examined carefully the proportions and capacity of cubes, cylinders, full and truncated cones, and spheroids.[51]

Thus it seems plausible that the additions at the end of chapter 6 of the new *Elements* were made at the emperor's request. Wooden models made for his study are still kept at the Imperial Palace Museum in Beijing today.[52] Some of them are made up of smaller parts, to show how solids can be decomposed to calculate their volume.[53] The use of models was recommended by Pardies in book V of his treatise, devoted to solids: 'I advise to make angles and figures with cardboard; and by this means one will easily understand these things.'[54] Among the students taught according to his pedagogy, Kangxi was probably the only one to commission the making of models in precious wood in order to check the statements of the textbook. His attitude as a student of geometry did not differ from his general attitude as a ruler: he always preferred to see and check for himself and to cross check various sources of information rather than to rely solely on the word of one or a few of those who served him.[55]

Beside the geometry treatise, proportions were also widely used in other mathematical textbooks produced for the emperor, which will be discussed in the next chapter; in that respect their importance is even greater in the Manchu and Chinese treatises than in their French source. These treatises incorporate one important notion of traditional Chinese mathematics into the discussion of proportions: whereas there is no generic word for them in French or in Latin, in Chinese the terms of a proportion are called *lü* 率; in the Manchu version they are called *shuwai*, which is a transliteration of the other pronunciation of the same Chinese character, *shuai*. *Lü* were first defined in Liu Hui's 劉徽 commentary (3rd century CE) to the *Nine chapters on mathematical*

---

50 Taipei NCL, Rare Books 6398, *juan* 6, items 71, 72, 90.
51 De Fontaney, Letter to de la Chaize, Ningbo, 15 February 1703; *LEC* 7: 202–203; see Jami 2002b, 33.
52 Liu 1998, no. 89.
53 This is visible on the picture in Rawski and Rawson 2005, 230 (no. 150).
54 Pardies (1671) 1673, 54.
55 Jami 2002b, 38–39.

procedures (*Jiuzhang suanshu* 九章算術, 1st century CE): 'In general, when numbers are associated with one another, they are called *lü*'.[56] Although neither the commentary nor the original text were known in the seventeenth century, the term remained common in Chinese mathematical texts. It was first used in the context of Euclidean geometry by Ricci and Xu Guangqi, in a comment to the definition of proportions:

> Any two quantities being compared, comparing this quantity to another quantity, this quantity is the former *lü*; the other quantity to which it is compared is the latter *lü*.[57]

This is not a definition of *lü*; instead the term, with which the reader is expected to be familiar, is used to clarify what the comparison of quantities is about. The term *lü* was also used from the onset in Jesuit arithmetic: the numbers processed in the rule of three, among others, were called *lü*, again without the term being defined.[58] In Gerbillon and Bouvet's textbook, on the other hand, *lü* was defined in item 2 of chapter 6:

> In general, when measuring or valuing two things to compare them, the measures or values yielded by the comparison are large and small; they are called a ratio (*bili* 比例). What is compared and the thing to which it is being compared are both called *lü* (*lü* is the name of homogenised numbers (*qishu* 齊數). So [they are called] hereafter). The thing compared is the former *lü*. The thing to which it is compared is called latter *lü*.[59]

The phrasing is quite similar to that of the Ricci-Xu translation; but here *lü* is defined concomitantly with ratio (*bili* 比例). The former term is thereby given a status in Euclidean geometry on a par with the latter, instead of merely serving as an aid to understanding it. In Pardies' work, the terms of a ratio are defined as follows:

> 3. The quantity that one compares to another is called antecedent, and this other one the consequent.[60]

In the Chinese translation, *lü*, once defined, is used in the compounds 'former *lü*' (*qianlü* 前率) and 'latter *lü*' (*houlü* 後率), which render 'antecedent' and 'consequent' respectively. The integration of *lü* into Euclidean geometry provided a bridge between this field and that of arithmetic. The former, usually regarded as entirely novel in China, had been introduced as separated from the latter, which was regarded as embedded in the Chinese mathematical tradition. Another difference in

---

56 凡數相與者謂之率 (my translation); see Chemla & Guo 2004, 167–168, 956–959.
57 Guo 1993, 5: 1235.
58 See the *Instructions for calculation in common script* (*Tongwen suanzhi* 同文算指, 1614), in Guo 1993, 4: 77–278.
59 Taipei NCL, Rare Books 6398, *juan* 6 item 2.
60 Pardies (1671) 1673, 58 (VI.3).

terminology between Pardies and the Chinese textbook is that whereas in the former the notion of 'proportion' is defined as the equality of two ratios,[61] the latter discusses more generally the 'principles (or pattern) of ratios' (*bili zhi li* 比例之理); the standard case is that of 'ratios with identical principles (or patterns)' (*tongli bili* 同理比例);[62] this is quite similar to the terminology used in the 1607 translation.[63] The term 'principle' (*li* 理), central to Neo-Confucian philosophy, thus took a technical meaning within Euclidean geometry.

8.5 Practical geometry

As mentioned above, Kangxi's ultimate purpose in studying Euclidean geometry was 'to understand practical geometry'. In this respect Pardies' work again proved adequate: it presented practical geometry as a final topic, in its ninth book. The seventh and last chapter of Gerbillon and Bouvet's treatise is devoted to 'the practice and application of what is said in the previous six chapters'.[64] It deals with geometrical constructions and gives a few techniques that overstep the limits of Euclidean geometry as traditionally defined; thus the technique for reproducing a geographical map closes the chapter.[65] Pardies' work also went beyond these limits when tackling trigonometric lines and their logarithms.[66] His book IX, entitled 'Problems, or practical geometry', opens with a definition:

1. A problem is a proposition that teaches how to do something and demonstrates its practice, whereas theorems are speculative propositions, in which one considers the properties of ready-made things.[67]

This was a traditional distinction; however in keeping with his choice not to distinguish explicitly between 'definitions, principles and propositions',[68] Pardies had not previously used the word 'theorem'; he defined it only in contrast with 'problem'.[69] The geometrical constructions given in his book IX include some given in Clavius' edition of the *Elements*. Gerbillon and Bouvet gave some more of these, with detailed justifications. For example, the construction of the midpoint of a segment constitutes proposition I.10 in Euclid's *Elements*, and most previous propositions are prerequisites for it:

---

61   'L'égalité de raisons s'appelle Proportion.' Pardies (1671) 1673, 59.
62   Taipei NCL, Rare Books 6398, *juan* 6 item 3.
63   Engelfriet 1998, 183–199, esp. 189.
64   Taipei NCL, Rare Books 6398, *juan* 7; this is written in both Manchu and Chinese on a slip of paper that is pasted on to the manuscript; Liu 1995, 10.
65   *Jihe yuanben*, Taipei NCL, Rare Books 6399, *juan* 7 item 53.
66   Pardies (1671) 1673, 154 ff. Logarithms are introduced in book VIII.
67   Pardies (1671) 1673, 134.
68   Pardies (1671) 1673, 'Preface' a x v.
69   Engelfriet 1998, 129.

propositions I.1, I.4 and I.9 are actually quoted in it; propositions I.3, I.5, I.7 and I.8 are used indirectly. Clavius gives the practical construction as part of the proposition, as do Ricci and Xu.[70] Pardies does not mention the construction as a separate item; he takes it for granted in his paragraph II.15 ('In all isosceles triangles, the angles that rely on the base and on the two equal sides are equal to one another'), and describes it as part of his paragraph IX.3 ('From a given point *d* draw a perpendicular to line *b a c*').[71] In Gerbillon and Bouvet's new *Elements*, on the other hand, the construction is given in the third item of the final chapter:

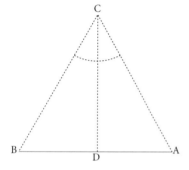

General method for dividing a straight line into two equal parts: Suppose there is a straight line AB. One wishes to [divide] it in the middle into two equal parts. On line AB, as in the 1st item of this chapter (⋆) construct an equilateral triangle ACB. Again, as in the previous item (⋆), construct CD, that bisects angle ACB. Then line AB must be divided into two equal parts in D. Why? In the two triangles CAD and CBD, the two superior angles ACD and BCD are equal. Moreover, as to the two sides CA and CB, because they are two sides of the equilateral triangle ABC, are also equal. Moreover, as to line CD, because it is a side common to the two triangles, the equality goes without saying. But if the superior angles and the two adjacent sides are all equal, the two bases, segments AD and DB, must also be equal as in chapter 2 item 5 (⋆). Thus line AB in D is divided in D into two equal parts.[72]

The construction is justified in great detail: items 7.1 and 7.2 are used in the construction itself, and item 2.5 in the justification; the figure corresponding to each item is reproduced within the text (the figures are marked by (⋆) in the translation above; see Illustration 8.3). These items correspond respectively to propositions I.1, I.9 and I.4 of the 1607 translation; there the first two, which are problems, are quoted in the construction itself and the third one, which is a theorem, in the proof.[73] In fact the construction given by Gerbillon and Bouvet is similar to the one in the 1607 translation, and is justified in the same way, except for the fact that in the 1607 translation triangle ABC constructed first is isosceles rather than equilateral. This similarity and the fact that the construction is absent from Pardies' work suggest that it was added to the imperial textbook at Kangxi's behest.

The two French tutors however, did not simply take up all the 1607 text of this proposition. They left out the 'praxis' (*yong* 用法) that concludes proposition I.10:

---

70 Jami 1996, 190–199.
71 Pardies (1671) 1673, 18–19 & 135; Jami 1996, 193.
72 Taipei NCL, Rare Books 6398, *juan* 7 item 3.
73 Guo 1993, 5: 1167–1170.

CHAPTER 8 | The imperial road to geometry: new *Elements of geometry*

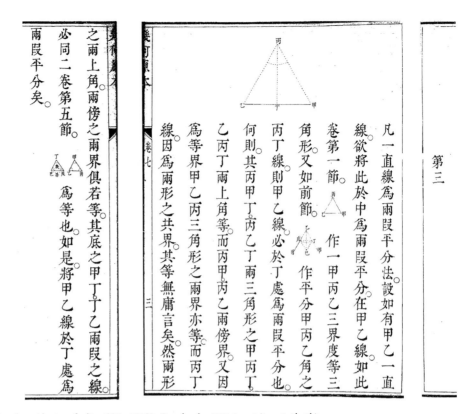

Illustration 8.3 Constructing the midpoint of a line (Taipei NCL, Rare books 6399, *juan* 7 item 3, 2b–3b).

Praxis: Take A as centre, and, using any measure—but it must be longer than half of line AB—above and below, construct a short [segment of] circumference. Next, using this original measure, taking B as centre, do likewise. The intersections are C and D. Finally construct the straight line CD, then it intersects AB in E.[74]

In this series of concise instructions, the fact that E is the midpoint of AB is not stated, even though in the main proposition the midpoint of AB was named D. Thus in the 1607 translation there was a disjunction between the proof and the praxis added by Clavius to the Euclidean text. Gerbillon and Bouvet, by contrast, follow Pardies in not separating justification from construction. They do not mention the compass construction in the passage quoted above, but instead refer to item 7.1 (the construction of an equilateral triangle on a given segment), where this use of the compass is described in the main body of the construction rather than as a separate part. In sum, the imperial

---

74 Guo 1993, 5: 1171.

175

geometry textbook bridged the gap between 'practice and speculation' that was apparent in the 1607 translation.

8.6 The emperor's role in the composition of the new *Elements*

From the two Frenchmen's diaries it is clear that Kangxi took an active part in the composition of the textbook. Further evidence of this can be found in the fact that he annotated two of the extant copies of the manuscripts. The Manchu manuscript kept in Hohhot has marginal annotations in imperial vermilion ink in the same language; indeed according to Gerbillon on 26 March 1690 Kangxi 'corrected a few words in red letters' in the definitions they had presented to him.[75] However, he did not always use vermilion ink: his Chinese handwriting has been identified in the marginalia in black ink on one of the manuscripts kept in Taipei, which is a draft of the translation from Manchu into Chinese done in 1691.[76] In this manuscript the text is not written continuously: some propositions start on new pages. This is consistent with the French Jesuits' account: they submitted their written work progressively, presenting a few propositions at a time. The Chinese translation was an opportunity for the emperor to revise geometry, in both senses of studying anew and editing a text. The identification of Kangxi's handwriting is corroborated by the use of *zhen* 朕, the imperial first person singular pronoun, in the annotations.[77] There is another handwriting: a dialogue took place between the emperor and someone who worked with the Jesuits. The latter pointed to changes made in the text in the margins; to this the former responded by approval or disagreement. Some annotations are written directly on the manuscript, some others are on strips of papers pasted onto it. The comments mostly bear on the conformity of the text with its Manchu source, which Kangxi checked in detail.[78]

One of the emperor's comments reveals a concern that mathematical terminology should be compatible with non-technical usage. Thus, in defining various types of ratios, what Pardies called *invertendo*, that is, the inversion of the first and the second terms and that of the third and fourth terms of a proportion,[79] was first rendered in Chinese as *fanli bili* 反理比例, (reverse principle ratio) while direct proportions were called *shunli bili* 順理比例 (direct principle ratio). However, the first term read literally 'ratio opposing principles'. Kangxi therefore edited the

---

75 Du Halde 1736, 4: 274; Landry-Deron 1995, 2: 50.
76 Taipei NCL, Rare Books 6398 *passim*; see Liu 1995, 10.
77 At *juan* 7 item 49.
78 Liu 1995, 12.
79 Pardies (1671) 1673, 61.

term into *fan bili* 反比例, and commented: 'Principles cannot be opposed, therefore remove this. '[Opposing] principle' goes against syntax'.[80] In other words, mathematical usage should not go against the received understanding of such a fundamental philosophical concept as *li* 理. One wonders whether the two Frenchmen were aware of its meaning at the time.

When the work was completed, the emperor wrote a preface for it, as Bouvet later reported to Louis XIV:

When we had finished explaining to the emperor all of practical and speculative geometry, following the same order as we had in the *Elements*, this prince, delighted to have become a good geometer, evinced full satisfaction. And to show how much these two works were to his taste he had them both translated into Chinese. He took the trouble to compose prefaces himself to put at the opening of each.[81]

This account seems to describe practical geometry, chapter 7 of both Manchu and Chinese versions, as a separate treatise. It is also possible that the second preface referred to is that of one of the other mathematical treatises that the Jesuits wrote for the emperor.[82] In any case, all the extant copies of the new *Elements*, both in Chinese and in Manchu, have a single preface. There are two slightly different versions of the Chinese preface: the one found in the draft kept in Taipei seems to have been edited into a slightly more elegant style in later versions, including the one kept in the imperial library. Beside minor changes in the text, one sentence in small characters has been added. This sentence is in italics in the translation below, in which the text in smaller characters is between brackets:

Jihe yuanben (the name of the origin of numbers. *In the one written by Matteo Ricci, because the method of composition was not clear, it was difficult to elucidate what comes first from what comes later; so we have done another translation*) is the foundation of measuring and counting the myriad things.[83] It is the source of astronomy, geography, and others. In general, those who study it must begin with what is easy and then reach what is difficult. Without skipping a step, they penetrate subtlety. Therefore the obvious shapes are given first, the complicated shapes afterwards. Within each proposition, the two are put together similarly: what is easy to learn first, what is complicated afterwards. The writing is in the right order so that the students achieve gradual progress. When the general idea can be understood by looking at the figures, we do not seek to give the meaning and to write detailed explanations.[84]

---

80 Taipei NCL, Rare Books 6398, *juan* 6 item 8. I follow the transcription and interpretation given in Liu 1995, 12, while proposing that the missing character in the last sentence might be *fan* 反.
81 Bouvet 1697, 144–145.
82 See pp. 198–199.
83 This first sentence echoes Xu Guangqi's preface to the 1607 *Elements*: '*Jihe yuanben* is the basis of number and magnitude.' (幾何原本者度數之宗); Wang 1984, 75.
84 Manuscript Gugong 律八三五 29, n.p. Comp. Taipei NCL, Rare Books 6398 & 6399.

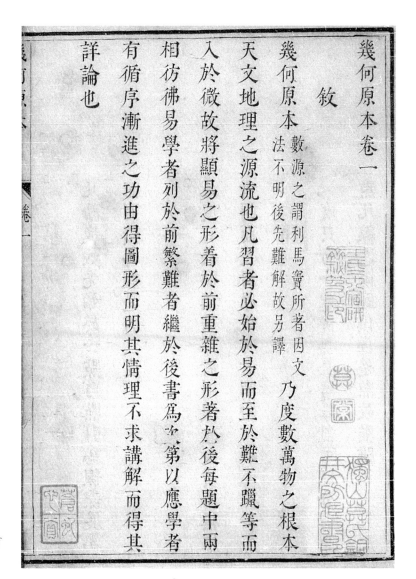

Illustration 8.4 Preface to the *Elements of geometry* manuscript (Taipei NCL Rare Books 6399, *juan* 1: 1a).

At this stage Kangxi intended to publish and circulate both the Manchu and the Chinese versions of the treatise. It is in this light that the added sentence should be interpreted: these new *Elements* were necessary because of the shortcomings of the 1607 translation that he deemed unclear. This was not an uncommon assessment: several Chinese scholars, including Mei Wending, had endeavoured to comment on and clarify it.⁸⁵ However the emperor alone was in a position to commission a new version of the work from the successors of Matteo Ricci. In his preface, Pardies claimed that 'in the intention [he] had to teach geometry with all possible ease, the way [he] followed seemed to [him] the most appropriate.'⁸⁶ Along the same lines, the emperor emphasised the progression from easy to difficult within the text: he had caused Euclidean geometry to be reordered according to pedagogical concerns, thus opening an imperial road to geometry.

---

85   See pp. 48–49; Martzloff 1980; Engelfriet 1998, 366–371, 383–431.
86   Pardies (1671) 1673, 'Preface', f. a xi r.

CHAPTER 9

Calculation for the emperor: the writings of a discreet mathematician

9.1 Antoine Thomas (1644–1709) and his *Synopsis mathematica*

Gerbillon and Bouvet, while giving lavish details on their own tutoring of the emperor, mentioned repeatedly that Antoine Thomas and Tomás Pereira formed a team similar to theirs, and that, while the French taught geometry and philosophy in Manchu, Thomas taught arithmetic—and much more—in Chinese. But there is comparatively little information on his tutoring. However he was the main author of several volumes of lecture notes, similar to Gerbillon and Bouvet's *Elements of geometry* (*Jihe yuanben* 幾何原本), but considerably longer. He was no less prolific in making observations for European savants: thus, about half of the content of the volume of *Observations* published in Paris in 1688 consists of observations contributed by him rather than by the Jesuits of the French mission who are mentioned in the work's title.[1] Moreover, whereas the French Jesuits' contributions to this volume mainly regard animals and their anatomy, Thomas' contributions mostly concern astronomy. He had longer experience of making and recording observations as he travelled, as well as of working within Chinese institutions. The contrast between him and his French confreres regarding visibility and actual production reflects the fact that he did not benefit from patronage as illustrious as that of his French confreres. This may be the reason why the textbooks he produced as an imperial tutor have long remained unnoticed, although they formed the core of imperial mathematics, and played a major role in shaping its structure.[2] This chapter is devoted to exploring these textbooks; as materials produced for the emperor's private use, they exist only in manuscript form.[3]

---

1   Goüye 1688, 120 ff. Hsia 2009, 107–108; see pp. 117–118.
2   See Han 2003a, Han & Jami 2003, Jami & Han 2003.
3   See the list of extant manuscripts in Bibliography 1, pp. 395–396.

CHAPTER 9 | Calculation for the emperor: the writings of a discreet mathematician

Thomas' life and career before his arrival in Beijing are relatively well known.[4] Born in Namur in 1644, he studied at the newly founded Jesuit college. In 1660 he entered the Jesuit novitiate in Tournai. Fifteen years of study and teaching ensued; during these years he repeatedly wrote to the Society's General, asking to be sent to the China mission, arguing that he was versed in mathematics and astronomy: it is possible that he acquired these skills partly by private study rather than as part of his Jesuit education.[5] In 1677 his request was finally granted, two years after he was ordained priest.[6] As was then usual he headed for Portugal; he studied Portuguese while teaching mathematics in Coimbra. From there he made what seems to have been his first published observation, that of the lunar eclipse of 29 October 1678.[7] The main work he produced while in Coimbra was a mathematical textbook published in 1685. Its title reveals Thomas as personifying the missionary–mathematician that China Jesuits are supposed to be: 'Outline of mathematics comprising various treatises of this science, laid out briefly and clearly by F. Antoine Thomas, S.J., for novices and candidates for the China mission'.[8] The work seems to have reflected Thomas' teaching at the Coimbra college; this teaching did not arouse great enthusiasm in him, nor possibly in his students.[9] One of the things that may have prompted him to write a mathematical work is the fact that in Coimbra he found himself 'destitute of all possible support of books': the college library seems to have been lacking in mathematical works.[10] However his textbook is intended for those aspiring to the China mission, that is, for Jesuits who had already completed the theology course, rather than for ordinary students: it uses as little technical language as possible, which might have suited both audiences. For example, in the section on spherical trigonometry, standard Latin language is used to state the formulae, rather than mathematical notations; no justification is given.[11] Consisting of more than one thousand pages in two in octavo volumes, the work was intended as a *vademecum*; Thomas hoped that it would provide an easy entry into the mathematical sciences.

The topics listed in the table of contents closely match those that the Jesuits taught and practised at the Beijing court. Volume 1 discusses arithmetic, elementary geometry, practical geometry, the sphere (cosmography), geography, hydrography, hydraulics and music.

---

4   De Thomaz de Bossierre 1977, 1–34.
5   Golvers 2004b, 34–35.
6   De Thomaz de Bossierre 1977, 4–5.
7   De Thomaz de Bossierre 1977, 13.
8   Hereafter *Synopsis mathematica*; Thomas 1685; the work was reissued in 1729; see Löwendahl 2008, no. 361.
9   Bosmans 1924, 172.
10 Golvers 2004b, 38.
11 Bosmans 1924, 174.

Volume 2 discusses optics, statics, clocks, spherical triangles, the astrolabe, calendar and astronomy. The circumstances under which the book was composed, as well as its purpose, are summarised in a notice to the reader:

> Having found some leisure in Portugal before the voyage into the Kingdom of the Chinese, I have written this Mathematical Outline, which previously in Belgium I had meditated to write in order to open some easy way into the mathematical sciences. For these arts are accustomed to terrify the beginner in his first entry [into the subject], unless some easy way is smoothed into the more difficult demonstrations [so that] beginning by these he does not immediately become tired and is scared away from the proposition. If now and then one comes across anything less balanced, or hurried, then I pray that you will excuse and ignore it, given that I am just about to be taken off on a voyage to the Oriental missions. Meanwhile I pray to God that this short work will be fruitful and that with the sole assistance of the Holy Spirit our efforts, with a mind of apostolic zeal, may be for the Divine Glory and may inspire many to succour China where the gates of the Gospel are most widely open.[12]

The *Synopsis mathematica* was dedicated to Maria-Guadalupe de Lencastre, Duchess of Aveiro (1630–1715), who had taken Thomas, amongst many missionaries, under her patronage. It seems that he sent her each chapter as he was writing the book; after he left Portugal, she recommended it for publication and paid for its printing.[13] This patronage as well as the work's title, content and purpose turned it into a piece of missionary mathematics, devised as it was especially for China. Surely the author of such a work was uncommonly well prepared to fulfil his evangelising task by the unusual means of becoming an imperial tutor.

Despite this preparedness, when Thomas sailed from Lisbon in April 1680, it was with the aim of attempting to enter Japan, where Christianity had been forbidden since 1614.[14] His first destination was Goa, as was usual for missionaries of the *Padroado*. From there he travelled on to Siam, and finally reached Macao in July 1682.[15] During the three years he spent there, it became clear that there was no possibility of entering Japan.[16]

Meanwhile in Beijing, the emperor was seeking to attract more Western 'experts' to his court:

> In the 24th year of Kangxi, on the 12th day of the 2nd month (16 March 1685), His Majesty ordered Ledehun 勒德洪 and Mingju:[17] 'Today Verbiest is already aged; I have heard that in Macao there is someone as versed as him in astronomy and younger. Together with the Ministry of Rites, enquire from

---

12 'Ad lectorem', Thomas 1685, vol. 1, n.p.
13 Bosmans 1924, 173; De Thomaz Bossierre 1977, 14.
14 De Thomaz de Bossierre 1977, 11–12; Miyazaki 2003, 12.
15 De Thomaz de Bossierre 1977, 15–23.
16 De Thomaz de Bossierre 1977, 32–35.
17 The two Manchu Grand Secretaries; Zhao 1977, 21: 6125.

Verbiest about that person's name, and report; and if there is someone skilled in medicine report it as well.'[18]

Verbiest was by then in his sixties. Kangxi sought an assistant and a successor for him among the Jesuits rather than among the officials of the Astronomical Bureau. He had decided to perpetuate the lineage of Westerners in his service and he did not regard the presence in Beijing of Claudio Filippo Grimaldi, who was already assisting Verbiest, as sufficient to ensure continuity. The requested report came promptly on the following day:

Verbiest said that there is but one person versed in the calendar, his name is Antoine Thomas (An Duo 安多); as for medicine, he does not know whether there is anyone skilled in it.[19]

This exchange is yet another illustration of the fact that Kangxi's interest in Western learning was not limited to the mathematical sciences, and that he was keen on gathering medical knowledge from all provenances.[20] On the day he received the reply, the emperor nominated Grimaldi as his envoy to Macao in charge of bringing Thomas back to the court:[21] the latter had the rare honour of entering China escorted by an imperial envoy, and went straight to Beijing, where he was to work for the rest of his life.[22] He started this new career as Verbiest's assistant at the Astronomical Bureau.

On 8 November 1685, the very day of his arrival in the capital, Thomas was given audience by the emperor. A few days later, he wrote to the Duchess of Aveiro, telling her that the emperor had heard of his *Synopsis mathematica*, and asking her to send him some well decorated copies.[23] Thus the story of Thomas' itinerary from his native province to Beijing epitomises the European representation of the evangelisation of China: the mathematical sciences were the means by which the Jesuits gained entry into Chinese territory and imperial favour, thereby protecting and furthering the missionary enterprise. Indeed he was one of the few albeit famous 'specialists' who worked in the service of the emperor. Also, like his French confreres, he relied on a European patron in order to establish links of patronage with the emperor, by presenting him with a gift that provided evidence of the skills he was putting in his service.

---

18 BnF Chinois 1330, 1a–1b.
19 BnF Chinois 1330, 1a–1b.
20 Dong 2004, 69–71; Puente Ballesteros, 2009, 50–182.
21 Jami & Han 2003, 146; Masini 2002, 187–188.
22 De Thomaz de Bossierre 1977, 34–36.
23 De Thomaz de Bossierre 1977, 36.

In March 1690, when Bouvet and Gerbillon started tutoring Kangxi in geometry, Thomas was already teaching him this subject. In the entry dated 8 March of his diary, Bouvet noted that:

> [...] Frs Thomas and Pereira continued to go [to the Hall of the Nourishment of the Mind] every day, the former to teach Euclid to the Emperor, the latter to serve as interpreter, the former being unable to express himself in Chinese with ease.[24]

Once the French took over Euclidean geometry, the remaining subjects of arithmetic,[25] algebra,[26] practical geometry, trigonometry and logarithms[27] continued to be taught by Thomas. Although it is not possible to reconstruct the chronology of his tutoring, a number of treatises he wrote for this purpose are extant. There is evidence that Thomas is the main if not the sole author of the two largest ones, the *Outline of the essentials of calculation* (*Suanfa zuanyao zonggang* 算法纂要總綱) and the *Calculation by borrowed root and powers* (*Jiegenfang suanfa* 借根方算法). Other, shorter treatises can also be ascribed to him. For some of them he may have relied in part on earlier material, since according to Bouvet,

> [...] Father Antoine Thomas explained to [the emperor] in Chinese the use of the main mathematical instruments and the practices of geometry and arithmetic that Father Verbiest had once taught [the emperor].[28]

So Thomas' textbooks may also have reflected what Verbiest had taught the emperor in the 1670s. Moreover, like all Jesuits writing for Kangxi, he must have received the help of some literati provided by the Imperial Household Department.

9.2 The *Outline of the essentials of calculation*

While Gerbillon and Bouvet rewrote the *Elements of geometry* for the emperor, Thomas undertook what was in effect a rewriting of another late Ming Jesuit mathematical work: the *Instructions for calculation in common script* (*Tongwen suanzhi* 同文算指, 1614), which was based in part on Clavius' *Epitome arithmeticæ*, while taking up material from the Chinese mathematical tradition. In teaching the emperor and in composing the *Outline of the essentials of calculation*, Thomas used his own *Synopsis mathematica*.[29] However, as its title indicates, his Latin work is a mere outline. Its books are divided into sections, which are in turn

---

24　BnF, Ms. Fr. 17240, f. 265 v.
25　See p. 140.
26　BnF, Ms. Fr. 17240, f. 268 r.
27　BnF, Ms. Fr. 17240, f. 265 v.–266 r., 269 v.
28　Bouvet 1697, 127.
29　Han & Jami 2003.

subdivided into items; most items describe a procedure or rule using a single numerical example; sometimes a second problem of the same kind is given. The *Outline of the essentials of calculation*, on the other hand, is a collection of problems arranged according to the methods used to solve them. Most of its chapters take up the numerical example of the corresponding chapter in the *Synopsis mathematica*, although the numerical data have often been changed. Together with it, a large number of problems solved by applying the same rule are given. A significant number of these additional problems are taken from Ricci and Li's *Instructions for calculation in common script*, whereas the examples taken up from the *Synopsis mathematica* are often rephrased so as to fit their Chinese context. Thus, in his Latin treatise, Thomas gives the following example for the simple direct rule of three:

5 terrestrial degrees contain 110 Belgian miles; how many do 8 [degrees] contain?[30]

As might be expected, the Chinese version uses Qing imperial units instead; it is the second problem of chapter 6:

Again supposing there are 2 terrestrial degrees, this corresponds to 500 *li* ; now for 7 degrees what must the number of *li* be?[31]

These figures are consistent with official standards: one terrestrial degree (a circle being divided into 360 degrees) corresponded to 250 *li*. Interestingly, Thomas later took part in an expedition whose goal was to measure a degree of the terrestrial meridian;[32] the standard *li* was then changed so that there were 200 *li* to one terrestrial degree, which was more convenient for calculations in the emperor's view.[33] This is one of many ways in which the *Outline of the essentials of calculation* reflects an imperial culture of mathematics, in which the Jesuits were involved as astronomers and tutors, rather than a specifically Jesuit culture.

Turning to the structure of the *Outline of the essentials of calculation*, a comparison between its table of contents and those of its two antecedents, Chinese and Latin, reveals strong similarities between their constructions.[34] Judging from the two Chinese works, the way in which the Jesuits taught the subject in China—and, it would seem, in their European colleges—remained stable throughout the seventeenth century. In fact, the Latin antecedents of the *Instructions for calculation in common script* and the *Outline of the essentials of calculation* have quite

---

30   Thomas 1685, 1: 24.
31   Gugong 13276–81, 25b.
32   See pp. 244–245.
33   Bosmans 1926, esp. 176.
34   Han & Jami 2003, 150–152.

Table 9.1 Contents of the *Outline of the essentials of calculation* compared with those of its possible sources

Synopsis Mathematica[1]		*Outline of the essentials of calculation* 算法纂要總綱[2]		*Instruction for calculation in common script* 同文算指[3]
I.1.1 De numeratione	1	定位之法 Method for determining places	I.1.1	定位 Determining places
I.1.2 De additione (I.2.2, I.2.3)	2	加法 Addition	I.1.2 (I.2.6)	加法 Addition
I.1.3 De subtractione (I.2.2, I.2.3)	3	減法 Subtraction	I.1.3 (I.2.6)	減法 Subtraction
I.1.4 De multiplicatione (I.2.2, I.2.3, I.2.4)	4	乘法 Multiplication	I.1.4 (I.2.6)	乘法 Multiplication
I.1.5 De divisione (I.2.1, I.2.2, I.2.3, I.2.4)	5	除法 Division	I.1.5 (I.2.6)	除法 Division
I.3. De regula aurea	6	三率求四率之法 Method for finding the fourth *lü* with three *lü*	II.1, II.2, II.3	三率準測法 變測法 重準測法 Direct, reverse and compound methods of the three *lü*
I.4.2 De regula alligationis	7	和較三率法 Method of the of three *lü* with sums of differences[4]	II.5	和較三率法 Method of the of three *lü* with sums of differences
I.4.1 De regula societatis	8	合數差分法 Method of differentiated distribution according to combined numbers	II.4	合數差分法 Method of differentiated distribution according to combined numbers
I.4.3 De regula positionis[5]	9	借衰互徵法 Method of borrowing grades for mutual comparison	II.6	借衰互徵法 Method of borrowing grades for mutual comparison
—	10	疊借互徵法 Method of repeated borrowing for mutual comparison[6]	II.7	疊借互徵法 Method of repeated borrowing for mutual comparison
I.5.1 De extractionis radicis quadratae	11	開平方法 Method of square root extraction	II.12 (II.13, II.15)	開平方法 Method of square root extraction
III.1 (2 & 3) De resolutione triangulorum rectilineorum	12	三角形總法 General methods of triangles	—	
III.5 De superficierum dimensione	13	算各面積總法 General methods for the areas of differen figurest	—	
I.5.2 De extractione radicis cubicæ	14	開立方法 Cube root extraction	II.16	開立方法 Cube root extraction
III.6 De dimensione solidorum	15	算體總法 General methods for solids	—	

1 The Roman number refers to the number of the book, the first Arabic number to that of the section, the second one to that of the item.
2 Gugong 13276–81.
3 I refers to *Qianbian* 前編, II to *Tongbian* 通編; the last number refers to the headings rather than to the *juan* 卷 (there are sometimes two or three headings in a *juan*).
4 In Latin *summa differentiarum*: this refers to the procedure used in the "rule of alligation" or "rule of mixtures"; Smith 1925, 2: 587–589.
5 This is the well-known "rule of false position". The *Synopsis* does not mention the rule of double false position.
6 Liu 2002a, 161.

similar structures themselves; this suggests that Thomas may have used Clavius' *Epitome arithmeticæ* when teaching arithmetic in Europe. In any case, the continuity between Clavius' textbook and Thomas' contrasts with the break in the teaching of Euclidean geometry.[35]

The scope of the *Outline of the essentials of calculation* is broader than those of both the arithmetical part of the *Synopsis mathematica* and the *Instructions for calculation in common script*. For chapters 12 (triangles), 13 (geometrical figures) and 15 (solids), Thomas drew on Book III of his *Synopsis*, devoted to practical geometry, (a subject also discussed in chapter 7 of the *Elements of geometry* composed by his two French colleagues). In integrating this material, a new structure was devised in comparison with the two sources: in the *Outline of the essentials of calculation* subjects are arranged according to their dimension, or to the degree of the equations used in solving the problems; practical geometry is thus embedded into calculation.

In another respect the *Outline of the essentials of calculation* differs from both its antecedents: these deal with operations on fractional numbers separately from those on integers. Thomas' Chinese textbook, on the other hand, gives the rule for operations on fractions at the end of each relevant chapter,[36] after having discussed operations involving units, including time units and angular units—these last are called 'astronomical' (*tianwen* 天文) operations —,[37] all of which are found in the *Synopsis mathematica*.

To my knowledge, eight copies of the *Outline of the essentials of calculation* in Chinese and one in Manchu have been preserved.[38] Comparing them gives further insight into the process of the composition. One can distinguish two main versions of the text, which correspond to two stages; an intermediate version between these two seems to have served as the basis for a Manchu translation, which is only extant in a draft form.[39] The later version is an augmentation of the earlier one in two ways: first, some more problems have been added; secondly and most strikingly, explanations and justifications have systematically been added for each problem.[40] In the earlier version, the first problem of each category opens with *sheru* 設如, 'supposing'; all the others open with *you sheru* 又設如, 'Again, supposing'; the

---

35 See pp. 162–166.
36 They correspond to parts of the chapters of the *Synopsis mathematica* and the *Instructions for calculation in the common script* between brackets in the table.
37 Gugong 13276–81, 7b–23b.
38 See the list in Bibliography 1, pp. 395–396.
39 Han & Jami 2003, 148–149, 153–154.
40 In what follows Gugong 13276–81 is used as representative of the first version, BML, Ms. 82–90 A as representative of the second one.

statement of the problem closes with *ruogan* 若干, 'how much' or 'how many'. The solution, which follows immediately, ends with the numerical result. In the later version, by contrast, this result is most of the time followed by a lengthy explanation, which opens with *heze* 何則: 'why?' and closes with *ke zhi yi* 可知矣, 'can be known'. In this explanation, explicit reference is sometimes made to the *Elements of geometry* and to another treatise, the *Elements of calculation* (*Suanfa yuanben* 算法原本), which will be discussed below; in both cases the reference is to a particular item of these texts on which the explanation relies.

Thus in the later version, the format of the text has been made similar to that of the Chinese version of Gerbillon and Bouvet's geometrical treatise, with the same terms used as markers within each item.[41] This similarity of format can hardly be coincidental. The evolution of the text can be summarised as follows: after a first version of the *Outline of the essentials of calculation*, of which a finely written copy (*jing xie ben* 精寫本) copy was produced for imperial use,[42] the work was revised as part of a process of homogenisation of the mathematical lecture notes. It was at this stage that explanations were added to all the problems, so that their structure would parallel that of the propositions of the new *Elements of geometry*. The systematic correction of *she yan zhi* 設言之 into *sheru* 設如 in the first chapter of the latter[43] suggests that the homogenisation may have been done simultaneously on both textbooks. This would imply that the *Outline of the essentials of calculation* was rewritten in 1691. It would also shed some light on the title of the work, 'Outline of the essentials of calculation', which hardly matches its bulk: it is possible that Thomas started from a translation of his *Synopsis mathematica*, and that subsequent rewritings turned the Chinese version into a full-blown textbook, with an abundance of solved problems explained in great detail.

In the case of the *Elements of geometry*, there is direct evidence of the emperor's role in the revision of the text.[44] Since the *Outline of the essentials of calculation* was produced in similar circumstances and during the same period, it is reasonable to suppose that the changes to the latter work were also made in response to Kangxi's demand: they resulted in a homogenisation of the styles of the two works. Another intervention of the emperor must have been to propose problems for inclusion. Thus according to Bouvet on 24 March 1690,

---

41 See p. 168.
42 Gugong 13276–81, written in *Song typeface* (*songtizi* 宋體字).
43 See p. 169.
44 See pp. 176–177.

His Majesty, before withdrawing, gave two arithmetic problems to solve, one to F. Thomas and the other to F. Gerbillon, which they carried out successfully.[45]

So even the earlier version of the *Outline of the essentials of calculation* probably contained problems proposed by Kangxi. It is possible for example that he contributed the famous problem of the 'chickens and rabbits in a cage' (*long nei ji tu* 籠內雞兔):

Again supposing there are chickens and rabbits inside a cage, one does not know their number, but one knows that there are 80 heads and 200 feet. How many chickens and rabbits are there?

If they were all chickens, 80 [of them], there should be 160 feet, 40 less; if they were all rabbits, 80 [of them], there should be 320 feet, 120 more. Add this number to the less number 40; one obtains 160, which is the 1st *lü*. The less number 40 is the 2nd *lü*. The number of heads 80 is the 3rd *lü*. Multiply the 2nd *lü* and the 3rd *lü*, divide by the 1st *lü*; one obtains the 4th *lü 20*, that is 20 rabbits. This number once obtained, then there are precisely 60 chickens.[46]

The method applied here is the rule of false position (one begins by reckoning what the number of feet would be if there were only chickens or only rabbits, and derives the actual number of each species from these), which is the main subject of the seventh of the nine chapters, 'Excess and deficit' (*Ying buzu* 盈不足). The problem itself is first found, with different numerical data and with pheasants (*zhi* 雉) instead of chickens in the *Mathematical classic of Master Sun* (*Sunzi suanjing* 孫子算經),[47] a work possibly dating back to the third century CE that was included in the *Ten mathematical classics* (*Suanjing shishu* 算經十書) at the beginning of the Tang dynasty.[48] In the *Instructions for calculation in common script*, the problem was given in the chapter on the rule of double false position ('Repeated borrowing for mutual comparison', *Diejie huzheng* 疊借互徵), also with different numerical data.[49] Thomas' textbook does not comment on the problem's origin, and puts it in a different category from the late Ming work: it now belongs in the 'method of the three *lü* with sums of differences'. As can be seen in this example, this method is named after the procedure that gives the first *lü* of the proportion: it is the sum of the two differences between the total number of feet given and that yielded by supposing that all animals are rabbits and chickens respectively. Interestingly, the

---

45 BnF, Ms. Fr. 17240, f. 266v.
46 Gugong 13276–81, 47a–b. In this translation, all numbers but one render numbers denoted using the characters of tens and hundreds. The *20* in italics renders 二〇, the place value notation used in the layout of written calculation. In most problems both notations are used together in the description of calculations within the text. The layout of operations is never shown. The same convention is used in the other translations given in this chapter.
47 Chapter 3, problem 31; Lam & Ang 2004, 157–158; 222.
48 Guo 1993, 1: 244; Guo & Liu 2001, 14.
49 Guo 1993, 4: 179–180.

same chapter of the *Outline of the essentials of calculation* closes with a problem that is emblematic of the Western mathematical tradition, known as the 'problem of Hieron's crown', the solution of which is traditionally attributed to Archimedes.[50] The problem is found in two earlier treatises in Chinese: in the *Instructions for calculation in common script*,[51] it is in the same chapter as the chickens and rabbits problem. In an anonymous mathematical work of the seventeenth century extant only in the form of a manuscript, the *Western mirror of Europe* (*Ouluoba xijing lu* 歐羅巴西鏡錄), the famous story of Archimedes' discovery of the solution at the bath is recounted.[52] The problem also appears in Thomas' *Synopsis mathematica*.[53] In his *Outline of the essentials of calculation*, it reads:

> Again supposing there is a gold vessel weighing 160 *liang* 兩: it contains an unknown amount of silver. How much gold and silver does it contain? Having filled up a wooden pail with water, measure the overflowing water in cubic *cun* (*jian fang cun* 見方寸). If there are 12 cubic *cun*, multiply the weight of one cubic *cun* of gold, 16 *liang* 8 *qian* 錢,[54] by 12 *cun*, one gets 201 *liang* 6 *qian*; compare it with the previous number, it is 41 *liang* 6 *qian* more. Again multiply the weight of one cubic *cun* of silver, 9 *liang*, by 12; one gets 108 *liang*; compare it with the previous number, again it is 52 *liang* less. Add this number with the 41 *liang* 6 *qian* more; one gets 93 *liang* 6 *qian*; this is the 1st *lü*. 12 cubic *cun* is the 2nd *lü*. The less 52 *liang* is the third *lü*. Multiply the 2nd *lü* by the 3rd *lü*; divide by the 1st *lü*; one gets the 4th *lü*, that is 6 cubic *cun* 666 cubic *fen* 分 reinforced, this is the quantity of gold.[55] Subtract this quantity from 12 cubic *cun*; one gets 5 cubic *cun* 334 cubic *fen*; weakened, this is the quantity of silver. Then the weight of gold is 111 *liang* 9 *qian* 8 *fen* 分, reinforced value. The weight of silver is 48 *liang* 2 *fen*, weakened value. The total weight of gold and silver is 160 *liang*, which matches the original number.[56]

The solution is similar to that of the chickens and rabbits problem, with an additional step at the beginning, where the volume of the vessel, which is not given in the problem, is assessed by dipping it into water. Archimedes is mentioned neither here nor in the *Instructions for calculation in common script*. In fact, it is unlikely that this problem would have appeared to Chinese readers as 'Western' in the absence of explicit mention of its origin, since Chinese mathematical works

---

50 Dijksterhuis 1987, 18–19.
51 Guo 1993, 4: 177–178.
52 Liu 2002a, 162–163.
53 Thomas 1685, 31–32.
54 This value is given in the section entitled 'Practice of the five metals' (*Wu jin yongfa* 五金用法) of chapter 15; Gugong 13276–81, 202a.
55 *Qiang* 強 (reinforced) and *ruo* 弱 (weakened) correspond respectively to approximation by deficit and by excess: these terms refer to the fact that the exact value would in principle be obtained by slightly increasing (in the first case) or decreasing (in the second case) the number obtained by calculation.
56 Gugong 13276–81, 50b–51b.

contained problems of the same type, exemplified by the problem of the 'chickens and rabbits in a cage': they mostly belonged to the category 'Excess and deficit'. The seventh of the *Nine chapters on mathematical procedures* contains a problem similar to that of Hieron's crown, except for the fact that instead of gold and silver one deals with jade and stone.[57] Interestingly, the *Instructions for calculation in common script* also gives this latter problem.[58] The problem, like all those of the *Nine chapters on mathematical procedures*, is found in the *Great classified survey of the Nine chapters on mathematical methods* (*Jiuzhang suanfa bilei daquan* 九章算法比類大全, 1450); however it is not in Cheng Dawei's *Unified lineage of mathematical methods* (*Suanfa tongzong* 算法統宗, 1592).[59] Wherever Ricci and Li Zhizao 李之藻 found it, they changed the total weight of the object made of stone and jade but not the rest of the data. In Thomas' work, on the other hand, all the data are different, including the weights of a cubic *cun* of stone and of jade. In short, although the methods given in the *Outline of the essentials of calculation* include those of the *Instructions for calculation in common script*, Thomas did not simply copy the latter: rather than taking up earlier categorisations, he rewrote the problems, and reordered them according to the methods he used to solve them.

9.3 Practical geometry and tables

'The use of the main mathematical instruments and the practices of geometry', taught by Thomas, are discussed in short works that are appended to some of the extant versions of the *Outline of the essentials of calculation*.[60]

1. The *Basics of the eight [trigonometric] lines* (*Baxian biao gen* 八線表根) is a ten page explanation of the geometrical meaning of trigonometric lines, which shows how to derive them from one another, using the Pythagoras theorem and proportions; calculations are carried out for angles of 30°, 45° and 15°. This could be the material for the tutorial that Bouvet dated 27 April 1690:

The emperor had us called and repeated the explanation previously given to him of a few geometry propositions that serve to calculate the sines of various angles, and, having shown to us that he had understood these propositions, he sent us to eat.[61]

---

57 Liu 2002a, 161–162.
58 Guo 1993, 4: 155.
59 Guo 1993, 1: 174–175; 2: 237.
60 Bouvet 1697, 127; Gugong and BML, Ms.82–90 C; the copies kept at the Gugong are accessible in a facsimile publication: Gugong bowuyuan ed. 2000, 403: 275–486; a few other titles, some of which pertain to astronomy rather than mathematics in the modern sense, are listed in Héraud 1993, Annexe A: 12–18.
61 BnF, Ms. Fr. 17240, f. 272 v.

2. The *Numerical tables* (*Shubiao* 數表) are of two kinds: logarithms of numbers from 1 to 10,000; sine, tangent and secant, with the logarithms of sine and tangent of angles from 1° to 90°, for each minute of arc. These are probably the tables that Thomas had copied in Chinese at the Palace in mid-March 1690.[62] Their source could be Vlacq: the format and layout are identical, but whereas the Latin tables give eight significant digits, and the Chinese ones only give six.[63] On the 8th of April these tables were completed: during the tutorial on that day the emperor had 'brought to him the tables of sine, tangents and secants with their logarithms that F. Thomas caused to be put in Chinese characters for him'.[64] Three years later, the emperor had these tables supplemented by de Fontaney and Visdelou, two of the 'King's Mathematicians' who had not settled in Beijing in 1688, but who were summoned back there in 1693. Visdelou recounted it with a touch of irony that contrasts with the reverent tone of Bouvet and Gerbillon's diaries:

> We have had the honour of seeing His Majesty twice in his Versailles.[65] The first time he showed us its beauties, then for our amusement he commissioned us to reduce in tables of absolute numbers and logarithms the proportions of lines, solids and regular bodies. Both times we presented our tables to him. He examined them and confronted them with those that F. Antoine Thomas had made, for the latter had not done those of the inscribed and circumscribed solids, which are much more difficult than the others.[66]

There is a hint of condescension here: Visdelou implies that he and de Fontenay carried out the most difficult part of a task that Thomas had left incomplete. So that some of the manuscripts discussed in the present chapter may have resulted from joint or successive effort by more than one Jesuit, not to mention the Chinese and Manchu who worked with them, whose names remain unknown to us.

3. The *Practice of numerical tables* (*Shubiao yongfa* 數表用法) has the same structure as the *Outline of the essentials of calculation*. Its problems are solved with the same methods, but logarithms as given in the *Numerical tables* are systematically used for all multiplications, divisions and root extractions. Again the content of this work is

---

62 BnF, Ms. Fr. 17240, f. 266 r.
63 Ozanam 1685 has the same format as Vlacq 1670 ; both are found in the catalogue of the Beitang Library catalogue; Verhaeren 1949 (Ozanam: no. 533; Vlacq: no. 3061). However Thomas could only have brought Vlacq's tables with him, since Ozanam's were published the year he arrived in Beijing; there is evidence that Vlacq's tables were used later under imperial patronage (Han 1991, 40).
64 BnF, Ms. Fr. 17240, f. 269 v.
65 The French Jesuits seem to have commonly referred to the Garden of pervading spring residence in this way among themselves; von Collani 195, n. 423.
66 Letter to Bouvet; von Collani 2005, 195.

mentioned several times in Gerbillon and Bouvet's diaries between March and May 1690; the latter reports that on 31 March:

[The emperor] having asked the way to use the table of logarithms that he had had put in Chinese recently, F. Gerbillon showed him the way to use them for division, which His Majesty understood perfectly after he had practised 2 or 3 examples, which gave him extraordinary joy.[67]

According to Gerbillon's diary for that date, it was the use of logarithms for multiplication that he explained to the emperor, who 'expressed esteem for this invention and pleasure in knowing its usage'.[68] The use of logarithms occupied the emperor for several days; on 2 April he 'encountered some difficulty in one of them because the logarithm answering exactly the number he was looking for was not in the table'.[69] On 2 May he was still doing 'practices of arithmetic using logarithm tables'.[70]

4. The *Explanation of the proportional compass* (*Biligui jie* 比例規解) bears the same title as a treatise written in 1630 by Johann Adam Schall von Bell and Giacomo Rho when they were working on the calendar reform. This earlier text was based on Galileo's *Le operazione del compasso geometrico e militare* (1606).[71] It opened with a description of the Galilean compass and of its construction, and explained the various graduations inscribed on the two legs that give the result of several operations.[72] The text written for Kangxi, on the other hand, only explains the use of the graduations. Like the *Outline of the essentials of calculation*, this book is written as a series of problems; the solution of each problem entails the use of the proportional compass instead of calculations. Here again Bouvet mentions the use of an instrument in various sessions of tutoring. On 13 April 1690,

[...] Having proposed a problem on solids to be solved by calculation and the proportional compass, His Majesty took the trouble to calculate himself, and some divergence was found between the two ways; so that the emperor, in order to see where the error was, had some kind of small beans brought that he had measured in his presence; and we took this opportunity to make him understand that the proportional compass could not approach the accuracy of calculation, which His Majesty did not suppose, because, he said, the compass was made according to the rules of geometry, which are in no way subject to error.[73]

---

67 BnF, Ms. Fr. 17240, f. 268 r.
68 Du Halde 1736, 4: 276.
69 BnF, Ms. Fr. 17240, f. 268 v.; Ozanam 1685, 40–41 explains how to deal with numbers greater than 10,000.
70 BnF, Ms. Fr. 17240, f. 276 r.–273 v.
71 Standaert 2001, 740.
72 Jami 1998, 665.
73 BnF, Ms. Fr. 17240, f. 270 r.

This is typical of the initiatives taken by Kangxi during the tutoring sessions to check the accuracy of what he was taught; it also reveals his high expectations concerning geometry as a field of certainty. Bouvet mentions various other explanations of the compass, including one that the emperor asked them to repeat to Zhaochang; these went on until February 1691.[74] A number of proportional compasses that must have been used during these sessions are kept at the Imperial Palace Museum. Some of them had been brought from Europe, others were made in the imperial workshops.[75]

5. The *Practice of instruments for measuring heights and distances* (*Celiang gaoyuan yiqi yongfa* 測量高遠儀器用法) explains how to assess heights and distances with surveying instruments, in different situations. This seems to have been one of the applications of mathematics that the emperor enjoyed and valued most; both Gerbillon and Thomas assisted him in operations of survey during his trips on which they accompanied him.[76] Gerbillon described how Kangxi practised the use of instruments at the Hall of the Nourishment of the Mind during one of the first tutoring sessions, on 24 March 1690:

> The Emperor [. . .] tried with us a divided circle one foot in diameter, which he had had made during his absence, to measure small heights and distances: that circle also had a divided geometric square inside it, so that one did not have to use sines to solve triangles. He then tried in the courtyard of that apartment a semi-circle, which the late Father Verbiest had once had made, and which His Majesty had had repaired recently and put on a good knee, done on the pattern of that of the semi-circle which I had presented to His Majesty, and he computed this operation straightaway on his Souan pan (*suanpan* 算盤, abacus), so fast that it took Father Thomas longer than him to compute it with our figures.[77]

Here again Kangxi took initiatives; his outdoing Thomas—who was after all in charge of teaching him arithmetic—in calculation thanks to the use of the Chinese abacus was one of many reminders to the Jesuits that theirs was no ordinary student, and that they should never assume that he knew less than they did.

6. The *Tables of distance between the horizon and the terrestrial sphere* (*Dipingxian li diqiu mian biao* 地平線離地球面表) are also an aid to surveying. They give the distances for up to 5 terrestrial degrees. Here each terrestrial degree corresponds to 200 *li*: these tables were calculated or revised in relation to the emperor's decision to set the length of

---

74 BnF, Ms. Fr. 17240, ff. 270 v., 275 v., 278 v., 288 v.
75 Liu 1998, no. 64–73.
76 See Du Halde 1736, 4: 302–528.
77 Du Halde 1736, 4: 274; Jami 2002b, 32–33.

the *li* so that there would be 200 *li* to one degree of terrestrial meridian. This suggests that the Jesuits' composition of mathematical lecture notes for the emperor extended over a decade.

9.4 The foundations of calculation: back to Euclid

Several versions of the manuscript *Elements of geometry* contain as an addendum a shorter treatise entitled *Elements of calculation* (*Suanfa yuanben* 算法原本), which forms a counterpart to the *Elements of geometry*. Both Manchu and Chinese versions of it are extant.[78]

With the *Elements of calculation*, the Jesuits' tutoring tackled themes that had not been previously discussed in their Chinese mathematical works. The treatise deals with what is today called 'number theory', which is also the subject matter of books VII to IX of Euclid's *Elements*. The text is organised like Gerbillon and Bouvet's *Elements of geometry*, although it consists of a single *juan* 卷: it contains 75 items (*jie* 節). Of these, items 1 to 49 closely correspond to the content of book VII of Euclid's *Elements*; they follow the same order, with several definitions often grouped together in one item.[79] Items 50 to 56 contain propositions from book VIII. Item 58 corresponds to the first proposition of book IX. Items 60 to 75 deal with arithmetical progressions.

Despite the similarity with some of the contents of Euclid's *Elements*, the *Elements of calculation* is not a straightforward translation. Again as in Gerbillon and Bouvet's treatise, most items consist of a general statement introduced by *fan* 凡 ('any', 'all'), an instantiation introduced by *sheru* 設如 ('supposing') and a justification introduced by *heze* 何則 'why') and often concluded by *kezhi yi* 可知矣 ('can be known').[80] As an example, let us compare item 1 with its counterpart in Euclid's *Elements*. The Chinese text reads as follows:

One is the root of numbers. A multiplicity of ones combined together is called a number. However there cannot but be inequalities in the amounts or sizes of numbers, and in order to homogenise these one must have a definite method; only then can one investigate their norms. Therefore if a small number added to itself repeatedly is equal to a large number, this small number is said to be a norm (*zhun* 準) that exhaustively measures the large number. Suppose the large number is 8 and the small number is 2; adding 2 to itself 3 times must be equal to 8; then the small number 2 is a norm that exhaustively measures the large number 8. If the small number added to itself several times cannot be equal to the large number, this small number is said not to be a norm that exhaustively measures the large number. If the large number is 8, the small number is 3, 3 added to itself once is 6, which is smaller than 8; added to itself twice it is 9, which again is larger than 8. In such a case it is not a norm exhaustively

---

78 The Manchu version is entitled *Suwan fa yuwan ben bithe*; see Bibliography 1, p. 396.
79 Han 1991, 29–31.
80 See Bibliography 1, p. 396.

measuring the large number. But when a small number can be a norm that exhaustively measures a large number, the small number is called one equal part of the large number. If a small number cannot be a norm that exhaustively measures a large number, the small number is called an unequal part of the large number. Therefore 2, which can measure 8, is one equal part of 8; 3, which cannot measure 8, is not one equal part of 8. The method of obtaining equal parts is the only one in searching for an exact, faultless accord in the measurement of the many by the few, of the large by the small.[81]

This is quite different from the first four definitions of book VII of Euclid's *Elements*, for which there had been no Chinese translation:

1. A unit is that by virtue of which each of the things that exist is called one.
2. A number is a multitude composed of units.
3. A number is a part of a number, the less of the greater, when it measures the greater;
4. but parts when it does not measure it.[82]

The concision of the Euclidean definitions contrasts with the profusion of details, including a numerical example, of the Chinese textbook. The notion of unit, which is defined at the opening of book VII of the Greek work, is absent from the *Elements of calculation*, where 1 takes on the status of a number. This is evident for example from the phrase used to render 'prime number': *yi wai wu shu du jin shu* 一外無數度盡數, literally 'number not exhaustively measured by any number except one'. Number (*shu* 數), on the other hand, is defined in a way similar to that of Euclid's *Elements*. The notion of 'measuring' is not defined there: it applies to both numbers and magnitudes. Neither does it appear in *juan* 6 of Gerbillon and Bouvet's *Elements of geometry*, where ratios are defined directly as stemming from comparison between quantities. In the *Elements of calculation*, on the other hand, the notion of 'measuring exhaustively' is defined, relying on addition as a prerequisite.

The *Elements of calculation* seem to have been derived from Euclid's *Elements* with the aim to produce a counterpart to the *Elements of geometry* and to provide foundations for arithmetic, or, rather for calculation—to use a more literal translation of *suanfa* 算法. The separation of these two fields was consistent with the view of some mathematics teachers in seventeenth century Europe that books VII to IX did not belong in the *Elements*. Thus, Claude-François Milliet Dechales, whose geometry textbook was entitled *Elemens d'Euclide*, did not regard the arithmetic books of Euclid as relevant:

I leave out the seventh, eighth, ninth and tenth books of Euclid's *Elements* because they are useless in most parts of Mathematics. I have often

---

81 BML, Ms. 82–90 C, 1a–1b.
82 Euclid 1956, 2: 277.

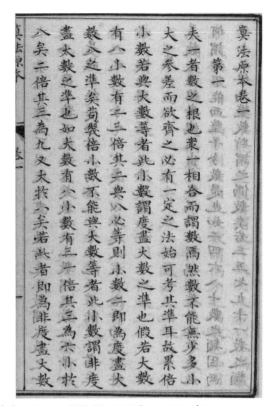

Illustration 9.2 Opening paragraph of the *Elements of calculation* (item 1; BML, Ms. 82–90 C, *juan* 1: 1a–1b).

wondered at the fact that they have been included among the *Elements*, since it is obvious that Euclid only composed them in order to establish the doctrine of incommensurables, which, being only a vain curiosity, should not have been placed among the *Elements*, but should instead form a separate treatise.[83]

Neither, it seems, did Gerbillon and Bouvet regard incommensurables, discussed by Pardies in the seventh book of his work, as a part of geometry.

The authorship of the *Elements of calculation* cannot be determined with the same certainty as that of most of the treatises discussed so far. However, several clues can be considered. The similarity of style with the *Elements of geometry* points in the direction of the French Jesuits, as does the fact that the *Elements of calculation* is mostly found appended to copies of the *Elements of geometry*. However it seems unlikely that Gerbillon and Bouvet's systematic report—or should one say advertising?—of their work for the emperor should not have

---

83 Dechales, 1677, 325; see p. 162.

mentioned this treatise among their production. Furthermore, the *Elements of calculation* relies on Euclid's *Elements*, which runs contrary to their choice of 'Modern' textbooks. On the other hand, the fact that Thomas was in charge of arithmetic makes him a plausible author. It is possible that the teaching of Euclid by Thomas to which Bouvet referred in March 1690 was in fact that of the arithmetical books of the *Elements* rather than that of the first six books.

A comparison between the Manchu and the Chinese versions of the manuscript does not contradict this hypothesis: given that the French were asked to translate Pereira's treatise on music from Chinese into Manchu,[84] they could equally well have been commissioned to translate the *Elements of calculation*. The type of corrections found in the Manchu version suggests that it was translated from Chinese. The term used to render 'to measure' in Chinese is *dujin* 度盡, literally 'to measure and exhaust'; it seems to have been coined for the *Elements of calculation* and is not used elsewhere. In the Manchu manuscript, there are three successive renderings for the same term: 'to measure exactly', 'to measure exhaustively', and 'to measure and exhaust'.[85] The corrections resulted in a closer match between Manchu and Chinese. Another point suggests more directly that the Manchu manuscript was translated from Chinese, rather than composed by someone familiar with European languages. Manchu can express the equivalent of 'larger' and 'smaller' as can Latin and French; Chinese, on the other hand, has no flexion, so that one simply uses *da* 大 (large) and *xiao* 小 (small). The use of the positive rather than comparative degree of adjectives in the Manchu version when two numbers are compared would therefore suggest that this version is based on a language in which there is no specific superlative form for adjectives, in this case Chinese rather than Latin or French.[86]

Both versions of the *Elements of calculation* include a relatively long preface. The parallel with the *Elements of geometry* suggests that the emperor must have written it. In contrast with the preface of the *Elements of geometry*, this one emphasises the importance of mathematics within classical learning, referring both to the relation between numbers (*shu* 數) and principles (*li* 理) and to the categorisation of Chinese society into warriors, peasants, craftsmen and merchants (*shi nong gong shang* 士農工商):

The principles of numbers are extremely subtle; the uses of numbers are extremely great. Therefore, although reading widely and examining thoroughly is in every respect a scholarly occupation, understanding numbers is a priority for intelligent gentlemen. For therein must lie the beginning of the

---

84 See p. 158.
85 Watanabe 2005a, 189–190.
86 Watanabe 2005a, 190.

investigation of things to extend one's knowledge (*gewu zhizhi* 格物致知).[87] If one can truly clarify numbers, then one's knowledge will necessarily be solid (*shi* 實). If knowledge is solid, then in dealing with affairs one will necessarily be reliable beyond doubt. When there is no doubt, all affairs are completed. If numbers cannot be clarified, then knowledge will necessarily be empty. If knowledge is empty, then when you encounter [real] entities there is much confusion and all affairs wane. Moreover, in the infinity of the universe, in the minuteness of daily life, there is nothing that does not dwell in numbers. In terms of the subtlety of their principles, there are always definite principles and definite numbers for the heavenly and earthly motions, the dimensions of the Earth, the Sun and Moon's periods, the celestial bodies' trajectories, the planning of human affairs, and the waxing and waning of all things. If one can clarify the principles of these numbers and explain in detail their methods of use, then by inference one can always process numbers according to a method, or determine a method from numbers. In terms of the greatness of their use, understanding numbers for the state enables one to know the national production and consumption, the incomes and expenses in money and cereals. Understanding numbers for commandment enables one to advance or retreat in a timely manner, and to deploy the appropriate number [of troops]. If peasants know numbers, then they get to plough and weave at appropriate times. If merchants know numbers, then they distinguish interest from capital in their expenses and revenue. As for craftsmen, only those who know the marking line can obtain standard squares and circles; only artisans familiar with the square and compass manufacture exquisite work. Thus for various necessary uses of innumerable kinds, discarding numbers is not possible. One must practise their methods; one must also know their principles in detail. If one does not carefully investigate and thoroughly understand the particulars so that the principles and methods are known in depth, one has 'entered the gate but not yet the room'.[88] Seldom do we not fail because of shallow striving in what our hand writes, what our eyes see, what our thought gathers, and what our mind penetrates. This is why we have written this book, the *Elements of calculation*, in order that those who must study should first practice the general principles, then discuss the principles of each method so that these are known solidly and are applied without doubt. If one truly can develop these through practice, then one will never strive without reaching the way. This, thus, is the preface.[89]

This is a strong plea for the necessity of mathematics, as a basis both for sound knowledge and understanding of the world and for success in state as well as individual affairs. It integrates its study into scholarly pursuits that literati of the time would have recognised as legitimate. Two terms repeatedly used here are essential to this integration: principles (*li* 理) again link mathematics to philosophy, while the emphasis on the solidity (*shi* 實, also rendered by real, concrete, practical) of the knowledge gained in mathematics echoes a preoccupation often expressed by scholars since the late Ming period.[90]

---

87 This well-known phrase is a quotation from the *Great learning* (*Daxue* 大學).
88 A quotation from the *Analects* (*Lunyu* 論語).
89 BML, Ms. 82–90 C, 1a–1b; comp. Watanabe 2005a, 183–184.
90 Cheng 1997, 531–533.

The whole text is strongly evocative of Kangxi's attitude to mathematics: beside his insistence on relentless and thorough study, the mathematics lessons of the 1690s seem to have reinforced his belief in the universal usefulness of numbers.

9.5 Cossic algebra: the *Calculation by borrowed root and powers* and its Summary

Algebra was the most novel subject in the Jesuits' tutoring of the emperor. It is first mentioned in Bouvet's diary in March 1690:

The 30th. Father Thomas gave Tchao laoye [Zhaochang] the solution of two problems of algebra, it seems, in order to present them to the emperor. F. Gerbillon, to whom Tchao laoye mentioned this, solved them as well so as to be ready in case His Majesty happened to talk about them. [. . .]
   The 31st [. . .] His Majesty would not undertake to solve the two problems that had been presented to him.[91]

A few days later the subject is mentioned again:

11th [April. His Majesty] gave F. Thomas a problem of algebra to solve.[92]

At this stage the emperor had already heard about some fundamentals of algebra. However his systematic study of the subject seems to have been undertaken only in 1692, when he 'desired to be taught algebra, i.e. the various parts of computation, and various parts of common arithmetic.'[93] During the next three years, Thomas was probably assisted in his mathematical tutoring and writing by an Italian Jesuit, Alessandro Cicero (1639–1703), who resided in Beijing between 1692 and 1695.[94] The composition of the *Calculation by borrowed root and powers* only started after the emperor was cured of an attack of malaria by the cinchona provided by the French Jesuits in 1693:[95] Thomas mentions this in his correspondence in 1694. The emperor ordered him to work on the text at the Hall of the Nourishment of the Mind and forbade him to show it to anybody, as with Gerbillon and Bouvet when they wrote on anatomy. Grimaldi's return to Beijing in August 1694 relieved Thomas from his charge of interim Administrator of the Calendar and made him available for writing out a full treatise.[96] By the end of 1695 Thomas reported that he had been continuously

---

91 BnF, Ms. Fr. 17240, f. 268 r.
92 BnF, Ms. Fr. 17240, f. 269 v.
93 ARSI, Jap. Sin. 105 II, *Annus Christi 1692*, Cap. 3 num. II; I am grateful to Dr Noël Golvers for generously sharing information drawn from Thomas' correspondence.
94 Pfister 1932–34: 392.
95 On this famous episode, see e.g. Von Collani 2005, 79–84, Puente Ballesteros 2009, 235–249.
96 Ricci 21st Century Round Table (accessed 10 February 2006) http://ricci.rt.usfca.edu/biography/view.aspx?biographyID=632.

CHAPTER 9 | Calculation for the emperor: the writings of a discreet mathematician

occupied with the treatise of algebra for over a year and that he expected to complete it within three months.[97] However, in early 1696 Kangxi started preparing for a military campaign against the Oirats headed by Galdan (1644–1697), which he decided to lead personally.[98] Together with Pereira and Gerbillon, Thomas followed him West.[99] During this campaign the emperor regularly observed the Pole Star and took note of the plants and animals he saw, but we do not know whether he found the leisure to study and amend Thomas' algebra treatise until his return to Beijing in July 1697.[100] It is also unclear whether there were further sessions of tutoring after that. In any case, it took at least another three years to finish the treatise: in October 1700, Thomas mentions it once more, complaining that completion of the work has so far been postponed.[101]

The *Calculation by borrowed root and powers* (*Jiegenfang suanfa* 借根方算法) is a treatise on cossic algebra, in which the unknown and its powers are represented by abbreviations in equations. A century after Viète first introduced letters to represent both known and unknown numbers, the emperor studied algebra as it had been practised and taught in Europe since the Middle Ages.[102] This may be because algebra was not part of elementary mathematical education: Thomas had not included it in his *Synopsis mathematica*.

The algebra presented in the *Calculation by borrowed root and powers* resembles that presented in Clavius' *Algebra* (1608). However, like the *Outline of the essentials of calculation*, Thomas' algebra seems to have been composed rather than directly translated from a particular European work: so far its possible sources have not been identified. It is the longest of all extant mathematical treatises written by the Jesuits for Kangxi: it contains about 110,000 characters, in 547 folios. The work is divided into three sections (*juan* 卷): initial (*shang* 上), middle (*zhong* 中) and final (*xia* 下); these are in turn subdivided respectively into three, two and three parts. The first part of each section contains a number of items (*jie* 節); the second parts each consist of one single item, in which what precedes is used to 'determine the fundamental principle of each problem' (*ding ge sheru zhi benli* 定各設如之本理, lit. each 'supposing that'); these parts are divided into as many paragraphs (*duan* 段) as there are types of problems. Part 3 of the initial section is a continuation of its part 2, whereas part 3 of the final section consists of

---

97   ARSI, Jap. Sin. 148, f. 201v, 203 v. 205 v. Jap. Sin. 105 II, *Annus Christi 1692*, Cap. 3, num. II.
98   Romanovsky 1998, Oyunbilig 1999; Perdue 2005, 184–190.
99   Spence 2002, 154–155; Du Halde 1736, 4: 386–423, 448–482.
100 Perdue 2005, 185.
101 ARSI, Jap. Sin. 148, f. 564 r.
102 Reich 1994.

three items in which the methods given previously are applied to problems on base and altitude (*gougu* 句股, the equivalent of right-angled triangles in the Chinese tradition), the calculation of areas and solids respectively. Like the *Outline of the essentials of calculation*, the work closes with applications to geometry,

The text is structured in a way similar to that of other treatises described above. In the version studied,[103] general statements are introduced by *fan* 凡, instantiations with abstract numbers (i.e. given without units) by *jiaru* 假如, and those that take the form of problems by *sheru* 設如; questions are concluded by *ruogan* 若干 ('how much?'). They are immediately followed by *yushi* 於是 ('thereupon'), which announces the solution; justifications open with *heze* 何則 and are concluded by *ke zhi yi* 可知矣. References are made to the *Elements of calculation* in justifications; notes in small characters at the end of some problems indicate which method of the *Outline of the essentials of calculation* can alternatively be used to solve them. Thus the *Calculation by borrowed root and powers* appears as the closing work in a set of treatises that construct mathematics as a coherent whole.

The first part of the initial section defines the notations and rules of the operations performed on the borrowed root and powers. The second and third parts altogether contain 53 paragraphs, each of which discusses problems of a particular type. The middle section is devoted to fractions of borrowed root and powers. Although in the first part the root and powers are defined as possible denominators, only numbers are taken as denominators in the problems given in the second part. In the final section, problems equivalent to quadratic and cubic equations are solved, the former by radicals, the latter by calculating the successive digits of the root.

The treatise opens with a foreword (*bianyan* 弁言), which could have been authored by the emperor. It gives the reader some hints of the wonderful novelties that he is about to discover, so as to arouse his curiosity on this subject, which Kangxi has ensured has hitherto remained a mystery to him:

This method, called *algebra* (*aerrebala* 阿爾熱巴拉) in the West, is the acme of calculation. It differs widely from ordinary calculation. The idea on which it is based is all in all to scrutinise the number sought by means of a borrowed root: one first examines the whole number equal to one or several roots; if one obtains this number, then one can obtain the number sought. As for its use in reckoning, it is easier than other methods. All that cannot be obtained through ordinary methods can be obtained through this one. Not only is it especially convenient to use, but it is also excellent in astronomy. Calculations by the rule of three, differential distribution and base and altitude each have their own calculation pattern, and their methods must be followed; only then can one

---

[103] BML, Ms. 39–43; the copy kept at the Gugong is similar.

CHAPTER 9 | Calculation for the emperor: the writings of a discreet mathematician

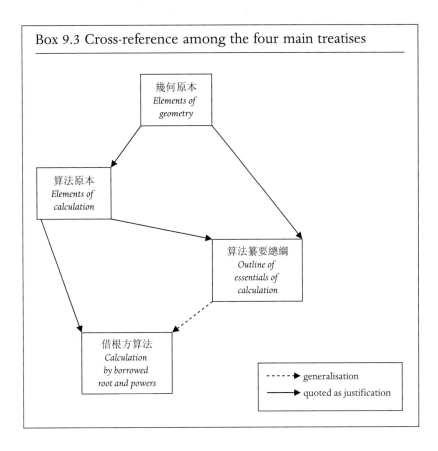

Box 9.3 Cross-reference among the four main treatises

master them. Yet these methods possess a common principle, which is infinitely adaptable and is not restricted to one method: one can always obtain the number sought. Thus the good hunter, when he sees what he is seeking, does not have to depend entirely on the usual encircling, but can get it by some ingenious idea of his own. The excellence of this method is very similar to that. The method is divided into three sections, initial, middle and final. The initial section clarifies the standards of this method. The middle section discusses the meaning of fractional borrowed root and powers. The final section explains the bases of the extraction of square, cube and other roots, and those with lengthwise [terms] in calculation by root and powers. For each of these there are problems (*sheru* 設如) so as to clarify things.[104]

The hunter's metaphor evokes the emperor's great taste for hunting. The notion of borrowing (*jie* 借) was not new in itself. It occurred in ancient Chinese texts in relation to a particular use of a counting rod in some procedure,[105] and in Jesuit works when an auxiliary number was

---

104   BML, Ms. 39–43.
105   Chemla & Guo 2004, 939.

Table 9.4 Contents of the *Calculation by borrowed root and powers*

	Initial section, part 1	
1.	定位之法	Determining rank
2.	定記號之名及用法	Determining the names and use of signs
3.	加法	Addition
4.	減法	Subtraction
5.	乘法	Multiplication
6.	除法	Division
7.	借根方公算法	Common calculation by borrowed root and powers
8.	以兩邊數平加減求與根方相等之真數法	Searching for an actual number equal to root and powers by addition and subtraction on both sides
9.	降位法	Lowering rank
10.	同乘法	Similar multiplication
11.	相除法	Joint division
12.	相比例法	Joint proportions
13.	參乘方開方法	Fourth root extraction
14.	肆乘方開方法	Fifth root extraction
15.	15- 伍乘方開方法	Sixth root extraction
16.	16- 陸乘方開方法	Seventh root extraction
17.	17- 柒乘方開方法	Eighth root extraction
18.	18- 捌乘方開方法	Ninth root extraction
19.	以本卷借根方算法算各所求之數	Calculating various numbers sought by the calculation by borrowed root and powers given in this section
	Initial section, part 2	
20.	以本卷借根方算法定各設如之本理（第一段至四十段）	Determining the basic principles of various problems by the calculation by borrowed root and powers given in this section (paragraphs 1 to 40)
	Middle section, part 1	
1.	有零數求同相比例之至小零數	Seeking the smallest fractional number in proportion with a given fractional number
2.	有兩零數之母不同變為相同之母數	Transforming the different denominators of two given fractional numbers into identical denominators
3.	加法	Addition
4.	減法	Subtraction
5.	乘法	Multiplication
6.	除法	Division
7.	零數之零數加法	Addition of a fraction of a fraction
8.	零數之零數減法	Subtraction of a fraction of a fraction
9.	開平方零數之法	Extraction of the square root of a fraction
10.	以本卷借根方算法算各所求之數用零數算各設如	Calculating numbers sought by borrowed root and powers given in this section, using fractions to calculate various problems [24 problems]
	Middle section, part 2	
11.	以本卷借根方算法定各設如之本理	Determining the basic principles of various problems by the calculation by borrowed root and powers given in this section [18 types of problems]

Final section, part 1

1.	論平方帶縱根數	Discussion of a square with a lengthwise root term [six types]
2.	有一平方或多根數或少根數與真數相等開方法	Extraction of one square plus a root term or minus a root term equals an actual number [$x^2 \pm mx = n$][1]
3.	有一平方與多根數或多真數或少真數相等開方法	Extraction of one square equals a root term plus an actual number or minus an actual number [$x^2 = mx \pm n$]
4.	有平方數帶縱開平方法	Square extraction of a square term with a lengthwise [term] [$mx^2 \pm nx = p$; $mx^2 = n \pm px$]
5.	開立方帶縱平方數之法	Extraction a cube with a lengthwise square term [$x^3 \pm mx^2 = n$]
6.	開立方帶縱平方及根數之法	Extraction of a cube with lengthwise square and root terms [$x^3 \pm mx^2 \pm nx = p$]
7.	開立方帶縱根數之法	Extraction of a cube with a lengthwise root term [$x^3 \pm mx = n$]
8.	有立方數帶縱開立方法	Cube extraction of a cubic term with lengthwise [terms] [$mx^3 \pm nx^2 \pm px = q$]
9.	有平方數帶縱少立方開方法	Extraction of a lengthwise square term minus a cube [$mx^2 - nx^3 = p$]
10.	平方之平方帶縱平方數開方法	Extraction of a square's square with a lengthwise square term [$x^4 = mx^2 + n$]
11.	有平方之立方帶縱立方數開方法	Extraction of a square's cube with a lengthwise cubic term [$x^6 \pm mx^3 = n$]
12.	以本卷之法算各所求之數	Calculation of various numbers sought by the methods of this section [26 problems]

Final section, part 2

13.	以本卷帶縱開平方開立方法定各設如之本理	Determination of the basic principles for various problems by cube and square root extractions with a lengthwise [term] given in this section [24 paragraphs]

Final section, part 3

14.	勾股弦各較算法	Relative calculation of various bases, altitudes and hypotenuses
15.	算各方面積較法	Relative calculation of various rectangular areas
16.	算各方體積較法	Relative calculation of various rectangular volumes

1 For the convenience of the reader, I have added a translation into symbolic algebraic notations between square brackets; it should be kept in mind that there are no negative numbers.

used in a procedure, as in the rule of false position.[106] Root (*gen* 根), on the other hand, had not hitherto been part of mathematical terminology: it is a literal translation of *radix* (Latin) or *racine* (French). In sum, the reader is given to understand that what can be learnt from this

---

106 See p. 28.

treatise subsumes all that is contained in the *Outline of the essentials of calculation*.

The first section, which introduces the root and powers and explains how to perform operations on them, opens with a table that gives the names and ranks of the root and powers used thereafter:

The 'short names of the root and power numbers' (*gen fang shu zhi jie ming* 根方數之捷名) correspond to the traditional Chinese names of powers: these are characterised by the number of multiplications from which they result; the elaborate forms of Chinese numerals are used for these abridged names. The 'names of the root and powers', on the other hand, are Chinese renderings of those used in cossic algebra (Table 9.5). Thus the notations used in Clavius' *Algebra* were derived from abbreviations of the names for powers, as in Table 9.6.

This table is quite similar to the one in Thomas' work. The names given to the successive powers in Chinese, however, are not direct translations of those used by Clavius: different European treatises on cossic algebra used different names for the successive powers. Carlo Renaldini (1615–1679) gave quite an extensive list of these in his *Opus mathematicum* published in 1657. The date of this work suggests that cossic algebra was still used and taught in Europe in the middle of the seventeenth century: Antoine Thomas could have studied it in Europe. Among the terms given by Renaldini, one finds plausible Latin antecedents for most Chinese terms coined by Thomas (Table 9.7).

Some of the Chinese terms are ambiguous: the character *er* 二 ('two') for example can indicate either an increase of rank by 2 (corresponding to the Latin *super*), as for ranks 7 and 8, or the squaring of a term (corresponding to the Latin *bi-* or to the iteration of a word), as for ranks 4 and 9. In Tables 9.5, and 9.7, *er* has been rendered by 'second' in the former case and by 'double' in the latter. In Thomas' treatise explanations in small characters are given only for the latter sense: a double square and double cube, which are respectively the square of the square and the cube of a cube.[107] But *er shang lifang* 二上立方, for example, could be read either as 'double upper cube' or as 'two upper cubes'. This must be the reason why this first item of the treatise closes on the following note:

We fear that using the names of the root and power numbers in the following calculations might be muddled and confusing. Quick names have been devised specially, with their number written in the elaborate form to distinguish them, so that hopefully it should be easy.[108]

---

107 The only other term explained in small characters is 'square of a cube', which is the same as the cube of a square.
108 BML, Ms. 39–43, *juan* I-1, 3b.

Table 9.5 Table of power numbers (transcription and translation), *Calculation by borrowed root and powers* (BML, Ms. 39–43, Initial Section 1a–1b)

九	八	七	六	五	四	三	二	一	定位數根方數	○真數方一數
五一二	二五六	一二八	六四	三二	一六	八	四	二		一
二立方	二立方之平方	二上立方	立方之平方	上立方	二平方	立方	平方	根	根方之名數	
捌乘方	柒乘方	陸乘方	伍乘方	肆乘方	參乘方	立方	平方	根	根方捷名數	

9	8	7	6	5	4	3	2	1	number fixing the rank	0 actual number
512	256	128	64	32	16	8	4	2	root and power numbers	1
double cube	second square of the cube	second super cube	square of the cube	super cube	double square	cube	square	root	names of the root and power numbers	
8 multiplication power	7 multiplication power	6 multiplication power	5 multiplication power	4 multiplication power	3 multiplication power	cube	square	root	short names of the root and power numbers	
x^9	x^8	x^7	x^6	x^5	x^4	x^3	x^2	x	modern notation	

However we are told that when multiplying or dividing the various powers by one another, it is the ranks that should be used, added or subtracted accordingly: neither the 'names' nor the 'quick names' allow for direct reading of the addition of ranks that yields the product of two powers.[109]

Symbols for addition, subtraction and equality are introduced next: they are slightly modified from those used in European works, with which we are still familiar today, so as to avoid the risk of confusion with the characters *shi* 十, *yi* 一 and *er* 二 (ten, one and two) respectively. For example, what we would now write $x^2 + 10x - 119 = 432$ was written as shown in Box 9.8.

---

109 BML, Ms. 39–43, *juan* I–1, 2b–3b.

Table 9.6 Abbreviations for the names of powers in Clavius' *Algebra* (1608) (BnF, Paris)

0	1	2	3	4	5	6	7	8	9	10
N	℞	ʒ	ce	ʒʒ	ſ	ʒce	ℬſ	ʒʒʒ	cce	ʒſ
1	2	4	8	16	32	64	128	256	512	1024

11	12	13	14	15	16	17	&c
Cſ	ʒʒce	Dſ	ʒℬſ	ceſ	ʒʒʒʒ	Eſ	&c
2048	4096	8192	16384	32768	65536	132072	&c

Clavius, *Opera Mathematica*[1]

Table 9.7 Terminology of cossic algebra in Latin and in Thomas' Chinese treatise

Rank	Latin term[1]	Thomas' 'names for the root and power numbers'	
0	Numerus simplex, & absolutus	actual number	真數
1	radix	root	根
2	quadratus	square	平方
3	cubus	cube	立方
4	biquadratus	double square	二平方
5	supersolidus	super cube	上立方
6	cubi-quadratus	square of the cube	立方之平方
7	supersolidus secundus	second super cube	二上立方
8	—[2]	second square of the cube	二立方之平方
9	cubi-cubus	double cube	二立方

1 Given by Renaldini 1657, 15–16.
2 A tentative reconstruction from the Chinese would be *cubi-quadratus secundus*; the various terms given by Renaldini, however, all refer to three iterations of squaring.

Let us give an example of how a problem is stated successively in general terms, then with abstract numbers:

There is an original number; one wants to divide this number into two parts. First, let us fix that the first part being multiplied by some number, and the second part being multiplied by some number, the two products obtained are equal (or one is a multiple of the other). One seeks how much these two part numbers are.

Suppose there is an original number one hundred; one wants to divide this number into two parts. Multiplying the first part by thirteen, and the second part by six, the two products obtained are equal. One seeks how much these two part numbers are. Thereupon borrow one root to be the first part number. [. . .][110]

The problem is solved by expressing the equality between the two parts in terms of the borrowed root. The calculations are described, each

---

110   BML, Ms. 39–43, *juan* I-3, 8b–9a (paragraph 44).

CHAPTER 9 | Calculation for the emperor: the writings of a discreet mathematician

---

Box 9.8 Equation and transcription

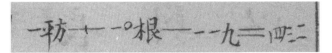
1 square + 10 roots − 119 = 432

Calculation by borrowed root and powers
BML, Ms. 39–43, juan III-2, 25b.

---

intermediary expression being added horizontally after it as been stated in the text. A note in small characters is added at the end:

(This method can also be derived similarly using the 'Method of repeated borrowing for mutual comparison'[111] in the *Outline*.)[112]

Finally, after this step by step description of the solution, the calculations are laid out as shown in Box 9.9.

A 'concrete problem' follows the one given with abstract numbers:

Supposing there are coins of gold and silver, altogether 1295. The value of each gold coin is 1 [copper] tael, 8 [copper] coins and 5 [copper] cents. The value of each silver coin is 7 [copper] coins and 4 [copper] cents. The total value of the silver coins is seven times the total value of the gold coins. One asks how many coins of gold and silver there are.[113]

This is solved in the same way as the question on abstract numbers given before. For every question and problem, the resolution is given rhetorically before being summarised in a layout. It looks as though all the oral explanations given during the tutorials have been transcribed and form the bulk of the text.

An abridged version of this algebraic treatise, the *Summary of calculation by borrowed root and powers* (*Jiegenfang suanfa jieyao* 借根方算法節要), in a single volume of 53 folios, was also composed.[114] It consists of excerpts from the initial and final sections of the longer treatise: items 1 to 12 and 19 of the initial section, and items 1 to 5, 12, and 14 to 16 of the final section. This abridgment leaves out fractions, as well

---

111 Item 10 of the *Outline of the essentials of calculation*.
112 BML, Ms. 39–43, juan I-3, 9b.
113 BML, Ms. 39–43, juan I-3, 10a. This problem must have appeared rather exotic in the Chinese context, as gold was not used as a currency and silver circulated in the form of ingots rather than coins. Copper taels should not be confused with silver taels, which were worth about one hundred times more. There were no coins worth a copper tael or a copper cent: these were just accounting units.
114 BML, Ms. 82–90 B.

The Emperor's New Mathematics

> **Box 9.9 Layout summarising the solution of a problem**
>
>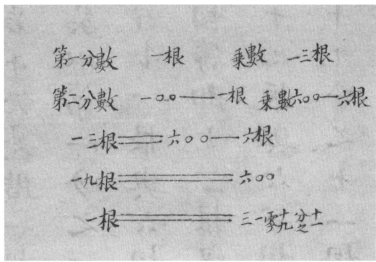
>
> | | | | |
> |---|---|---|---|
> | 1st part | 1 root | product | 13 roots |
> | 2nd part | 100−1 root | product | 600−6 roots |
> | | 13 roots = 600−6 roots | | |
> | | 19 roots = 600 | | |
> | | 1 root = 31 11/19 | | |
>
> Supposing there is an original number 100, one wishes to divide into two parts. Multiplying the first part by 13, and multiplying the second part by 6, the two products are equal. Find how much the two parts are.
>
> *Calculation by borrowed root and powers*
> BML, Ms. 39-43, *juan* I-3, 9b.

as roots and equations above the second degree. No evidence has been found so far as to the particular circumstances in which it was produced.

9.6 Symbolic linear algebra: a treatise within the treatise

Towards the end of the initial section of the *Calculation by borrowed root and powers*, in the last five paragraphs, a different notation is introduced into the solution of linear problems; it is neither discussed nor even mentioned in previous explanations. The problem to be solved is the following:

One wishes to seek two numbers. First, let us fix that multiplying the first number by some number, the second number by some number, the sum of the two products is fixed as being a certain amount. Again multiplying the first number by some number, the second number by some number, the sum of the two products is fixed as being a certain amount. One seeks how much is each of these two numbers.[115]

---

115 BML, Ms. 39–43, *juan* I-3, 28a.

This is a general statement of a linear problem in two unknowns; it is given in terms similar to all previous, more specific problems. Its form does not seem to call for a solution different from those given just before. In the solution of the numerical instantiation, a new notation is nevertheless proposed:

Suppose one wishes to seek two numbers. If the first number is multiplied by 7, the second number is multiplied by 11, the sum of these two products is 130. Again, the first number is multiplied by 12, the second number is multiplied by 5, the sum of these two products is 112. One seeks how much each of these two numbers is. (This method uses the characters *jia* 甲, *yi* 乙, *bing* 丙, *ding* 丁, not the root number.) Use *jia* as the first number, *yi* as the second number. [...][116]

In the following paragraphs, problems involving up to seven numbers sought are solved, using the cyclical characters to denote the numbers sought (as in Box 9.10).[117] These are processed as the root has been previously: they are added and subtracted from one another, multiplied and divided by numbers.

Introducing symbols for linear equations within a treatise of cossic algebra was not Thomas' idea: it was common practice in European treatises;[118] Clavius had done so in his. In any case, one could hardly imagine a less spectacular way to usher in symbolic algebra, so widely celebrated as one of the great European inventions that laid the foundations of modern mathematics, and regarded as one of the branches of mathematics sorely missed by China until the nineteenth century. The fact is, however, that its introduction was quite unremarkable, and one can hardly be surprised that Thomas' symbolic notation went unnoticed. Fundamental as this notation appears to us today, it was but a minor point in the Jesuits' tutoring to Kangxi.

The analysis of the mathematical lecture notes shows that the main contributor to them was Antoine Thomas rather than the two French Jesuits, although his work was much less advertised than theirs at the time, and has as a consequence remained little known to the present day. Unlike Gerbillon and Bouvet, Thomas represented continuity with Clavius' mathematics and with the Jesuits' late Ming heritage: he was closer to the Ancients than to the Moderns in his teaching of the emperor. After all, he worked with Verbiest for three years, and derived his authority as an imperial tutor from his role as a continuator of Verbiest's science, in contrast with the French who made a point of representing new, 'French,' science at the court.

---

116 BML, Ms. 39–43, *juan* I-3, 28b.
117 BML, Ms. 39–43, *juan* I-3, 85a–86b; see Jami & Han 2003, 101.
118 I am grateful to Prof. Kenneth Manders for drawing my attention to this point.

> Box 9.10 Layout of the linear problem in seven unknowns using symbolic notation (BML, Ms. 39–43, I-3, 85a)
>
>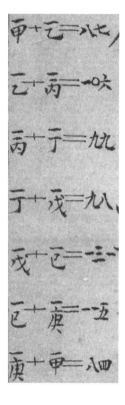
>
> $$1A + 1B = 87$$
> $$1B + 1C = 106$$
> $$1C + 1D = 99$$
> $$1D + 1E = 98$$
> $$1E + 1F = 131$$
> $$1F + 1G = 115$$
> $$1G + 1A = 84$$
>
> Suppose seven persons each have silver. One does not know how much each has, but one knows that the total silver that the first and second have is 87 taels, the total silver of the second and the third is 106 taels, the total silver of the third and the fourth is 99 taels, the total silver of the fourth and the fifth is 98 taels, the total silver of the fifth and the sixth is 131 taels, the total silver of the sixth and the seventh is 115 taels, the total silver of the seventh and the first is 84 taels. One asks how much silver each person has.

Taken as a whole, the lecture notes reflect what a mathematics teacher of a European Jesuit college would have transmitted to his students, rather than the discipline as it was practised in the small milieu of the *Académie Royale des Sciences*. In fact the case of mathematics illustrates the fact that the transmission of scientific knowledge did not proceed from the practice of increasingly professional milieus such as that of the *Académie,* but from the practice of college teachers. In Beijing their audience was reduced to one student: the emperor himself. But, one might ask, what use did the latter make of his knowledge of geometry and calculation? Did he carry out the plan he mentioned to the Jesuits to publish their sciences in his empire? The answer to these questions must be sought outside the Palace.

CHAPTER 10

Astronomy in the capital (1689–1693): scholars, officials and ruler

10.1 Mei Wending in Beijing

While the emperor studied with the Jesuits, interest in the sciences also grew outside the Palace. In 1689, the officials in charge of compiling the *Ming History* (*Mingshi* 明史)[1] asked Mei Wending to work on the draft of the chapters to be devoted to astronomy; it is likely that his name had been suggested by Shi Runzhang, whom Mei had assisted in the compilation of the local gazetteer of their common home district in the 1670s; in the meantime Shi had been appointed to work on the *Ming History*.[2] Mei, now in his mid-fifties, travelled to the capital, where he remained for four years. At first he found lodgings with Mei Xuan 梅鋗, a nephew of his and an official.[3] There seems to have been a consensus at the Office for Ming History (Mingshi guan 明史館) that no one there or elsewhere in the capital was sufficiently versed in astronomy to write the chapters on this subject for the *Ming History*. Indeed Mei regarded the official historiographers as ignorant of astronomy:[4]

> Those who spoke valued Guo Shoujing's Season granting (*Shoushi* 授時) [system] and joined in despising the Great concordance (*Datong* 大統) [system], without knowing that the Great concordance was [the same as] the Season granting.[5]

The Season granting system had been used in China since 1281; it is usually taken to represent the advanced state of astronomy under the Mongol Yuan dynasty (1279–1367).[6] At the beginning of the Ming dynasty (1368–1644), the same system was taken up, with minor changes, under the name Great concordance. The idea that the Ming

---

1 See p. 74.
2 See p. 83; Zhao 1977, 13328.
3 Qian 1998, 121. Mei Xuan passed the metropolitan examination in 1667; Han 1996, 433; Han 1997, 18; *Jiangnan tongzhi* 江南通志, *SKQS* 528: 470; *Rongcun ji* 榕村集, *SKQS* 1324: 731.
4 Mao Jike 毛際可, 'Biography of Mr Mei' (*Mei xiansheng zhuan* 梅先生傳), in Mei 1939, 2.
5 Mao Jike, in Mei 1939, 2.
6 On this system see Yabuuti & Nakayama 2006; Sivin 2009.

214

system was inferior to the Yuan system, though, suited both Chinese scholars and the Manchu state. For the former, the fall of the Ming resulted from a general neglect of practical learning (*shixue* 實學) during this dynasty. For the latter, the Mongol Yuan dynasty served as a precedent that legitimised the rule of a non-Chinese dynasty. Mei Wending was one of the very few scholars of the time who was in a position to make his own judgement on the two astronomical systems. His account of the prevailing view of astronomy among scholars in charge of the *Ming History* corroborates the emperor's poor assessment of Chinese scholars' competence in this field.

The first draft of the astronomical chapters of the *Ming History* had been composed by the famous Ming loyalist scholar Huang Zongxi. Taking up this draft, in which he claimed to have spotted more than fifty mistakes, Mei Wending composed his own *Draft for the Astronomical Records of the Ming History* (*Mingshi lizhi nigao* 明史曆志擬稿) in 1690.[7] The fact that his work for a Qing imperial project drew on that of a Ming loyalist illustrates the crucial role of the compilation of the *Ming History* for the winning over of Chinese elites by the Manchu rulers. For these elites, writing the history of the fallen dynasty for the one that had conquered it was a way of reconciling loyalty to the Ming and service of the Qing, which they had long regarded as incompatible.

The compilation of the *Ming History* gave Mei Wending an opportunity both to enter the capital's circles of officials engaged in imperial scholarly projects and to apply his specialised competence to a text-oriented project, rather than to the kind of technicalities that the Astronomical Bureau dealt with. Shortly after his arrival in the capital, he acquired a powerful patron in the person of Li Guangdi. Taking a scholar specialised in astronomy under his protection was very likely the latter's response to Kangxi's criticism of him at the Nanjing Observatory.[8] Li had already studied mathematics with Pan Lei 潘耒 (1646–1708), another compiler of the *Ming History* whom Mei Wending had met in Nanjing years before.[9] Pan was a disciple of the famous scholar Gu Yanwu 顧炎武 (1613–1682),[10] under whom he had studied phonology, and whose works he edited.[11] His skills in the mathematical sciences have been assessed diversely;[12] in any case Li Guangdi had been far from satisfied with his pedagogy:

---

7 Mei Wending, *Wu'an lisuan shuji* 勿菴曆算書記, SKQS 795: 966.
8 Han 1996, 431–432.
9 See p. 84.
10 Hummel 1943, 421–426; Durand 1992, 208.
11 Hummel 1943, 606–607.
12 The *Qingshi gao* 清史稿 gives a positive assessment (Zhao 1977, 13343), the *Chouren zhuan* 疇人傳 (1799) a negative one (Ruan 1982, 448).

I am rather dull-witted. I used to study [Napier's] calculating rods with Pan Lei. He is short-tempered; when I did not understand, he would give up and scold me: 'It takes no longer to understand this than [to have] a meal. How can you be so muddled?' He used to mumble indistinctly; he turned out to be unable [to teach me] and quit. Only now that I have Mr Mei who is gentle and good at guiding me, do I get to understand.[13]

Mei Wending's writings were in part intended to help scholars like Li Guangdi grasp the basics of mathematics; he had composed a treatise on Napier's rods, entitled *Calculating rods* (*Chousuan* 籌算).[14] The fact that he was Li Guangdi's protégé, rather than his peer like Pan Lei, may also explain the greater patience shown by Mei. Li Guangdi's description of what he achieved under his new master, though, remained modest:

I have studied *Calculating rods*, and looking at the book I can understand; [but] I close the book, and then I forget. There is no other reason to that than my nature.[15]

In addition to Li Guangdi and Pan Lei, Mei Wending also met scholars such as Lu Longqi, who had met Verbiest years earlier,[16] as well as Liu Xianting 劉獻廷 (1648–1695), Zhu Yizun 朱彝尊 (1629–1709) and Dai Mingshi 戴名世 (1653–1713).[17] He asked Dai to write a preface for his collection of mathematical works, but the latter never got round to doing it.[18] Pan Lei, on the other hand, did write a preface to the *Discussion of rectangular arrays* (*Fangchenglun* 方程論) in 1690.[19] Mei's network extended to the imperial family: Fuquan 福全 (1653–1703), one of the emperor's brothers, also invited him.[20] The emperor's interest combined with the compilation of the *Ming History* to make the mathematical sciences a subject of curiosity and perhaps even a fashion in the capital.

Beside officials and members of the imperial family, Mei Wending also visited Antoine Thomas. He recorded a conversation they had on 14 January 1691:

On the day following the full moon in the month of winter sacrifices of year *gengwu* 庚午,[21] I met Mr An 安 [Antoine Thomas] from the Far West and discussed mathematics (*suanshu* 算數) with him. He said that in measuring fields it was possible not to use the [method for calculating] taxation according to the land owned (*lümu* 履畝). On first hearing this I strongly disagreed, but later on, thinking about it, I obtained this method.[22]

---

13 Li 1995, 2: 775.
14 See p. 85; *SKQS* 795: 983–984.
15 Li 1995, 2: 776.
16 See pp. 68–69.
17 Hummel 1943, 521–523, 182–185, 701–702; Li 1998, 530–532.
18 Durand 1992, 109–110.
19 *Meishi congshu jiyao* 1761, *juan* 11, 1a–2a.
20 Mao Jike, in Mei 1939, 2; Hummel 1943 249–251.
21 The date is expressed in a literary style: it is Kangxi 29/12/16.
22 Mei Wending, *Lisuan quanshu* 曆算全書, *SKQS* 795: 337.

Mei's reminiscence is recorded as an appendix to a 'Method for measuring a plane area at a distance' (*Yao liang pingmian fa* 遙量平面法): sighting from two different standpoints, one measures the angles between the four corners of a distant quadrilateral, and then derives its area.[23]

While he never met Verbiest, Mei discussed with Antoine Thomas at the time when the latter was tutoring the emperor in mathematics. Mei's recollection suggests that Thomas gave him some hint of the surveying method, and that Mei worked out what he meant afterwards. This may have been due to Thomas' allegedly poor Chinese; but maybe he refrained from giving full explanations to someone who did not belong to the imperial court. Did Antoine Thomas know about Mei's works, and in particular about the *Discussion of rectangular arrays*? If so, his treatise on cossic algebra, and especially the part in which he uses symbols to deal with linear problems, could be a response to Mei's denigration of the Jesuits' skills in calculation.[24] On the other hand, the secrecy that surrounded Thomas' writing makes it unlikely that Mei Wending learned much about the former's algebra from their dialogue. If symbolic algebra was indeed a response to the *Discussion on rectangular arrays*, the arbiter for whom it was intended was the emperor. In any case, the encounter between Thomas and Mei Wending was that between the promoters of two competing methods, the former's 'borrowed root and powers' (*jiegenfang* 借根方) forming part of mathematics as an imperial monopoly, while the latter's rectangular arrays (*fangcheng* 方程) was circulated among and approved by scholar-officials such as Pan Lei, as a legacy retrieved from Chinese antiquity.

Mei Wending was not the only visitor that the Jesuits received during that period: according to Bouvet's diary, on the day before Mei's discussion with Thomas, Changning 常寧 (1657–1703), another of Kangxi's brothers, came to their residence.[25] The prince was received by Bouvet and Gabiani; he expressed interest in geographic maps, and was accordingly shown some; he also asked a number of questions about the Christian religion.[26] In contrast to the exchange between Mei and Thomas, the conversation reported by Bouvet had no technical content. What mattered to the latter was to find the emperor's brothers so well disposed towards the Jesuits. He also reported that Kangxi's sons, and especially the eldest, Yinti, were similarly well disposed towards them. Among all these princes, it was the favour of the second son and heir apparent, Yinreng 胤礽 (1674–1725), that was most

---

23 *SKQS* 795: 334–337.
24 See pp. 92–93.
25 Hummel 1943, 69–70.
26 BnF, Ms. Fr. 17240, f. 286 r. Landry-Deron 1995, 2: 147–148.

valuable.²⁷ In the light of court politics, the imperial princes' attitude towards the Jesuits was probably designed to please the emperor. Thus Changning's visit took place shortly after he had returned from the relatively unsuccessful campaign against Galdan led by Fuquan in 1690, during which Tong Guogang, the emperor's uncle who had led the Qing delegation to Nerchinsk, was killed. At the time all leading officers, including the two brothers of the emperor, had been degraded or fined.²⁸ The two princes' display of interest in the mathematical sciences may well have been part of their strategy to regain imperial favour, as was the case with Li Guangdi.

10.2 A patron's commission: the *Doubts concerning the study of astronomy*

The *Doubts concerning the study of astronomy* (*Lixue yiwen* 曆學疑問), a work produced by Mei Wending during the years he spent in the capital, was to play a crucial role in the construction of a consensus between the emperor and his Chinese scholar-officials regarding astronomy. In 1689, shortly after Mei's arrival in the capital, Li Guangqi suggested to him that he should write an essay on astronomy:

Astronomical methods have reached great perfection under our Dynasty. Masters of the Classics are dumbfounded by this: there is no quick exposition expressing their meaning. You should produce a concise work, roughly following the style of the *New writing on the image of [the hexagram] alteration* (*Gexiang xinshu* 革象新書) by Zhao Youqin 趙友欽 of the Yuan dynasty,²⁹ so that everyone might grasp the essential points. If more people apply themselves to it, this branch of learning will be increasingly in the public eye.³⁰

The terms in which this commission was given indicate that it was a direct response to the emperor's discontent with the poor knowledge of astronomy among his officials in general and by Li Guangdi in particular. Although Mei Wending 'pondered' over Li's suggestion,³¹ it is less than obvious that he found it attractive. He had other plans: he set out to write a *General study of astronomical methods* (*Lifa tongkao* 曆法通考), in fifty-eight chapters. The work was at least drafted: Mei wrote a preface for it.³² But it aroused little enthusiasm in Li, who pointed out:

---

27 Bouvet 1697, 232–235.
28 Hummel 1943, 251; Spence 2002, 154.
29 On Zhao Youqin 趙友欽 (1271–c.1335) and his *New writing on the image of [the hexagram] alteration*, see Arai 1996.
30 *SKQS* 795: 970.
31 *SKQS* 795: 970.
32 Mei 1995: 49–52.

Your work forms quite a voluminous book, difficult to complete, and not easy to have printed either. You had better organise a treatise item by item, so as to clarify [each item's] subtle meaning, and to enable the families of masters of the Classics to complete their study [of the subject].³³

Mei Wending's *opus magnum* in fifty-eight chapters was never completed. Instead, after moving to Li Guangdi's residence in the summer of 1691, he started to work on the required popularisation work. Li Guangdi kept close watch on its progress. Mei Wending later recalled that:

When coming back from the court Li Guangdi would enquire eagerly about what I had been discussing on that day. If there was a draft, he would correct it with his own hand. This went on for several months.³⁴

Thus Li Guangdi's patronage of Mei Wending, like the emperor's patronage of the Jesuits, entailed the close supervision of the work commissioned; this must have helped improve Li Guangdi's understanding of the subject. In any case, by April 1693, the *Doubts concerning the study of astronomy* was completed to the satisfaction of the patron, as witnessed by the fact that the latter wrote a preface for it.³⁵ Its author, however, did not regard the work as complete: although Li Guangdi had it printed after Mei left Beijing, the latter still intended to supplement it and to append maps.³⁶

The *Doubts concerning the study of astronomy* is a dialogue between two scholars: the questioner, while ignorant of the technicalities of astronomy, upholds orthodox views. His interlocutor, who on the contrary is well versed in these technicalities, explains them and clears up all the technical and ideological doubts expressed in the questions. The work is a staging of Mei's teaching of Li, rather than a record of actual discussions between them. It is divided into three chapters (*juan* 卷), which consist of sixteen, seventeen and nineteen dialogues respectively. Whereas chapters 2 and 3 mainly contain clarifications of technical matters, the first chapter focuses on assessing and comparing three astronomical systems: Chinese (*zhong* 中), Western (*xi* 西) and Muslim (*huihui* 回回). It touches upon matters that were controversial at the time. The first dialogue is entitled: 'Astronomy was coarse in the past and is accurate nowadays' (*li gushu jinmi* 曆古疏今密). This title suggests that scholars of the present time know better than the ancients: this would at the very least arouse suspicion on the reader's part.

---

33 Mei Juecheng, Preface to the *Fundamentals of astronomical phenomena* (*Lixiang benyao* 曆象本要), 1742; quoted in Han 1996, 433.
34 *SKQS* 795: 970.
35 *Meishi congshu jiyao* 1761, *juan* 46, Preface: 2a; the date does not appear in the *Siku quanshu* version; *SKQS* 794: 5; on the *Doubts concerning the study of astronomy* see Chu 1994, 190–200; Jami 2004: 714–725.
36 Mei 1995, 22–23; Chu 1994, 186.

The dialogue opens with the questioner's articulation of the received narrative of the loss of perfect ancient knowledge in the burning of books by the first emperor; since then, astronomers have striven to recover this perfect knowledge, and today one is closer to it than ever before. Mei Wending's refutation of this viewpoint is based on the assumption that change is inherent to the Way of the heavens (*tiandao* 天道). The Sages, foreseeing that the calendar would need adjustments in order to match the heavens, appointed specialists to deal with it. He also argues that, because of the length of astronomical cycles, change is only perceptible by accumulating records of observations over generations. This is why the moderns, relying on all the data left by their predecessors, are now closer to the actual patterns of the heavens. The Sages' wisdom, Mei Wending goes on, consisted in understanding that there could be no fixed method. They foresaw astronomical changes to come, but did not hold astronomical knowledge themselves. Instead, they assigned specialists to record these changes, thereby making it possible for the astronomical system to approach perfection in the present time. In sum, a good ruler should entrust the care of astronomy to specialists and accept that they should make changes in the astronomical system. This first dialogue establishes two kinds of legitimacy. On the one hand, it is licit and indeed commendable for an emperor to have the astronomical system changed so as to better match celestial phenomena: this is a justification of the Qing astronomical reform based on traditional values. On the other hand, determining what the astronomical system should be is the prerogative of specialists such as Mei Wending himself: this establishes his own authority, for teaching astronomy and for writing the *Doubts concerning the study of astronomy*.[37] It should be emphasised that Mei's argument has very little in common with the notion of progress developed in early modern Europe. In his view, the moderns are far from knowing better than the ancients. They only know more, and this is possible only thanks to the ancients' superior wisdom. There is no improvement or progress of the human mind along history, and the accumulation of human knowledge is merely a token of the ancients' superior merit.

His authority in matters of astronomy once established, Mei Wending goes on to discuss similarities and differences between the Western system and the Chinese system, and to propose how to make use of the best elements in the former to supplement the latter. Beyond the fact that similar phenomena are dealt with in different terms in the two systems, Mei sees a difference in approach between them:

[...] what the Chinese system sets out is motion as it should be (*dangran zhi yun* 當然之運), whereas what the Western system derives is the origin

---

37 Jami 2004, 717–718.

wherefore it is so (*suoyiran zhi yuan* 所以然之源); this is what can be adopted from it.³⁸

As a follower of Neo-confucian philosophy, Mei Wending assigns great importance to the intelligibility of what he asserts.³⁹ Whereas the 'wherefore it is so' is a major contribution of Western methods, and is lacking in the Chinese methods, he finds other aspects of the former useless or unnecessary: in all that pertains to human convention, such as when a day or a month should begin, or the names of constellations, there is no point in adopting Western ways. In sum, 'it is for its precision in measurement and calculation that the Western method should be used, and not for the love of its difference'.⁴⁰

In the *Doubts concerning the study of astronomy*, Mei does not explicitly mention the Jesuits, whom he had strongly criticised in his *Discussion of rectangular arrays*, written decades earlier;⁴¹ instead, he assesses 'Western methods'. The patronage of Li Guangdi, which was part of the latter's effort to regain imperial favour, made this change necessary: Western learning was not only the basis of official astronomy, but also a field of learning patronised by the emperor, in which the Jesuits were his tutors. Mei himself had become closer to the centre of power since he worked on the *Ming History*: he had a personal reason for being more moderate than he had been in the past.

Mei goes on to discuss Muslim astronomy, that is, the system formerly used at the Muslim section (*huihuike* 回回科) of the Astronomical Bureau, which had been closed down in 1657. He notes that Muslim and Western astronomy have the same origin, but that Western astronomy is more accurate.⁴² Moreover, like its Chinese counterpart, 'Western astronomy too was coarse in the past and is accurate nowadays' (*xi li yi gu shu jin mi* 西曆亦古疏今密), due to the accumulation of data; this is why the Muslim system, being older, is less accurate.⁴³ The historicity of astronomy does not apply to China alone, but is universal: appropriating Western learning entails historicizing it.

A different use of history is made when Mei Wending needs to convince his questioner on another matter: 'it is credible that the Earth is round' (*di yuan ke xin* 地圓可信). This was by no means a trivial assertion to make at the time. In received knowledge, following traditional cosmology, earth was represented as square and heaven as circular; this representation was dissociated from technical

---

38 *SKQS* 794: 7.
39 See p. 16.
40 *SKQS* 794: 8.
41 See pp. 92–93.
42 *SKQS* 794: 8–11.
43 *SKQS* 794: 12–13.

astronomy.[44] While giving empirical evidence in support of the Earth's rotundity, such as the variation of the altitude of the celestial North Pole according to latitude, Mei also claims that the idea is present in a number of ancient sources. First, he mentions the ancient *Spherical heaven (huntian* 渾天*)* theory. He goes on to cite authorities outside astronomy, including the medical classic *Inner Canon of the Yellow Emperor* (*Huangdi neijing* 黃帝內經) and the Song philosopher Cheng Hao 程顥 (1032–1085). He further argues that 'the theory of the Earth's rotundity certainly does not come from Europe: it originated in the Western Marches' (*xiyu* 西域), that is, the route by which Muslim astronomers reached China under the *Pax Mongolica*. The evidence he gives for this is the fact that among the instruments made by these astronomers under the Yuan dynasty, there was one that represented the Earth as a globe. 'This', he concludes, 'was the ancestor of the Western theory'.[45] History thus provides two arguments for winning over his questioner to the idea that the Earth is round: on the one hand the idea has always been present in China, on the other hand its present form did originate in the West. The appropriation of foreign knowledge ceases to be problematic if this knowledge proves not to be foreign after all. At the same time, evidence of this knowledge in early Chinese sources invalidates the Westerners' claim that they know any better. This double rhetoric was then assuming more and more importance in discussions of Western learning, whether by its promoters or by its opponents. In the eighteenth century it became a powerful rationale underlying the application of the methodology of evidential scholarship (*kaozheng xue* 考證學) to ancient sources dealing with mathematics and astronomy; more generally, it has played a major role in the historiography of Chinese science to the present day.

10.3 Three solar eclipses

During the years Mei spent in Beijing, three solar eclipses were visible there: 3 September 1690, 28 February 1691 and 17 February 1692.[46] In the emperor's view, eclipse prediction remained a touchstone for astronomical systems, and he applied the test to all sources of astronomical knowledge available. Bouvet recorded that in July 1690,

A Lama, by imperial commission, calculated a solar eclipse on particular tables that he has in a book written in foreign characters. Meanwhile two mandarins of the Tribunal of Kin tien Kien (Qintianjian 欽天監) calculated on our

---

44 Chu 1999.
45 *SKQS* 794: 14–15.
46 Zhang Peiyu 1990, 1043.

European tables, in order to compare the calculations together. And the emperor after having [seen] them both himself and having found the calculation of the Lama was quite remote from the truth, gave it to Fr Thomas for closer examination; the latter reported in writing. His Majesty was quite pleased [...].[47]

The 'foreign characters' might have been Tibetan. While Kangxi used the Jesuits' skills in the sciences and discarded their religion, he did the opposite with the lamas: Tibetan Buddhism played an important role in imperial religion, but he found that its priests had little to offer regarding astronomy.[48] The Jesuits were only one of several groups of non-Chinese 'specialists' working for the emperor, each group being used for specific purposes.

The next time an eclipse occurred, the Astronomical Bureau duly predicted its time and location in the sky for the various provinces of China. It then sent in a report of the observation in the standard format on the day:

Your subjects, the Director of the Astronomical Bureau Gemei 戈枚 and others, respectfully memorialise concerning the matter of their observation of the solar eclipse:

We humbly note that, apart from the full report on the the time and position of the solar eclipse of the day *jiwei* 己未, 1st of the 8th month of this year, already [submitted], we went to the Observatory together with Director Li Huandou 李煥斗 and Vice-Director Deng Tiandong 鄧天棟 of the Bureau of Sacrifices of the Ministry of Rites. Together with Observatory Attendant Anbai 安拜 and others of the Section of Heavenly Signs, we used the instruments to measure that at 3 *ke* 刻 and 8 *fen* 分 of the true *mao* 卯 [hour], waning began to the north-west.[49] The eclipse peaked at 2 *ke* and 5 *fen* of the initial *chen* 辰 [hour].[50] Return to [complete] roundness occurred at 2 *ke* and 9 *fen* of the true *chen* 辰 [hour], at the north-east.[51]

We respectfully note that the writings on divination (*Zhanshu* 占書) state:

Solar eclipse in the 8th month: military equipment is expensive; Zheng suffers a great disaster.

Solar eclipse in [the lodge] *Zhang* 張 ('Spread'):[52] In the past a prince failed in the rites at the ancestral temple.

Solar eclipse on a *jiwei* 己未 day: Ministers are uneasy, and there is conspiracy among the people to invade territory [...].[53]

The astrological comments bear on three elements that characterise the eclipse: the name of the day in the sexagesimal cycle of heavenly stems

---

47 BnF, Ms. Fr. 17240, f. 279 v.–280 r.; Landry-Deron 1995, 2: 100.
48 On the role of Tibetan Buddhism at court, see Rawski 1998, 243–263.
49 I.e. first contact at 6.53 a.m.
50 7:35 a.m.; a modern calculation gives 7.42 a.m.; Zhang Peiyu 1990, 1043.
51 I.e. last contact at 8.39 am.
52 This lodge includes stars of Hydrae.
53 Cui & Zhang 1997, 134, no. 4.

and earthy branches (*ganzhi* 干支), on the month,⁵⁴ and the location according to the division of the sky into twenty-eight lodges (*xiu* 宿). These are of variable width: between 1.5° and about 30°.⁵⁵ The time at which the eclipse is to occur has no bearing on the divinatory comments. Precision was of little consequence in this respect.

Gerbillon also recorded his observation of this eclipse:

> On the third of September Father Bouvet and I observed a solar eclipse: it started at forty-seven minutes past six, and approximately forty or forty-five seconds, and ended at ten minutes past eight, and approximately thirty seconds. It was of approximately three fingers.⁵⁶

The Jesuits recorded the 'magnitude' of the eclipse rather than the time of its culmination. Moreover Gerbillon's record of the time of last contact differs from that of official astronomers by about half an hour. Bouvet, who observed the eclipse with his confrere, gave slightly different figures for the seconds: allowance has to be made for the difference between two observers working together. However, Bouvet recorded the eclipse under the entry of 1 September of his diary instead of 3 September. As mentioned above he did not write up his diary every day,⁵⁷ and may have confused the beginning of the eighth Chinese month with that of the month of September in the Western calendar.

Around the time of the eclipse, ominous events of the kind reported in the divinatory interpretation indeed took place on the battlefield: the battle against the Oirats during which Tong Guogang was killed, and which was at first misrepresented as a victory in reports to the emperor, had occurred two days earlier; the bad news must have reached him very shortly after this report was handed in.⁵⁸ I know of no evidence as to whether any correlation was established between the two.

A few months later, another solar eclipse was visible in Beijing, as noted by Gerbillon:

> The 28th [of February 1691], which was the first day of the second Chinese month, there was a solar eclipse of more than four fingers. As I was at the Palace, I was unable to observe it exactly. We prepared the instruments necessary for the emperor to observe it. And indeed he observed it in front of the noblemen of his court, to whom he wanted to give some evidence of the fruit he had reaped from his studies.⁵⁹

---

54 The Chinese calendar was luni-solar; the first day of a month corresponded to the New Moon, so that solar eclipses would always occur on such a day.
55 The lodges play a role similar to that of the twelve zodiac constellations in Western astronomy: in modern terms, they define slices of right ascension; Cullen 1996, 17–18.
56 Du Halde 1736, 4: 284; Landry-Deron 124.
57 Landry-Deron 1995, 1: 7; see another example of this pp. 148–149.
58 Spence 2002, 154.
59 Du Halde 1736, 4: 299; Landry-Deron 1995, 2: 171–172.

Gerbillon goes on to report a crisis to which this eclipse gave rise at the Astronomical Bureau:

> The Tribunal of Mathematics, having observed this eclipse, consulted the book called Chen Chou (*Zhanshu* 占書) in which is written what should be done, what must happen, and what is to be feared in relation to eclipses, comets and other celestial phenomena, and they found that on such an occasion there was a bad man on the throne and that he had to be removed and substituted by a better one.
>
> The Tartar president of the Tribunal did not want this comment to be inserted in the memorial that, according to custom, was to be presented to the emperor on this eclipse. His lieutenant argued with him for very long, and claimed on the contrary that one should insert what was in the book; that this was the rule of the Tribunal, and if they followed it, no one could disapprove of their behaviour.[60]

The identity of the 'lieutenant' who insisted on including the unseemly comment in the memorial presented to the emperor is unclear; he could have been a Chinese or a Bannerman.[61] In any case, the Director, Gemei,[62] of whom little else than his successive posts at the Bureau is known, won the argument, for the quotations from the *Zhanshu* in the memorial submitted on the very day of the eclipse read:

> Solar eclipse in the 2nd month: Beans are expensive and oxen die; Lu 魯 suffers a great disaster.
> Solar eclipse in [the lodge] *Wei* 危 (Rooftop):[63] Risings of soldiers.
> Solar eclipse on a *dingsi* 丁巳 day: Beneath are many soldiers.[64]

As eclipses went, this one was no more threatening than any other, and there is certainly nothing at which the emperor might take offence in this report. One wonders whether the emperor heard of the argument between his astronomers; in any case, a few months later, on 14 August 1691, he spoke to the Jesuits about the interpretations that these astronomers dutifully submitted after observing celestial phenomena:

> The conversation happening to be on the Tribunal of Mathematics, His Majesty showed to us great contempt for those who superstitiously believe that there are good and bad days, and fortunate hours: he told us clearly that he was very convinced, not only that these superstitions were false and vain, but also that they were detrimental to the good of the State, when those who govern give them credence; that he knew that this had in the past cost several innocents' lives, of whom he named a few, among them some Christians of the Tribunal of

---

60 Du Halde 1736, 4: 299; Landry-Deron 1995, 2: 172.
61 He was probably one of the officials mentioned in Bo 1978, 88, for the 29th year of Kangxi.
62 Gemei was Director (*jianzheng* 監正) at the Astronomical Bureau from 1690 to 1695; Bo 1978, 88–89; Qu 1997, 50–51.
63 This lodge includes stars of Aquarii and Pegasi; Cullen 1996, 18.
64 Cui & Zhang 1997, 135, no. 5.

Mathematics, who were tried at the same time as F. Adam, and who were sentenced to death and executed for not having, it was said, chosen appropriately the time of the funeral of one of the Emperor's sons, which had brought misfortune to the imperial house.[65]

One cannot help wondering exactly in what terms Kangxi spoke. Clearly he was aware that the baby prince, his mother and the Shunzhi emperor had all died of smallpox.[66] One of the reasons why Kangxi was chosen among Shunzhi's sons to succeed him is that he had survived this disease: his face was marked with smallpox scars.[67] Smallpox was a major threat for the Manchus during and after the conquest of China, as they were much more susceptible to it than the Chinese.[68] The disease probably provided a sufficient rationale for Kangxi to account for the deaths that Adam Schall and his Chinese collaborators at the Astronomical Bureau had been accused of bringing about during the calendar case.[69] While he abided by Chinese cultural imperatives, the emperor did not share all the beliefs that underlay these imperatives. In his management of the various groups serving him, he typically presented a facet of himself that matched their own views. He valued the Jesuits' services enough to apply this general policy to them. Gerbillon reports that the emperor went further in deriding Chinese cosmological beliefs:

That the people and even noblemen should lend credence to these superstitions, he said, is a mistake without further consequence. But that the sovereign of an Empire should let himself be fooled could cause terrible harm. I am so convinced of the falsity of this kind of superstition, he added, that I do not have the slightest consideration for them. He even joked on what the Chinese say about all constellations ruling over the Chinese Empire, so that they have nothing to do with others.[70] On which His Majesty added that he had sometimes said to some Chinese, when they told him that sort of tales: 'At least leave a few stars to take care of neighbouring Kingdoms'.[71]

Gerbillon's rendering of Kangxi's words might evoke the European Enlightenment to us. However the emperor is not denying the existence of links between heaven and earth in general, but rather challenging the traditional Sino-centric view that the whole of the heavens

---

65 Du Halde 1736, 4: 342.
66 Dennerline 2002, 118.
67 Bouvet 1697, 11; see also the portrait kept at the Metropolitan Museum of Art in New York, which is reproduced on the cover of the 1975 Vintage edition of Spence 1974.
68 Chang 2002.
69 See pp. 51–52.
70 A 'map of the world' conveying the type of representation that Kangxi is mocking is reproduced in Smith 1991, 69.
71 Du Halde 1736, 4: 342–343.

is mapped onto China alone. Such a view denies both the land of his ancestors and that from which the Jesuits came any significant cosmological status: he connives with them at gazing at China from an outsider's viewpoint.

The words reported by Gerbillon are consistent with the biting response made to Li Guangdi when the latter had proposed a divinatory interpretation of the fact that the Old Man Star was visible:[72] Kangxi's attitude cannot be reduced to duplicity. Rather, he drew a line between the way he made judgements and decisions in governing, for which purpose he used all resources available, and his role within Chinese cosmology, which entailed active response to celestial phenomena. He probably saw no contradiction between the two: after all, he was both the universal ruler of the Qing empire and the Confucian emperor of China.[73] His response to the prediction by the Astronomical Bureau of the third eclipse of this period, due to occur on New Year's Day, illustrates how seriously he took this second role. After receiving the report, he sent an edict to the Ministry of Rites, more than a month before the eclipse was due:[74]

Since antiquity sovereigns have respected Heaven and been assiduous in government affairs. Whenever heaven displays warnings, one must actually improve people and affairs so as to respond to heavenly admonitions. Now the Astronomical Bureau has reported calculations [according to which] an eclipse will occur on New Year's Day of the 31st year of Kangxi. A solar eclipse is a change in heavenly phenomena; moreover this one will be seen at the beginning of a year. We, vigilant and conscientious, are striving to examine Our conscience and seek perfection as appropriate. But if Our subjects of various ranks are very pure, then each must fulfil his duty in his employment appropriately. We personally undertake to clarify the meaning [of all this]. On this New Year's Day, when Rites are carried out, all banquets should be cancelled.[75]

The predicted eclipse thus served as a reminder to officials. The decision to cancel the feasts that usually followed the performance of rites no doubt contributed to perpetuating the correlation between eclipses and unpleasant events. On the other hand, the emperor ostensibly took his share of the moral load, as was expected from him. The austerity prescribed also affected the Imperial Household, although it did not entail the cancellation of all feasting there:

His Majesty visited the Tangse shrine to perform rites, and returned to the Palace. Having completed the worship of the Gods, He led the Honourable Princes, the Grand Ministers of the Imperial Household Department and the Grand Secretaries to visit the Palace of the Empress Dowager, where

---

72 See p. 129.
73 Elvin 1998, 220–221 emphasises the latter aspect.
74 Dated to the 24th day of the 11th month of the 30th year of Kangxi (11 January 1692).
75 *Shengzu renhuangdi shengxun* 聖祖仁皇帝聖訓, *SKQS* 411: 261.

He performed rites.⁷⁶ Congratulations to the Emperor and banquets were called off.

At noon there was a solar eclipse.

The King of Korea Yi Sun 李焞⁷⁷ sent his vassal Min Am 閔黯⁷⁸ and others to convey his congratulations for the Winter Solstice, the New Year and the Imperial Birthday. They offered and presented the Yearly Tribute Gifts. A banquet was bestowed on them according to regulations.⁷⁹

The ritual of allegiance of Korea to Qing emperors went through despite the austerity called for by the eclipse; there is no evidence, on the other hand, that the emperor took part in the banquet offered to the Korean envoys. On the next day, he asked that measures be taken in order to ensure sufficient cereal reserves in areas of the empire that had not had a good harvest the previous year: he appropriately opened the year's work by an act of fatherly care for the people.⁸⁰

A few days after the eclipse, the Astronomical Bureau sent in its report of observation with the usual divinatory interpretation:⁸¹

We respectfully note that the writings on divination state:

Solar eclipse on New Year's Day: a high official dies, or else is cashiered from office.

Solar eclipse on a *xinhai* 辛亥 day: favourite officials are anxious.

Eclipse in [the lodge] *Xu* 虛 ('Barrens'):⁸² The warning relates to officials in charge of taxes on goods, and of fabrics, utensils and treasures.⁸³

The emperor responded by a rescript:

Since antiquity, many are the unworthy high officials who have been executed; this is tied up with human affairs. When those who do prognostication are on duty and write these divinatory words, nowadays they often surmise the current trends in stating the interpretations. One can learn from this.⁸⁴

There is a hint here that rulers are not to take astronomers' prognostications literally. It is not the whole of Chinese cosmology, but only the

---

76 On these rituals see Rawski 1998, 236–240; on the Tangse (Tangzi 堂子) see Di Cosmo 1999, 379–381, 394–396.
77 King Sukjong 肅宗 (r. 1674–1720) is referred to by his family and given names; this reflects the fact that the Qing regarded the Korean Chosŏn dynasty as vassals.
78 Min Am (1636–1694), a Korean official, had already taken part in two earlier embassies to China, in 1675 and 1678; I am grateful to Prof. Jun Yong Hoon for this information.
79 *SL* 154: 1b.
80 *SL* 154: 2a.
81 On the 17th day of the same month (4 March 1692).
82 This lodge includes stars of Aquarius and Sagittarius.
83 Cui & Zhang 1997, 135, no. 6.
84 Jiang 1980, 260.

way contemporary official astronomers work that is targeted. Bringing together all the materials quoted above, it appears that while Kangxi abided by Chinese 'orthopraxis', duly adjusting rituals in response to celestial phenomena, he made it plain to his officials that he did not allow prognostication to influence his decisions. In this particular case, the content of the prognostication made it unlikely that any official might complain about the emperor's ignorance of it.

10.4 An imperial pronouncement on mathematics and classical scholarship

The emperor's first address to his officials after the New Year eclipse was accompanied by a display of his skills in music, mathematics and astronomy. This was duly recorded in the entry of the *Imperial Diary* (*Qiju zhu* 起居注) for the 4th day of the 1st month. Twenty officials, Bannermen and Chinese in equal numbers, were summoned to the Gate of Heavenly purity (Qianqing 乾清), and welcomed with these words:

As there is a little leisure for the beginning of the year, We would like to have a chat with you on harmonics and mathematics, and to consult those who understand their meaning.[85]

Thereupon the officials were let in before the throne. The ensuing discussion was recorded by Zhang Yushu and Li Guangdi as well as by the Imperial Diarists.[86] As often, the emperor's words during this 'leisurely chat' were duly handed down to posterity: the account of the scene in the *Veritable Records* (*Shilu* 實錄), compiled after Kangxi's death on the basis of the *Imperial Diary*, is also found as an entry of the *Sagely Instructions* of the Kangxi emperor, in the chapter devoted to 'Sagely studies' (*shengxue* 聖學): this was fitting as harmonics was essential to rites.[87]

Kangxi emphasised the need to know mathematics in order to make sense of harmonics as set out in the *Complete collection on nature and principle* (*Xingli daquan* 性理大全, 1415) an early Ming compilation of Song philosophers; it was one of the works studied to prepare for civil service examinations. Kangxi had written a preface to it for an edition that he had previously commissioned.[88] Contrary to the *Imperial Diary*, Zhang Yushu's recollection shows that in his harangue the emperor addressed several of those present individually:

---

85 Taipei NPML, no. 104000027; I am grateful to Prof. Chu Pingyi for making this material available to me.
86 The episode is also mentioned in the chronicle of Wang Xi's 王熙 (1626–1703) life; Han 2003d, 438–439.
87 *Shengzu renhuangdi shengxun*, SKQS 411: 205–206.
88 Cheng 1997, 498; Elman 2000, 116–117. The *Complete collection on nature and principle* was commissioned in 1414 by the Yongle 永樂 Emperor. Kangxi's preface is dated 1673.

His Majesty said to the Grand Secretaries and others:

'Do you think that the method of the three parts decreased by one [part] (*sanfen sunyi* 三分損一), and that of [taking] 1 as diameter and 3 as perimeter (*jing yi wei san* 徑一圍三), mentioned in the *Complete collection on nature and principle*, are practical? During the Ming someone wrote a discussion on musical tones. Previously I ordered Xiong Cilü to read it. Yesterday, Xiong Cilü, having completed the reading, reported that the meaning [of this Ming work] was mainly the same as that of Cai Yuanding's theory. I asked him whether Cai Yuanding's theory was accurate, he said that it seemed close enough.'[89]

The passage of the *Complete collection on nature and principle* under discussion is an extract from the *New Writing on the [standard] pitchpipes* (*Lülü xinshu* 呂律新書), a work by the Southern Song scholar Cai Yuanding 蔡元定 (1135–1198) included in it. The Ming work, on the other hand, is Zhu Zaiyu's *New explanation of the study of the [standard] pitchpipes* (*Lüxue xinshuo* 律學新說, 1584).[90] The issue here is the way to construct the twelve standard pitchpipes (*lülü* 律呂), which were to set the tone of ritual music. In Xiong Cilü's eyes, the two works were in agreement on this subject, and both were accurate. The emperor, however, went on to refute this conclusion:

In Our view, using the method of taking 1 as the diameter and 3 as the perimeter, it could not but be impossible for calculations to match. If one used it in administering the calendar, there would necessarily be many mistakes. Now if you try to use this method to calculate solar and lunar eclipses, the error can be seen immediately. There is also the theory (*shuo* 說) of multiplying and dividing by the precise lü (*milü* 密率): if the diameter is 1, then the perimeter is 3 and a fraction; if the diameter is 7, then the perimeter is 22. This is used in all derivations, but you can only calculate small quantities, you cannot calculate large ones. If you calculate small quantities, the difference is insignificant; but as for large quantities the error will be tens to hundreds, or thousands to myriads.[91]

The first criticism addressed to Song musical theory here is its use of 3 as the ratio between the diameter and the circumference of the circle in the calculation of the section of pitchpipes. Neither does Kangxi deem 22/7,[92] a value found in ancient Chinese sources, a satisfactory approximation. He indicated that 'if the diameter is 1 *chi* 尺, then the circumference should be 3 *chi* 尺 1 *cun* 寸 4 *fen* 分 1 *li* 釐 with a fraction'.[93] But he is not entirely satisfied with Zhu Zaiyu's work either:

---

89 Zhang Yushu 張玉書, *Zhang Wenzhen ji* 張文貞集, SKQS 1322: 517.
90 Needham 1954–, 4-1: 220–224.
91 SKQS 1322: 517.
92 That is 3.1429 to five significant figures.
93 In modern terms: the ratio between the diameter and the circumference of the circle is 3.141... Taipei NPML, no. 104000027; *SL*, 154: 3a.

[. . .] In the late Ming, Prince Zaiyu 載堉 discussed musical tones. He emphasised that the mutual production at intervals of eight, with three parts decreased or increased, was not right. But his own theory cannot be perfect either. It is obvious that the general calculation method cannot allow the slightest error. One can test this against facts. Although even people who are not well versed in the meaning of a text may be able to discern the right from the wrong in it, they cannot get away with empty verbiage.[94]

Thus Zhu Zaiyu's theory was dismissed as not allowing for sufficient precision and Xiong Cilü was chastised for his superficial reading. After agreement was duly expressed by two of the officials present, including Zhang Yushu himself, the emperor went on:

Recently there is a man of a Mei 梅 clan from Jiangnan; We have heard that he thoroughly understands mathematics. We have sent someone to test him. What he says on the measurement of shadows does not correspond [to reality] at all. At all times, with the method for measuring shadows, on any day, at any time, for any position of the sun, one can distinguish very minutely. This man sets a gnomon that is so short that one cannot go beyond the *cun*. If for one *cun* there is a difference of one *miao* 秒, then when it gets a *chi* there will be a difference of one *fen*, and when it gets to a *zhang* there will be a difference of one *cun*.[95] Because his calculation method is not accurate, he uses a short gnomon to measure shadows, so that his cheating people will not be visible.[96]

This is the earliest known reference to Mei Wending made by the emperor. The latter had checked on the former, either spontaneously or possibly because Li Guangdi had recommended Mei to him. The test set to the latter was similar to that set to Verbiest and his opponents over two decades earlier, when the Jesuit had triumphed. Mei Wending's failure to meet imperial standards suggests that observation may not have been his strong point. Astronomical instruments were not entirely strange to him: he made some sundials.[97] However instruments do not come out as a major centre of interest in his writings, whereas they were crucial to official astronomy; in any case Mei did not have the means to make a gnomon tall enough to meet the emperor's standards. In other words, Mei Wending's scholarly and book oriented approach to astronomy was entirely irrelevant to Kangxi at the time. In Zhang Yushu's account, the emperor concluded by a diatribe against today's scholars in general and Xiong Cilü in particular, for their ignorance of mathematics, still dwelling on Xiong's inept acceptance of 'the method of taking 1 as the diameter and 3 as the

---

94 *SKQS* 1322: 517–518.
95 Here there seems to be some confusion between length and angular measures: a *miao* is a second of an arc (*fen* refers both to a minute of an arc and to the tenth of a *cun*); *miao* was also used sometimes as an equivalent of *si* 絲.
96 *SKQS* 1322: 518.
97 Zhu Yizun 朱彝尊, *Pushuting ji* 曝書亭集, *SKQS* 1318: 334.

perimeter'. Li Guangdi's reminiscence of the emperor's words is less detailed than that of Zhang Yushu; but it conveys the tension that underlay the whole scene:

In the year Renshen 壬申, His Majesty asked Xiong Cilü about mathematics and astronomy: which was correct – the New writing on [standard] pitchpipes or Zhu Zaiyu's book? In fact Xiong Cilü did not know; he casually replied that Cai Yuanding's book was right. His Majesty was greatly indignant, and said: 'Could such musical tubes be used? Or is this just about writing books? If one measures the main points of heaven and earth, a difference in *miao* 秒 and *hu* 忽 is just acceptable;[98] here we have small tubes, the difference can be calculated: how could these be usable? 1 as diameter and 3 as perimeter, that is a hexagon, how can one achieve calculations with that?' Therefore on the next day he called the Manchu Grand Secretary Alantai 阿蘭泰,[99] the Chinese Bannerman Yu Chenglong 于成龍,[100] the Chinese officials Xiong Cilü, Chen Tingjing 陳廷敬,[101] Zhang Yushu and [Li Guangdi] to go up to the Hall for a consultation. On the previous day, His Majesty, without clarifying the theory of mutual production at intervals of eight and the three parts decreased or increased, had also asked Zhang Ying about it.[102] Zhang Ying had mistakenly brought into 1 as diameter and 3 as perimeter into the theory of mutual production at intervals of eight. His Majesty had again expressed strong disagreement; one had not understood his reasons. On that day he asked Zhang Yushu, who was unable to reply. Xiong Cilü had wildly asserted that Cai Yuanding was correct, and could not explain why. His Majesty said to Yu Chenlong:

'You are usually a sensible man, what do you think about this?'

Yu said:

'What do I know?'

Then the emperor said:

'You Chinese know absolutely nothing about mathematics. There is only one called Mei, in Jiangnan, who knows a little; [but] he is completely muddled.'[103]

Thus in Li Guangdi's reminiscence, the reprimand was not targeted at Xiong Cilü alone. All the Chinese officials present had failed to reply satisfactorily to the emperor's queries on the mathematical aspects of harmonics. The latter's judgment as reported by Li bore on Chinese scholars in general; it is an expression of the conqueror's contempt for the conquered. Li Guangdi, who at the time was closely following Mei Wending's writing of the *Doubts concerning the study of astronomy*, must

---

98 These are sexagesimal units: *miao* are seconds and *hu* are fourths.
99 Alantai belonged to the Bordered Blue Banner; his carrier was mainly a military one; at the time he was a Grand Secretary from 1689 to 1699; Zhao 1977, 6128; 9703–9704; *Qinding baqi tongzhi* 欽定八旗通志, *SKQS* 667: 268–270.
100 Yu Chenlong (1638–1700) was a Left Censor-in-chief (*zuo duyushi* 左都御史) at the time; Hummel 1943, 938–939.
101 Chen Tingjing (1639–1712) was the Minister of Justice (*xingbu shangshu* 刑部尚書) at the time; he was also one of those in charge of the *Ming History*; Hummel 1943, 101; Zhao 1977, 6434–6436.
102 On Zhang Ying see p. 131.
103 Li 1995, 2: 815.

have felt targeted by the emperor's comment on his protégé. One can imagine that on that day Li's queries about Mei's progress when he returned from audience were especially eager.

The *Imperial Diary* records that during this scene the emperor displayed the skills in music, mathematics and astronomy that he had acquired from the Jesuits. First, he had his musicians play, and recognised all the notes of the scale: this was the result of Pereira's tutoring in music. Then he addressed Li Guangdi in particular and explained how one could calculate the quantity of water flowing through a sluice gate. The audience duly expressed their admiration:

> Today your subjects depend on Your sagely instruction. We have heard what we had never heard before and seen what we had never seen before. We are overwhelmed with joy to the utmost.[104]

Having received this obligatory praise, the emperor set up the most spectacular evidence of his skills:

> [The emperor] ordered that a sundial be fetched; and, drawing a mark with the imperial brush, he said:
> 'This is the place that the sun's shadow will reach at noon.' Thereupon, having set up [the sundial] exactly in the middle of the Gate of Heavenly purity, [He] ordered all the ministers to watch out. His Majesty went back into the Palace. When it got to noon, the sun's shadow coincided exactly with the place marked by the imperial brush. It did not deviate by a breadth of a hair.[105]

Kangxi's physical presence at this demonstration of imperial efficacy was not required:[106] it was enough that twenty officials, waiting outside in the cold, witnessed the accuracy of his prediction. The prediction of noon shadow length was the test that Mei Wending had failed, but which had vindicated the Jesuits and their astronomy in early 1669. The witnesses of the emperor's success were of course well aware of the source of his expertise.

10.5 The officials' responses

The collected papers of Zhang Yushu include a 'Memorial for the ordering of musical tones and of mathematics' (*Qing bianci yuelü suanshu shushu* 請編次樂律算數書疏), which he submitted in his capacity as Minister of Rites.[107] The memorial opens with an account of the imperial pronouncement that is closer to the official proceedings than to Zhang's personal recollection. It goes on:

---

104 Taipei NPML, no. 104000027; *SL* 154, 4a.
105 Taipei NPML, no. 104000027; *SL* 154, 4a.
106 On a similar use of calligraphy as a way to prolong imperial presence, see Hay 2005, 329–330.
107 Han 1996, 436–437.

When Your Majesty, with His supreme heavenly talent, establishes methods and determines systems in the study of the investigation of things, He must be sure to test all the marked effects of their applications. After that it will be absolutely without the slightest doubt that these methods can be used to measure the shadow of a sundial, to distinguish minutes and seconds, to plot annual difference,[108] to check eclipses, to measure heights and depths, to finalise musical tones; adapting these to various uses, everything will coincide exactly. [. . .] But the study of musical tones and mathematics has long been lost. There are inherited mistakes, and we do not point them out. There are hidden meanings and subtle points, and we do not explore their content. When Your subjects bowed and listened to Your teaching, it was as though in a moment we had woken up to reality and recovered all that was hazy in mathematics. How come Your subjects in and outside China respectfully pray Your Majesty to grant us the special favour of deciding to order this into books and to issue them everywhere? For general teaching and study, for the correction of the mistakes in mathematics that have accumulated through the ages, for the bequeathing to all generations of a good method for harmonising sounds. Learning and government will equally benefit [from this]; Your subjects will be truly happy, as will the generations to come.[109]

This seems to be the earliest mention of an imperial publication on mathematics and music. It was fitting that a Minister of Rites should write it: music played a central role in rites. Beneath the received rhetoric that turned the emperor into a Confucian teacher, however, this memorial can also be read as a plea that he should share the knowledge he held with his subjects. In the light of Kangxi's insistence on secrecy regarding everything he learned from the Jesuits, this was not merely a formal request. It can be read as a criticism of the emperor's attitude: a teacher (the emperor) should not be content with blaming his disciples (his officials) for being ignorant of knowledge he kept to himself; it was his duty to instruct them. Secrecy did not fit in well with scholarly ethics: if mathematics was indeed a tool for understanding cosmology and harmonics, then it was the emperor's duty to make it widely available to his subjects. Thus he was charged to put an end to the contradiction between his insistence on the necessity for his officials to know these disciplines, and his continued withholding of what he himself knew about them. The counterpart of accepting the relevance of Western learning to classical learning was that the former could no longer be monopolised personally by the emperor.

A further response to Kangxi's pronouncement on harmonics and mathematics came from Mao Qiling 毛奇齡 (1623–1716), who had passed the special examination in 1679, and thereafter worked on the *Ming History*. Although he was not listed in the *Imperial Diary* among those who attended the imperial pronouncement, Mao wrote an essay

---

108 Equivalent to what is called precession in modern terms.
109 Zhang Yushu, *Zhang Wenzhen ji*, SKQS 1322: 411–412.

entitled *Explanation of the foundations of music [according to] the Sagely Edict* (*Shengyu yueben jieshuo* 聖諭樂本解說), in which he discussed the subjects tackled by Kangxi on that day. An outline of the work was submitted to the emperor in 1692. The first chapter (*juan* 卷) was written four years later, and the completed work, in two chapters, was presented to the emperor during the 1699 Southern tour by its author, who had by then retired.[110] The work is a non-technical discussion, composed as a commentary, of the memorial quoted above.[111] Interestingly, it refers to the memorial above as having been presented by Iccanga 伊桑阿, a Manchu Grand Secretary, who had attended the imperial pronouncement. In retrospect authorship of the memorial was clearly a source of prestige.

The idea that the emperor should commission works on mathematics and harmonics may well have come from the emperor himself. According to the Jesuits' accounts, he had always intended to publish at least part of what they taught him.[112] The memorial and its erudite commentary could well have been a response to an intention he had expressed; it was not uncommon for an emperor's initiatives to be thus dressed up as a benevolent response to a request from his subjects. On the other hand, there is no indication that at a time when the Jesuits continued to produce lecture notes for him in a confidential manner, Kangxi was ready to share their content with his officials.

---

110 Hummel 1943, 563–565; Editors' note to the *Explanation of the foundations of music [according to] the Sagely Edict*, SKQS 220: 197.
111 It dates the memorial to the Kangxi 31/5/15 (29 June 1692).
112 See p. 158.

PART IV

Turning to Chinese scholars and Bannermen

CHAPTER 11

The 1700s: reversal of alliance?

Two decades elapsed between the memorial begging the emperor to publish works on mathematics and music, and his initiation of this publication. During this period, while the Jesuits remained active at court, the situation of the mission changed for the worse. 1692 had been a year of triumph: on 20 March, the emperor endorsed a memorial proposing that 'churches built in the provinces should be allowed to remain, and that those who burn incense and worship there should be allowed to do it as usual: it should not be forbidden'.[1] This text, which became famous in Europe under the name of 'the Edict of Toleration',[2] was a result of the court Jesuits' intercession to help the Hangzhou missionaries and their parishioners, who were facing strong hostility from local officials.[3]

The 1700s witnessed a gradual change of attitude on Kangxi's part. On the one hand, he was increasingly interested in and supportive of Chinese scholars versed in the mathematical sciences. On the other hand he distanced himself from the missionaries. This is usually thought to be a consequence of the visit of a papal legate to Beijing in 1706, but it should also be understood in the light of the general change of atmosphere at the court during the 1700s.[4] It became more and more doubtful whether Prince Yinreng, Kangxi's second son and the heir apparent, was fit to succeed his father; at the same time the emperor seems to have grown generally suspicious of his sons and officials.[5] When Yinreng was deposed from his status of heir apparent, first for a few months in 1708, then permanently in 1712, none of his brothers was nominated to replace him.[6] Fierce competition among them for succession to the throne ensued and continued until the emperor's death. Factions, which played an important role in politics throughout the reign, then crystallised around a number of contenders.[7] While missionary work became more and more difficult,

---

1 Han & Wu 2006, 359; Lecomte 1990, 499–500 (there is a typo on p. 499: 'inutiles' should read 'utiles') and Fu 1966, 105–106 give extensive translations of this edict.
2 Le Gobien 1698.
3 Witek 1988, 90–93; Lecomte 1990, 494–498.
4 Standaert 2001, 683.
5 Spence 2002, 164–165, 169.
6 Wu 1979, 112–120; Hummel 1943, 924–925.
7 Wu 1979, 156–176; Durand 1992, 191–218.

239

some Jesuits remained in Kangxi's service and then in that of his successors, both at the Astronomical Bureau and at court, until the news of the dissolution of the Society of Jesus by the Pope (1773) reached the capital in 1775.

11.1 Locating the Beijing Jesuits *c*.1700

In addition to their contribution to official astronomy and their tutoring of Kangxi in the sciences, the Jesuits were also active as court craftsmen and artists. While they continued to make and maintain clocks and astronomical instruments, some new crafts were added to those already put in the emperor's service. In 1696 glassworks were established in a building adjoining the French Jesuits' residence inside the Imperial City. Kilian Stumpf (1655–1720), a Bavarian Jesuit newly arrived in Beijing who mastered the craft of glassmaking, was put in charge of it.[8] For the first time European techniques for producing transparent glass were put in practice in China; snuff bottles were among the objects produced there.[9]

An officer of the French *Compagnie de la Chine* who visited Beijing in 1702 left an account of the employments of the Jesuits, whom he portrays as servants of the emperor, paid as such.[10] The mission comprised eleven Frenchmen out of a total of eighteen Jesuits: Bouvet and then de Fontaney had gone back to France and brought back more missionaries.[11] Among the Beijing Jesuits, the officer listed thirteen fathers and five brothers. They lived in three houses; although the Northern Church (Beitang 北堂) had initially been intended by the French as their residence, the emperor assigned the Jesuits to live where it was convenient for him: Stumpf resided at the Northern Church because the glassworks were next door. One of the two Portuguese residences, the Eastern Church (Dongtang 東堂) hosted two fathers in charge of astronomical observations: Antoine Thomas and Jean-Baptiste Régis (1663–1738), and two physician-apothecaries, both brothers: Bernard Rhodes (1646–1715) and Giuseppe Baudino (1657–1718).[12] This residence, located to the East of the Imperial City, was relatively close to the Imperial Observatory. The other residence, the Southern Church (Nantang 南堂) adjoined one of the three branches of the Astronomical Bureau; it was there that work on the calendar reform had started in 1629.[13] It contained workshops where instruments were made.

---

8 Reil 1978, 61–63; Curtis 2004.
9 Olivova 2005, 229; see Curtis 2001 & 2009.
10 Madrolle 1901, 149–150.
11 Twenty-nine Frenchmen joined the China mission between 1698 and 1701. Dehergne 1973, 402–403.
12 Dong 2004, 28–29.
13 See p. 33.

Tomás Pereira probably resided there: he was in charge of the emperor's arms, automata and clocks, with the help of two other Jesuits for the latter. The Jesuits' position in the emperor's service and their craftsmanship was visible at their residence. A Russian envoy who visited Beijing in 1693 had been duly impressed by it:

> [...] the cloister [...] was encompassed with a high Stone-Wall, and provided with two exquisite regular Stone-Gates, after the *Italian* Manner. On the left side of the Entrance, under Shelter of a Roof, made for that Purpose, in the Court, stand the Cœlestial and Terrestrial Globes, of an extraordinary size, each being about a Fathom[14] in diameter. From hence we proceeded on to the Church, which was a very beautiful *Italian* building, furnish'd with an Organ, made by Father *Thomas Pereyra*: And the Church it self suitable to the *Roman* Catholick Usage, was richly adorn'd with fine Images, and Altars; and was withal large enough to contain Two or Three thousand People. On the Top was a Clock and Chimes. Having seen the Church, the Fathers brought me into the *Musæum*, which was stored with all sorts of *European* Rarities.[15]

By 1700, the French mission significantly contributed to the emperor's service as well: music now fell to the lot of Louis de Pernon (1664–1702), who made, tuned and played instruments; he was one of a trio who regularly performed for Kangxi. The fine arts were represented by a sculptor, Brother Charles de Belleville (1657–1730), and by an Italian layman painter brought to China by Bouvet in 1698, Giovanni Gherardini (1655–1723?). The latter worked for the emperor, but also decorated the inside of the Northern Church, which opened in 1703. On the ceiling Gherardini painted a trompe-l'œil dome, probably in the manner of those still extant in the Jesuit churches of Rome, which greatly impressed Chinese visitors.[16] Thus the French Jesuits too used the skills available both in the service of the emperor and to enhance their own image within the city.

The Jesuits had a number of apprentices. They could be no less versatile than their masters. For example, the painter Jiao Bingzhen 焦秉貞 (c.1660–1726) started his career at the Astronomical Bureau, where he is said to have acquired a good knowledge of mathematics.[17] He also became acquainted with Western techniques of painting, including perspective. There is no evidence as to who taught these to him; however we know that some Chinese artists worked under Ludovico Buglio's (1606–1682) supervision. Jiao, who was a convert,

---

14 About two metres.
15 Ides 1706, 79; a depiction of the church as it was in 1711 does not show the celestial and terrestrial globes; Brockey 2007, 196.
16 Madrolle 1901, 145–149; Vissière & Vissière 1979, 146; Corsi 1999, esp. 107; Pirazzoli-t'Serstevens 2007, 11. A painting kept at the Cabinet des estampes de la Bibliothèque nationale de France (Paris, photograph 76 C 76801) most likely represents the Northern Church; Golvers 1993, 12–13 & 502.
17 Zhao 1977, 13911.

The Emperor's New Mathematics

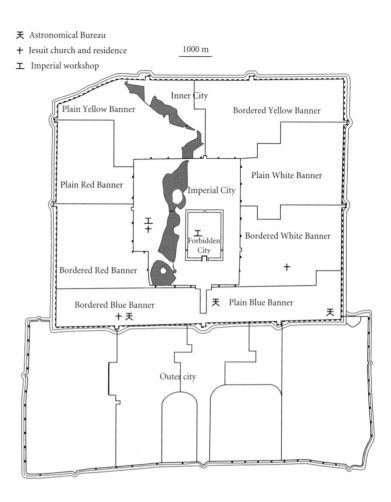

Map 11.1 The Jesuits in Beijing *c.*1700 (Jami 2008a, 58).

could have been one of them.[18] By 1689, when he held the rank of Supervisor of the Five Offices (*wuguanzheng* 五官正) at the Astronomical Bureau, Jiao was already known as a painter who mastered Western techniques.[19] His most famous work, commissioned by the emperor, is a series of illustrations for a new version of Lou Shu's 樓璹 (fl. 1133–1155) *Pictures of tilling and weaving* (*Gengzhitu* 耕織圖) for which Kangxi in person wrote poems. In these forty-six paintings, bound to form a book, Jiao used perspective. The paintings were completed by the spring of 1696, when Kangxi wrote a preface for the work. A xylographic edition was published in the same year. Other versions of the hand-painted *Pictures of tilling and weaving* were produced

---

18 Golvers 1993, 116–117; Han 2005a, 199.
19 Hu (1816) 1963, 1.

thereafter, including one by Jiao Bingzhen himself.[20] Whereas we do not know about Jiao's work as an astronomer nor about his skills in the mathematical sciences, we do know that the geometrical bases of perspective were taught by the Jesuits to at least one Chinese official, Nian Xiyao 年希堯 (1671–1738), who later wrote a treatise on the *Study of vision* (*Shixue* 視學, 1735).[21] There seems to be no evidence as to exactly when and from whom Nian learnt about perspective; he must have come in contact with it at the court.[22]

The technical and artistic work done by the Jesuits and their 'apprentices' in the service of the emperor took place under the auspices of the Imperial Workshops (*Zaobanchu* 造辦處), managed by the Imperial Household Department.[23] To the end of the reign, these remained located at the Hall of the Nourishment of the Mind, where the Jesuits taught, made and maintained scientific instruments; Yinti, Kangxi's eldest son, supervised their activities. The glassworks, located outside the Forbidden City, were also attached to the Imperial Workshops.[24] Not only the Jesuits' sciences, but all their technical and artistic skills, came under the jurisdiction of the Imperial Household Department.

An account of Western curios in Kangxi's private apartments at the Garden of pervading spring was given by Gao Shiqi, then a retired official, after the emperor invited him to visit him there several times in June 1703; this was a very rare favour. Among the valuable objects that Gao was thus able to admire were 'Chinese antiques and European musical instruments'.[25] A number of musicians performed for him. Kangxi talked about harmonics, then played himself *Pu'an's Incantation* (*Pu'an zhou* 普唵咒)[26] on a 'Western iron-wire zither with one hundred and twenty strings, made in the Palace workshops', most likely a harpsichord.[27] Having thus made a European instrument resound with a Buddhist tune, he argued that he was thus retrieving an ancient Chinese artefact:

The *konghou* 箜篌 harp existed during the Tang and Song dynasties; it has long been lost. Now we have its method back.[28]

---

20   Kangxi & Jiao 2003, 28–33; Bray 2007, 529; Strassberg 2007, 89–96.
21   See Corsi 2004, 163–192.
22   Ruan 1982, 505–506; Li 2000, 557–558.
23   Wu 2004.
24   Curtis 2001, 181–184; see p. 151; on Yinti's role see Spence 1975, 72–73; Jami 2008b, 202.
25   Gao 1994, 31.
26   Pu'an was a Chan Master of the Northern Song period.
27   Rather than a piano (Fu 1966, 113): the description suggests that the instrument had sixty keys, that is, a range of five octaves. It could have been made by Pereira or by de Pernon; on keyboard instruments in China see Lindorff 1994.
28   Gao 1994, 32.

Other Western *exotica* were revealed to Gao's admiring gaze. In one hall, the *Gengzhitu* was engraved on the pillars with poems and a preface in Kangxi's own calligraphy; this was probably a reproduction of Jiao Bingzhen's album. In the emperor's private theatre (*guanjuchu* 觀劇處), Kangxi drew Gao's attention to some European paintings and to the glass windows. While claiming that 'China [could] now produce a sort of glassware superior to the Western product', the emperor gave him 'some twenty articles of glassware and a mirror-screen five *chi* 尺 high, imported from Europe'. On another day, Gao received more presents, including some snuff bottles and 'three European paintings'. On account of his old age, he was even granted the amazing permission to contemplate the portraits of two imperial concubines (*guifei* 貴妃), possibly painted by Gherardini;[29] with these Kangxi meant to show him an example of the Westerners' capacity to paint 'marvellous noble faces'.[30] Among the riches he saw, Gao Shiqi was of course most struck by those that only the emperor could collect, thanks to his Jesuit craftsmen. Gao recorded the details of his visits so that his fellow officials would eventually know of the emperor's Western *exotica*, but also of the honour that Gao had been granted.

At the time, the Jesuits' tutoring of Kangxi had ceased, for, as Bouvet wrote to Leibniz in 1701, the emperor 'had finished satisfying himself several years ago about what he wanted to know of the theory of our sciences'.[31] This does not imply that they no longer wrote for him: by 1700 Antoine Thomas was still working on his algebra treatise.[32] Throughout his reign, they continued to use the mathematical sciences in his service. This service often led them outside the capital: Thomas was sent to measure the length of a degree of terrestrial meridian, together with the emperor's Third Son, Prince Yinzhi 胤祉 (1677–1732), who had studied mathematics with Thomas. According to the Jesuit's account, fixing a new standard for the *li* 里 was part of Kangxi's plan for correcting the maps of his empire that had been drawn earlier.[33] Some parts of the empire had already been surveyed; thus in 1690, after the Nerchinsk treaty with Russia, the lower Amur River (Heilongjiang 黑龍江) area was surveyed by Manchu troops, which resulted in a map of that area.[34] In 1698, after defeating Galdan, the emperor sent Thomas and Gerbillon to 'Western Tartary' (today's Inner Mongolia) to do some surveying. Soon after their return, he

---

29 Corsi 1999, 117.
30 Gao 1994, 32.
31 Widmaier 1990, 159.
32 See p. 201.
33 Bosmans 1926, 160.
34 The map is in Manchu; Chengzhi (Kiccenge) 2009.

CHAPTER 11 | The 1700s: reversal of alliance?

noticed a discrepancy between the length of the *li* used in their map and that used in an earlier one. The project of correcting the maps of the empire must not have seemed particularly urgent to him at the time, for it was only in the late autumn of 1702 that Yinzhi, Thomas, and a group of officials that included Mingtu 明圖, the Manchu Director of the Astronomical Bureau, were sent to measure a degree of meridian in order to fix a standard *li*, such that one degree of terrestrial meridian would contain exactly 200 *li*.[35]

11.2 The mathematical sciences in history

A few days after the meridian expedition set off, the emperor addressed Zhang Yushu and Li Guangdi, praising the techniques used by the expedition:

As to the measurement of distances using instruments (*yong yiqi celiang yuanjin* 用儀器測量遠近), there are fixed principles with absolutely no error. If for 10,000 there is an error of one, this is [due to] difference in the practice (*yongfa* 用法), and not [to] inaccurate numbers. By this means, in the calculation of geographical features and of field areas, all can be instantly discerned. But one must be meticulous and diligent, so that squares are accurate. In general the test for this is nothing but triangles. Although the word triangle (*sanjiaoxing* 三角形) was not used formerly, calculations methods in past dynasties must have had some foundations. For example, the base and altitude (*gougu* 句股) method is pretty close to triangles. In fact this method must have been transmitted from antiquity. It has not been found in books, therefore we do not know when it started.[36]

The first words evoke the title of one of the treatises attributed to Antoine Thomas, the *Practice of instruments for measuring heights and distances* (*Celiang gaoyuan yiqi yongfa* 測量高遠儀器用法), a treatise whose methods were put into practice for the measurement of the degree of meridian.[37] The main points made here are the importance of trigonometry for surveying, the assertion that it underlies the traditional base and altitude method and that it must have been known in Chinese antiquity. Here as with music instruments, the emperor upheld the idea that Western learning retrieved ancient Chinese knowledge, even though, in this case, he acknowledged the lack of explicit reference to triangles in ancient sources.

The implications of redefining the standard length unit reached beyond cartography. Historians of metrology have identified three main standards for the *chi* 尺 during the Ming dynasty,[38] and listed

---

35 Bosmans 1926, 167.
36 *Shengzu renhuangdi yuzhi wenji* 聖祖仁皇帝御製文集, *SKQS* 1299: 44; dated Kangxi 41/10/24, 12 December 1702.
37 See pp. 194–195; Bosmans 1926.
38 Qiu *et al.* 2001, 406–409; Luo 1957, 50–57.

245

several extant standards for the Qing dynasty. There is plenty of evidence that the standardisation of units remained a slogan: the sheer size of the empire would have made its implementation quite difficult. Moreover the names of the three Ming *chi* reveal that standards varied according to trades.[39] While Yinzhi and Thomas were out on the meridian expedition, Kangxi further discussed the issue of standard length and weight units with Li Guangdi. The emperor claimed that he had used a Western method to ascertain that 'the builder's *chi* of the early Ming is the ancient *chi*':[40] he measured the propagation of the sound of cannon using a second pendulum, assuming that for every seven seconds sound travelled five *li*.[41] Accordingly, there should be 250 *li* to a degree of terrestrial meridian.[42] Li Guangdi then questioned the reliability of the pendulum; again Kangxi had tested it, and retorted that the pendulum was swifter when the oscillations were wider, and was slower when they were narrower, so that its period was constant. The ratios of 250 *li* per terrestrial degree and of 5 *li* per 7 seconds for the propagation of sound, as well as the determination of the length of the second pendulum all appear in Thomas' *Outline of the essentials of calculation*.[43] Thus the emperor once more vindicated Western learning as a scholarly tool by means of which he investigated some technical aspects of Chinese history. Some days later, however, Kangxi informed Li Guangdi that the outcome of the expedition was that there should be 200 *li* to a degree of meridian. He upheld that the Zhou dynasty's *chi* was 'correct', arguing that there were in fact two different standards at the time. The unit determined by a Manchu prince and a Jesuit for the survey of the empire was thus validated against the knowledge held by the dynasties of antiquity, the correctness of which could be verified.

At the time, the emperor's opinion of Chinese scholars' skills in mathematics had not improved. He derided Zhang Ying[44] on this matter: during the discussion of a passage of the *Ritual of Zhou* (*Zhouli* 周禮) concerning the division of the kingdom into fiefs, it turned out that the Grand Secretary believed that there were 10 squares

---

39 These were the builder's *chi* (*yingzao chi* 營造尺), the tailor's *chi* (*zaiyi chi* 裁衣尺) and the surveyor's *chi* (*liangdi chi* 量地尺).
40 Li 1995, 2: 813.
41 There are some inaccuracies in Li Guangdi's account; Jami 2007b, 157.
42 In modern units, this *li* would then be about 444 m long.
43 The two ratios are found in the longer version of the *Outline of the essentials of calculation*, in two different problems on the simple direct rule of three; on the former, see p. 185. The determination of the length of the second pendulum is found in the same version, in the section on the indirect simple rule of three (BML, Ms. 82–90 A, vol. 1 ff. 26b–27a, 34b–35a; see Jami 2007b, 154–156). With 1 *li* = 444 m, this corresponds to 317 m/s for the speed of sound (modern value: 331 m/s at 0°C).
44 Zhang Ying, who had been present at the Nanjing Observatory in March 1689, was now a Grand Secretary; see pp. 131 & 232.

of side 100 *li* in a square of side 1000 *li*.⁴⁵ For the emperor this would be yet another illustration of the need for mathematical knowledge in order to understand the Classics. Li Guangdi, perhaps piqued by Kangxi's dismissive comments, such as the one he made on this occasion, and possibly in response to the emperor's assertion that trigonometry came from Chinese antiquity, mentioned to him Mei Wending's *Doubts concerning the study of astronomy*. Kangxi's reply sounded almost like a warning: 'For many years We have paid great attention to astronomy and mathematics; We are able to assess right and wrong.'⁴⁶ Li then presented the work to him—as he had probably planned to do when first commissioning Mei to write it more than a decade earlier. Kangxi's first assessment of the work was positive, with a small reservation: 'There are no mistakes, but the calculations are not yet complete. Probably Mei has not yet finished his work.'⁴⁷ Interestingly, this comment concurs with Mei's own unwillingness to have the work printed: he was not entirely satisfied with it either.⁴⁸

An exchange of books produced under their respective patronage seems to have taken place between the monarch and Li Guangdi: a few months later, in the spring of 1703, Kangxi in turn gave the latter copies of Gerbillon and Bouvet's *Elements of geometry* (*Jihe yuanben* 幾何原本) and of the *Elements of calculation* (*Suanfa yuanben* 算法原本). Li humbly acknowledged that he could not fully understand them.⁴⁹ This is perhaps not surprising: beside Li's self-confessed lack of skill in mathematics, both works were produced in conjunction with oral teaching. They may have been difficult to study without a teacher. Li once more undertook to remedy his own ignorance. He was then governor of Zhili 直隸, and there were a number of young scholars among his retinue in Baoding 保定; these included Wei Tingzhen 魏廷珍 (1669–1756), Wang Lansheng 王蘭生 (1679–1737), Chen Wance 陳萬策 (1667–1734), Wu Yongxi 吳用錫, and Li's eldest son Li Zhonglun 李鍾倫 (1663–1706). Li invited Mei Wending to teach mathematics and astronomy; Mei's grandson Mei Juecheng joined the group.⁵⁰ Mei Wending mentions a 'new translation (*xin yi* 新譯) of the *Elements of geometry*' in the bibliography of his own writings:⁵¹ he and

---

45 Li 1995, 2: 814; the expression 'The royal domain [covers] one thousand *li*' (*wang ji qian li* 王畿千里) occurs in the *Book of documents* (*Shangshu* 尚書) and in the *Ritual of Zhou*.
46 Li Guangdi, *Rongcunji* 榕村集, SKQS 1324: 722; Zhao 1977, 1667.
47 SKQS 1324: 723; see Han 1996, Han 1997.
48 See p. 219.
49 Li 1825, *juan xia*, 18b–19a; Li (1826) 1999, 581–582; on the two works, see pp. 166–179 & 195–200.
50 Han 1996, 437; Li 1998, 539; Li 2006, 46–47.
51 Mei Wending, *Wu'an lisuan shumu* 勿菴曆算書目, SKQS 795: 986; Martzloff 1993a, 168.

his students were the first Chinese scholars to use some of the court Jesuits' treatises, and to study imperial mathematics.

A year after receiving the *Doubts concerning the study of astronomy*, the emperor returned it to Li Guangdi:

In the evening of the 23rd day of the 8th month of *guiwei* 癸未,[52] we saw the copy of the *Doubts concerning the study of astronomy* that His Majesty had read. [It bore] small grain-like circles and large fly-leg like marks. Comments were written in vermilion ink in miniscule characters, on separate pieces of paper. It was to be feared the book had been assessed as bad. And some discussant had inserted tiny bits of Korean paper that had been slipped in so as to mark some [passages]. Mr Anxi[53] 安溪 asked His Majesty for a comment in vermilion [on the book itself]; He replied:

'He has spent tens of years of effort to complete one book, how could I undertake to criticise it lightly all by myself? Yet he is one of your 'cultivated talents' (*xiucai* 秀才):[54] he can only discuss principles, and does not know how to do computations. After all, in astronomy calculations are best; empty discussions are irrelevant.'[55]

It is not known whether Kangxi had shown the *Doubts concerning the study of astronomy* to anyone; in particular there is no evidence that he consulted the Jesuits before making a judgment on the work. He was less dismissive of Mei Wending than he had been ten years earlier, but his basic standpoint has not changed: he did not consider the *Doubts concerning the study of astronomy*, which was written in a discursive mode, as evidence of what he regarded as technical competence. It is ironical that of all of Mei's writings, Li had presented this one to Kangxi: some of Mei's earlier works on mathematics and astronomy are far more technical, and certainly full of calculations, which probably made them much more interesting in the emperor's eyes. Thus Li Guangdi's patronage, which oriented Mei towards a more literary style, was a mixed blessing if the ultimate goal was to draw the emperor's attention to Mei's skills in mathematics and astronomy.

However irrelevant to imperial astronomy Kangxi may have deemed Mei Wending's 'empty discussions', the *Doubts concerning the study of astronomy* probably inspired him to write an essay of his own on some of the matters discussed in it. This essay was written about a year after the emperor returned the *Doubts concerning the study of astronomy* to Li Guangdi. The *Discussion of triangles and computation imperially composed* (*Yuzhi sanjiaoxing tuisuanfa lun* 御製三角形推算法論) is a short text (about 620 characters); it was dictated

---

52 28 September 1703.
53 Li Guangdi.
54 i.e. a District Graduate.
55 Li 1995, 2: 814–815.

by Kangxi in Chinese, then translated into Manchu, and given to read to a number of officials and to some of the emperor's sons in December 1704. This was an imperial response to a mistake in eclipse prediction by the Astronomical Bureau.[56]

The *Discussion* opens in a most conventional manner, with a quote from Mencius: 'The compass and the carpenter's square are the culmination of squares and circles; the sage is the culmination of humanity.'[57] It praises the knowledge and achievements of high antiquity in mathematics and astronomy, while blaming later Chinese astronomers for not having kept up this knowledge, 'because the vulgar easily fear trouble, put great value on official honours and positions, and little value on respecting the heavens and transmitting the calendar'.[58] This is a criticism of official astronomers. The emperor then goes on to remind how he was prompted to study astronomy by the realisation of his officials' ignorance of it during the reinstatement of the Jesuits at the Astronomical Bureau in 1668–1669.[59] He then gives his view on the history of astronomy:

Astronomy originated in China and was transmitted to the Far West. Westerners received it, did measurements without fail, revised it year after year without end, and therefore obtained precision in differences (*chafen* 差分). It is not that they have different procedures. Although there are differences between designations of items, these really have nothing to do with the origin of astronomy; actually they are linked to the yearly revisions and checks made to [improve] precision. Reckon the multitude of angles in squares and circles, classify the measurements of latitude and longitude, then astronomical methods prevail for one thousand years: why reject them? With triangles (*sanjiao* 三角), the subtleties in the multitude of angles in squares and circles are

---

56 Jin 1989, 20, translates the date in Manchu as Kangxi 43/10/23. According to Prof. Watanabe Junsei (personal communication), the date is Kangxi 43/11/23, that is, 19 December 1704; the document was translated into Manchu and circulated on the next day. See the original bilingual text at: http://archive.wul.waseda.ac.jp/kosho/ni05/ni05_00410/ni05_00410.html. What makes this date significant is that there was a solar eclipse on Kangxi on 43/11/1; the emperor observed it, and rebuked the officials of the Astronomical Bureau for presenting a mistaken prediction; Zhao 1977, 266 & 1416; see Lü 2007, 140.

The date of the essay has another implication: over the last ten years, there has been a controversy amongst historians of science concerning the date of the imperial essay, and therefore on whether Mei Wending and the emperor developed their ideas independently. Against Wang 1997, who relied on Jin 1989 for the date, Han 1998, 187, and Chu 2005, 74, have argued that the essay must have been written at least ten years earlier, and that there was a convergence of views between the two men. Wang 2006a responded by referring back to Jin 1989, (the only one among these scholars who could read Manchu), who noted that the date was given in the Manchu version and only there.

57 Translation Lau 1970, 118.
58 *SKQS* 1299: 156.
59 See pp. 74–75.

easy to understand. If one rejects them and searches by another way, one is bound to end in confusion, and astronomy cannot succeed. Yixing of the Tang, Guo Shoujing of the Yuan[60] did nothing more than borrow the Islamic system, adding a bit of touching up; by coincidence it worked temporarily, that is all. It could not have worked for very long. Obviously this is due to the fact that their viewpoints were not rooted in mathematics. Astronomy starts with measurement and ends with computation. It is granted to the people in the form of the seasons [for their work]; it is tested through [observation of] conjunctions and eclipses. How could one evade what all eyes can see?[61]

In this final section of the essay, the emperor extended to astronomy what he had said about trigonometry in his address to Zhang Yushu and Li Guangdi two years earlier. He asserted that his own role of intermediary between the cosmos and the human realm was fulfilled through mastery of the mathematical sciences. He concurred with Mei Wending concerning the history of astronomy, insofar as he too asserted that Western astronomy originated in China. Unlike Mei, however, he dismissed two major astronomers of the past millennium. The overall historical narrative articulated here, which was to become the received one in the eighteenth century—with some variants—[62] conveniently legitimised his choice of the Jesuits as imperial astronomers over the Muslims who were previously in office at the Bureau; so much so that one might wonder how far the emperor actually believed in it. In any case, he explained away differences in terminology as resulting from the changes made by Western astronomers over the centuries during which they relentlessly improved on what they had received from China. In his rhetoric, the Chinese origin of astronomy and the superiority of Chinese antiquity served to criticise Chinese astronomers of the more recent past. Using the Ancients against one's recent predecessors was a common rhetorical device; Mei Wending used it as well to promote his own innovations.[63] Comparing those predecessors disadvantageously with their Barbarian counterparts was certainly less common. The excellence of the latter's methods, Kangxi argued, lay in their use of triangles. This point was already made in the address to Zhang Yushu and Li Guangdi; but there he had emphasised the application of trigonometry to surveying, while here in the essay, he

---

60   Yixing is the creator of the the Nine planets system (*Jiuzhi li* 九執曆), officially adopted in 729; Guo Shoujing is the creator of the Season granting system (*Shoushi li* 授時曆), officially adopted in 1281 (Yabuuti & Nakayama 2006; Sivin 2009); the former was influenced not by Islamic astronomy (which did not yet exist in his time), but by Indian astronomy imported into China in the wake of Buddhism; Cullen 1982.
61   *SKQS* 1299: 156–157.
62   On this much studied topic, see, among others, Jiang 1988; Wang 1997; Chu 2003; Liu 2002b.
63   See p. 94; Jami 2004, 717–718.

discussed the more sensitive field of astronomy. The fact that the last statement in the essay echoes its title suggests that there laid the main point the emperor intended to make: in his own time Western astronomy outdid Chinese astronomy because the former used triangles, that is, geometry and trigonometry. While it is not surprising that the emperor defended the mathematical sciences taught by the Jesuits, it is quite remarkable that he articulated so clearly what his contemporaries versed in astronomy (and after them, historians of science) perceived as the main changes brought about by Western learning.[64]

Once Kangxi acknowledged that Mei Wending's work could not be lightly underestimated, Li Guangdi continued to promote his protégé. In 1705, when the emperor was starting on another Southern tour, Li mentioned Mei's name among those of eminent scholars. On his way back, the emperor sent for Mei Wending; the latter gratefully recorded:

On the 20th day of the 4th [intercalary] month,[65] I received a summons to an audience on the Imperial boat in Linqingzhou 臨清州. I took the little boat that had come for me, and on the 22nd and the 23rd, I was granted audience, bestowed food and offered a seat.[66] When by midnight we stopped, His Majesty lit the way for me, and a Palace Attendant was ordered to escort me back. On the 24th, during the imperial stopover in Yangcun 楊村, I went to thank Him.[67]

This was the crowning of Mei Wending's career, the ultimate accolade, coming when he was already in his seventies. During the audience he presented one of the works he had once planned to include in his synthesis of Chinese and Western mathematics,[68] and which Li Guangdi had had printed in Baoding, the *Essentials of trigonometry* (*Sanjiaofa juyao* 三角法舉要).[69] The treatise, divided into five chapters, is entirely based on Western learning, although it retains the traditional Chinese term *gougu* 句股 (base and altitude) to refer to right-angled triangles. Opening with elementary definitions, it guides the reader from simple triangles on to polygons inscribed in or circumscribed around circles, and finally to general measurement (*celiang* 測量) problems similar to those found in the Jesuit textbook *Practice of instruments for*

---

64 Sivin 1995, VII: 62 characterises the response of seventeenth-century Chinese astronomers to Western learning as 'a scientific revolution' (in the Kuhnian sense of a change of paradigm), arguing that 'geometry and trigonometry largely replaced traditional numerical or algebraic procedures'.
65 Mei Wending writes '4th month' (Mei 1995, 325). However, according to Li Guangdi, the audience took place in the 4th intercalary month of the same year (Li (1826) 1999, 595–596; Mei 1995, 353). The *Veritable Records* (*Shilu* 實錄), which mention all the stops in the emperor's Southern tour, corroborates Li's dating (*SL* 220: 18b).
66 13 and 14 June 1705 (rather than 14 and 15 May), still following Li Guangdi and the *Veritable Records*.
67 Mei 1995, 325.
68 See p. 86.
69 *SKQS* 795: 985.

measuring heights and distances (*Celiang gaoyuan yiqi yongfa* 測量高遠儀器用法).⁷⁰ The *Essentials of trigonometry* certainly suited Kangxi's taste better than the *Doubts concerning the study of astronomy.* The presentation of the former work looks like a response to the reservations that the emperor had expressed to Li Guangdi concerning the latter work, and to the praise of trigonometry as a tool for surveying and astronomy in Kangxi's essay. The conversation between Mei and the emperor must have been much more technical than those in which Li Guangdi took part with either of them. Still, Mei was delighted that the emperor approved of the main point of his narrative of history of astronomy expounded in the *Doubts concerning the study of astronomy,* as suggested by a poem written to commemorate the audiences:

His sagely holiness and heaven-granted talent are the continuation of Yao and Shun.
He has watched the Heavens and, in His leisure time, the Morning Star shining.
With His completed discussion of triangles, [Sagely] Rules and Advice are handed down to posterity.
Ancient and modern, Chinese and Western are all consistent.

(The *Discussion of triangles imperially composed* says: 'Western learning in fact originated in Chinese methods.' How great are the words of the King!)⁷¹

Only in the Classics did Mei Wending find the appropriate words to praise the emperor's approval of his own idea: 'How great are the words of the King!' is taken from the *Book of documents*, the very work that recounts how the (mythical) sovereigns Yao and Shun established the calendar, among other institutions.⁷² This indicates the importance that Mei attached to the 'Chinese origin of Western learning'. It does not imply, though, that this idea was the main, let alone the sole, topic that he discussed with the emperor, nor indeed that the emperor thought that the historical discussion of astronomy was Mei's most important achievement.⁷³ The emphasis put by most historians of science on this aspect of their shared interest probably results from the fact that the main work written by Mei Wending after this audience was a *Sequel to the Doubts concerning the study of astronomy* (*Lixue yiwen bu* 曆學疑問補) in which, among other things, he further explored the history of astronomy and the evidence for the

---

70 Mei Wending, *Lisuan quanshu* 曆算全書, *SKQS* 795: 350–515; see p. 194.
71 Mei 1995, 325–326 & 329; Han 1997, 27.
72 Legge 1960, 3: 218.
73 Chu 1994, 189–205; Han 1997, 26–28 emphasise solely this point.

Chinese origin of Western learning.[74] Li Guangdi recorded Kangxi's praise of Mei:

> As to astronomy and mathematics, We are most attentive to their study. We now realise that the likes of Mei Wending are rarely seen. He is also a refined scholar; what a shame he is so old![75]

This was an indirect acknowledgement that he had previously underestimated Mei, whom it was now too late to recruit to his service. A few days after receiving Mei, the emperor sent him a calligraphy: 'Erudite learning partaking in the subtle' (*jixue canwei* 績學參微).[76] The title of Mei's non-technical writings, *Collected poems and prose from the Hall of Erudite Learning* (*Jixuetang shiwenchao* 績學堂詩文鈔), echoes this great honour. For both men the encounter had been significant. Kangxi was convinced that there were some Chinese scholars competent in the mathematical sciences, while Mei Wending exemplified the fact that specialisation in them could open the way to imperial recognition, if not to an official career.

11.3 From favour to distrust: the papal legation

The first decade of the eighteenth century saw a decline of fortune for the Catholic missions in China. By 1700, there were about 140 missionaries in China and 200,000 Christians; the latter number had probably doubled in about fifteen years.[77] Enthusiastic reports of the 'Edict of Toleration' spurred zeal for the China mission in Europe, and especially in France. However, those very numbers seem to have been a cause of concern for the emperor. Having come across many missionaries and converts during his 1703 Southern tour, he made his displeasure about it felt to the Beijing Jesuits early the following year, as the latter reported to Rome.[78] Another element must have harmed the Jesuits' interests at the time: Songgotu, the minister who had been one of their main protectors, was arrested in July of that year, and died in prison the same year. He was accused of being part of a conspiracy in favour of the emperor's second son Yinreng, then still the heir apparent, who was his grand-nephew: Songgotu's whole family was involved in the affair.[79] His fall was but one episode in the factional politics that dominated throughout the reign; however, from then on,

---

74 Chu 2005, 60–73. The 'Chinese origin of Western learning' in astronomy seems to fascinate historians of science; see, among others, Jiang 1988; Han 1997; Wang 1997; Han 1998; Han 2002; Wang 2006b.
75 Li (1826) 1999, 596; Mei 1995, 353.
76 Mei 1995, 353.
77 Standaert 2001, 307–308, 383–384.
78 Rouleau 1962, 296.
79 Wu 1979, 77–82; according to Wu, Gao Shiqi may well have denounced Songgotu when he visited the imperial apartments; see pp. 243–244.

crises related to the issue of Kangxi's succession seem to have dominated court politics, with the emperor growing generally more and more distrustful of those around him.[80] This must have had some impact on his attitude towards the Jesuits, insofar as links of patronage associated them with members of one or another of these factions.

This being said, the issue of the authority of Rome over China missionaries played a major role in the decline in the court Jesuits' fortune during the 1700s. Two different albeit related questions came into play: the general problem of unifying catholic missions worldwide under papal control, and the famous Rites Controversy, which is probably the facet of the Jesuit enterprise in China best known in Europe. Since Ricci's time, the interpretation of Confucian rites prevalent among the China Jesuits was that they were purely social rites, and were therefore compatible with Christianity. Over the years a number of dissenting voices were heard on this matter within the Society of Jesus. However, it was the arrival of the mendicant orders in China that brought the issue to Rome in the 1640s. The Controversy reached its peak during the Kangxi reign: in 1693 Charles Maigrot (1652–1730), a priest of the *Missions étrangères de Paris*, Vicar Apostolic of Fujian, indicted Confucian rites as incompatible with the true religion; the debates that followed in Europe resulted in a papal decree that forbade Chinese Christians from taking part in sacrifices to ancestors and to Confucius. A legate was sent to China to make this decree known, and to unify all missionaries working in China under the authority of one single superior. The legate, Charles-Thomas Maillard de Tournon (1668–1710), was granted an audience by Kangxi for the first time in December 1705. Although the emperor agreed to the idea of having a superior for all missionaries, there was a major bone of contention: in Kangxi's view, the superior of the China mission had to be one of the court Jesuits of his own choice, while for Tournon this superior was to be nominated by himself, to whom the Pope's authority had been delegated. A second audience took place in June 1706, in the presence of Maigrot. The two priests were to explain their position on Confucian rites to the emperor; Maigrot apparently turned out to be incapable of speaking in Chinese, let alone of citing the Classics in support of their viewpoint.[81] The emperor was irritated by what in his view amounted to outside interference in the affairs of the empire. Maigrot was summoned again on his own, and eventually expelled. On his way back to Rome, Tournon was kept a prisoner in Macao by the Portuguese, who needed to stay on good terms with the Beijing court, and

---

80 Spence 2002, 160–169.
81 It is possible that Maigrot, who had by then been in Fujian for more than a decade, spoke a local language rather than Mandarin.

would have defended the Jesuits and the *Padroado*; he died there a few years later.⁸²

Following this legation, it was generally perceived in the capital that 'the emperor seemed to get sick of Europeans',⁸³ including the court Jesuits, who of course did not share Tournon and Maigrot's views and were dismayed by the whole affair. Aware that his image as a Confucian monarch was at stake, Kangxi made sure that his change of disposition was known among officials, as once more witnessed by Li Guangdi:

On the 19th day of the 10th month of the year *bingxu* 丙戌,⁸⁴ Xiong Cilü and I jointly presented Master Zhu's works.⁸⁵ His Majesty ordered all the eunuchs to withdraw; calling me and Xiong Cilü near him, He said:

'Do you know that the Westerners are gradually getting into mischief? They have even insulted Master Confucius [Kong fuzi 孔夫子]. The reason why We treat them well is solely in order to use their skills. Sure enough they are good at astronomical and mathematical studies. You share this with scholars; see to it that local officials outside know the reason [why We treat the Westerners well], so that they can understand Our intention.'⁸⁶

It was all the more necessary for the emperor to advertise his disapproval of Westerners at this juncture as he was intent on continuing to use their skills. In particular, one of the projects he thereafter commissioned from them was of strategic significance, while requiring the cooperation of local officials.

11.4 The Jesuits as imperial cartographers

This project, doubtlessly the longest and most exacting task that the emperor entrusted to the Jesuits, was the survey of his empire. It has often been said that it was the Jesuits who gave Kangxi the idea of having a map of his empire made.⁸⁷ This, however, may well be the result of the French Jesuits' successful communication policy in Europe. There is evidence that the emperor had long intended to have accurate maps and geographical data concerning his whole empire: according to Thomas, Kangxi had already mentioned this plan to Verbiest.⁸⁸ Since the beginning of the reign, there were a number of

---

82   On the Tournon legation and the Rites controversy, see Rouleau 1962, Mungello 1994, Standaert 2001, 497–498, 682–684, Rule 2008.
83   ARSI, Jap. Sin. II 154, 1.
84   23 November 1706.
85   This was probably a draft of the imperial edition of the *Zhuzi quanshu* 朱子全書 (Complete works of Master Zhu, 1713), which Li Guangdi and Xiong Cilü supervised together with Wu Han 吳涵 (d. 1709); see p. 269.
86   Li 1995, 2: 643.
87   See e.g. Pfister 1932–34, 530; Yee 1994, 180.
88   Bosmans 1926, 160.

complaints about the inaccuracy of maps from officials. Local surveys were sometimes commissioned, not only in Manchuria but also in areas that suffered from floods.[89] This may have prompted the emperor to create, in 1686, an office in charge of compiling a *Gazetteer of the Great Qing unification* (*Da Qing yitong zhi* 大清一統志, 1746). The work, which took sixty years to complete, was to include maps as well as details of the physical geography of the empire. Whenever they became aware of mistakes in existing official maps, the Jesuits also pointed them out; it seems that they did offer their services to correct these.[90] The decision made by the emperor in 1708 to commission them to survey the empire and draw a general map 'using the surveying methods of Western learning' (*yong xixue liangfa* 用西學量法) came after some trials done in 'Tartary', in the surroundings of Beijing and along portions of the Great Wall.[91] An Office for the Complete View of the Imperial Territory (*Huangyu quanlan guan* 皇輿全覽館) was created, which seems to have functioned independently from the one in charge of the *Gazetteer*.[92] Three French Jesuits, Bouvet, Régis and Pierre Jartoux (1669–1720) started on what was to be the first campaign of the general survey on 4 July 1708, with the aim of producing a complete map of the Great Wall. They returned to Beijing on 10 January 1709, and presented a map 'of more than 15 feet' that included the details of the relief and main rivers, as well as the nearby military forts.[93] The following expeditions included Jesuits from the Portuguese mission; despite the fact that they were later published in France, the maps were not exclusively the work of the French.[94]

Unlike the tutoring in mathematics, the survey and production of the map, which took ten years to complete, was not merely a matter of a few Jesuits and scholars working for the emperor under the jurisdiction of the Imperial Household Department: instead it was a state enterprise on a major scale. Officials of the Ministry of Personnel, of the Ministry of War, of the Astronomical Bureau took part in the expeditions, while local officials were in charge of providing horses as well as any material required for the work. Each governor was to collect the map of the province he was in charge of as soon as it had been drawn, and to send a trusted servant to present it directly to the emperor; he was also to report on the surveyors' activities on the

---

89   Foss 1988, 223.
90   *LEC* 22: 316–317; Yee 1994, 180; Leibundgut 2007, 35–36.
91   Han & Wu 2006, 366, Hostetler 2001, 71; on these cartographic expeditions, see Fuchs 1943, De Thomaz de Bossierre 1977, 67–69, Foss 1988, 223, Widmaier 1990, 159.
92   This Office is mentioned in 1713, on the occasion of the emperor's birthday ceremony; *Wanshou shengdian chuji* 萬壽盛典初集, *SKQS* 653: 125; the date of its creation remains to be ascertained.
93   Fuchs 1943, 61.
94   Landry-Deron 2002, 143–149.

territory he was in charge of.[95] The survey of some frontier zones was done by officials without the Jesuits.[96] The latter, while providing the techniques that the emperor had deemed the best, were merely a few among a very large staff involved in a project that was of major consequence to the Qing state, rather than the sole authors of the general map that came to be known in Europe as 'the Kangxi Atlas'. Looking back at the Jesuits' accounts of their tutoring of the emperor in the light of this project, Kangxi's emphasis on applied geometry and on surveying instruments suggests that during the 1690s, when the tutoring in mathematics was the most intensive, he already had a strong interest in cartography.

11.5 Assessing mathematical talent: Chen Houyao's interview

Following the imperial recognition of Mei Wending's competence, Li Guangdi continued to recommend scholars versed in mathematics to the emperor. In 1708 he mentioned the name of Chen Houyao 陳厚耀 (1648–1722), who had passed the Metropolitan examination in 1706; the latter was then summoned to Beijing.[97] The following year, in the late spring of 1709, he was admitted into the Garden of pervading spring, at the Southern Study,[98] and questioned by the emperor on his mathematical abilities through the intermediary of Li Yu 李玉, a Palace Eunuch, and in the presence of two Grand Secretaries. Part of the dialogue took place in writing. In this indirect manner, without Chen being admitted into the emperor's presence, the exchange continued as the latter left the capital for his summer residence in Jehol, where Chen followed in the company of Kuixu 揆敘 (1674 ?–1717), then Administrator of the Hanlin Academy (*Hanlin yuan zhangyuan xueshi* 翰林院掌院學士).[99] Chen left an account of his exchanges with the emperor.[100] Asked what mathematics he had studied, Chen began his reply by asserting that he had been trained in philosophy in the orthodox way, using the same Ming compilation of Song philosophers that Kangxi had criticised on the subject of harmonics:

---

95 For an example of such a report, by the governor of the Sichuan province, see *Zhongguo diyi lishi dang'anguan* 1985, 5: 674–676.
96 Wang 1995, 134–135.
97 *Jiangnan tongzhi* 江南通志, *SKQS* 511: 585–586.
98 The office of the emperor's personal secretaries at the Garden of pervading spring was named *Nanshufang* 南書房, after the one located in the Forbidden City.
99 Kuixu was the son of Mingju and the younger brother of the poet Singde 性德 (1655–1685). After Yinreng's deposition in 1708, Kuixu was among those who proposed that Yinsi 胤禩, Kangxi's 8th son, be promoted heir apparent; Hummel 1943, 430–431; Wu 1979, 141–142.
100 Jiao 1998 121–123; reproduced in Han 2003c, 459–463.

> In my youth I read *Nature and principle* (*Xingli* 性理)[101] and researched calendrical astronomy (*lifa* 曆法). Because I did not know mathematics, I also studied it. Gradually I came to understand the 'nine chapters'. Then again I focused on triangles; their basic principles are in the division of the circle (*geyuan* 割圓); their use is surveying. All this is very deep and profound, with infinitely subtle meaning. I only possess rudiments, and its complete meaning remains unknown to me.[102]

A variety of topics are then discussed. On measurement at a distance, Chen states that he has no surveying instruments, but that one can manage without them. This is certainly no news to the emperor, as a method of doing this is found in the *Practice of instruments for measuring heights and distances*. Chen also knows that the length of the Sun's shadow varies according to the North Pole's altitude. The emperor has him calculate the diameter of the Earth according to the new standard *chi*. Here, as in discussions with Li Guangdi, the emperor brings out matters relevant to his particular interests at the time, in this case the survey undertaken by the Jesuits.

The lengthiest and most detailed exchange concerns methods of calculation, especially ways of determining the rank of a product or a quotient, and root extraction. Like the emperor himself,[103] Chen is more familiar with the abacus than with written calculation. The example of the method he uses for multiplication that he presents to the emperor is taken from the Song philosopher Shao Yong's exposition of the cycles of time: calculation is firmly embedded into classical scholarship rather than in technical matters. The exchange on calculation techniques ends with Kangxi offering to 'teach calculation methods' to Chen. The latter is also asked whether he knows Mei Wending; he replies that he has visited the elderly scholar in order to seek his advice, and that he deems himself much inferior to Mei. The only mathematical work mentioned in the dialogue is Cheng Dawei's *Unified lineage of mathematical methods* (*Suanfa tongzong* 算法統宗, 1592).[104] The emperor asks Chen if he knows it, to which the answer is affirmative; although this work is seldom mentioned in sources of that period, both the emperor and the scholars versed in mathematics were familiar with it. A 'method for measuring piles' (*duoji zhangliang zhi fa* 垛積丈量之法) is mentioned in the following question, and Chen answers that he knows it too. Interestingly, among the works attributed to him, one is entitled *Piles* (*Duoji* 垛積) or *Proportions of borrowed root and powers for piles* (*Duoji jiegenfang bili* 垛積借根方比例);

---

101 The *Complete collection on nature and principle* (*Xingli daquan* 性理大全, 1415), had by then become a standard work. Kangxi was somewhat critical of it; see pp. 229–230.
102 Jiao 1998, 121.
103 Jami 2002b, 32.
104 See pp. 17–22.

this second title was most likely chosen later. One wonders how much Chen had worked on the subject at the time; perhaps his study of the subject was a response to Kangxi's suggestion.[105] Another subject discussed is the various values used for the ratio between the diameter and the circumference of the circle: Chen uses 113 for the diameter and 355 for the circumference, which he correctly ascribes to Zu Chongzhi 祖冲之 (429–500).[106] This could not but satisfy Kangxi, who deemed 7 and 22 insufficiently accurate.[107]

Finally, the sphericity of the Earth is discussed in the presence of several Literary Officials (*cichen* 詞臣). Chen Houyao quotes evidence for this, first from the Chinese Classics, then from the Jesuits' geographical writings; and lastly, he mentions the results of observation. None of the officials present is aware of such a notion: they all adhere to the classical idea that 'the heavens are round and the earth is square' (*tian yuan di fang* 天圓地方).[108] The emperor then sends them a pile of books dating back to the Yuan and Ming dynasties, which we are told discuss the notion that the cosmos is similar to an egg in which the heavens surround the Earth; the latter is likened to an egg yoke.[109] In them the emperor finds evidence that scholars of the Song dynasty already discussed the subject. This is an example of Kangxi's use of the idea that 'Western learning originated in China' in his campaign to win over his officials to the cosmography proposed by the Jesuits. The other arguments in favour of the sphericity of the Earth put forward by Chen Houyao, which rely on a work of Western learning and on direct observation, had less authority in their eyes and were only accessible to specialists.

Chen Houyao was the first official whose career was to change as the result of imperial recognition of his mathematical talent. By 1710, the emperor, while continuing to use the Jesuits' skills, had made it plain to his officials that he was seeking similar talents among literati. A few scholars under Li Guangdi's patronage had already accessed the hitherto monopolised textbooks on geometry and calculation, and Kangxi was now considering sharing them more widely with his officials. This enterprise was to be carried out during the last decade of his reign.

---

105 Chen's writing on piles is mentioned in Li 2000a, 433–434; on the coining of the phrase 'proportions of borrowed root and powers', see pp. 336–337.
106 On Zu Chongzhi see Ang 1997.
107 See p. 230.
108 Han 2003c, 466; Cullen 1996, 128–137.
109 This corresponds to the ancient Spherical heaven (*huntian* 渾天) theory; See Cullen 1996, 57–66.

CHAPTER 12

The Office of Mathematics: foundation and staff

For the emperor to share his mathematical knowledge with Chinese literati entailed a thorough rewriting of the Jesuits' lecture notes into an imperial textbook. For this purpose—as for all editorial projects carried out at the court—the first step was to recruit scholars versed in this particular type of knowledge. In the field of astronomy, however, there was no equivalent to the mathematical lecture notes. The circumstances under which, in 1711, the emperor undertook to remedy this absence were markedly different from those of the 1690s: he had come to treat the Jesuits with ostensible harshness.

12.1 The Summer solstice of 1711

On 22 June 1711, two days after praising his officials for the unprecedented accuracy of the cartographic work done using his new standard unit,[1] Kangxi 'calculated [...] and observed himself' the summer solstice from Jehol,[2] and '[...] found, or thought he had found, that it must have happened before noon. And yet the Tribunal of Mathematics, in the Empire's calendar, had marked it 56 minutes after noon.'[3] It was not unprecedented for the emperor to check the Astronomical Bureau's calculations and observations against his own: when he had found a mistake in the Bureau's prediction for the solar eclipse of 27 November 1704, he had put the blame on the 'calculators who truncated fractions too much in their calculations', and the officials of the Bureau had duly acknowledged their mistake. It seems that at the time astronomers routinely submitted as observations the figures that they had obtained from calculation, and that the emperor was aware of it.[4] But in 1711, things took a different turn; this was recounted in detail by Jean-François Foucquet (1665–1741), a French Jesuit who was closely involved.[5]

---

1 *SL* 246, 9a–10b.
2 The city today called Chengde 承德 was named Rehe 熱河 by Kangxi, commonly transcribed as Jehol in European sources; the latter spelling is used in what follows; the city was renamed Chengde by Yongzheng in 1733; Forêt 2000, esp. 16.
3 ARSI, Jap. Sin. II 154, 1.
4 Zhongguo diyi lishi dang'an guan 1997, 135 no. 9; Lü 2007, 140.
5 ARSI, Jap. Sin. II 154 (Foucquet 1716); the content of this document will be discussed in detail in chapter 13.

The Jesuits who had followed the emperor to Jehol as well as those of the Astronomical Bureau argued that the difference between Kangxi's result and what the official astronomers had predicted could not be called a mistake, and pointed out:

1. That the observation of shadows was subject to some error.
2. That the calendar being for the people, there was no need to look at it so closely.
3. That His Majesty had used small instruments and that the Tribunal was using large ones, which should cause some difference.[6]

The emperor angrily commented that this was 'an answer of vile and lowly souls.'[7] However, a few months later, on 25 October, when addressing his Grand Secretaries, he referred to this incident in mild terms:

> We are usually quite careful about astronomy. There are no mistakes in the main points of Western astronomy. But with measurements, on the long term there are bound to be errors. For the summer solstice this year, We heard from the report of the Astronomical Bureau that it had occurred at three quarters past noon.[8] We observed the Sun's shadow carefully; in fact it occurred at eleven fifty-four.[9] If at this time there is an error, it is to be feared that after several decades the difference will be greater. Just like with money and cereals, minute amounts, although each is insignificant, produce significant sums when added up. This fact can be verified, unlike in scholars' compositions, where one can work perfunctorily, using empty words.[10]

The final cutting remark on Chinese literati was a reminder that the emperor knew what he was talking about better than any of them when it came to astronomy. The tense situation that the Jesuits experienced at the time did not show through here. Something more fundamental was implied: the error stemmed from calculation rather than from observation. In other words, the system employed by the Jesuits was in need of revision. However Kangxi still trusted their skills: a few months later, in April 1712, he commissioned them to write out 'the principles of astronomy' for him.[11] That is to say, he wanted from them a textbook similar to those written by Pereira on harmonics and by Thomas, Gerbillon and Bouvet on mathematics.

---

6   ARSI, Jap. Sin. II 154, 2.
7   ARSI, Jap. Sin. II 154, 2.
8   午正三刻.
9   午初三刻九分; the times mentioned by the emperor do not match those given by Foucquet.
10 *Shengzu renhuangdi shengxun* 聖祖仁皇帝聖訓, SKQS 411: 211.
11 ARSI, Jap. Sin. II 154, 4–5.

12.2 Selecting talented men

However, the emperor did not simply repeat for astronomy what he had done for mathematics. Instead, he summoned some scholars versed in the mathematical sciences to the court. As Foucquet saw it:

[. . .] he made himself a school of some sort. A few chosen people came to him everyday; and he explained to them himself a few propositions from Euclid, enjoying the pleasure of appearing skilled in these abstract sciences, and savouring the praise that his disciples would not fail to give him, often without understanding. But this school did not last and was only the beginning of a sort of academy that the emperor then set up. He had enquired in Peking and in the Provinces about people skilled in various parts of mathematics among the Tartars and Chinese. The Viceroys and other high mandarins, to pay court to him, introduced to him the elite of the best minds, and the most qualified for the sciences. They were brought to him from all sides. It is among this elite of minds that he again made a choice, mainly among young people, to compose the academy we are talking about. He put more than a hundred of them, that is officers who preside, calculators, geometers, musicians, astronomers, students for all these faculties, to say nothing of a considerable number of craftsmen, who are working on the instruments. He assigned for this troop a vast area full of buildings, within the walls of the Tchang tchun yuen [Changchunyuan] [. . .].[12]

This description, phrased in somewhat dismissive terms, provides an accurate account of Kangxi's recruitment of scholars versed in the mathematical sciences—if not of the latter's skills. The Ministry of Rites was ordered to select by examination 'staff to serve in mathematics' (*xiaoli suanfa renyuan* 效力算法人員), whom he would examine personally. The exact date of this examination is not known; it was probably held some time in 1712. The fact that the Ministry of Rites was in charge of it suggests that it was deliberately modelled on the triennial Metropolitan examination, for which the emperor in person interviewed the successful candidates. Among the forty-two men thus selected, only one is named in the chapter of the *Draft Qing History* (*Qingshi gao* 清史稿) devoted to official astronomy: this was Gucong 顧琮 (1685–1755), the grandson of Gubadai 顧八代 (1640–1709), a Manchu official who had been quite favourable to the Jesuits and was interested in astronomy.[13] Although we do not know how much knowledge the grandfather gained from the missionaries, what interests us here is that Gucong came from a family with an interest in the

---

12 ARSI, Jap. Sin. II 154, 3; see also Ripa 1996, 2: 62.
13 In 1692 the name of Gubadai, then the Minister of Rites, appeared first among those who signed the memorial approved by Kangxi that came to be known as the 'Edict of toleration'; Wu 1966, 1789–1701; Han & Wu 2006, 358–359.

mathematical sciences.[14] He may well have been the most socially prominent rather than the most brilliant mathematical scholar thus selected. In the biographical section of the *Draft Qing History*, we are told that among three hundred candidates sent from all provinces, a scholar from Jiangsu named Gu Chenxu 顧陳垿 (1678–1747) who had passed the provincial examination in 1706 ranked first in the mathematical examination;[15] another biography of Gu states that he was accordingly called the 'Principal Graduate in mathematics' (*suan zhuangyuan* 算狀元), apparently an informal title.[16] The author of an essay on harmonics, the *Numbers laid out for bells and pipes* (*Zhonglü chenshu* 鐘律陳數), and one on phonology, the *Commentary on the terms for the Eight kinds of arrows with illustrated explanation* (*Bashi zhuzi tushuo* 八矢注字圖說), Gu is absent from histories of mathematics.[17] His modest career after 1712 suggests that the informal title he held had little to do with that of plain Principal Graduate (*zhuangyuan* 狀元)[18] earned once every three years by the scholar who ranked first in the Metropolitan Examination: clearly the Jesuit characterisation of 'the elite of the best mind' sent to the capital does not correspond to the criteria then prevalent in China. Gu Chenxu must have been the only 'Principal Graduate in mathematics': the empire-wide special examination only took place once. Nonetheless it contributed to enhancing the status of the mathematical sciences among literati; moreover it was the occasion of an unprecedented gathering of literati versed in this field in the capital.

By the summer of 1712, the emperor had surrounded himself with young men who could be interlocutors in discussing mathematics and astronomy. When he set off for his yearly stay in Jehol, his retinue included a number of such men. These included Chen Houyao, whom Kangxi had interviewed three years earlier, but also four men connected with the Astronomical Bureau: He Guozhu 何國柱 and He Guozong 何國宗 (d. 1766) were the sons of He Junxi 何君錫, a Supervisor of the Five Offices (*wuguanzheng* 五官正). The latter seems to have been well versed in 'the old calendrical method' (*gu li fa* 古曆法) and is believed to have been a disciple of Yang Guangxian, the Jesuits' great enemy in the 1660s.[19] He Guozong had just been granted the title of Metropolitan Graduate and appointed a Hanlin Bachelor.[20]

---

14 Zhao 1977, 1668; on Gubadai and Gucong, see Hummel 1943, 271.
15 Gu's biography gives 72 as the total number of men who passed the mathematical examination; the figure 42, mentioned above, seems more compatible with that of 45 given in a later edict; see p. 271.
16 Zhao 1977, 13367; Ruan 1982, 85–86.
17 I am grateful to Prof. Han Qi 韓琦 for drawing my attention to Gu Chenxu.
18 For the translation of official terms related to examinations, I mostly follow Hucker 1985; Elman 2000 uses slightly different translations.
19 Zhao 1977, 1666; Han 2003b, 85.
20 *Jifu tongzhi* 畿輔通志, *SKQS* 505: 501; 599: 647.

Minggantu 明安圖 (d. 1765?) was then an Official Student (*guanxuesheng* 官學生) at the Bureau: Cengde 成德 had been a Vice-Director (*jianfu* 監副) there.[21] Mei Wending's grandson Mei Juecheng was also summoned to join them. All of them were 'allowed to query and ask [the emperor] difficult questions, like disciples with their master'.[22]

A quick glance at the profiles of these six men reveals several interesting features. Apart from Chen Houyao, all of them belonged to a new generation, whose career would span the reigns of three emperors: Kangxi, Yongzheng (1723–1735) and Qianlong (1736–1795). Those whose family background we know about were the sons or grandsons of men themselves versed in the mathematical sciences, which suggests that the transmission of knowledge in this field remained a family affair. Last but not least, they fell into two categories: officials of the Astronomical Bureau and their families, which included two Bannermen (Minggantu and Cengde), and protégés of Li Guangdi. Only one of them, Chen Houyao, earned the title of Metropolitan Graduate through the standard examination process. The emperor's opinion regarding the mathematical skills of high officials had probably not changed; instead he had taken the necessary steps to search for mathematical talent where it could be found, that is, outside the top ranks of literati elites.

In Jehol, these disciples studied several mathematical works. In July, the emperor ordered officials of the Hall of Military Glory (Wuyingdian 武英殿), the Imperial Printing Office located inside the Forbidden City, to prepare a clean, corrected copy of Cheng Dawei's *Unified lineage of mathematical methods* (*Suanfa tongzong* 算法統宗, 1592) as 'this work [was] useful'.[23] This appears to have been the first acknowledgement of the value of Chinese mathematics on Kangxi's part. Such imperial praise could well have triggered a new edition of the work by Cheng Dawei's descendants: it was duly mentioned in Cheng Shisui's 程世綏 preface to the 1716 edition of the work.[24] Despite this praise and the subsequent widened availability of the work on the book market, it was to the 'imperial writings' on mathematics that the students were to apply themselves: a few days after the emperor's order concerning the *Unified lineage of mathematical methods*, they completed the study of the *Elements of geometry* and of the *Elements of calculation*.[25]

---

21 Zhao 1977, 1668; on Minggantu see Jami 1990.
22 Zhao 1977, 1668.
23 Zhongguo diyi lishi dang'anguan 1996, no. 1980.
24 Guo 1993, 2: 1217–1218; the preface is dated to Kangxi 55/8/16 (1 October 1716).
25 Zhongguo diyi lishi dang'anguan 1996, no. 1986.

12.3 The mathematical staff at the emperor's sixtieth birthday

The following year, on 12 April 1713, an imperial birthday ceremony (*wanshou shengdian* 萬壽盛典) was held to celebrate Kangxi's entering his sixtieth year. This great event was duly recorded in the *First collection of the imperial birthday ceremony* (*Wanshou shengdian chuji* 萬壽盛典初集, 1716).[26] This work gives a depiction of the imperial procession from the Due South (Zhengyang 正陽) Gate, through the Forbidden City to the Garden of pervading spring (which had long been Kangxi's favourite residence), with a list of those in attendance and of gifts presented, as well as all the congratulatory poems written on this occasion. Some of the men who served the emperor in fields related to the mathematical sciences took part in this major event. Inside the Forbidden City, in front of the Gate of Heavenly purity, and next to officials of the Hall of the Nourishment of the Mind stood

- Stumpf, Suarez, Bouvet and Parrenin and other Westerners from the Three Churches (*Santang xiyangren* 三堂西洋人);[27]
- Dongtai 董泰 and others from the Cabinet of Mathematics (*Suanfa fang* 算法房);
- Zhang Wei 張位 and others from the Office for the Complete View of the Imperial Territory (*Huangyu quanlan guan* 皇輿全覽館);[28]
- Qiu Qiyuan 邱起元 and others from the Office of the *Quintessence of the Masters and Histories* (*Zishi jinghua guan* 子史精華館) [...][29]

This list indicates that, after the special examination, a Cabinet of Mathematics existed alongside offices devoted to projects such as the map of the empire and the *Quintessence*, an encyclopaedia of terms and phrases found in the works of Masters and in the Histories. By then, most of the Jesuits who had tutored Kangxi before 1700 had died: Gerbillon in 1707, Pereira in 1708, and Thomas in 1709; Grimaldi had passed away in 1712. Their colleagues and successors were present at the ceremony as a group.

Western learning was represented among the gifts presented to the emperor as well as among the men who attended the procession. Kilian Stumpf (1655–1720), who had succeeded Grimaldi as Administrator of the Calendar after working as the latter's assistant, joined with the Bureau's Manchu Director Mingtu to present a 'Perpetual calendar of the Kangxi era of the Great Qing' (*Da Qing Kangxi wannian*

---

26 *SKQS* 653 & 654.
27 On the three churches of Beijing see pp. 240–242. A fourth church was established in 1723; Standaert 2001, 584. On Jose Suarez (1656–1736), see Pfister 1932–34, 399–402.
28 Zhang Wei was an Imperial student; *SKQS* 654: 667.
29 *SKQS* 653: 125.

li 大清康熙萬年曆) and 'Tables of squares, cubes and roots for 10,000 numbers' (*Ping li fang gen wan shu biao* 平立方根萬數表).[30] Among the gifts presented by the Jesuits, there was a calculator (*suanfa yunzhou* 算法運軸, lit. 'calculating revolving spools'), which may have been a variant of the famous Pascal calculator; several such calculators, made in the imperial workshops, are still preserved at the Palace Museum;[31] there was also a box of wine (*putaojiu* 葡萄酒) and two bags of cinchona.[32] The most spectacular set of Western gifts was not presented by the Jesuits, but by Wang Hongxu 王鴻緒 (1645–1723),[33] a compiler of the *Ming History* who was then the Minister of Revenue (*Hubu shangshu* 戶部尚書); this set included two surveying instruments: a level (*diping yi* 地平儀; see Illustration 12.1) and an 'instrument for estimating distances' *chaliang yuanjin yiqi* 察量遠近儀器.[34] Thus the Jesuits were neither the only nor the best providers of Western curios.

The 'Manchu and Han staff serving in mathematics' (*suanfa man han xiaoli renyuan* 算法滿漢效力人員) who watched the procession most likely included some of those who had passed the mathematics examination.[35] They contributed poetic eulogies (*gesong* 歌頌), which form a chapter of the record of the celebration. There they are referred to as 'the subjects at the School for tutoring the young' (*Mengyangzhai zhu chen* 蒙養齋諸臣): this suggests that at the time they had already settled at the Garden of pervading spring.[36] Altogether twenty-four of the men who worked on mathematics in the emperor's service at the time can be identified (see Table 12.2).

It is not easy to understand the hierarchy among them. On both lists, Metropolitan Graduates are named first. Among the thirteen men whose name appears on both lists, two Provincial Graduates are named first; neither of them seems to have precedent over the students of various ranks. Quite the contrary: Mei Juecheng, appointed an Imperial Student by Kangxi, seems to have taken precedent over He Guodong 何國棟, another son of He Junxi who became a Provincial Graduate that same year.[37] Whatever hierarchy there was does not seem to follow that of the civil service ranks; we have no indication of how each of these individuals' skills in mathematics and harmonics were rated.

---

30 *SKQS* 654: 79; Mingtu was the Manchu Director of the Astronomical Bureau from 1711 to 1739; he was favourable enough to the Jesuits to write an inscription for the Southern Church on the occasion of its renovation; Dudink 1993, 14–15.
31 Liu 1998, 96–98.
32 *SKQS* 654: 51–52; on cinchona, see Puente Ballesteros 2007.
33 On Wang Hongxu, see Hummel 1943, 826.
34 *SKQS* 654: 83–84.
35 *SKQS* 653: 632–633.
36 *SKQS* 654: 480.
37 *Jifu tongzhi*, *SKQS* 505: 609.

CHAPTER 12 | The Office of Mathematics: foundation and staff

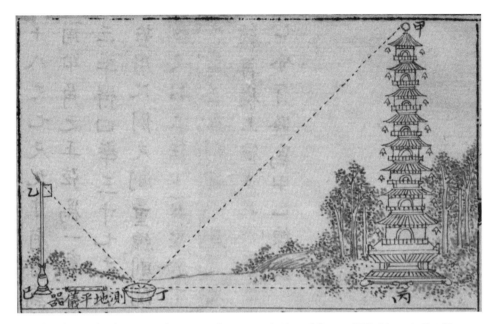

Illustration 12.1 Use of a level as illustrated in the *Practice of instruments for measuring heights and distances* (BML, Ms. 75–80 D, 16b).

Eight of the twenty-four men can be identified as Manchu or Mongol; at least one third of the staff belonged to the Banners. This confirms the importance that Kangxi attached to promoting the mathematical sciences among the conquest elites.

At the time of Kangxi's sixtieth birthday, Cheng Shisui, the descendant of Cheng Dawei, was an Imperial Student.[38] However, he does not seem to have been connected to court mathematical activities: having an ancestor versed in the field did not imply that one would be among the 'staff'. It would be interesting to know how far Cheng Shisui mastered the content of his ancestor's work.

12.4 An editorial project in the mathematical sciences

The last decade of the Kangxi reign saw the flourishing and completion of a number of imperial editorial projects; after the rhyme dictionary *Rhyme Repository of the Adorned Literature* [Studio] (*Peiwen yunfu* 佩文韻府, 1711), the *Kangxi Character Dictionary* (*Kangxi zidian* 康熙字典, 1716) was compiled, as well as the *Quintessence of the Masters and*

---

38 *SKQS* 654: 700; Cheng became a Provincial Graduate in 1714; *Jiangnan tongzhi* 江南通志, *SKQS* 510: 884.

267

Table 12.2 Manchu and Han staff serving in mathematics

Name[1]	Rank or position[2]
Dongtai 董泰	Formerly Hanlin Bachelor (*yuan ren Hanlinyuan shujishi* 原任翰林院庶吉士)[3]
Boerhe 博爾和	Imperial Student (*jiansheng* 監生)
Mushitai 穆世泰	Imperial Student
Mucengge 穆成格	Provincial Graduate (*juren* 舉人)
Yulun 余掄	Imperial Student
Pan Yunhong 潘蘊洪	District Graduate (*shengyuan* 生員)
Shen Chenglie 沈盛烈	Tribute Student (*gongsheng* 貢生)
Zhu Song 朱崧	Imperial Student
Feng Rui 馮罃	Tribute Student
Ye Changyang 葉長揚	Provincial Graduate
Wang Chong 王翀	Imperial Student
Xu Juemin 徐覺民	Tribute Student
Wang Yuanzheng 王元正[4]	Confucian Apprentice (*tongsheng* 童生)
Chen Shiming 陳世明	Imperial Student
He Guozong 何國宗	Hanlin Bachelor (*Hanlinyuan shujishi* 翰林院庶吉士)
Chen Houyao 陳厚耀	Secretariat Drafter (*zhongshu sheren* 中書舍任)
Mei Juecheng 梅瑴成	Imperial Student
He Guodong 何國棟	Provincial Graduate
Lundali 倫達禮[5]	Member of the staff selected by imperial order to serve in mathematics (*qin qu xiaoli suanfa renyuan* 欽取效力算法人員)
He Guozhu 何國柱[6]	Manager of the calendar in the Five Bureaus (*wuguan sili* 五官司曆)
Gu Chenxu 顧陳垿	Provincial Graduate 1705[7]
Cengde 成德	Former Supervisor of the Five Offices (*yuan ren wuguanzheng* 原任五官正) as of 1712[8]
Minggantu 明安圖	Official Student (*guangxuesheng* 官學生) as of 1712[9]
Zhaohai 照海	Manchu Provincial Graduate 1708[10]

1 The first fourteen persons on the list attended the procession together; the next six attended it in another quality; the last four are not mentioned in the *First collection of the imperial birthday ceremony*.
2 Except for the last four persons, the rank is taken from the *First collection of the imperial birthday ceremony*; *SKQS* 653 & 654.
3 *SKQS* 654: 448.
4 Wang later became a Calendar manager of the Five Bureaus at the Astronomical Bureau, in 1720–21 (Qu 1997, 53).
5 A member of the Gioro 覺羅 clan; *SKQS* 653: 670.
6 He Guozhu attended the ceremony as member of a mission about to depart for Korea; *SKQS* 653: 174; see pp. 277–279.
7 Huang (1955) 1982, 85.
8 Zhao 1977, 1668.
9 Zhao 1977, 1668.
10 *Qinding baqi tongzhi* 欽定八旗通志, *SKQS* 665: 838.

CHAPTER 12 | The Office of Mathematics: foundation and staff

Histories (*Zishi jinghua* 子史精華, 1727) mentioned above. Li Guangdi, who had been one of the editors of *Rhyme Repository*, was thereafter involved in a number of other editorial projects. Some of those articulated imperial orthodoxy: the main ones were an edition of the *Complete works of Master Zhu* (*Zhuzi quanshu* 朱子全書, 1713), a compilation of the *Book of change* (*Yijing* 易經) with various commentaries entitled *Middle way to the Change of Zhou* (*Zhouyi zhezhong* 周易折中, 1715), and the *Essential significance of nature and principle* (*Xingli jingyi* 性理精義, 1717), which presented the essentials of the teachings of the Cheng-Zhu school.[39] Other projects supervised by Li Guangdi related to astrology, ritual and music, fields that were one way or another connected to the mathematical sciences. These included the *Investigation of the origins of planetary ephemerides* (*Xingli kaoyuan* 星曆考原, 1713), which was thereafter used for divinatory calculations,[40] the *Summary of the Monthly ordinances* (*Yueling jiyao* 月令輯要, 1715), and the *Clarification of the subtleties of phonetics* (*Yinyun chanwei* 音韻闡微, 1725).

Just as Li Guangdi supervised more than one project at a time, scholars who worked under him also took part in several of these projects. Thus the names of He Guozhong and Mei Juecheng appear among the compilers of the last three works listed above. Together with them, protégés of Li Guangdi who had studied mathematics in Baoding a decade earlier also took part in some of these projects; these included Wei Tingzhen, Wang Lansheng and Chen Wance, as well as Li's third son Li Zhongqiao 李鍾僑 (1679–1732). Wang Lansheng, a mere District Graduate, was recommended to the emperor by Li Guangdi in the autumn of 1712, and assigned to correct the *Complete works of Master Zhu* and the *Middle way to the Change of Zhou* a few months later. Not all those who worked on these compilations were Li's protégés, though: thus Wu Xiaodeng 吳孝登, a Chinese Bannerman, was among the Imperial Correctors (*yuqian jiaodui* 御前校對) for all three works mentioned.

It was amidst these multiple compilations that the emperor decided to employ the 'staff serving in mathematics' on a similar project in the mathematical sciences. On 23 July 1713, while spending the hottest days of the year in Jehol, he gave instructions to his Third Son Yinzhi:

---

39 See pp. 15–16.
40 Smith 1991, 50–52.

All the writings on harmonics and mathematics ought to be edited. The writings on harmonics and mathematics that We have composed are now being sent down. You shall lead the Bachelor He Guozong and others in establishing an Office in the Summer Palace[41] to edit them.[42]

Rather than Li Guangdi, who had provided him with a number of scholars versed in mathematics and was already working with them on editorial projects, the emperor chose his Third Son, who had studied and worked with the Jesuits, to supervise an editorial project that focused on fields connected to Western learning. The imperial writings on mathematics mentioned in this document must have been the abundant lecture notes produced by the Jesuits in the 1690s.[43] Similarly, those on harmonics must have included the treatise on Western musical theory authored by Tomás Pereira, the *Essentials of the [standard] pitchpipes* (*Lülü zuanyao* 律呂纂要).[44] The appropriation of the Jesuits' sciences as imperial scholarship was undertaken under Manchu rather than Chinese supervision, and excluding those who had been the emperor's source of knowledge in these fields. This exclusion was a deliberate choice: at the time, four Jesuits and two other missionaries followed Kangxi to Jehol. They included Joachim Bouvet, who had tutored Kangxi in Euclidean geometry, and Teodorico Pedrini (1571–1746), a musician; however it was not Bouvet, but another Jesuit, Luigi Gonzaga (1673–1718), who was there in the capacity of mathematician.[45]

About a month after ordering the setting up of the Office, Kangxi granted the title of Provincial Graduate to both Wang Lansheng and Mei Juecheng.[46] One might wonder whether this special promotion was a reward for their skills in mathematics, or for the part they took in the completion of the *Complete works of Master Zhu*. A few days later, the emperor further clarified how he saw the relevance of the former subject:

In Our writings on harmonics and mathematics, We have probed for their origins. These are now clear: We know that the Yellow Bell is most essential. Altogether both mathematics and standard measure units are closely related to it.[47]

---

41   Here *xinggong* 行宮 refers to the Jehol *Bishu shanzhuang* 避暑山莊.
42   *SL* 255: 13b–14a; date Kangxi 52/6/2, i.e. 23 July 1713. The entry of the *Draft Qing History* for that date tersely mentions the 'Compilation of writings on harmonics and mathematics' (*xiu lü suan shu* 修律算書).
43   See chapters 8 & 9.
44   See Wang 2002, Jami 2008b, 191–192; see p. 72.
45   Ripa 1996, 2: 119.
46   Date: Kangxi 52/7/5 (25 August 1713); Wang 1836, *juan* 1, 29b.
47   *SL* 255: 22b.

These words hammered in the new project's relevance to classical learning; accordingly, they were later included in the *Sagely instructions* (*Shengxun* 聖訓), in which the utterances of the sovereign worthy of being passed on to posterity were collected.[48] The edict stresses the importance of harmonics; this seems to have been the first of what were to be frequent direct interventions of the emperor in the project.

On 11 November 1713, about four months after ordering Yinzhi to set up an Office in Jehol, Kangxi gave him further instructions concerning its organisation:

> For the edition of the works on harmonics and mathematics, establish the Office at the School for tutoring the young (*Mengyangzhai* 蒙養齋). Moreover, check on music instruments for shrines, temples and the palace. As to the forty-five men who study mathematics—the Provincial Graduate Zhaohai 照海 and the others, set them another examination. Those who excel most in study shall be ordered to proceed to the place where the books are to be edited.[49]

This second process of selection was carried through, and some 'Mathematics Graduates' were thus eliminated from the editorial team: these included Gu Chenxu, who had ranked first in the first examination, and Chen Shiming, who had attended the birthday ceremony. The two of them then sought the patronage of Chen Menglei 陳夢雷 (1651–1723). The latter, who had been Yinzhi's tutor, was then working on his famous *Compendium of books and illustrations ancient and modern* (*Gujin tushu jicheng* 古今圖書集成), which was to include, among many others, works on mathematics, harmonics and astronomy.[50] This suggests that the gathering of mathematical talents in the capital had some impact beyond the narrow circles of the Great Interior. A different order, dated to the same day, is quoted in Wang Lansheng's collected papers:

> The Sixteenth Son received a decree: 'You shall lead He Guozong, Mei Juecheng, Wei Tingzhen, Wang Lansheng, Fang Bao 方苞 and others in editing Our Imperial writings on astronomy, harmonics and mathematics. You shall also make musical instruments. It is ordered that an office (*ju* 局) should be opened [for this purpose] at the School for tutoring the young, inside the Eastern Gate of the Memorial [Office] (*Zoushi dong men* 奏事東門) of the Garden of pervading spring.[51]

A new member of Li Guangdi's circle appears here: Fang Bao (1668–1749), a brilliant scholar, had been involved in the Dai Mingshi case. After the latter was executed for allegedly treasonable writings, Li Guangdi obtained Fang Bao's pardon on the grounds of the latter's

---

48 *Shengzu renhuangdi shengxun*, SKQS 411: 211.
49 SL 256: 8b; date: Kangxi 52/9/20 甲子 (7 November 1713).
50 Chen 1995, 153; Han Qi 2003, 79; on Chen Menglei see Hummel 1943, 93–95; Durand 1992, 202.
51 Wang 1836, 1: 29b–30a.

exceptional literary talent.⁵² Fang was then assigned to work at the School for tutoring the young. The chapters devoted to astronomy in the *Draft Qing History* also describe the setting up of the project, without mentioning the Office:⁵³

In the fifth month of the fifty-second year: Compilation of the writings on harmonics and mathematics. The Sincere Prince of the Blood Yunzhi 允祉,⁵⁴ the Fifteenth Son Yunwu 允禑 and the Sixteenth Son Yunlu 允祿 served as compilers by decree. He Guozong and Mei Juecheng served as editors (*huibian* 彙編). Chen Houyao, Wei Tingzhen, Wang Lansheng, Fang Bao and others served as proofreaders (*fenjiao* 分校). The writings compiled were presented everyday, and His Majesty personally added corrections.⁵⁵

This is the institution described by Foucquet as an 'academy'. In fact, it resembled neither the academies set up in Europe in the seventeenth century, in which monarchs did not intervene personally, nor the Chinese scholarly academies (*shuyuan* 書院) where scholars taught and exchanged viewpoints.⁵⁶ The function of the Office seems to have been strictly limited to the compilation of works on particular topics. Its members submitted written materials to Kangxi who edited them. This is exactly how the Jesuits had worked in the 1690s when teaching him mathematics. This time, however, it was the emperor who had the status of a teacher, and those who wrote the materials were regarded as his students. In publicising knowledge that he had hitherto monopolised, Kangxi took up the role of the sovereign as a master, thus conforming to the Confucian tradition.

Music seems to have been first of the three subjects on which the staff of the Office were to work, as witnessed by an order issued in September 1713:

---

52 Hummel 1943, 235–237; 701; on the Dai Mingshi affair, see Durand 1992.
53 The only mention of Kangxi's Office of Mathematics in the *Draft Qing History* is in the biography of Gucong; Zhao 1977, 10637.
54 All the sons of Kangxi were given a name the first character of which was *yin* 胤 at birth. After one of them succeeded their father, the character *yin* became taboo as part of the emperor's personal name, so that the first character of all the other brothers' names was changed to *yun* 允, even in documents recording events dating from the Kangxi reign. 'Sincere Prince of the Blood' (Cheng qinwang 誠親王) was Yinzhi's official title.
55 Zhao 1977, 1668.
56 This is why I translate this institution's name (*Suanxue guan* 算學館) literally, rather than using 'academy' an English term evocative of institutions—European and Chinese—of a very different nature. Han Qi has suggested that the foundation of the Office of Mathematics was inspired by Foucquet's mention of the Paris *Académie*, which he called *Gewu qiongli yuan* 格物窮理院 (lit. Academy of the investigation of things and fathoming of principles), in his *Dialogue on astronomy* (*Lifa wenda* 曆法問答) (Han 1999b, 316–317); but, despite the Jesuits' claim that they influenced the emperor, there is no evidence in Chinese sources that he ever considered imitating European institutions.

Edict to Li Xu 李煦 and Cao Yong 曹顒:[57] We have spent decades to achieve imperially composed writings for the *Origins of pitchpipes and the calendar* (*Lüli yuanyuan* 律曆淵源). They are now almost completed. However instruments makers and good bamboo are lacking. Summon the old man called Zhou 周 from Suzhou; in his home there should be men who can make instruments. Moreover, choose large quantities of good bamboo and send them in. Also ask him whether anyone knows about pitchpipes; if so send that person over with him. However, if he is too old to make the journey, he must send the right man. That will be alright after all.[58]

This edict contains the earliest mention known to me of the title of the work that the staff of the Office was to produce. It does indicate that astronomy forms part of the project. Associating pitchpipes and the calendar in the title was an obvious reference to precedents in early China: the two matters, traditionally linked, were, for example, dealt together in the 'Monograph on harmonics and calendrical astronomy' (*Lülizhi* 律曆志) of the *Han history* (*Hanshu* 漢書), completed in the late first century CE.

But the main point brought out by this edict is that the endeavour to fix tones according to the Classics entailed using the competence of craftsmen to make properly pitched instruments. In music, Western learning was one of several sources for imperial scholarship, but by no means the main one: the instrument makers mentioned by Foucquet as working at the Office were musical craftsmen as well as makers of astronomical and mathematical instruments.

12.5 The imperial princes

The records of the edict in the *Veritable Records* (Shilu 實錄), in Wang Lansheng's writings and in the *Draft Qing History* quoted above mention different names. The latter two list the entourage of Li Guangdi, most of whom worked under his supervision for other editorial projects; there is no evidence that any of them took the mathematics examination. The three documents also differ regarding whom was in charge of the Office of Mathematics. Both official sources indicate that it was Yinzhi. On the other hand, Wang Lansheng's writings mention the Sixteenth Son, Yinlu 胤祿 (1695–1767) instead of Yinzhi. The document quoted in Wang's writings, which were compiled long after the Kangxi period, may have been altered; more generally, not all sources available nowadays fully reflect Yinzhi's role in the compilation project. The reason for this is the strong enmity that the

---

57   Li Xu and Cao Yong were textile commissioners (*zhizao* 織造) respectively in Suzhou and Nanjing; Spence 1966, esp. 263–270.
58   Zhongguo diyi lishi dang'anguan 1985, 5: 164, no. 1428; see Spence 1966, 117 and Chiu 2007, 132.

Yongzheng emperor (r. 1723–1735), who was Kangxi's Fourth Son (Yinzhen 胤禛) and his successor, felt towards Yinzhi. After Yinreng's final deposition from his title of heir apparent in 1712, Yinzhi appears to have been a plausible candidate to the succession. It is likely that after Yongzheng's accession, a number of documents concerning other contenders, including the Third Son, were altered or destroyed. Yinlu, who was on good terms with both Yongzheng and his successor Qianlong,[59] seems to have received more credit than Yinzhi for the editorial project, whereas extant sources indicate that the latter was the main leader of the project. The mention of Yinlu's name alone on the document quoted in Wang Lansheng's collected writings contradicts both official records and the Jesuits' accounts, and matches the victor's account. Yinzhi's fate as an editor evokes that of his protégé Chen Menglei: the name of the general editor of the *Compendium of books and illustrations ancient and modern* was simply erased from the work when it was printed in 1728.[60] Yinwu 胤禑 (1693–1731), the Fifteenth Son, does not seem to have taken side in the succession struggle.[61] Thus the Office of Mathematics cannot simply be associated with one particular faction in this struggle.

By the time this Office was set up, the strife among Kangxi's sons had become public knowledge. In 1708, Yinreng had been demoted from his position of heir apparent after what appears to have been a spell of insanity; he had been put in the custody of his eldest brother Yinti. At the time, Yinti was among those who proposed to the emperor that he should nominate Yinsi 胤禩 (1681–1726), his Eighth Son (whose accession to the throne, Yinti argued, had been predicted by physiognomists) as heir apparent in replacement for Yinreng; they were all rebuked by Kangxi. Moreover, following Yinzhi's accusation that Yinti had cast a spell on Yinreng with the help of a 'Lama sorceress', Yinti was placed under house arrest, where he remained for the rest of his life, and Yinreng was reinstated. When in 1712 Yinreng was once more demoted and placed in perpetual confinement, Yinzhi remained the eldest of the emperor's sons to retain the status of an imperial prince.[62]

Yinzhi's appointment to the head of the Office of Mathematics was a sign of his favour with his father: it resulted from his enhanced status after this crisis, but also from his knowledge of the mathematical sciences: he was the best versed in the mathematical sciences of Kangxi's sons.[63] According to Foucquet, the prince had 'applied himself to mathematics in his youth and [he knew] about it. His master was

---

59 See Hummel 1943, 925–926.
60 Hummel 1943, 94.
61 Hummel 1943, 331.
62 On the complex rivalries and alliances between the brothers, see Hummel 1943, 916–917; 924–931; Wu 1979, 132–138 (where the Eldest Son's name is transliterated as Yinshi).
63 Wu 1979, 165–167.

Father Antoine Thomas, the Flemish Jesuit who taught him the principles of numbers and of geometry.'⁶⁴ As mentioned above master and student measured the length of a degree of terrestrial meridian together in 1702.⁶⁵ Bouvet also reported that Yinzhi studied the *Elements of geometry* (*Jihe yuanben* 幾何原本), which he and Gerbillon had composed, under the supervision of his father in 1693, when the prince was still in his teens.⁶⁶ However, Yinzhi was not the only one of the emperor's sons who had studied the mathematical sciences. Kangxi closely followed their education, and saw to it that they became acquainted with these disciplines. The future Yongzheng emperor, who does not seem to have been particularly interested in them, nonetheless took part in astronomical observations in his father's lifetime, as he later remembered:

In former years when there was a solar eclipse of four or five *fen*, the sunlight was hardly visible. Our Late Father in person led Us and Our brothers to the Palace of Heavenly purity with a telescope. We used paper spread all around it. When the sunlight was blocked, we would check how many marks were obscured. Thus We are familiar with such experiments.⁶⁷

Yongzheng remembered well the technique used to assess the magnitude of an eclipse; like his father before him, he used this knowledge to rebuke the officials of the Astronomical Bureau about their alleged negligence when a solar eclipse occurred on 15 July 1730.⁶⁸ This confirms that several of Kangxi's sons had some knowledge of astronomy. The mention of Yinlu's name in the document found in Wang Lansheng's collected writings reflected the former's involvement with the Office of Mathematics. According to Foucquet,

To help the third [prince] govern the new academy, [the emperor] added the twelfth, the fifteenth, and the sixteenth [princes]. Of these four princes, the twelfth knows little mathematics, the 15th and 16th, both still young, study it daily, the 15th is only 25 years old, and the 16th about 22. As for the 3rd, [he] is now instructing his two younger brothers. But the custom is that in the morning he takes them to the emperor, who in turn instructs the three of them, and makes them report about some work he has prescribed to them. It is said that he thus keeps them busy to prevent them from moving.⁶⁹

Foucquet was evidently aware of the tensions and rivalries between Kangxi's sons. The fact that the three sons who are mentioned in the *Draft Qing History* seem to have taken different stands in the succession

---

64 ARSI, Jap. Sin. II 154, 3.
65 See pp. 244–245.
66 Von Collani 2005, 75.
67 *Shizong xianhuangdi shengxun* 世宗憲皇帝聖訓, *SKQS* 412: 127.
68 *SKQS* 412: 127; Cui & Zhang 1997, 136.
69 ARSI, Jap. Sin. II 154, 3–4.

strife leads one to wonder whether the emperor did not deliberately bring together sons that he knew belonged to different factions. Since we do not know what stand Yintao 胤祹 (1686–1763), the Twelfth Son mentioned by Foucquet, took, it is unclear whether the presence of a prince who 'knew little mathematics' at the Office can be interpreted along those lines. Yintao's name also appears in Wang Lansheng's collected writings, in connection with a work on phonetics to which Wang contributed, presumably the *Clarification of the subtleties of phonetics imperially commissioned* (*Yuding Yinyun chanwei* 御定音韻闡微, 1725). Two other sons of Kangxi, as well as Li Guangdi, are mentioned on the list of contributors to that work. It is possible that, like some of the latter's protégés, the Twelfth Son was at some stage involved in several editorial projects, not all connected with Western learning, although his name does not appear on the resulting works.[70]

Yinzhi regularly reported to the emperor on the activities of the Office of Mathematics. Thus in January 1714 he informed him that logarithmic tables were being compiled, and that they were editing the first three sections of the *Elements of geometry*.[71] While a good part of the communication concerning the compilation undertaken at the Office took place in Manchu rather than in Chinese, there is no evidence that at that time Kangxi still pursued his earlier plan to publish works on mathematics and music in his ancestors' language.[72]

Kangxi also had his sons trained in Western music. At the time, Teodorico Pedrini, an Italian Lazarist sent to Beijing by the Pope and who effectively succeeded Pereira as tutor in music, taught Yinzhi, Yinwu and Yinlu.[73] The emperor insisted that the purpose of this tuition should be to learn the fundamentals of musical theory rather than just playing the keyboard.[74] There was a close link between his sons' instruction in music and the elements of Western musical theory that were to be included in the imperial compilation on music.

Kangxi's eagerness to impress on to his sons his views on the importance of the mathematical sciences and of Western learning is apparent in the *Maxims of fatherly advice* (*Tingxun geyan* 庭訓格言) that were edited by Yongzheng. As well as moral admonitions, the emperor's opinions on learning in general and on these disciplines in particular are quoted in this work. After evoking how, following the reinstatement of the Jesuits at the Astronomical Bureau in 1669, he had devoted constant effort to the mathematical sciences and how

---

70 Wang 1836, *juan* 1, 33a; *Yuding yinyun chanwei* 御定音韻闡微, SKQS 240: 2–3.
71 Zhongguo diyi lishi dang'anguan 1996, no. 2310, 2324.
72 Zhongguo diyi lishi dang'anguan 1996, *passim*.
73 Ripa 1996, 2: 297; on Pedrini see Allsop & Lindorff 2007; Gild-Bohne 1991, 43–55; 1998.
74 Chiu 2007, 81; the edict quoted is dated to 22 June 1714 (Chiu 2007, 110), that is, Kangxi 53/5/11.

he 'took pains to investigate their difficulties' so that those who studied after him would find them easy, the emperor went on to explain his approach to music, stressing that tuning instruments correctly was indissociable from retrieving the music of antiquity. This was possible because the fundamental pitch was found in nature:

> The original pitch of heaven and earth is continuous in ancient and modern times and does not change. It unites China and foreign countries into Great Harmony. Within the Six Directions, beyond the Four Seas, the sounds are the same, the principles are the same. Therefore if one ignores the music of the past to indulge in the music of the present, one will not only ignore the music of the past, but also the music of the present. If one insists on returning to the music of the past and regards the music of the present as not worthwhile, in the end one will not return to the past either.[75]

Past and present are thus indissociable and complementary. One draws on both instrument makers and Western learning to nourish one's understanding of ancient Chinese learning: the latter is not simply a foundation of Chinese culture, but is in fact the true knowledge about something that is universal. This rationale is slightly different from that expounded in the *Discussion of triangles and computation*, where the assumption is that Western learning is historically derived from ancient Chinese learning.[76] Positing that the object studied is natural and thereby universal provides an alternative, less 'sinocentric' rationale for the relevance of Western learning. One might wonder whether astronomy and music had different epistemological statuses in the emperor's eyes or whether he adjusted his discourse to his audience: his sons may well have been less concerned than Chinese literati about the foreignness of Western learning.

12.6 Mathematicians, astronomy and cartography

The staff of the Office of Mathematics did not devote all their time and energy to the edition of imperial writings. While He Guozong and Mei Juecheng took part in various literary projects, others carried out in the emperor's service tasks for which their skills were required; not surprisingly, the cartographic survey of the empire was one of these tasks. In 1713, shortly after the sixtieth birthday celebration, He Guozhu set out for Korea as a member of an embassy sent to King Sukjong 肅宗 (r. 1674–1720) which was to gather material for a map of Korea.[77] Official Korean sources include a record of observations carried out by

---

75 *Shengzu renhuangdi tingxun geyan* 聖祖仁皇帝庭訓格言, *SKQS* 717: 650; see Chiu 2007, 134.
76 See pp. 249–250.
77 Ledyard 1994, 299; on cartography and border trespassing see Kim 2007.

He Guozhu in Seoul to determine the geographical coordinates of the place:

> In the 29th year of the Sukjong reign, The Calendar Manager of the Five Bureaus (*ogwan-saryŏk* 五官司曆) He Guozhu, sent by the Qing, observed the altitude of the North Pole in Seoul [...], and obtained 39°37′15″; with reference to the meridian of Beijing's Shuntian 順天 prefecture, Seoul is 10°30′ to the east.[78]

Together with the materials provided by Korean officials, these coordinates were used for the imperial map, and subsequently for European maps.

He Guozhu's visit to Seoul provided an opportunity for some Korean scholars versed in mathematics to pay him a visit. One of them, Hong Chŏng-Ha 洪正夏 (b. 1684) recorded his dialogue with He and inserted it in the last chapter of his treatise *Writings of Nine and one* (*Kuilchip* 九一集, c.1714).[79] The 'Calendar Manager' (*saryŏk* 司曆) first proposed a series of problems for which we are given the answer and sometimes the method used. The first ones are elementary: multiplication, division, and extracting the square root of 225. Then He Guozhu showed Hong how to assess the weight of an object on non-graduated scales, which the latter apparently did not know how to do.[80] Aqitu 阿齊圖, the ambassador, who was present, then suggested that Hong should in turn propose problems to He Guozhu to test the skills of the latter, who, he claimed, was 'the fourth best mathematician in the empire'.[81] At this point the problems get more difficult. Some of them, which puzzle He Guozhu, are solved by Hong using counting rods; the Korean mathematicians, on their side are puzzled by problems that, according to He, can be understood thanks to the *Elements of geometry* (*Jihe yuanben* 幾何原本) and the *Complete meaning of measurement* (*Celiang quanyi* 測量全義, 1631).[82] The association of these two titles suggests that He was referring to the 1607 translation of Euclid's *Elements*. This is consistent with the fact that the dialogue took place before Kangxi handed over Gerbillon and Bouvet's *Elements of geometry* and the *Elements of calculation* (*Suanfa yuanben* 算法原本) to Yinzhi so that the staff of the Office should work on it. The fact that the last character of *Celiang quanyi* is wrongly written as *yi* 儀 (instrument) instead of *yi* 義 (meaning) in Hong's work suggests that he did not know about this work. On the other hand, He Guozhu expressed puzzlement at the sight of his Korean visitor performing calculations

---

78 Quoted by Guo & Li 2004, 211.
79 I follow the translation given by Jun 2006, 481; on the dialogue see Horng 2002, Guo & Li 2004.
80 On problems involving non-graduated scales, see pp. 349–351.
81 Hong 1983, 478–479.
82 See pp. 25, 34.

with counting rods (*suanzi* 算子), which had been used in China since antiquity and at least until the fourteenth century, but had some time thereafter fallen into disuse. Hong then presented to He forty rods from his own set to take back to Beijing.[83] In China, it was Mei Wending who first reconstructed the use of counting rods for elementary operations.[84] Decades later, Mei Juecheng rediscovered the meaning of the notations employed in the procedure of the celestial element (*tianyuan shu* 天元術) developed during the Song and Yuan dynasties.[85] It is generally assumed that knowledge in general circulated from China to Korea in the eighteenth century; the Korean kings regularly sent astronomers to Beijing to learn about the latest methods used at the Astronomical Bureau.[86] However, the fact that He Guozhu and Mei Juecheng worked together on the compilation of the *Origins of pitchpipes and the calendar* after the former returned from Korea leads one to wonder how much Chinese scholars' rediscovery of their own mathematical tradition owes to their knowledge of the practice of mathematics in Korea in their time.

Observations similar to those made by He Guozhu in Seoul were needed both for the map of the empire and for the imperial astronomical treatise: in order for that treatise to be usable in any part of the empire, observations made from provincial capitals were needed. In November 1714, the emperor ordered that measurements of the North Pole's altitude and of the obliquity of the ecliptic should be made every day, arguing that 'these are the most crucial points in astronomical methods'. The resulting values were '39°59′30″ for the altitude of the North Pole at the Garden of pervading spring, that is, 4′30″ more than at the capital's Observatory', and 23°29′30″ for the obliquity of the ecliptic.[87] The Garden of pervading spring was to be taken as the point of reference for all measurements for the imperial compendium on astronomy. The next month, Yinzhi pointed out that measurements should be made throughout the empire. One 'student in mathematics' (*xuexi suanfa yi ren* 學習算法一人) was sent to each of the main provinces (Guangdong, Yunnan, Sichuan, Shenxi, Henan, Jiangnan and Zhejiang) to measure the height of the Pole and the Sun's shadow, with the help of the local administration; thus He Guodong, the third He brother, was sent to Guangdong.[88] As had been the case for Korea, to improve his map of

---

83 Hong 1983, 493.
84 Jami 1994c, 165–173.
85 *Meishi congshu jiyao* 1761, *juan* 61: 8b–11b.
86 Shi 2008a, 207–211.
87 *SL* 260, 10b–11a.
88 *SL* 260, 4b–5a; *Qinding daqing huidian zeli* 欽定大清會典則例, *SKQS* 625: 145–146.

Tibet the emperor did not rely on the Jesuits, who could enter neither country. Instead

[...] he chose two Lamas who had studied Geometry and Arithmetic in an Academy of Mathematics, established under the protection of his third son. He commissioned them to make the map from Si ning [Xining 西寧] in Chen si [Shenxi 陝西] province to Lasa residence of the Great Lama, and from there to the source of the Ganges, with an order to bring him some water of this river.[89]

So far these lamas have not been identified. The Jesuits, naturally, were skeptical about their work:

In the year 1717, this map was put in the hands of the Missionaries Geographers by order of the emperor, to be examined; they found it without comparison better than the one given to them in 1711. It nevertheless did not appear to them to be devoid of faults: but out of respect for the school from which these Lamas came, they were then content with correcting the most obvious ones, that would have shocked the emperors' eyes. They even left Lasa above the 30th degree of latitude, where the Lamas had put it, showing more respect for the actual measurement used by these Lamas than for astronomical observation.[90]

Thus the tasks set to imperial mathematicians entailed both working on texts and making observations; in fulfilling both tasks they were direct competitors of the Jesuits.

12.7 The mathematical sciences in examinations

As mentioned above, the examination in mathematics took place only once: like the 1679 special examination for the recruitment of 'Profound scholars of vast learning' (*boxue hongru* 博學宏儒),[91] it was aimed at staffing a particular project, albeit one much less prestigious and sensitive than the *Ming History*. The 1713 recruitment provided the emperor with some staff who seem to have been satisfactory for his purpose. But it was also an acknowledgement that high officials were not and would not be as versed in the mathematical sciences as he was and wished them to be. Looking back at this issue from our twenty-first century standpoint, it might seem obvious that adding the mathematical sciences to the curriculum of imperial examinations would eventually solve this problem. But no evidence points in this direction; in any case, what interests us here is how things looked seen from Beijing in the last decade of the Kangxi reign. The complexity of the examination system, supervised by the Ministry of Rites, and the close links between its curriculum and literati culture needed to be taken into account. There appears to be no evidence that Kangxi made any attempt to

---

89   Du Halde 1735, 4: 460; quoted by Fuchs 1943, 67.
90   Du Halde 1735, 4: 460; quoted by Fuchs 1943, 67.
91   See p. 74.

include the mathematical sciences or any other part of Western learning into the curriculum.

It has been argued that on the contrary he actively took steps to eliminate the sciences from examination questions, where they had sometimes appeared until then.[92] This assertion seems to result from the misreading of an official document, which is the sole evidence provided for this supposed elimination. A close look at the sources is necessary here. To the best of my knowledge the emperor mentioned the works on the mathematical sciences he had commissioned only once in the context of examinations, a few months after the completion of the work devoted to harmonics, during a formal audience that took place on 2 March 1715:[93]

At the *chen* 辰 hour,[94] His Majesty held court at the Residence of Tranquillity (*Danningju* 澹寧居) at the Garden of pervading spring. At that time he called in the Grand Secretaries (*daxueshi* 大學士), the Secretaries (*xueshi* 學士), the Nine Ministers, the Supervisor of the Heir Apparent's Household (*zhanshi* 詹事), and the Supervising secretaries and Censors (*kedao* 科道), and told them:

'Today We have called you especially about the matter of imperial examination compounds. The questions set at the examination compounds are of crucial importance. There is already a decree concerning the examination questions in the books on the Classics produced by local associations. You must avoid those strictly, and must not issue questions that have been thoroughly practiced or often imitated. As to the works containing what We often teach on the *Book of change*, those on astronomy, harmonics and mathematics that are being compiled, and others, everyone expects questions modelled on these works. You are to choose the officials in charge of examinations: although none has been nominated yet, they will be strictly forbidden to set questions from these works. Indicate that questions cannot be taken either from any compiled works or granted works and the like. Otherwise people will have the fortune to succeed because of modelled questions, and how will we find real ability? [. . .]'[95]

The work containing imperial teachings on the *Book of change* mentioned here is probably the *Middle way to the Change of Zhou*, which was by then just completed but had not yet been circulated.[96] The emperor's concern here is to avoid examination questions with which some candidates may be particularly familiar: he justifies this in terms of the efficacy of the selection process rather than of fairness. He goes on to refer to some recent cases of badly chosen examination questions and of fraud. He further accepted Li Guangdi's suggestion that questions

---

92 Elman 2000, 481–485; the assertion in taken up in Elman 2005, 167–169.
93 The completed *Correct interpretation of the [standard] pitchpipes* was presented to the emperor on Kangxi 53/11/17 (23 December 1714).
94 From 7 to 9 a.m.
95 Zhongguo diyi lishi dang'anguan 1984, 3: 2214.
96 The imperial preface to the *Middle way to the Change of Zhou* is dated to Kangxi 54/3/18 (21 April 1715); *Yuzuan Zhouyi zhezhong* 御纂周易折中, SKQS 38: 1.

related to 'nature and principle' (*xingli* 性理) should not be set for the next two years; this was probably related to the fact that the compilation of the *Essential significance of nature and principle* was in progress at the time. Evidently works on the mathematical sciences were treated on a par with imperial compilations on subjects of classical scholarship, and were deemed potentially relevant to imperial examinations. The emperor was certainly not demanding that the subjects of astronomy, harmonics and mathematics should be entirely banned from examination questions,[97] no more than he did so for questions related to the *Book of change* (the latter would have been a ludicrous suggestion). His wording suggests on the contrary that at the time candidates might have expected to have to answer questions on these technical subjects, all the more so because the imperial emphasis on them was well known.

Kangxi's pronouncement on the setting of examination questions was of considerable weight, although it was not easily implemented.[98] More or less truncated versions of it, in which the fact that officials responded to him during the audience is no longer visible, appear in various compilations. The usual abridged transcription in the *Veritable records* of his reign leaves out about half of the original text.[99] Even shorter versions appear in his collected prose, and in his *Sagely instructions* edited by Yongzheng,[100] and, finally, in the chapters devoted to imperial examinations of a collection of political documents, the *General survey of documents of our Dynasty* (*Huangchao wenxian tongkao* 皇朝文獻通考). The main point of Kangxi's pronouncement still comes out clearly there:

As to the works containing what We often teach on the *Book of change*, those on astronomy, harmonics and mathematics that are being compiled, and others, you officials in charge of examinations are strictly forbidden to set questions from these works.[101]

Here again, it is not the mathematical sciences and the *Book of change* as subjects for examination questions that are the object of the interdiction, but merely questions drawn from particular works on these subjects. Thus there is no evidence that there ever was a 'ban on natural studies' in examination questions.[102] If indeed questions on

---

97 Contrary to what is asserted in Elman 2000, 484–485 and Elman 2005, 168.
98 *Qinding huangchao wenxian tongkao* 欽定皇朝文獻通考, *SKQS* 633: 237.
99 *SL* 262: 3a–5a.
100 *Shengzu renhuangdi yuzhi wenji* 聖祖仁皇帝御製文集, *SKQS* 1299: 425–426; *Shengzu renhuangdi shengxun*, *SKQS* 411: 280.
101 朕常講易及修訂天文律呂算法諸書，爾等考試官斷不可以此諸書出題。*Qinding huangchao wenxian tongkao*, *SKQS* 633: 237; quoted in Xi 1969, *juan* 191: 7b; this is the passage quoted by Elman as evidence that Kangxi banned questions on natural studies; Elman 2000, 484–485.
102 Elman 2000, 485 & Elman 2005, 167.

astronomy, harmonics and mathematics did not appear in examination questions after this date (such a negative conclusion is difficult to ascertain), one would have to turn to the officials who set the questions for an explanation: did they believe that some mastery of the mathematical sciences was needed in order to fulfil the tasks entailed by their own positions, and those of the junior colleagues they were recruiting? The issue here is whether the emperor convinced Chinese elites that they should emulate him in studying the mathematical sciences. On the whole, the evidence gathered so far in the present book does not suggest that by 1713 he had succeeded in doing so. The compilation of *Origins of pitchpipes and the calendar* at the Office of Mathematics was commissioned in order to turn the disciplines that the work was to discuss into 'a matter for scholars', to use Mei Wending's words.[103] This entailed the exclusion of the Jesuits from the project. At a time when they were losing the emperor's favour, they could not but perceive this exclusion as a supplementary threat.

---

103   Mei Wending, *Lisuan quanshu* 曆算全書, *SKQS* 794: 103; see Jami 2007b, 149.

CHAPTER 13

The Jesuits and innovation in imperial science: Jean-François Foucquet's treatises

The Office of Mathematics was not the only threat to the Jesuits' monopoly as imperial experts in the arts and the sciences. As the Rites Controversy was gaining momentum, the *Propaganda Fide* ([Sacred Congregation for the] Propagation of the Faith), established by Pope Gregory XV in 1622, sent to China missionaries of various other religious orders, some of whom entered the service of the emperor. In 1711, Guillaume Fabre-Bonjour (1669–1714), a French Augustinian, the Italian Lazarist Teodorico Pedrini and Matteo Ripa (1682–1746), an Italian secular priest, arrived in Beijing.[1] They worked as cartographer, musician and painter respectively; while the Jesuits rightly perceived them as competitors, all Catholic missionaries had to collaborate in the emperor's service. Czar Peter the Great of Russia (r. 1682–1723) also sent an Orthodox mission to Beijing; Russia thus became another potential provider of Western learning for Kangxi.[2] This new element forms part of the background against which Jesuit engagement with astronomy during the last decade of the Kangxi reign should be understood.

13.1 Foucquet in Beijing: from the *Book of change* to astronomy

The French Jesuit Jean-François Foucquet had been in China since 1699, working as a missionary first in Fujian and then in Jiangxi. He had taught mathematics at the Jesuit College of La Flèche from 1695 to 1697;[3] however, since his arrival in China he had devoted himself to 'the study of ancient Chinese books'.[4] In 1711 he was summoned to the capital where he became involved in astronomy for several years. He recounted this involvement in a lengthy report, which focuses on his attempt to present to the emperor what in his view were important

---

1 Standaert 2001, 340–341; 349–350; 353.
2 Standaert 2001, 369–370; *LEC* 17: 351; ARSI, Jap. Sin. II, 154, 82.
3 Pfister 1932–34, 549; De Dainville 1954, 111.
4 ARSI, Jap. Sin. II 154, 5.

updates that European astronomy had undergone in the previous fifty years.[5] In so doing he provided Kangxi with potential astronomical counterparts to the lecture notes on mathematics and harmonics produced by Thomas, Gerbillon, Bouvet and Pereira in the 1680s and 1690s. The tone of Foucquet's report, some letters written by his confreres, and the fact that he was recalled to Europe in 1720 suggest that his character as well as his ideas played a role in the conflicts of this period.[6] However biased his account may be, evidence from other sources indicates that the stakes of the controversies that surrounded him were not only personal.

As we have seen, while the Office of Mathematics set out to construct imperial learning in the mathematical sciences, the Beijing Jesuits continued to provide Kangxi with expertise and materials in these and many other fields. In 1716, Cyr Contancin (1670–1732), who was then the superior of the French residence in Beijing, listed the activities of the missionaries under his authority: Bernard Rhodes, a brother whose death Contancin lamented, had been a surgeon and an apothecary for seventeen years. Jean-Baptiste Régis (1663–1738) had spent the last few years surveying the empire; so had Pierre Vincent de Tartre (1669–1724) and Pierre Jartoux (1669–1720). The latter also served the emperor as a mathematician, while Jean-François Foucquet wrote treatises on astronomy for the monarch. Jacques Brocard (1664?–1718), another brother, made and repaired clocks.[7] Contancin, who does not seem to have had any scientific or technical skill, humbly closed this enumeration by describing himself as 'a useless servant'.[8] Evidently the Jesuits, especially those of the French mission, still understood such skills to be the means to protect the mission, but the optimism conveyed to their European audience in the 1690s had given way to a defensive mode. The issue of what to teach to the emperor and how to teach it was controversial, but this was only one aspect of the manifold conflicts in which Western learning was caught at the time: between the missionaries on the one hand and Yinzhi and his mathematicians on the other hand, between the Jesuits of the Portuguese mission and those of the French mission, and among the latter. The interpretation of the Chinese classics as well as views on the best way to put the sciences in the service of the mission divided

---

5 'Relation exacte de ce qui s'est passé à Péking par raport à l'astronomie européane depuis juin 1711 jusqu'au commencement de novembre 1716', manuscript, ARSI, Jap. Sin. II 154; another copy of this document, also by Foucquet's hand, is kept at the Vatican Library (BAV, Borg. Lat. 566, ff. 144–183). A very useful if not completely reliable transcription is found in Witek 1973, 479–678.
6 Witek 1982, 238–244.
7 Brocard came from Dauphiné, the clock-making region of France; Standaert 2001, 844–845.
8 ARSI, Jap. Sin. 177, f. 70 r.–70 v. letter dated 1 September 1716.

the missionaries. In the imperial family, the dependence of official astronomy on European informants who were no longer fully trusted had become an increasing concern. The atmosphere at court was generally tense, especially after the final deposition of Yinreng as heir apparent in October 1712. Matters related to the sciences thus intertwined with a variety of sensitive issues.

It was due to Bouvet that Foucquet was called to the court in 1711. After returning to China from France in 1698, the former had developed a strong interest in the *Book of change* (*Yijing* 易經). He became the first and foremost upholder of the so-called Figurist interpretation of the Chinese classics, which, relying on some traditions of Christian theology, read 'signs' referring to God's revelation into the *Book of change*, other classics, and Daoist texts. By 1706, when the papal legate visited Beijing, Bouvet had composed two treatises in Chinese, the *Original meaning of heavenly studies* (*Tianxue benyi* 天學本義) and the *Mirror of ancient and modern revering of Heaven* (*Gujin jingtian jian* 古今敬天鑒).[9] The former work quoted extensively from the *Explanation of the meaning [of the Classics] during daily tutoring* (*Rijiang [...] jieyi* 日講...解義) by Kangxi's Daily Tutors (*rijiang guan* 日講官).[10] In other words, the Jesuit relied on imperial interpretations to make his case. While Bouvet had a number of followers among the China Jesuits, including Foucquet, his views were strongly opposed by his superiors. After Tournon's explicit condemnation of the *Original meaning of heavenly studies* (which he could not read, as he did not know Chinese) during his stay in Beijing, most of Bouvet's confreres were worried that his interpretation might further weaken the Society's position in Europe, a position already endangered by the Rites controversy.[11]

Therefore they were quite concerned when, in 1711, after Bouvet reported to Kangxi that Foucquet was versed in the Classics, the emperor, who had some interest in Bouvet's investigation of the *Book of change*, summoned Foucquet to Beijing to assist him. Foucquet arrived in the capital on 7 August 1711, that is, a few weeks after the emperor's remonstrance to the Jesuits concerning the summer solstice.[12] After he had collaborated with Bouvet on the Classics for a few months, an imperial command made him turn to other tasks.

On the morning of the 8th [April 1712], all the Europeans being gathered in the Palace, an eunuch came and told, upon His Majesty's request, that he wanted the principles of astronomy explained to him, and by the word principles he meant a doctrine by means of which a man who knows geometry and arithmetic could make astronomical tables. The order added that Europeans

---

9 Han 1998; Han 2004.
10 See p. 73.
11 Standaert 2001, 668–672.
12 See pp. 260–261.

should choose two persons capable of fulfilling this function well, and that above all explanations should be done with figures likely to make the doctrine more intelligible.[13]

The Jesuits decided to appoint two men for this task: Foucquet and Franz Thilisch (1670–1716). The latter, recently arrived in China and sent to the court as a mathematician, was in Jehol with the emperor at the time of the summer solstice of 1711; the estimate of the time of the solstice that he calculated for Kangxi differed from that of the Astronomical Bureau. To explain this, Thilisch told the emperor that he had used different tables from those used by his colleagues of the Bureau. The fact that different tables were used simultaneously in Europe seems to have been news to Kangxi,[14] who expected all the astronomy done under him to follow one single system. The terms of his order indicate that he wanted to be in a position to decide which tables were to be used by the Jesuits in his service.

13.2 The *Dialogue on astronomical methods*: a reform proposal?

Foucquet and Thilisch never worked together. Beside the fact that the former resided at the Northern Church (Beitang 北堂) and the latter at the Eastern Church (Dongtang 東堂), they seem to have disagreed profoundly as to what the work they were to write for the emperor should contain. In agreement with Kilian Stumpf, now the Administrator of the Calendar, Thilisch planned to produce a text that would uphold the practice of the Astronomical Bureau; however, he did not yet know Chinese, and there is no evidence that he ever wrote anything on astronomy. Foucquet, on the other hand, was well versed in written Chinese from his lengthy study of the Classics; he undertook to compose

[a] work in the form of a dialogue, on the idea of the epitome or abridgement of Copernican astronomy, composed by Kepler, an idea which he had proposed to the emperor and which the emperor had approved. Father Foucquet had preferred this to other manners, because he knew that a few books written in this form by Tchou uen cong [Zhu Wengong 朱文公],[15] a famous philosopher, are held in high esteem by the Chinese, and because he hoped, in questions and answers, to be able to explain more easily and more clearly the difficulties and subtleties of astronomy.[16]

---

13   ARSI, Jap. Sin. II 154, 4–5.
14   ARSI, Jap. Sin. II 154, 1.
15   I.e. Zhu Xi; Witek 1973, 487; the *Classified conversations of Master Zhu* (*Zhuzi yulei* 朱子語類, 1270) as well as the Analects themselves are written in the form of dialogues. On 'recorded conversations' as a mode of discourse of Song dynasty philosophers, see Gardner 1991.
16   ARSI, Jap. Sin. II 154, 12.

The form of a dialogue, associated not only with Confucius and the *Analects*, but also with Zhu Xi and thereby with imperial orthodoxy, had also been used by Mei Wending. The book that Foucquet proposed to write would make clear to the emperor 'all the defects of Tychonic astronomy which [was] followed at the Tribunal'.[17] This, the French Jesuit claimed, was the only way to prevent the greater harm of having these defects revealed by Yinzhi, whose purpose in writing a new astronomical sum was to run down the books used at the Bureau.[18]

Foucquet's claim that he modelled his treatise on that of Kepler in order to expose the shortcomings of Tycho Brahe's astronomy has led some historians to see in him the first European who attempted to introduce heliocentrism into China.[19] Foucquet does discuss the model put forward by Copernicus in several passages of this work, but his intention was clearly not to propose heliocentrism as the foundation of imperial astronomy.[20] As a whole, his *Dialogue on astronomical methods* (*Lifa wenda* 曆法問答) is mainly based on the Tychonic model, while proposing a number of corrections that rely on recent observations. Foucquet listed what he regarded as the most significant updates he proposed compared with the *Books on calendrical astronomy according to the new method* then used at the Astronomical Bureau:

- taking the obliquity of the ecliptic as 23°29′ instead of 23°31′30″;
- taking refraction into account for all angles instead of only those of up to 45°;
- correcting the values of solar parallax, and of the distance and diameters of the planets;
- presenting a new method for calculating eclipses;
- correcting geographical tables using lunar and solar eclipses.[21]

These were mainly corrections of numerical values that functioned within the astronomical system then in use. None of them challenged the geocentric universe depicted in the *Books on calendrical astronomy according to the new method*. Foucquet's innovations aimed at securing a long-term need for Jesuit astronomers in Beijing, in response to the

---

17 ARSI, Jap. Sin. II 154, 12.
18 ARSI, Jap. Sin. II 154, 73.
19 Witek 1982, 181–189; Martzloff 1989, 986. Speculations on this issue are mainly relevant to the controversial role of the Society of Jesus in the development of astronomy in early modern Europe. There have also been debates on what might or should have happened in China if the Jesuits had introduced the heliocentric system at an early stage. Different viewpoints are held, among others, by Needham 1954–, 3: 437–458, Sivin 1973 and Libbrecht 1996; however the aim of the present study is to account for what did happen rather than for what did not.
20 For an outline of the mentions and uses of the heliocentric theory in Foucquet's work, see Hashimoto & Jami 1997, 179–180.
21 ARSI, Jap. Sin. II 154, 32–33.

threat that the Office of Mathematics posed to the missionaries' control of imperial astronomy:

> [I]n defending the known defects of Tychonic astronomy, one could not but succumb. If on the contrary, the Europeans were able to agree and substitute it with the theory based on recent observations, if they declared for example that the path of planets is elliptic or approaching the ellipse [...], nothing more would be needed to disconcert the Regulo's [Yinzhi] men, and to reduce them again to the condition of students of the Europeans; with the first six books of Euclid and a few propositions from Archimedes on solids that have been added to them since, no matter how well they knew them, it was not to be feared that they could by themselves penetrate the secrets of conic sections, or of higher geometry; thus the modern theories of the Sun, of the Moon and of other planets would be a sealed language for them; by following this path the Europeans could not fail to be employed to a reform of astronomy and hold the place of masters in it, should the emperor undertake it; then neither the emperor nor the empire could be ignorant of the fact that one was indebted to them for that [...][22]

Thus Foucquet's plan was to use innovation as bait, and to maintain Chinese astronomers in a state of dependency upon the Jesuits. Between 1713 and 1719, he gradually presented to the emperor the five treatises that form the *Dialogue on astronomical methods*. The titles of these works are almost identical to those of the treatises on the same subject written by the Jesuits between 1629 and 1635 that were included in the *Books on calendrical astronomy according to the new method*:[23] *Astronomical guide for solar motion* (Richan lizhi 日躔曆指, 1713), *Lunar motion* (Yueli 月離, 1713–1714),[24] *Astronomical guide for eclipses* (Jiaoshi lizhi 交食曆指, 1714–1715), *Astronomical guide for the fixed stars* (Hengxing lizhi 恆星曆指, 1717) and *Astronomical guide for the Five Planets* (Wuwei lizhi 五緯曆指, 1718–1719).[25]

According to Foucquet, Kangxi read at least part of the treatise on solar motion that he submitted in May 1713, and asked him 'why F. Verbiest had not explained this doctrine to him'. Foucquet replied that 'since Father Verbiest one had found many things that [he] intended to offer to His Majesty'.[26] However, other Jesuits' views on how to best use the sciences to retain the emperor's favour differed.

---

22   ARSI, Jap. Sin. II 154, 64; it is likely that here Foucquet uses the word 'theory' in the technical sense of 'methods of calculation', as in Newton's *New and most accurate theory of the moon's motion* (1702).
23   See pp. 33–34.
24   The treatise on the moon in the *Books on calendrical astronomy according to the new method* is entitled *Astronomical guide for lunar motion* (Yueli lizhi 月離曆指).
25   Copies of these treatises are kept at the Vatican Library and at the British Library; from them one can reconstruct two different stages of composition; Hashimoto & Jami 1997, 172–176; Hashimoto 1998, 800–805; the latter gives a full transcription of the table of contents. The dates of submission of the treatises to Kangxi are given in ARSI, Jap. Sin. II 154, 12, 26, 29, 30; BAV Borg. Cin. 377, f. 57; see Witek 1982, 186–187, n. 89.
26   ARSI, Jap. Sin. II 154, 13.

Thus, Pierre Jartoux (1669–1720), another French Jesuit who also taught Kangxi at the time, was extremely critical of the way Foucquet presented his writings to the emperor:

> I first gave [Foucquet] the employment of giving in my place Astronomy lessons in front of His M^{ty}, who had appointed me to this. There he had a splendid opportunity to win the Emp^r for the good of Religion [...]. One had to take a little pain to appear in front of His M^{ty} every two or three days, and to make him read in this way successively by separate lessons all of one's work: he [Foucquet] preferred to do long treatises comfortably in his room for ten years; the Emp^r, put off by their length, has not even read them, and they are today in the hands of the 3rd Prince, who in order to do without the Jesuits presently has some Chinese calculate all the tables of the Heavens, of sines, of logarithms etc., so that with a few changes either in the Theory that he has redone, or the practice, he can put forward his mathematics and astronomy instead of European Astronomy, that he will call the Astronomy or the Mathematics of Kamhi.[27]

Unlike Foucquet, Jartoux had been in Beijing since his arrival in China in 1701; he had worked on the general survey of the empire commissioned in 1708. He knew that the role of a missionary turned courtier entailed regular and sometimes daily attendance at the Palace; this was essential to retaining the emperor's favour. Indeed Foucquet seems to have failed in this respect. Whereas in 1712 he was in the emperor's retinue for the summer sojourn in Jehol, the following year he was excused from going on the grounds of ill health.[28] He perceived that this caused some change in the emperor's attitude to him. However, somewhat dismissive of the uses and obligations of the court, he emphasised that in astronomy as in the study of the Classics, he was defending the truth. Much to his dismay, a consensus gradually formed against him among his confreres that this was neither sufficient nor necessary in order to serve the emperor in the best interest of evangelisation. He finally realised this to the full in February 1716, when Yinzhi asked his father for an arbitration between the methods used at the Astronomical Bureau and those proposed by Foucquet:

> The Third Regulo, the 12th Count,[29] the fifteenth and the sixteenth sons of the emperor beg His Majesty to give his orders on the affair that regards the composition of the books of astronomy. We have noticed that the books of the new astronomy have been in use for a long time, and, although it has been years, no one has examined them. By the kindness of our august Father who has condescended to instruct us, using gnomons everyday starting in the 53rd year of Cang hi, we have observed the height of the sun and we have found that the obliquity of the ecliptic was slightly less than what the books of the new

---

27 Jartoux, letter written from Beijing, dated 23 September 1720; ARSI, Jap. Sin. 178, 384–385 & 182, Jap. Sin. 398–409, f. 398r.
28 ARSI, Jap. Sin. II, 154, 12–13, 19.
29 Foucquet renders 'Pei-tsu' (*beizi* 貝子) in French by 'comte'; Witek 1973, 534.

astronomy have determined. Moreover, the dialogues on astronomy that Fou Ching Tche [Foucquet], this European newly arrived at the court, has offered to Your Majesty, differ from the new astronomy in the calculations and the tables. As it is very important, regarding the composition of the books on astronomy, to know which way shall be followed, we beg Your Majesty to give orders on this.

Order from the Emperor: Let the Europeans Kiligan [Stumpf], Fou Ching tche [Foucquet], Yang Ping y [Thilisch], Sou Lin [Suarez] compare the two astronomies, examine them with care, what they have in common, how they differ, let them determine each point carefully, and let them report to me unanimously. Let them execute this order without delay.[30]

As when the dispute between the Jesuits and Yang Guangxian 楊光先 was being settled in 1669, the emperor asked all the astronomers to come to an agreement so as to serve him best. However, in this new controversy, both parties were Jesuits. Aware that the Europeans present in Beijing did not all agree and that in their respective native lands different tables were used, he needed a choice to be made for imperial science. The calendar and the map of the empire were to follow one single set of standards.

According to Foucquet, at first the French Jesuits supported him: what he had expounded in his writings was after all 'the doctrine approved by the Royal Academy of Sciences'.[31] However, arguments on the interpretation of ancient Chinese works continued within the French mission. Thus, against Foucquet, Jartoux 'persisted in feeling that Tchouang Tse (Zhuangzi 莊子) [...], far from talking about Trinity, taught atheism instead'.[32] Foucquet's French confreres feared that if he got the emperor's ear thanks to his astronomical writings, he might further promote Figurist ideas at court. The Jesuits argued among themselves as to what they should reply to the emperor's orders for several months, until in July 1716 Yinzhi wrote to them from Jehol, demanding a prompt response to the order issued by the emperor in February.[33] The French then made their support of Foucquet conditional on his ceasing to promote Figurist ideas; he refused. At that stage the tensions between the French mission and the Portuguese mission mattered less than the lack of discipline and therefore of reliability of Foucquet.[34] More importantly, all the Jesuits believed that there was a

---

30 ARSI, Jap. Sin. II 154, 30–31; so far I have not been able to locate the original Chinese or Manchu memorial.
31 ARSI, Jap. Sin. II 154, 38–39.
32 ARSI, Jap. Sin. II 154, 40; for Jartoux's view, see Jap. Sin. 182, 399r.-v.
33 ARSI, Jap. Sin. II 154, 56.
34 A letter from Jartoux, dated 23 September 1720, gives an account of continuous quarrels between Foucquet and the other French Jesuits; ARSI, Jap. Sin. 178, 384 & Jap. Sin. 182, 398–409.

genuine risk that Kangxi would do without their services as astronomers, and staff the Astronomical Bureau with members of the Office of Mathematics instead; this would deprive the mission of a protection that had proved vital.³⁵ This suffices to explain why the French Jesuits chose solidarity with a German Administrator of the Calendar, who belonged to the Portuguese mission, over deference to the Paris Academy of Sciences in the matter of astronomical constants. Apparently there was a consensus amongst all but Foucquet that any change in the astronomy that the Jesuits had practised since the advent of the Qing would be detrimental to the mission. Moreover, the continual improvements that were the *raison d'être* of the Paris Academy had no reason to be a priority for China missionaries. The joint reply sent to the emperor on 24 August 1716³⁶ mentioned the points on which Foucquet had proposed innovations:

> [Foucquet's] *Dialogues on astronomy* differ from the *Books on the new astronomy* [i.e. the *Books on calendrical astronomy according to the new method*] that are used at the Tribunal regarding all the following points: 1° refractions, 2° parallaxes, 3° diameters of the sun and moon, 4° obliquity of the ecliptic, points that have never been determined with precision. [. . .]. In each kingdom, although large gnomons have been erected for observation, observations always differ from each other by minutes and seconds, and therefore the tables of each kingdom somewhat differ. As to the astronomy used by the Tribunal of Mathematics nowadays, its rules have been approved by Your Majesty. Its calculations and observations are without mistakes, and compared to the astronomy of past dynasties, that of the present dynasty is of the highest perfection. We, your subjects, although devoid of wit and of lights, think that today's astronomy, ruled and determined by Your Majesty, has been in use for a long time, and that it should not be changed easily. The European astronomers omit nothing to reach precision in the foundations of the 4 above-mentioned points. Moreover, Your Majesty has had magnificent instruments made, and teaches subtle and learned methods himself; he has students trained by him, who are very skilled and very accomplished. We are about to see observation reach ultimate perfection; then it will be time to work on astronomical reform. [. . .].³⁷

Beside justifying the differences between the various European tables, the Jesuits defend the *status quo* for the astronomy practiced at the Bureau on the grounds that it has been used for a long time. Thus they embrace the interests of an institution that is in charge of ensuring continuity rather innovation. This is somewhat reminiscent of some of the arguments used against the calendar reform initiated by Xu Guangqi in 1629 and of Yang Guangxian's attack against Schall's

---

35 ARSI, Jap. Sin. 182, 399 v.
36 That is, Kangxi 55/7/8.
37 ARSI, Jap. Sin. II 154, 79.

'innovations' (*xin* 新) in the 1660s.³⁸ However, the Jesuits' plea here is not for ruling out reform entirely, but rather for postponing it until a sufficient number of observations are made in China. This implies an acknowledgement that imperial science is centred in Beijing. Foucquet, on the other hand, proposes that it should rely on data centralised in Paris, which in his view cannot fail to be the most accurate.³⁹ This is what Stumpf means when he writes that Foucquet's treatises provide the opportunity 'to glorify French mathematics' long awaited by the French Jesuits.⁴⁰ Having read the Jesuits' reply, the emperor took their advice and decided that there should be no significant changes to the astronomical system in use.

Foucquet, although disavowed, did not cease to work on astronomy at that point. Between 1717 and 1719 he presented to the emperor treatises on fixed stars and on the Five Planets; again these were handed over to Yinzhi, who asked Foucquet a few questions on them.⁴¹ This confirms that the prince, while pursuing his editorial task, continued to consider whether and how to reform imperial astronomy: as both Stumpf and Jartoux reported, he did not agree with the emperor's decision.⁴² Indeed the prince held the tables used at the Astronomical Bureau to be faulty.⁴³ One might also wonder to what extent his eagerness to reform astronomy was linked to his ambition to succeed his father. In any case, Foucquet's effort to promote astronomy as approved by the Paris Academicians could have reinforced court astronomy based at the Garden of pervading spring against official astronomy based at the Astronomical Bureau. In the conflict between the French and the Portuguese mission, the former ended up siding with the court, which fostered change in the mathematical sciences at this juncture, and the latter with the civil service, which defended the status quo.

There is little evidence, however, that Foucquet's innovations directly influenced the astronomical writings compiled at the Office of Mathematics. In his *Astronomical guide for the Five Planets*, the orbits of the planets were 'declared to be elliptic', again departing from the Tychonic model used by official astronomers, as he had argued should be done in his 1716 report. At that stage the little knowledge on conic sections available in Chinese might probably have hindered rather than prompted demand for further teaching from Foucquet.⁴⁴

---

38 See p. 50.
39 See Witek 1982, 185, no. 87.
40 ARSI, Jap. Sin. 177, 121; quoted by Witek 1982, 186, n. 87.
41 BAV Borg. Lat. 577, 57 v.
42 Witek 1982, 186, n. 87; ARSI, Jap. Sin. 182, 398.
43 ARSI, Jap. Sin. II 154, 4.
44 On conic sections in the *Books on calendrical astronomy according to the new method*, see Jami 1998, 668–674.

On the other hand, the mathematicians working under Yinzhi's supervision did not simply take up the methods used at the Bureau. As regards the obliquity of the ecliptic, which in Foucquet's view was one of the major shortcomings of 'Tychonic astronomy', rather than choose between the value given in the *Books on calendrical astronomy according to the new method*, 23°31′30″, and that proposed by Foucquet, 23°29′, they used the value deduced from their own measurements, that is, 23°29′30″.[45] Thus the work of the Office of Mathematics promoted the autonomy of imperial astronomy vis-à-vis Western learning: the Office was subservient neither to Tycho Brahe and the Astronomical Bureau, nor to the Paris Academy of Sciences.

13.3 Foucquet's writings on mathematics

Beside the *Dialogue on astronomical methods*, Foucquet also wrote two mathematical treatises, one on algebra and one on logarithms. The former was commissioned during the summer of 1712, when Kangxi's retinue in Jehol included Foucquet, as well as the emperor's six 'disciples' in the mathematical sciences:[46]

One day, in Jehol, when [Foucquet] was in front of the emperor shortly before the autumn hunt, His Majesty started of his own initiative to talk about algebra, looking at [him] and wanting to know what he thought about it. The Father told him that there was a new one, much easier and of much broader use than the old one. This was enough for the emperor, who was awaiting the opportunity, to immediately order the father to explain to him what this new algebra was.[47]

At the bottom of a page of one of his notebooks, Foucquet jotted down in Chinese what could be the first problem that he solved for the emperor using symbolic algebra:

On the 21st day of the 7th month of the 51st year of Kangxi, in Jehol, His Majesty asked: '36 bolts of silk and satin cost together 100 silver taels. Each bolt of silk costs 2 silver taels. Each bolt of satin costs 4 silver taels. How many bolts of each of these are there?'[48]

Foucquet's solution is written to the left of the question (see Box 13.1).

This simple problem is solved using the substitution method. The only peculiar feature in this solution is the signs used for 'plus', 'minus' and 'equal': respectively the *yang* and *yin* strokes as they appear in hexagrams, and the cross. Foucquet probably made these up there and then, as a reference to his views on the Christian message hidden in the

---

45   *SL* 260, 10b–11a.
46   See pp. 263–264.
47   ARSI, Jap. Sin. II 154, 21.
48   BAV, Borg. Lat. 516, f. 141v; the date corresponds to 22 August 1712.

Box 13.1 Foucquet's solution of the problem set by the emperor. (BAV, Borg. Lat. 516, f. 141 v.)

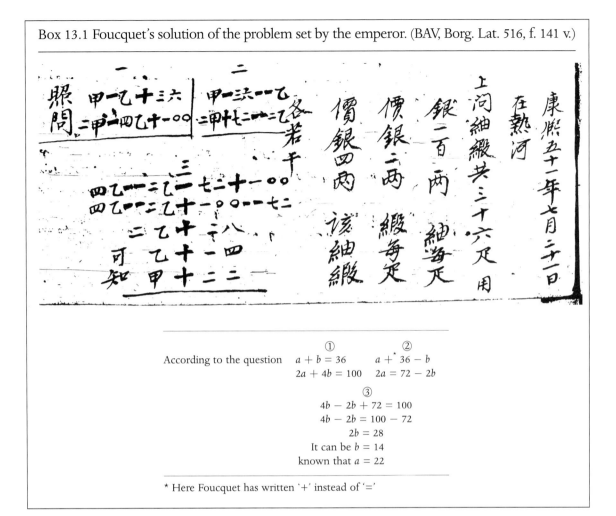

According to the question   ① $a + b = 36$ ② $a +^* 36 - b$
                                       $2a + 4b = 100$ $2a = 72 - 2b$

③ $4b - 2b + 72 = 100$
$4b - 2b = 100 - 72$
$2b = 28$
It can be $b = 14$
known that $a = 22$

* Here Foucquet has written '+' instead of '='

Classics, and in particular in the *Book of change*. The fact that in the second group of equations he wrote a 'plus' sign instead of an 'equal' sign suggests that he was not used to it at that stage; as his French manuscripts witness, he was in the common habit of using the cross '+' as meaning 'plus', and a single horizontal stroke '—' as meaning 'minus'.

This first exchange between the Jesuit and the emperor concerning symbolic algebra took place just a few days after the latter ordered Cheng Dawei's *Unified lineage of mathematical methods* (*Suanfa tongzong* 算法統宗, 1592) to be edited. A problem quite similar to the one above is given as a rhyme in the section on rectangular arrays (*fangcheng* 方程) of chapter 16 of that work, which is devoted to 'difficult problems' (*nanti* 難題).[49] The emperor probably gave a copy of that

---

49   Guo 1993, 2: 1402; see pp. 20–22.

work to Foucquet at some stage: in the latter's papers one finds the first four problems of chapter 11, devoted to rectangular arrays; they are solved using a symbolic notation similar to the one described above.[50] We do not know, however, whether Foucquet indicated that his 'new algebra' could be a substitute for the method of rectangular arrays, and whether he had studied Cheng Dawei's method, or merely copied the problems without looking at his solutions. In any case, Kangxi's curiosity was aroused by Foucquet's notation:

> The emperor [...] eventually said that as he was going to hunt, he wanted [...] one fascicule every other day on this algebra. He ordered at the same time to enquire whether among the other Europeans someone else could also work on it. F. Jartoux, a French Jesuit who understands analysis well, had just come back from Western Tartary, where he had done some surveying. The emperor wanted Jartoux to follow him on the hunt, to explain to him the fascicules that F. Foucquet would send. The latter therefore started to write; after giving the rules of operations on letters and signs, he had just started on the first principles of equations when F. Jartoux had an attack of a very dangerous disease. The emperor, having no one to explain to him this abstract doctrine, ordered that one should cease to send him fascicules until further notice.[51]

Kangxi found the subject interesting at first: during the hunt he informed Yinzhi that he had a new method of algebra, superior to the old one.[52] However after Jartoux's explanations were discontinued, Foucquet heard nothing more about the matter for a year. In the summer of 1713, at the time when he ordered Yinzhi to recruit men to compile works on mathematics and harmonics, the emperor also commissioned the Jesuits to 'write clearly and neatly the principles of the doctrine' of logarithms.[53] At that stage, Foucquet noted:

> [The emperor] has long had Vlacq's great logarithms, and the others that have been reduced for easy use and convenience of space.[54] He knows their practice and he has taught it himself to his children; but because he very much esteems this work and can feel its usefulness everyday, he wanted to penetrate its subtle doctrine.[55]

That Foucquet, who was not a wholehearted admirer of Kangxi and his sons, penned these words, confirms there was more than a rudimentary

---

50 BAV Borg. Cin. 318 (4). The order to edit Cheng's work is dated Kangxi 51/7/13 (14 August 1712); Zhongguo diyi lishi dang'anguan 1996, no. 1980.
51 ARSI, Jap. Sin. II 154, 21–22.
52 Zhongguo diyi lishi dang'anguan 1996, no. 3872; the document is not dated, but the reference to the emperor being in Mulan for the hunt indicates that it was written at the time when Foucquet was working on his treatise; Kangxi left Jehol on 2 September and returned there on 12 October (SL 250: 18a, 251: 2b).
53 ARSI, Jap. Sin. II 154, 19; see pp. 269–270.
54 First published respectively in 1628 and 1636; Gillispie 1970–1980, 14: 51–52; see Verhaeren 1949, no. 3061–3063. The tables in question were probably those translated by Thomas in 1690; see p. 148.
55 ARSI, Jap. Sin. II 154, 19.

grasp of mathematics in the imperial family. This contradicts the assessment by Matteo Ripa, who was then working at court as a painter:

> The emperor supposed himself to be an excellent musician and a still better mathematician; but though he had a taste for the sciences and other acquirements in general, he knew nothing of music and scarcely understood the first elements of mathematics.[56]

Just as the panegyrics of the emperor and of his understanding of the sciences such as the one presented by Bouvet to King Louis XIV in the 1690s must be taken with a pinch of salt,[57] so Ripa's negative assessment should be considered with care. First, one might wonder whether he knew enough mathematics himself to be in a position to judge another person's level. Secondly, the tone of his account of the court tends to be generally dismissive: he also claimed that the staff of the Office of Mathematics 'understood little or nothing' of the *Elements of geometry* on which the emperor set them to work.[58] This is however incompatible with the editorial achievement of the Office.[59] Similarly, the emperor's annotations to some of the manuscript lecture notes produced by the Jesuits and his discussions with Yinzhi leave little doubt as to the fact that he knew what he was talking about; there is no reason to believe that he understood less of the Jesuits' teaching in mathematics than he did of his Chinese teachers' tutoring in the Classics.[60] Ripa's dismissive view of imperial mathematics may have been motivated by his hostility to the Jesuits.

Let us return to Foucquet's narrative: since Luigi Gonzaga, who was in the retinue of the emperor in the capacity of a mathematician during the summer of 1713, seemed incapable of explaining the construction of logarithmic tables in writing, the task once more fell to Foucquet, who had remained in Beijing.[61] His short treatise on the subject seems to have given full satisfaction.[62] A manuscript of about twenty pages, which could be the whole or part of this treatise, is kept at the Vatican Library; it is entitled *Dialogue on numerical tables* (*Shubiao wenda* 數表問答).[63] It consists of four fascicules (*duan* 段, in

---

56 Ripa 1844, 63; quoted a.o. in Hummel 1943, 331; Kessler 1976, 152.
57 Bouvet 1697.
58 Ripa 1996, 2: 62.
59 See chapters 14 & 15.
60 See pp. 176–179. Jami 2002b, 40–41.
61 Zhongguo diyi lishi dang'an guan 1996, no. 2173 (translated in ARSI, Jap. Sin. II 154, 20), 2178, 2180, 2192.
62 ARSI, Jap. Sin. II 154, 20; some tables were 'translated' at the same time (Zhongguo diyi lishi dang'anguan 1996, no. 2192).
63 BAV Borg. Cin. 319(4), ff. 71–89; another manuscript bearing the same title is kept at the Palace Museum Library in Beijing; but I have not been able to consult it (Online union catalogue of Chinese rare books: http://202.96.31.45/libAction.do;jsessionid =CDCF8EE9B8935ABFE8972B6F20167D7B?method=goToBaseDetailByNewg id&new-gid=12386). It is possible that the Vatican version is not complete, as the memorial

French 'cahiers'), and opens with a brief account of how Napier (Naiboer 㮈伯尔, 1550–1617) had first constructed tables one hundred years earlier, and entrusted their completion to a friend of his.[64] The first two fascicules contain basic explanations of 'ratios of similar difference' (*yitong bili* 異同比例, arithmetic progressions) and of 'ratios of similar multiple' (*beitong bili* 倍同比例, geometric progressions); these terms differ from those used in Thomas' textbooks. Constructing tables, then, consists in writing out numbers in 'ratios of similar multiple' and numbers in 'ratios of similar difference' in matching columns. One column contains 'actual numbers' (*zhenshu* 真數), while the other contains 'false numbers' (*jiashu* 假數): these terms had been in use since Smogulecki first introduced logarithms in the 1650s; they parallel those used by Napier.[65] Foucquet had in his hands some already existing tables in Chinese, entitled *Extensive use of logarithms* (*Duishu guangyun* 對數廣運), which he studied in detail, but his *Dialogue on numerical tables* did not refer to these in particular.[66] The purpose of logarithmic tables, he explained in this work, was to simplify calculations, by replacing multiplication and division by addition and subtraction.[67]

Kangxi was satisfied with Foucquet's explanation of logarithms; it was easy to understand, and the subject was uncontroversial. However, the treatise on symbolic algebra was less successful: in September 1713, the emperor, who was in Jehol for his usual summer visit, gave his view on it:[68]

Since We started on Our journey, We have examined the *New method of algebra* (*Aerrebala xin fa* 阿爾熱巴拉新法) every day with Our sons. It is extremely difficult to understand. He [Foucquet] says it is easier than the old method, but looking at it, it is more difficult than the old method. There are many difficult and intricate points, and several muddled points. Previously, We happened to commission the Westerners in Beijing to write on the foundations of numerical tables. They wrote with utmost clarity. Copy out this edict and send it to the capital with the book. Order the Westerners to examine it together closely, and to delete all that is badly written. One more word: *jia* 甲 multiplied by *jia*, *yi* 乙 multiplied by *yi* is after all no number. And if one succeeded in multiplying, one

---

reporting the progress of the *Dialogue on numerical tables* mentions that 'eight paragraphs of the first section' have been written (Kangxi 52/6/23, i.e. 13 August 1713); less than a month later (Kangxi 52/7/17, i.e. 6 September 1713), the text, in one section comprising 48 paragraphs, appears to be complete; Zhongguo diyi lishi dang'anguan 1996, no. 2201, 2241.

64 BAV Borg. Cin. 319 (4), 71a; Henry Briggs, the 'friend' in question, is not named.
65 See p. 43.
66 BAV, Borg. Cin. 518 (15); two copies of the tables in question are still extant (Online union catalogue of Chinese rare books: http://202.96.31.45/libAction.do;jsessionid=CDC-F8EE9B8935ABFE8972B6F20167D7B?method=goToBaseDetailByNewgid&newgid=12389); a facsimile reprint is found in Gugong bowuyuan 2000, 401: 251–326.
67 For a detailed analysis of the *Dialogue on numerical tables*, see Jami 1986, 45–59.
68 ARSI, Jap. Sin. II 154, 21.

would not know its value. Considering that, this man's calculation method is mediocre—this is too little; in one word, it is laughable.⁶⁹

Three months later, the emperor ordered Stumpf that the Jesuits should speed up the revision of the work, and that it should not be done by Foucquet himself.⁷⁰ At this point Thilisch undertook to write another treatise on symbolic algebra, relying on Claude-François Milliet de Chales' *Cursus seu mundus mathematicus*: evidently the French Jesuits were not the only ones to use French mathematical works, especially when they were written in Latin.⁷¹ Both Foucquet and Jartoux found Thilisch's extracts from Dechales' work incomprehensible: they agreed that he had left out too much of the explanation. Foucquet nonetheless translated what Thilisch sent him into Chinese. The emperor does not seem to have liked the resulting text either; it is not known whether any of the Jesuits explained it to him.⁷² In any case, in April 1714, they were ordered to send all the books on algebra in 'Western characters' (*xiyang zi* 西洋字) in their possession to Yinzhi, together with Thilisch's writings.⁷³ Although there is no evidence that anyone at court knew enough 'Western characters' to read these books, the figures and tables that they contained could be checked; some members of the Office of Mathematics may also have looked at algebraic notations. Be that as it may, the superiority of symbolic algebra over cossic algebra, which Foucquet took for granted, was not obvious to anyone at court. The *New method of algebra* was not integrated into imperial mathematics. Foucquet held Yinzhi responsible for this: it was he who had not understood symbolic algebra, and insisted on giving priority to astronomy over other subjects.⁷⁴ If indeed the prince was hostile to the new algebra, it could also have been because he defended what he had been taught in his youth out of loyalty to his late master Antoine Thomas.

---

69   BAV, Borg. Cin. 439 A (a); Zhongguo diyi lishi dang'anguan 2003, 1: 52. In Foucquet's French translation of this edict (ARSI, Jap. Sin. II 154, 21), he has left out the last eight characters of the edict: 太少二字即可笑也. It may be simply because they are disobliging; but it is also possible that he first received a version of the edict that did not contain these words, which appear as an addition in the vermilion ink version kept at the No. 1 Historical Archive. Zhongguo diyi lishi dang'anguan 2003, 1: 52.
70   BAV, Borg. Cin. 439 A (dated Kangxi [52]/10/18, that is, 5 December 1713); see ARSI, Jap. Sin. II 154, 24–25.
71   Dechales 1690, vol. 1; both the 1674 and the 1690 editions of the work are found in Verhaeren 1949, no. 1259 to 1263; on Dechales see p. 162.
72   ARSI, Jap. Sin. II 154, 25.
73   BAV, Borg. Cin. 439 A & B.
74   ARSI, Jap. Sin. II 154, 22.

13.4 Symbols in the *New method of algebra*

An examination of Foucquet's treatise sheds further light on its failure.[75] The copy of the incomplete *New method of algebra* that he brought back from China is the only one extant; it consists of two sections (*juan* 卷), each further divided into items (*jie* 節).[76] The handwriting is neat, but the manuscript is not as beautifully copied as the lecture notes of the 1690s. The pages of the first section are numbered continuously from 1 to 21; those of the second section are not numbered.[77] The items in this second section are not in the same order in the text as in the table of content at the beginning of the section. Foucquet's treatise does not seem to be a translation, but an original composition; he first wrote an outline in French.[78] Like Antoine Thomas, having taught mathematics at a college before leaving Europe, he was capable of putting together lecture notes on a mathematical topic.

The text has the form of a dialogue, but the questions and answers do not correspond to the traditional problem–solution layout of Chinese mathematical works. The first section introduces the symbolic notation, and the rules for the four basic operations on expressions containing symbols. It opens with an 'Explanation of the differences between this new algebra and the old one' (that is, Thomas' *Calculation by borrowed root and powers*),[79] which, while paying lip-service to the 'old algebra', points out that 'where flexibility (*tongrong* 通融) is concerned, it has some imperfections.' The new method remedies these by introducing a new notation:

The rules of the old method and of the new method are about the same. The difference is that the signs (*jihao* 記號) used in the old method have the form of digits. The signs used in the new method are signs that can be adjustable; thus in the West one uses 22 letters. In China one can use the 22 characters for heavenly stems and earthly branches in their place, the advantage being that people are familiar with all of them. Therefore when using them one will make no mistakes.[80]

Literal notations should be used both for unknowns and for coefficients, because, as argued further, this enables one to retrace all the steps of the calculation.[81] This notation is introduced using a parallel between

---

75 To the best of my knowledge the extracts from Dechales' work prepared by Thilisch and translated into Chinese by Foucquet are no longer extant.
76 A copy of Foucquet's incomplete *Aerrebala xin fa* 阿爾熱巴拉新法 is kept at the Vatican Library: Borg Cin 319 (4), 1–69; the *Shubiao wenda* is part of the same volume.
77 This section comprises 47 double pages altogether. For a detailed analysis of the treatise, see Jami 1986, 60–107.
78 'Abrégé d'algèbre', BAV, Borg. Lat. 516, ff. 132–161v.
79 See pp. 200–212.
80 BAV, Borg. Cin. 319 (4), 1a.
81 BAV, Borg. Cin. 319 (4), 2a.

numbers and signs for the calculation of the area of a square of given side: the numerical value of this side is 12, its literal value is *jia* 甲 plus *yi* 乙. The two multiplications are given in parallel, with the corresponding squares (see Box 13.2).

This is the first occurrence of symbols in the text; it is expressed in a rather confusing manner. The analogy brought out by the two figures is one between squaring the number 12 = 10 + 2 and squaring the sum of *jia* (corresponding to 10) and *yi* (corresponding to 2). On the other hand, in the multiplication layouts, the two numbers that are in parallel with *jia* and *yi* respectively are 1 and 2. Moreover, in these

---

Box 13.2 Numerical and literal multiplication in the *New method of algebra*. (BAV, Borg. Cin. 319(4), 2a–2b.)

		甲	+	乙		1	2	
		甲	+	乙		1	2	
		甲乙	+	乙乙		2	4	
		甲甲	+	甲乙		1	2	
甲甲	+	2甲乙	+	乙乙		1	4	4

layouts, place value is used for numbers where addition is used for symbols. In other words, the juxtaposition of two digits is to be read as a number in place-value notation, while the juxtaposition of two cyclical characters is to be read as their product. So, is the analogy drawn between numbers and 'adjustable signs', or between the latter and the digits of a number? This ambiguity contributed to the puzzlement expressed by Kangxi concerning the meaning of two juxtaposed symbols: '*jia* multiplied by *jia*, *yi* multiplied by *yi* is after all no number'. Thomas already used the heavenly stems in his *Calculation by borrowed root and powers*, when dealing with systems of linear equations;[82] Foucquet also used them in his *Dialogue on numerical tables* to denote the successive terms of a continued proportion.[83] The only novelty here compared with what Kangxi had previously studied was precisely the multiplication of cyclical characters used as symbols by one another, and Foucquet's text is not exactly enlightening on that point.

This point stopped Kangxi when he re-examined the *New method of algebra* with his sons in 1713; however one may assume that during the 1712 autumn hunt he had gone further into the text with Jartoux, seen how the four basic operations are performed on symbols, and studied the resolution of linear problems in several unknowns, which is the object of the second section of the text. He may well have understood all this then thanks to Jartoux's explanations, and not remembered it the following year. This is not uncommon in the study of mathematics; moreover, around the time when he commissioned Foucquet to write on algebra, Kangxi complained that, as he was reaching his sixtieth year, he could no longer memorise what he read in books.[84] Judging from the rest of the treatise, the emperor's complaint that it was unclear may have resulted partly from the fact that Foucquet uses neither published Chinese mathematical books nor the textbooks composed by his confreres in the 1690s as references. The only work he refers to when justifying some statements is the 1607 *Elements of geometry*, rather than the work by Gerbillon and Bouvet; neither does he refer to the very work on which he claims to improve, that is Thomas' *Calculation by borrowed root and powers*.[85] His terminology is partly created anew, ignoring the common Chinese usage at the time: he uses *guichu* 歸除, which commonly referred to division itself, to denote the quotient of a division (usually called *shang* 商). This probably passed unnoticed by Jartoux, but must have confused readers as familiar with

---

82   See pp. 210–212.
83   BAV, Borg. Cin. 319 (4), 74 a–b.
84   SL 250: 16a–16b.
85   See e.g. BAV, Borg. Cin. 319 (4), 2a, where the reader is referred to the Pythagoras theorem; references to Ricci's translation of the *Elements* are also found in the *Dialogue on numerical tables*.

Chinese mathematical terms as the emperor and his sons. For powers, Foucquet introduces a new term, *ceng* 層 (lit. 'layer'), as an equivalent of 'degree'. He calls the symbols that denote unknowns and coefficients *tongrong jihao* 通融記號, which can be rendered literally as 'adjustable signs' (in French he calls them *signes généraux*—general signs). For addition, subtraction and equality, he uses the same signs (*jihao* 記號) as in his first Chinese transcription of symbolic algebra presented to Kangxi on 22 August 1712. However in the treatise, the signs are written in outline, probably to avoid confusion with the characters for one, two and ten (*yi* 一, *er* 二 and *shi* 十); for example, $z - x = 2$ and $y = 1 + x$ are written respectively as follows:[86]

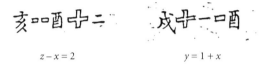

$z - x = 2$ $\qquad$ $y = 1 + x$

Illustration 13.3 Equations, *New method of algebra*. (BAV, Borg. Cin. 319(4), n.p.)

At the time when Foucquet wrote the *New method of algebra*, Kangxi was interested in the links between the *Book of change* and mathematics: as he told Li Guangdi: 'Calculation and the numbers of the *Change* match each other (*wenhe* 脗合).'[87] He deemed these links to be of a complex nature. Indeed Foucquet was trying to suggest such links. But to the emperor who had been discussing them in a sophisticated manner with scholars for years, the Jesuit's attempt must have seemed at best crude; his association of the Christian cross with *yin* and *yang*, if it was noticed at all, cannot have helped convince anyone acquainted with Chinese classical culture that his notations made any mathematical sense.

The second section of the *New method of algebra* deals with first-degree problems. The central object is defined as an 'equality formula' (*xiangdeng zhi shi* 相等之式) between two terms made up of 'what can be added and subtracted' (*ke jia ke jian zhe* 可加可減者); examples of such equations are given in pairs, one with a literal constant term and one with a numerical constant term.[88] Problems involving first degree equations (*di yi ceng zhi xiangdeng* 第一層之相等) are then classified into three categories: determined (*yiding* 一定), undetermined (*wuding* 無定), and over-determined (*guoyuding* 過於定). Only the first category is discussed in detail. Their solution includes three stages:

- The construction (*cheng* 成) of equations from the data of the problem entails sorting out what needs to be calculated (*ying suan* 應算) from the 'idle words' (*xianyan* 閒言) in the problem.

---

86   BAV, Borg. Cin. 319 (4), 47b.
87   *SL* 251: 21a.
88   BAV, Borg. Cin. 319 (4), 24a–24b.

- The former must be 'given a name' (*ming* 名), so that the ratios (*bili* 比例) between them can be determined.
- The preparation (*bei* 備) of these equations consists in simplifying the equalities obtained as far as possible.
- The solution (*jie* 解) consists in a sequence of nine rules (*jiuze* 九則); the successive forms of the equations used are laid out in three different columns (departing, returning and final columns: *wanghang* 徃行, *fanhang* 反行, *jiehang* 結行). The method used is that of substitution. Both the procedure and the layout are identical to those used in the first problem solved by Foucquet.[89]

The procedure that takes one from the problem to its solution is expounded in great detail, over more than thirty double pages of text, which are divided into eight fascicules. This division suggests that if Jartoux followed the text in his tutoring, each of the sessions that took place during the 1712 imperial hunt focused on one particular step of the procedure, rather than on solving a particular type of problem, as had been the case with Thomas' *Calculation by borrowed root and powers*. This lengthy explanation without any practice may explain why Kangxi was at a loss to figure out what the proposed method consisted in when he revisited Foucquet's treatise a year later.

The composition of the *New method of algebra* stopped just as Foucquet was about to start on second degree equations. Therefore what the emperor had in his hands a year later was text written using idiosyncratic terminology and notation, introducing a new way of combining cyclical characters, and whose lengthy explanation concerned only first-degree problems; it was for these that Thomas had briefly introduced literal symbols—only for the unknowns—some years earlier, without any explanation or comment. Foucquet did not have the opportunity to go far enough in his treatise to show where his main innovation lay: he intended to present a single notation allowing solution of linear problems such as those discussed by Mei Wending in his *Discussion of rectangular arrays* (*Fangcheng lun* 方程論),[90] and those of higher degree, for which Thomas used the cossic notation. Even if some scholars of the Office of Mathematics managed to understand the symbolic algebra introduced in Foucquet's treatise, they would not see in it anything that surpassed Mei Wending's method or the methods presented in the series of earlier Jesuit textbooks on which they were working at the time: they would therefore have had little reason to adopt it.

In his assessment of Foucquet's treatise, Kangxi did not mention Foucquet's curious signs for operations. Jartoux, on his part, cannot

---

89   BAV Borg. Cin. 319 (4), 24a–69b; this particular layout may well have been borrowed from a French or Latin treatise, but so far I have not identified a source for it.
90   See pp. 95–101.

have felt entirely comfortable with the association of *yin* and *yang* with the Christian cross, a smuggling of Figurist associations into a mathematical text. These signs are the only trace found in Foucquet's writings on mathematics of his controversial views on the *Classics*. They reveal the extent of his unconventionality, which in his confreres' eyes amounted to breach of discipline and of the respect due to their predecessors.

13.5 The Jesuits and the Office of Mathematics

Foucquet's writings on astronomy were put at the disposal of the Office of Mathematics as possible material for the work they were to compile, just like the writings by Pereira on harmonics, and those by Thomas, Gerbillon and Bouvet on mathematics. At the time Kangxi was neither the first, nor the only reader and editor of the Jesuits' new lecture notes. The latter were in direct contact with the members of the Office of Mathematics. Foucquet mentions several of these:

[On 28 November 1715] I was called to the Tchang Tchun Yuen [Garden of pervading spring] to explain La Hire's machine at the Regulo's place, which I did for two or three days, so that a young mathematician called Ho [He] could understand well what it was about, and explain it afterwards to the Regulo.[91]

The machine in question was 'the ecliptic machine of Mr de la Hire, but all accommodated to the method of the Chinese, whose luni-solar year is different from the European one'.[92] Foucquet had just presented it to the emperor with the second part of his *Astronomical guide for eclipses*.[93] The young mathematician to whom he gave explanations may well have been He Guozong; perhaps Foucquet had met him before, as both men were in the emperor's retinue during his stay in Jehol in the summer of 1712. The flow of circulation of knowledge had reversed: whereas the textbooks written in the 1690s remained a monopoly of the emperor for more than ten years before Chinese scholars had a chance to study them, it was now a member of the Office of Mathematics—probably the most prominent one—who was first instructed by Foucquet on how to use an astronomical machine, so as to eventually pass this knowledge on to Yinzhi; the experts were in direct contact. Foucquet was also asked to repair the two machines of Roemer that had been presented to Kangxi by the five French Jesuits who arrived in China in 1687. After the actual repairs were done by Jacques Brocard, Foucquet 'put in writing the theory of both' machines.[94] Beside the conversion of European dates into Chinese ones, the notice written for the planetary machine stated that the orbits of the Five Planets were not circular, but

---

91 BAV Borg. Cin. 577, f. 57r.
92 I have not been able to identify this machine.
93 ARSI, Jap. Sin. II 154, 30.
94 ARSI, Jap. Sin. II 154, 30.

elliptic (*you si danxing* 有似蛋形), as Foucquet later argued in his *Astronomical guide for the Five Planets*.⁹⁵ The mended machines and the instructions were sent directly to Yinzhi.

Relations between the Jesuits and the Office of Mathematics were not friendly:

In the beginning of July 1716, the Regulo having called F. Thilisch to calculate with his astronomy students a few difficult problems on right-angled triangles, these young men, hostile and arrogant, treated the father very haughtily. They claimed to have embarrassed him, and to know methods that he did not know.⁹⁶

It is possible that the 'difficult problems' submitted to Thilisch dealt with the base and altitude (*gougu* 勾股), the last of the 'nine chapters'. The links between right-angled triangles and base and altitude had long been discussed by Chinese mathematicians.⁹⁷ Since some writings on the base and altitude by Chen Houyao have been preserved,⁹⁸ one may suppose that such discussions also took place at the Office of Mathematics. Thilisch who, according to Foucquet, could not read Chinese very well, may have been puzzled by the terminology and techniques used in problems on base and altitude, which do not refer to the objects of Euclidean geometry.

Thus, whereas there were some direct contacts between the Jesuits who wrote treatises for the emperor and the staff of the Office, relations were hardly of the nature of those between teachers and students, which had earlier been essential in defining the Jesuits' place in Chinese literati society and at the court.⁹⁹ Yinzhi, who had been Thomas' student, avoided placing himself in the position of Foucquet's student; Thilisch seems to have been submitted to a rather humiliating treatment. In short, the status of teachers of mathematics was denied to the court Jesuits just as the discipline they had long taught was integrated into imperial scholarship. Perhaps this integration could not have been achieved without that denial.

Direct transmission of knowledge from the Jesuits to the members of the Office seems to have been the exception rather than the rule; but it did sometimes occur. Decades later, when Mei Juecheng published his grandfather's work in the *Selected essentials of Mr Mei's collection* (*Meishi congshu jiyao* 梅氏叢書輯要, 1761), he appended to it a short text that he had written himself, entitled *Lost Treasures from the Red Waters* (*Chishui yizhen* 赤水遺珍). In it Mei explained a 'simple method for

---

95 BAV, Borg. Cin. 438 B(e).
96 ARSI, Jap. Sin. II 154, 64.
97 In particular, Mei Wending endeavoured to explain Euclidean geometry in terms of base and altitude; Martzloff 1981b, Engelfriet 1998, 407–427.
98 Chen Houyao, *Meaning of base and altitude methods* (*Gougu fayi* 勾股法義), n.d.
99 Jami 2002a.

measuring the altitude of the North Pole' (*ce beiji chudi jianfa* 測北極出地簡法), which he attributed to Karel Slaviček (Yan Jiale 嚴嘉樂, 1678–1735),[100] a Jesuit versed in mathematics who arrived in Beijing in January 1717.[101] This method differed from the one used by the staff of the Office for the measurements of the altitude of the North Pole in various cities undertaken in 1714.[102]

The *Lost treasures from the Red Waters* also gave a 'quick method for seeking the precise lü of the diameter and the circumference' (*qiu zhou jing mi lü jie fa* 求周徑密率捷法), which we would nowadays write down as the series giving π; this can be derived from the power series expansion of the arc sine function. According to Mei, this was a translation of a method given by Jartoux (Du Demei 杜德美). There followed 'quick methods for seeking the chord and the sagitta' (*qiu xian shi jie fa* 求弦矢捷法), which are equivalent, in modern terms, to the power series expansions of sine and versed sine.[103] A posthumous work by Minggantu, the *Quick methods for the circle's division and precise lü* (*Geyuan milü jiefa* 割圜密率捷法, 1774) gave proofs for these three methods, which came to be known as 'the three procedures of Mr Jartoux' (*Du shi san shu* 杜氏三術) and derived six more.[104] Although the circumstances in which Jartoux might have taught these formulae to Mei and Minggantu are not known, the fact that both of them knew about them and ascribed them to Jartoux, who died in 1720, suggests that the transmission occurred while the two men worked at the Office of Mathematics. Minggantu's proofs of the validity of 'the three procedures of Mr Jartoux' provided a more complex counterpart, for trigonometric tables, to Foucquet's explanation of the construction of logarithmic tables, and were therefore directly relevant to the interests of the Office of Mathematics.

These are 'lost treasures' insofar as they were not included in the imperial compendium on which Mei Juecheng worked at the Office. They bear witness to the fact that rivalries and tensions did not entirely prevent communication between the two competing groups of specialists in the mathematical sciences who served the emperor.

As in the 1690s, the emperor also commissioned writings on medicine around the time he created the Office of Mathematics: Dominique Parrenin (1665–1741) composed a treatise on anatomy in Manchu. The emperor expected that 'The public must draw great advantage from this book, as it must contribute to save, or at least to lengthen, life.'[105]

---

100 *Meishi congshu jiyao* 1761, 61: 3b–4b.
101 Pfister 1932–34, 655; Zhongguo diyi lishi dang'anguan 2003 1: 15–17.
102 *SL* 261: 4b–5a.
103 *Meishi congshu jiyao* 1761, 61: 23a–27b; the versed sine of an angle is equal to 1 minus its cosine.
104 Jami 1988, Jami 1990.
105 *LEC* 17: 350.

Accordingly, he planned to have the work translated into Chinese. He appointed 'three Mandarins of the most skilled, two writers whose hand was excellent, two painters very capable of drawing figures, line drawers, designers etc.' to work with the French Jesuit, who was fluent both in Manchu and in Chinese.[106] It is possible that they used some earlier writings by Bouvet and Gerbillon, at least to construct the terminology. However the *Complete record of the body imperially commissioned* (*Dergici toktobuha Ge ti ciowan lu bithe*) was a translation of a work by the surgeon Pierre Dionis (1643–1718), who had taught at the *Jardin du Roy*.[107] Like Gerbillon, Bouvet and Foucquet, Parrenin drew on French royal science to serve the emperor, who carefully edited the work.[108] Kangxi eventually decided not to circulate the text: its illustrations in particular might be regarded as shocking.[109] The fact that the text was in Manchu sufficed to ensure that the few copies that were to be kept in imperial libraries would only be accessible to few readers.[110] The emperor must still have had in mind the Hanlin academicians' reaction to the idea put forward by Verbiest that the brain was the siege of memory; moreover Parrenin's work contained allusions to the 'invisible soul' and to the 'Lord who created things'.[111]

It has been argued that the representations of the human body underlying Chinese medicine are incompatible with that yielded by anatomy.[112] However, it would be worth reconsidering Kangxi's commission in a slightly different perspective: the nature of the knowledge of the human body accumulated in Chinese forensic medicine, a separate profession that had its own literature, was not so distant from that of European anatomists. An updated edition of the Song dynasty 'classic' of the profession, the *Collected records on the washing away of wrongs* (*Xiyuan jilu* 洗冤集錄, 1247) was published by the Ministry of Justice (*Xingbu* 刑部) in 1694 to be circulated throughout the empire.[113] Thus anatomy was relevant to a specialised profession of the civil service as well as satisfactory to the curiosity of the emperor. But in the present state of our knowledge of the relevant sources, no evidence supports the hypothesis that Kangxi

---

106   Walravens 1996, 365.
107   Dionis 1691; Walravens 1996, 364.
108   Walravens 1996, 366; Stary 2003a.
109   They were drawn from Thomas Bartholinus, *Anatome quartum renovata*. Lyon: Jean Huguetan, 1677; there was a copy of this work at the French residence (Verhaeren 1949, no. 948).
110   LEC 17: 350; Jami 2002b, 43; Jami 2010 on medical literature in Manchu see Hanson 2006.
111   See pp. 80–81. Walravens 1996, 370.
112   Walravens 1996, 364.
113   In 1770, plates of the human skeleton were added to this official version of the work. Will 2007, esp. 70–71; Will forthcoming, no. 502; on the *Xiyuan jilu* see Needham 1954–, 6–6: 178–179.

considered using yet another branch of Western learning as a tool for statecraft.

13.6 Tables and the standardisation of mathematical sciences

As shown above, tables and their use were central to the emperor's concerns and to the activities of the Office of Mathematics. The tables that Kangxi had had copied out in the 1690s may already have been based on Vlacq's tables; they were also used at the Office of Mathematics.[114] The emperor closely supervised the preparation of these tables for publication. In early January 1714, the staff of the Hall of Military Glory (the Imperial Printing Office), completed the copying of a volume of trigonometric tables.[115] The emperor proposed a different layout, with the degree at the top of the page, and the minutes listed in columns below. He ordered Yinzhi to use this layout for the tables produced at the Office.[116] At the same time, the aging emperor repeatedly insisted that the character strokes for the numbers in these tables should be thicker, so as to be more legible.[117]

Astronomical tables were more controversial. Foucquet appended a Chinese version of La Hire's tables, adapted to the Beijing meridian, to his *Astronomical guide for eclipses*.[118]

A few months later, after Foucquet's *Dialogues on astronomical methods* was disowned by his confreres, Yinzhi borrowed the original tables from him, which Foucquet interpreted as a sign of distrust rather than of interest.[119] There had been earlier instances of such reluctant loans:

This dangerous prince, having had several books from Europe brought to him, which he retained, out of authority or rather out of violence, noticed tables of square and cubic numbers which are at the end of the second volume of Prestet.[120] He took it to his Father who had ordered that a similar one be calculated, and used this opportunity to make him understand, that either the

---

114 BnF, Mss. Fr. 17240, f. 266 r.-v.; Han 1992.
115 These were the *Yuzhi shubiao jingxiang* 御製數表精詳, which are reproduced in Gugong bowuguan 2000, 401: 327–391; Zhongguo diyi lishi dang'an guan 1996, no. 2308, 2312.
116 Zhongguo diyi lishi dang'anguan 1996, 2326, 2328; in the Chinese translation of the Manchu documents, the character *du* 度 (degree) is wrongly written as *du* 讀; see Jami 2002b, 40–41.
117 Zhongguo diyi lishi dang'anguan 1996, 2335, 2339, 2343.
118 La Hire 1702; see Verhaeren 1949, no. 1944 and 1945. ARSI, Jap. Sin. 182, 398r. Foucquet was not the only provider of data for imperial astronomy. Jartoux presented tables for the calculation of sunrise and sunset (*suan ri churu hun ke* 算日出入昏刻); Zhongguo diyi lishi dang'anguan 2003, 1: 55.
119 ARSI, Jap. Sin. II 154, 81.
120 Prestet 1689; see Verhaeren 1949, no. 584.

Europeans were ignorant of their own tables, or they had kept these hidden from him.[121]

Evidently, both the emperor and his son could read tables written using Arabic numerals: unlike texts, tables could be used at the Office of Mathematics without the help of a translator.[122] It seems that in general borrowed books were eventually returned: in February 1717, Kangxi sent to Stumpf eleven of them, which included the first French edition of Dechales' *Elements de géométrie*.[123]

Whereas presenting symbolic algebra was Foucquet's initiative, the demands to which his other writings responded all concerned methods for constructing tables, or one might say the 'why it is so' (*suoyiran* 所以然) of the tables. Then as in the 1660s, Kangxi wanted to understand the criteria according to which decisions should be made. His demand that all astronomers should refer to the same tables follows the Chinese custom that the ruler issues quantitative standards for all the empire: beside the calendar, these included the definition of weight and length units. This made the emperor's concern legitimate and relevant in the context of the political culture within which he ruled. At the same time, his demand that the Jesuits should agree amongst themselves as to which astronomical parameters, and therefore which tables, were to be used also echoes concerns of standardisation of measurement displayed by academies of science in Europe: indeed different tables were used in Europe at the time, but the 'King's Mathematicians' had been given precise instructions so as to ensure that the observations they were to send to France could be compiled with those done in other parts of the world. Thus the 'Chinese tradition' provided a rationale for a feature that is regarded as a characteristic of 'European modernity'. In Beijing as in Paris, standardisation was rendered necessary by the unprecedented scope of the knowledge centralised, as the territory under the control of the Qing dynasty expanded. That this knowledge was drawn from multiple sources, between which a common measure had to be found, is a feature characteristic of empire throughout the ages.

Foucquet's mathematical and astronomical writings in Chinese had little or no direct influence on the compilation undertaken at the Office of Mathematics under Yinzhi's supervision. This very failure (it was indeed a failure in Foucquet's eyes) is revealing in several respects. First, by the 1710s the court Jesuits worked amidst multiple tensions: among the French, Figurism had become a bone of contention; and, almost three decades after the arrival of the 'King's Mathematicians' in China, the rivalry between the French and the Portuguese mission had by no

---

121 ARSI, Jap. Sin. II 154, 4.
122 On Kangxi's interest in European script, see APF Informazioni 118, f. 439.
123 Verhaeren 1949, no. 172.

means eased away. Secondly, the Jesuits saw Kangxi's setting up of the Office of Mathematics and the compilation of new reference works in the mathematical sciences as an attempt to deprive them of the credit they had gained by providing the Qing dynasty with an astronomical system. Indeed the mastery of Western learning by a sufficient number of officials was a prerequisite to doing without the Jesuits. However, although Kangxi by then displayed more esteem for the members of the Office than for the Jesuits, there is no evidence that he ever intended to do entirely without the latter. Beyond conflicts of ideas and of interest between individuals and institutions, Foucquet's attempt to introduce some innovations into imperial science gives us a glimpse of the complex process of validation of knowledge in the mathematical sciences at the court during the last decade of the reign. The emperor had earlier acted as intermediary and an arbiter between the Jesuits and the Chinese literati. He was now using the latter, who were constructing the mathematical sciences as a branch of imperial scholarship, to control the former, who, by working on the general map, continued to ensure the imperial monopoly on Western learning as a tool for statecraft.

PART V

Mathematics for the Empire

CHAPTER 14

The construction of the *Essence of numbers and their principles*

The mathematical compendium resulting from the work done at the Office of Mathematics under Prince Yinzhi's supervision, the *Essence of numbers and their principles imperially composed* (*Yuzhi shuli jingyun* 御製數理精蘊) was printed in 1722, a few months before Kangxi's death. The first striking feature of the work is its size: the text spans almost 4900 pages, each of which comprises nine columns of twenty characters. Many of these pages include cartouches with figures or calculation layouts, which take up the space of the five top characters for most of the page; the same illustration is sometimes reproduced on more than one page, so that the reader always has it under his[1] eyes when studying the text that refers to it; the tables then cover almost another 3000 pages.[2] This is the longest and the most lavishly illustrated single mathematical work ever printed in imperial China.[3] The details of its printing history are unclear; however it seems that there was at least one edition in moveable type done during the Kangxi period.[4]

14.1 Outline

The *Essence of numbers and their principles* consists of fifty-three chapters, divided into three parts, of very different lengths and layout. The first part (*shangbian* 上編, five chapters) is devoted to 'Establishing the

---

[1] It went without saying that the book, composed by males, was intended for male readers: scholars and officials.
[2] For a list of the various editions kept in Chinese libraries, see Li 2000a, 507–509. All the copies printed in the Kangxi era I have seen have the same layout; the *Complete library of the four treasuries* (*Siku quanshu* 四庫全書) copy reproduced this layout (*SKQS* 799–801); unless otherwise specified, the present description is based on the facsimile of a Kangxi edition in Guo 1993, 3, which does not include the tables. The format of the two other works included in the *Origins of pitchpipes and the calendar* is identical. For the tables I have used the *SKQS* copy, vol. 800–801, in which again the original layout has been preserved.
[3] As a point of comparison, the Kangxi edition of the *Unified lineage of mathematical methods* reproduced in Guo 1993, 2: 1217–1453, comprises about 950 pages of 10 columns of 21 characters, that is, less than one third of the size of the imperial compendium.
[4] According to Fan 1999, 88–89, there were two; for one of them chromo-typography was used: the punctuation was printed in red.

structure to clarify the substance' (*Ligang mingti* 立綱明體); the text is divided into items (*jie* 節), which are numbered.⁵ The second part (*xiabian* 下編, forty chapters), entitled 'Dividing into groups to put to use' (*Fentiao zhiyong* 分條致用), is written in the form of problems, each accompanied by a method (*fa* 法) for deriving the solution. Finally, there are eight chapters of tables (*biao* 表). In the first and second parts of the work, notes in small characters are sometimes inserted in the text, in the way that commentaries were traditionally added to classical texts; these give either supplementary explanations or cross-references to another item in the work. No individual author signs any passage of the work.

Each part of the *Essence of numbers and their principles* begins with a table of contents. Part I opens with a chapter on the origins and foundations of mathematics; it then gives 'elements' of both geometry and calculation. Part II consist of five sections. The 'Initial section' starts with a list of standard units; it then explains the four basic operations of arithmetic and the handling of fractions; written calculation is the only technique discussed here; operations are laid out with numbers written horizontally. The 'Line section' discusses ratios, which are used among others in methods akin to the rule of false position; its last chapter takes up Mei Wending's rectangular arrays (*fangcheng* 方程) method for solving systems of linear equations.⁶ The 'Area section' deals with square root extraction and the solving of second degree equations; it also discusses plane geometrical figures, starting from problems on the base and altitude (*gougu* 句股) in the Chinese tradition, going on to triangles and from there to all kinds of figures. The 'Solid section' has a structure parallel to the previous one, going on from cube root extraction and third degree equations to solids; its final chapter contains some problems on the formation of complex solids from stacks (*duiduo* 堆垛) of an elementary shape, in the Chinese tradition. The 'Final section' includes six chapters devoted to cossic algebra. This is followed by a chapter of 'difficult problems' (*nanti* 難題), and another on the principles that underlie the construction of logarithm tables; finally, two chapters are devoted to the use of the proportional compass. There are four kinds of tables: trigonometric lines, decomposition of all numbers up to 100,000 into two multiplicative factors (with lists of prime numbers (*genshu* 根數)), logarithms, and logarithms of trigonometric lines.

---

5 The character *jie* 節 does not appear in the numbering itself, but in cross-references given in small characters.
6 See pp. 90–101.

Table 14.1 Contents of the *Essence of numbers and their principles*

立綱明體		上/I	Establishing the structure to clarify the substance
數理本原 河圖 洛書 周髀經解		I.1	Origins of numbers and their principles River Diagram Luo Writing Explanation of the Classic of the Gnomon of Zhou
幾何原本一之五		I.2	Elements of geometry 1 to 5
幾何原本六之十		I.3	Elements of geometry 6 to 10
幾何原本十一、十二		I.4	Elements of geometry 11, 12
算法原本一、二		I.5	Elements of calculation methods 1, 2
分條致用		下/II	Dividing into groups to put to use
首部一	度量權衡 命位 加法 減法 因乘 歸除	II.1 Initial section 1	Measures and weights Assigning places Addition Subtraction Multiplication Division
首部二	命分 約分 通分	II.2 Initial section 2	Assigning parts Simplifying parts Homogenising parts
線部一	正比例 轉比例 合率比例 正比例帶分 轉比例帶分	II.3 Line section 1	Direct ratios Inverse ratios Compound ratios Direct ratios with parts Inverse ratios with parts
線部二	按分遞折比例	II.4 Line section 2	Ratios according to parts repeatedly rebated[1]
線部三	按數加減比例	II.5 Line section 3	Ratios according to added and subtracted numbers[2]
線部四	和數比例 較數比例	II.6 Line section 4	Ratios with sums Ratio with difference s[3]
線部五	和較比例	II.7 Line section 5	Ratios with sums of differences
線部六	盈朒	II.8 Line section 6	Excess and deficit
線部七	借衰互徵 疊借互徵	II.9 Line section 7	Borrowing grades for mutual comparison Repeated borrowing for mutual comparison
線部八	方程	II.10 Line section 8	Rectangular arrays

(Continued)

Table 14.1 Continued

面部一	平方 帶縱平方	II.11 Area section 1	Squares [roots] Squares bearing a length[4]
面部二	勾股	II.12 Area section 2	Base and altitude
面部三	勾股	II.13 Area section 3	Base and altitude
面部四	三角形	II.14 Area section 4	Triangles
面部五	割圓	II.15 Area section 5	Division of the circle
面部六	割圓	II.16 Area section 6	Division of the circle
面部七	三角形邊線角度相求	II.17 Area section 7	Seeking the sides and angles of a triangle from each other
面部八	測量	II.18 Area section 8	Mensuration
面部九	各面形總論 直線形	II.19 Area section 9	General discussion of various plane figures Rectilinear figures
面部十	曲線形	II.20 Area section 10	Curvilinear figures
面部十一	圓內容各等邊形 圓外切各等邊形	II.21 Area section 11	Various regular polygons inscribed in a circle Various regular polygons circumscribed to a circle
面部十二	各等邊形 更面形	II.22 Area section 12	Various regular polygons More polygons
體部一	立方	II.23 Solid section 1	Cubes [roots]
體部二	帶縱較數立方 帶縱和數立方	II.24 Solid section 2	Cubes bearing a side with difference Cubes bearing a side with sum[5]
體部三	各體形總論 直線體	II.25 Solid section 3	General discussion of various solids Rectilinear solids
體部四	曲線體	II.26 Solid section 4	Curvilinear solids
體部五	各等面體	II.27 Solid section 5	Various regular polyhedra
體部六	球內容各等面體 球外切各等面體	II.28 Solid section 6	Various regular polyhedra inscribed in a sphere Various regular polyhedra circumscribed to a sphere

Table 14.1 *Continued*

體部七	各等面體互容 更體形	II.29 Solid section 7	Various regular polyhedra inscribed in one another More solids
體部八	各體權度比例 堆垛	II.30 Solid section 8	Ratios between the weights of various solids Stacks
末部一	借根方連比例	II.31 Final section 1	Continued proportions of borrowed root and powers
末部二	借根方連比例 開諸乘方法（諸乘方表）	II.32 Final section 2	Continued proportions of borrowed root and powers: extraction of all roots
末部三	借根方連比例 帶縱平方	II.33 Final section 3	Continued proportions of borrowed root and powers: squares bearing a length
末部四	借根方連比例　線類	II.34 Final section 4	Continued proportions of borrowed root and powers: category of lines
末部五	借根方連比例　面類	II.35 Final section 5	Continued proportions of borrowed root and powers: category of areas
末部六	借根方連比例　體類	II.36 Final section 6	Continued proportions of borrowed root and powers: category of solids
末部七	難題	II.37 Final section 7	Difficult problems
末部八	對數比例	II.38 Final section 8	Logarithmic ratios
末部九	比例規解	II.39 Final section 9	Explanation of the proportional compass
末部十	比例規解	II.40 Final section 10	Explanation of the proportional compass
表		表/III	**Tables**
八線表上		III.1	Tables of the eight [trigonometric] lines I
八線表下		III.2	Tables of the eight [trigonometric] lines II
對數闡微上		III.3	Minute clarification of logarithms I
對數闡微下		III.4	Minute clarification of logarithms II
對數表上		III.5	Tables of logarithms I
對數表下		III.6	Tables of logarithms II
八線對數表上		III.7	Tables of logarithms of the eight [trigonometric] lines I
八線對數表下		III.8	Tables of logarithms of the eight [trigonometric] lines II

1 These include five cases: dividing into two parts so that the ratio between the shares is 2 to 8, 3 to 7, and 4 to 6; dividing into several parts so that the shares are in geometric progression; dividing into several parts so that each share is double or half the previous one.
2 These include four cases: dividing a given number into a given number of shares that are in arithmetic progression of a given ratio; dividing a given number into a given number of shares among which are related by various given ratios; dividing a given number into a given number of shares in arithmetic progression, given the difference between the first and last share; dividing an unknown number into a given number of shares in arithmetic progression, given the smallest and largest shares.
3 Calculating e.g. shares of profit, given the capital sum invested by each or the difference between the capital sums invested.
4 In modern terms, second degree equations.
5 In modern terms, third degree equations with a positive and negative constant term respectively.

14.2 Sources

The core material of the editorial work on the *Essence of numbers and their principles* was the lecture notes that the Jesuits had prepared for Kangxi in the 1690s.[7] However, a number of printed mathematical works were also available to the staff of the Office; they were used to various extents and in various ways. It is not possible to establish an exhaustive list of these works; however the imperial compendium explicitly refers to some of them.

Among the *Ten mathematical classics* (*Suanjing shishu* 算經十書) compiled in the seventh century CE, two are used. The most venerable of them, the *Mathematical Classic of the Gnomon of Zhou* (*Zhoubi suanjing* 周髀算經, probably 1st century CE)[8] is discussed in chapter I.1 as evidence on the study of numbers (*shuxue* 數學) in Chinese antiquity. Quotations from this work also open chapters II.12, II.15 and II.18, devoted respectively to base and altitude, the division of the circle and mensuration.[9] This is a way of asserting that the subject matter of these chapters stems from (in the first case) or are germane to (in the second and third cases) knowledge found in Chinese antiquity. The other work included in the *Ten mathematical classics* that is mentioned in the imperial compendium is the *Mathematical classic of Master Sun* (*Sunzi suanjing* 孫子算經, 3rd century CE); it is quoted twice in the passage on weights and measures, as evidence on ancient units.[10] The *Nine chapters on mathematical procedures* (*Jiuzhang suanshu* 九章算術, 1st century CE), nowadays regarded as the founding work of the Chinese mathematical tradition, seems to have been unknown to the compilers as to other scholars versed in mathematics in their time.[11] There are only six references to the 'nine chapters' (*jiuzhang* 九章) classification in the whole compendium: two are historical, and the others link the methods given in chapters II.6 and II.7 to variants of the method that gives its title to the third of the 'nine chapters', here called 'Differential distribution' (*chafen* 差分).[12] Nowhere is the complete list of the 'nine chapters' given; apparently the compilers did not deem this list particularly significant. No other mathematical work by a Chinese author is

---

7   See p. 271.
8   Cullen 1996.
9   Guo 1993, 3: 457, 526, 607.
10 The work is mentioned under the title *Mathematical procedures of Master Sun* (*Sunzi suanshu* 孫子算術); Guo 1993, 3: 181; on the work see Lam 2004.
11 On this work see Chemla & Guo 2004; the text was later reconstructed by Dai Zhen 戴震 when he worked on the compilation of the *Complete library of the four treasuries*; among the scholars versed in mathematics of the Kangxi reign, only Mei Wending seems to have had access to the first chapter of the work; see p. 85.
12 Guo 1993, 3: 12, 186, 294, 306, 317, 328; chapter 3 of the *Nine chapters on mathematical procedures* is entitled *Cuifen* 衰分 (see p. 19); *chafen* occurs in many later works.

mentioned. However, as we shall see, the *Unified lineage of mathematical methods* (*Suanfa tongzong* 算法統宗) must have served as an inspiration in some respects. Of Mei Wending's works, at least the *Discussion of rectangular arrays* (*Fangcheng lun* 方程論) served as a direct source for chapter II.10.[13]

The compilers also had at their disposal the late Ming works by the Jesuits, including the works that formed the *Books on calendrical astronomy according to the new method* (*Xinfa lishu* 新法曆書). These include a treatise entitled *Explanation of the proportional compass* (*Bili gui jie* 比例規解), the title of which was taken up in one of the emperor's manuscript textbooks, and then in chapters II.39 and II.40 of the imperial compendium. These two chapters are much more detailed than the manuscript textbook: for each line on the compass, its use in solving problems is explained as well as its construction. Chapter II.40 also gives explanations of how to draw the lines on a sundial and how to construct logarithmic scales (simple ones as well as those of sines and tangents) on rulers.[14] As we have seen, calculating devices constructed on these principles were made in the imperial workshops.[15] Again, the practice of mathematics at court was not solely textual.

The compendium mentions two of the first Jesuit works: the 1607 translation of Euclid's *Elements* and the *Instructions for calculation in common script* (*Tongwen suanzhi* 同文算指).[16] The former was not used directly in the *Essence of numbers and their principles*, where Gerbillon and Bouvet's textbook is the sole reference for geometry. On the other hand, the names given to some methods in the *Instructions*, which are different from those used in the compendium, are mentioned in part II when those methods are introduced: this is a form of acknowledgement of the Jesuit work on arithmetic.

It is also worth considering whether any of the contributors to the *Essence of numbers and their principles* had previously produced mathematical writings that were used at the time of the compilation. To my knowledge no mathematical work by any of them was printed prior to the imperial project: most of them were relatively young when it started. Chen Houyao was an exception: he was already in his fifties. Two collections of mathematical writings attributed to him have been preserved in the form of manuscripts. One is entitled *Exploring the mystery of mathematics* (*Suanyi tan'ao* 算義探奧).[17] Some of the matters discussed there are also found in the

---

13 See pp. 93–101.
14 Guo 1993, 3: 1219–1235.
15 See pp. 155–156. Liu 1998, no. 65–67, 71, 73, 74, 75.
16 Guo 1993, 3: 16.
17 Kept in Beijing at the Library of the Institute of History of Natural Sciences (Li 2000a, no. 530).

imperial compendium; however one particular essay, the *Significance of an intricate method* (*Cuozong fayi* 錯綜法義),[18] deals with combinatorics, which is not included in the imperial compendium: this was most likely Chen's personal work, although we do not know whether it was completed before or during his participation in the compilation. The term *cuozong* 錯綜 ('intricate') is used in the compendium to describe a particular type of continued proportions; this is taken up from Gerbillon and Bouvet's textbook and bears no relation to Chen's essay, in which the term is used only in the title, in a non-technical sense.[19] The other extant collection of his writings is entitled *Mathematical writings of Chen Houyao* (*Chen Houyao suanshu* 陳厚耀算書).[20] It contains manuscript copies of the *Elements of calculation* (*Suanfa yuanben* 算法原本)[21] and of chapters II.25, II.30 and II.34 of the *Essence of numbers and their principles*, but also an *Illustrated explanation of the base and altitude* (*Gougu tujie* 句股圖解), which opens with a 'Method for seeking the base and altitude from the area taught by the Emperor' (*qin shou ji qiu gou gu fa* 欽授積求句股法). On the whole it is unclear whether this second collection contains any original work by Chen rather than texts produced or copied as part of the compilation enterprise. The content of the writings traditionally attributed to Chen Houyao's mathematical writings cautions us as to the difficulty of ascribing manuscripts to individual authors in the context of court mathematics. Another example of this is the traditional attribution of Antoine Thomas' *Outline of the essentials of calculation* (*Suanfa zuanyao zonggang* 算法纂要總綱) to Nian Xiyao, on the grounds that there was a copy of it in his papers.[22] Nian Xiyao is best known as the author of a treatise on perspective, the *Study of Vision* (*Shixue* 視學, 1735). As this work is based on Andrea Pozzo's *Perspectiva pictorum et architectorum* (1693–1700),[23] it is reasonable to infer that Nian had contacts with some court Jesuits during either the Kangxi reign or the Yongzheng reign, probably with the painter Giuseppe Castiglione (1688–1766).[24] However there is no evidence either that this happened as early as the 1690s, when Thomas composed the *Outline of the essentials of calculation*,[25] or that Nian took any part in the compilation of *Essence of numbers and their*

---

18   Guo 1993, 4: 685–688.
19   Guo 1993, 3: 70; Taipei NCL, Rare Books 6399, 6: 14a–15a.
20   In Prof. Li Peiye's 李培業 collection (Li 2000a, 292); I am grateful to Prof. Qu Anjing 曲安京 for providing pictures of the title pages of the fascicules.
21   I have not been able to ascertain whether this is chapter I.5 of the *Essence of numbers and their principles* or the original textbook discussed above, pp. 195–200.
22   Li & Qian 1998, 6: 529; Li 2000a, 557.
23   On Nian Xiyao and his treatise on perspective, see Corsi 2004, esp. 77–88; there are two editions of Pozzo's treatise in the Beitang Library (Verhaeren 1949, no. 2511 & 2512).
24   On Castiglione see Pirazzoli-t'Serstevens 2007.
25   See pp. 184–191.

CHAPTER 14 | The construction of the *Essence of numbers and their principles*

principles. In sum, the presence in some scholars' papers of material used for the compilation of the imperial compendium is not sufficient to draw conclusions as to whether and how these scholars contributed to it.

14.3 The historical narrative of mathematics

The first and second parts of the *Essence of numbers and their principles* both open with historical materials. The 'Origins of numbers and their principles' (*Shuli benyuan* 數理本原), which serves the purpose usually fulfilled by a preface, simply retells the received account:

Investigating high antiquity, [one finds that] 'the [Yellow] River gave forth its Diagram, and the Luo [river] gave forth its Writing'. With the [former], the eight trigrams were produced;[26] with the [latter], the Nine Divisions [of the Great Plan (*Hongfan* 洪範) of the *Book of documents* (*Shangshu* 尚書)][27] were

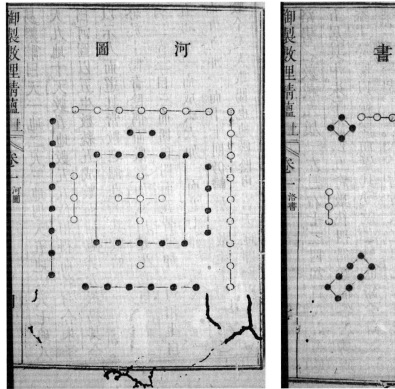

Illustration 14.2 River Diagram and Luo Writing, *Essence of numbers and their principles*. (Bodleian, Sinica 361 / 29, 4a, 7a; see Guo 1993, 3:13, 14.)

---

26 The trigrams are traditionally attributed to the mythical sovereign Fuxi 伏羲.
27 The Great Plan is a chapter of the *Book of documents* traditionally attributed to Yu 禹 the Great; see Legge 1960, 3: 320–321.

ordered. The study of numbers also began thereupon. Now the Diagram and the Writing are auspicious omens in response to Heaven and Earth; they first came forth in response to Sages. The study of numbers fathoms the principles of the myriad things; it is from Sages that it attains clarification.[28]

The quotation in the first sentence is from the Appended Explanation (*Xici* 繫辭) section of the *Book of change* (*Yijing* 易經), traditionally attributed to Confucius. By the Qing period, the River Diagram and the Luo Writing were assumed to be two layouts of numbers from one to ten and from one to nine respectively (the latter is laid out like a 3×3 magic square).[29] The mythical association of the former chart with Fuxi 伏羲, the inventor of the hexagrams, and that of the latter chart with Yu 禹 the Great, to whom the Great Plan chapter of the *Book of documents* is attributed, only appear in commentaries since the Han dynasty. Since the Song dynasty, the two charts gained new importance with the development of the 'figures and numbers' (*xiangshu* 象數) tradition of commentaries on the *Book of change*; the pictorial representations of the two charts date back only to that time.[30] The imperially commissioned *Middle way to the Change of Zhou* (*Zhouyi zhezhong* 周易折中) on which several of the compilers of the *Essence of numbers and their principles* worked under Li Guangdi's supervision give both the charts and the narrative:[31] it is no surprise that the imperial compendium on mathematics paraphrased the orthodox commentary on the *Book of change*. In fact, none of this was particularly original or innovative: this narrative of the origins of numbers had been quite commonplace since the Song dynasty. The *United lineage of mathematical methods* is typical in this respect.[32] There are close similarities between its narrative and that of the *Essence of numbers and their principles*; but the latter goes further by stressing that the study of numbers (*shuxue* 數學) is not a human invention, but stems from patterns produced by the cosmos and revealed to Sages who alone were able to convey their meaning. Moreover it traces back mathematics to the River Diagram and Luo Writing and does not assume the *Book of change* to be its foundation:[33] mathematics stems from the same source as the *Book of change*, rather than deriving from it. The two approaches to numbers are thus put on the same footing: this is quite an unprecedented claim as to the status of mathematics.

---

28 Guo 1993, 3: 12.
29 Nielsen 2003, 103–105, 169–171.
30 Smith *et al.* 1990, 120–122. The explanation of the two charts given in the *Essence of numbers and their principles* invoke the authority of Zhu Xi and Shao Yong.
31 *Yuzuan Zhouyi zhezhong* 御纂周易折中, SKQS 38: 468–469; see p. 269.
32 See p. 18.
33 As is done for example by Liu Hui in his preface to the *Nine chapters on mathematical procedures* (Guo 1993, 1: 96; Chemla & Guo 2004, 126–127).

CHAPTER 14 | The construction of the *Essence of numbers and their principles*

Just like the origins of mathematics, its history since the Three Dynasties of antiquity is recounted in the received way:

> In former times the Yellow Emperor ordered Li Shou 隸首 to make counting rods (*zuo suan* 作算),[34] and the meaning of the 'nine chapters' was expounded. Yao 堯 ordered Xi 羲 and He 和 to settle the calendar, respectfully delivering the seasons to the people, and the tasks of the year were thus completed.[35] The *Officers of Zhou* (*Zhouguan* 周官) taught the gentry the Six Arts; numbers were one of them. The discourse of Shang Gao 商高 in the *Gnomon of Zhou* may be examined. From the Qin and Han onwards, there has been no lack of successors, such as Luoxia Hong 落下閎, Zhang Heng, Liu Zhuo 劉焯 and Zu Chongzhi, who have each left writings.[36] The Tang and Song set up examinations for [the title of] Classicist in mathematics (*suanxue mingjing* 算學明經). The [set] books were proclaimed in the College (*xuegong* 學宮), and they ordered the Erudites (*boshi* 博士) and their students to practise them.[37] From this one knows that the study of reckoning is really an important affair in the investigation of things and the extension of knowledge (*gewu zhizhi* 格物致知).[38]

The names of the experts of the imperial period mentioned here all occur in dynastic histories: the evidence given relates to official involvement in astronomy and mathematics rather than to their study in general. Despite the emperor's high opinion of Cheng Dawei's work, neither his name nor the tradition he strove to unify are mentioned in the imperial compendium. Whereas Mei Wending (who is not mentioned either) had argued that the study of numbers was the affair of scholars, the editors of the imperial compendium argue that it is first and foremost the affair of the state: scholars are to put their competence in the service of the dynasty.

After the 'Origins of numbers and their principles', the River Diagram and the Luo Writing, chapter I.1 turns to an 'Explanation of the Classic of the Gnomon of Zhou' (*Zhoubi jingjie* 周髀經解), focusing on the famous passage in which Shang Gao, a learned man of the recently defeated Shang dynasty, passes on his knowledge about the origins of numbers to the Duke of Zhou (Zhou Gong 周公), a particularly virtuous member of the new imperial clan.[39] As an introduction to this, more recent history is evoked:

---

34 The counting rods, the calculating instrument widespread in China at least until the Song dynasty, had fallen into oblivion in China by the late Ming; Li Shou was traditionally regarded as the inventor of calculation and of the musical scale.
35 So far the text refers to various founding sovereigns and heroes mentioned in the *Book of documents*.
36 Luoxia Hong (fl. 110 BCE), Zhang Heng, Lui Zhuo (544–610) and Zu Chongzhi all were renowned astronomers.
37 The books in question are the *Ten mathematical classics*; see Siu & Volkov 1999.
38 Guo 1993, 3: 12.
39 See the translation of this passage in Cullen 1996, 174.

The study of numbers has long been lost. Since the Han 漢 and Jin 晉, what has been preserved was thin as a thread. Thereafter, the likes of Zu Chongzhi and Guo Shoujing[40] devoted their entire minds to figures and numbers (*xiangshu* 象數). They established the methods of the precise *lü* [of the circle] and waning and waxing (*xiaozhang* 消長) as the basic rules of the practice of calculation. But their methods measured the infinite by means of the finite; they only talked about heavenly motion, and did not tackle the Earth. Therefore when surveying there were changes, when measuring there were increases and decreases; for there were accumulated remainders that they had not exhausted. During the Wanli period (1573–1620) of the Ming, Westerners began to enter China. Among them there were one or two exercised in calculation, like Ricci, Smogulecki etc. They wrote the *Elements of geometry*, the *Instructions for calculation in common script* and various other works. Although they provided the main substance (*da ti* 大體), they really did not clarify the subtleties of principles and numbers. Since our dynasty chose its capital, men from afar admired it, and more and more of them have arrived. Schall, Verbiest, Thomas and Grimaldi have succeeded one another as Administrators of the Calendar. In the meantime they have clarified the study of calculation, and the principles of measures and numbers have gradually gained more details. However, when asked about their origin, they all said it had originally been transmitted from the Chinese land.[41]

While acknowledging the contribution of Jesuit astronomers to mathematics (at least those who were dead: Killian Stumpf, who was Administrator of the Calendar from 1711 to 1720, is not mentioned), imperial mathematicians claimed that these Jesuits acknowledged that Western learning had originated in China. However, apart from Kangxi's knowledge that the word *algebra* was of Eastern origin,[42] which his Jesuit tutors must have told him, I know of no evidence that they said anything in support of his idea. But then, neither does the dismissive claim that Zu Chongzhi and Guo Shoujing indulged in 'figures and numbers', which caused their astronomical methods to be inaccurate, rely on any evidence.[43] The narrative that follows is highly conventional; it is quite similar to the one found in Mei Wending's preface to his *Brush calculation* (*Bisuan* 筆算, 1693).[44] Astronomy spread abroad during the Three Dynasties. In China it was supposedly lost in the first Qin emperor's 'burning of books', while the knowledge that had previously spread from China to the West continued to be handed down from generation to generation. The famous dialogue between Shang Gao and the Duke of Zhou is reproduced and explained in order to 'clarify the lineage of the study of numbers, and to make it known to those who study them that in China and abroad there are not different principles'.[45] The universality of mathematics is asserted

---

40 See p. 250.
41 Guo 1993, 3: 16.
42 Jami 2002b, 36.
43 Kangxi professed a low opinion of Guo Shoujing; see pp. 124–125, 250.
44 See Jami 1994c, 162–164.
45 Guo 1993, 3: 16 '[以]明數學之宗。使學者知。中外無二理也焉爾。'

on historical grounds, which also serve to legitimise imperial mathematics as stemming from Chinese antiquity.

The definition of measures and weights (*duliang quanheng* 度量權衡) at the beginning of the second part of the compendium is preceded by a wealth of quotations from ancient texts: commentaries of the Classics, histories, ancient dictionaries and the *Mathematical Classic of Master Sun*. This textual evidence leads to the conclusion that there is no single uniform system for weights and measures, which justifies the definition of a system that makes no reference to any earlier one: imperial mathematics is not subservient to historical precedents. Only in one case is an old unit mentioned in the definition of the system:

In the old appellation, for one celestial degree there were 250 *li* 里 on earth; verifying with today's *chi* 尺, for one celestial degree there are 200 *li* on earth. For the old *chi* is 8/10th of today's *chi*.[46]

This is a reminder of the new standard defined for the *li* in 1702, after Thomas and Yinzhi's measurement of one degree of terrestrial meridian.[47] When it comes to defining imperial units, no suggestion is made that ancient ones should be re-enforced: the investigation into ancient sources serves the opposite purpose. The new *li* is defined by a ratio between heavenly units, that is, the degrees as imported by the Jesuits, and earthly units: the emperor thus fulfils his role of ensuring harmony between Heaven and Earth, or in other words between the cosmos and the human realm, which his predecessors, served by the like of Zu Chongzhi and Guo Shoujing, had allegedly failed to do. At the same time it can be said that there is an element of modernity[48] in this attitude towards antiquity, insofar as the choice to set new standards is asserted and justified: the emperor's mastery of cosmology is unprecedented.

14.4 The structure of imperial mathematics

The 'Origins of numbers and their principles' also explains how to deal with numbers and their principles:

[...] in discussing numbers, one establishes the parts of *jihe* 幾何, so as to set up methods for finding them [i.e. numbers] from one another. In addition, subtraction, multiplication and division, no number is lost, be it for amount, weight, price, or excess and deficit. In discussing principles, one establishes the shapes of *jihe*, so as to clarify the reason why calculations are set out. In ratios,

---

46 Guo 1993, 3: 128.
47 See pp. 244–245.
48 In the restricted sense of a way to position oneself in rupture rather than continuity vis-à-vis tradition.

compound or with parts, no principle is lost, be it for shape, size, distance, or height.[49]

The meaning of *jihe* in this passage and elsewhere in the compendium deserves close attention: the term cannot be rendered by a single word in all its occurrences. To understand how it is used here, one should take into account the fact that it occurs in the construction of an original duality between numbers and principles. Probably inspired both by Ricci's preface to the 1607 translation of Euclid's *Elements* and by Mei Wending's view on the structure of mathematics, it nonetheless differs from both.[50] For Ricci, the two relevant objects were number (*shu* 數) and magnitude (*du* 度), the two instances of quantity (*quantitas*, which he translated by jihe 幾何).[51] He considered the parts (*fen* 分) of things, contrasting two approaches to these parts, each characterised by a question: respectively, how many (*jihe zhong* 幾何眾) and how large (*jihe da* 幾何大).[52] For Mei on the other hand, numbers (*shu* 數) formed the general category, indissociable from principles, and two different fields were relevant to them: calculation (*suanshu* 算術) and measurement (*liangfa* 量法). In his view, calculation had primacy over measurement; principles underlay both, and his explicit ambition in writing the *Discussion on rectangular arrays* was to 'clarify the principles of calculation' (*ming suan li* 明算理) from within the field of calculation.[53] In contrast with both these views, the imperial standpoint is that in mathematics such clarification is done by the means of shape, that is, what we now call geometrical objects. The 'shapes (*xing* 形) of *jihe*', which pertain to principles, underlie the methods by which one operates on the 'parts (*fen* 分) of *jihe*'. Numbers and principles do not correspond to two types of mathematical objects, but rather to two processes: establishing (*li* 立) and clarifying (*ming* 明), which apply to parts and shapes respectively. These two processes address two aspects of mathematics that evoke the philosophical duality between 'what must be so' (*suodangran* 所當然), and 'why it is so' (*suoyiran* 所以然), a duality that Mei Wending associated respectively with the Chinese tradition and the Western tradition in astronomy.[54] Here, however, dividing knowledge according to its origin is irrelevant, as one is dealing with the unified, all encompassing field of imperial mathematics.

---

49 Guo 1993, 3: 12.
50 Verbiest also referred to the number–principle duality in relation to astronomy; see p. 80.
51 Engelfriet 1998, 138–141; it has often been said that *jihe* is a phonetic transliteration of *geo*; however the evidence brought together by Engelfriet does not point in that direction.
52 See pp. 26–27.
53 See pp. 94–95.
54 See pp. 220–221.

Parts and shapes are thus the two aspects of *jihe* considered in mathematics. Here *jihe* cannot refer to geometry, that is to say the subject matter discussed in chapters I.2 to I.4.[55] Rather, *jihe* is akin to Ricci's Chinese rendition of *quantitas*. However, *jihe* is not the object studied in mathematics, as it was for Ricci; this object is instead numbers as clarified according to principles. In this context, *jihe* is a generic object considered as composed of parts, and as having a shape; it is these parts and shapes that mathematics addresses. *Jihe* could be translated by 'amount' in order to avoid 'quantity', which evokes the ten Aristotelian categories, a reference that Ricci had in mind but that is no longer relevant here. In part II of the compendium, *jihe* is used in most problems as a marker of interrogation that can be rendered by 'how much' or 'how many'; this confirms that the term had not yet taken the sense of 'geometry' that it has since come to have. The difficulty of proposing a single English translation for *jihe* throughout the compendium is not due to an inconsistent use of the term, but rather to the flexibility of classical Chinese: in this language, the same word can be used with different grammatical functions in a sentence, to a much greater extent than in English.

After this explanation of how to handle numbers and their principles, the latter are evoked again, in a way that links origins to practice:

Tracing back to the origins, addition and subtraction actually come from the River Diagram, and multiplication and division certainly come from the Luo Writing. One odd and one even face and support one another. You add and subtract progressively without exhausting multiplicity. Odd and even are each divided; vertical and horizontal mutually harmonise. You multiply and divide crosswise without obstructing transformation and communication.[56]

The two mythical charts are each given a technical meaning; thus, the River Diagram expounds the alternate sequence of odd and even numbers from one to ten, the former associated to Heaven (and *yang*) and the latter to Earth (and *yin*), and each number generating the next one.[57] In this context, *suanfa* 算法, which to Cheng Dawei and his predecessors covered the whole of mathematics, represents only one aspect of it, literally 'calculation methods':

The rudiments of calculation methods are addition, subtraction, multiplication and division; but they are also their entirety. Although one reaches the ever changing, one remains within those. But there are different ways of using the methods. If one needs to take a sum, then one uses addition; if one needs to take a difference, then one uses subtraction; if one needs to gather and assemble a product, then one uses multiplication; if one needs to take its

---

55 Nonetheless for clarity's sake *Jihe yuanben* is rendered by *Elements of geometry* in this chapter as in previous ones.
56 Guo 1993, 3: 12.
57 Guo 1993, 3: 13–15.

scattered parts, then one uses division. There is also addition followed by subtraction, subtraction followed by addition, multiplication followed by division, division followed by multiplication. There is also the combined use of addition and subtraction with multiplication and division; in the old appellation, [kinds of] calculation were designated according to the 'nine chapters', from 'Rectangular fields' to 'Base and altitude'. Numbers can be complex or simple; principles can be obvious or obscure; methods can be shallow or deep; calculations can be difficult or easy. But which of them is not obtained by addition, subtraction, multiplication and division? Therefore if one talks about them in a shallow way, then they are the rudiments of calculation methods; if one talks about them in a thorough way, they are really the whole body of calculation methods.[58]

This is the sole reference to the traditional nine-fold division of mathematics in the compendium. The 'nine chapters' thus evoke calculation methods that are combinations of the four basic operations; according to the dual character of mathematics, they address 'the parts of *jihe*', but not 'the shapes of *jihe*' through which one clarifies principles.

The structure of the *Essence of numbers and their principles* reflects that of mathematics as it emerges from the opening statements of the work:

Here a book has been compiled, in which categories follow each other. Point, line, area and solid form the outline (*gang* 綱); sum, difference, direct and inverse divide up sections (*mu* 目). As to methods, taking no account of the magnitude [of the task], only the excellent (*shan* 善) ones have been chosen. From the shallow to the profound, and grasping the simple to manage the complex, one harmonises principles and numbers, with the aim of benefiting the empire and the state, and so as to pass on all of this to the generations to come.[59]

The closing sentence epitomises the enterprise pursued at the Office of Mathematics very neatly: the philosophical concepts that underlie the orthodox worldview are used to reconstruct mathematics as learning in the service of statecraft. Geometrical objects define categories into which the compendium is organised: geometry plays a structuring role. Progression in difficulty is incorporated into these categories. The four basic operations and ratios then define a finer order within the categories of problems defined by points, line, area and solid: in this way geometry and calculation define the structure of mathematics in the particular sense of the Chinese term *gangmu* 綱目, which refers to a multilayer structuring of data in a book.[60]

14.5 Rephrasing the Jesuits' textbooks

The part of the *Essence of numbers and their principles* that is most straightforwardly taken from the Jesuits' lecture notes is that devoted

---

58 Guo 1993, 3: 184.
59 Guo 1993, 3: 12.
60 As in the title of the famous *Bencao gangmu* 本草綱目 (*Systematic materia medica*, 1593).

to the *Elements of geometry* and to the *Elements of calculation*; these cover chapters I.2 to I.5. They seem to have been the first part of the text on which the staff of the Office worked: the 'disciples' of Kangxi studied them in Jehol during the summer of 1712.[61] One of them, Mei Juecheng, had probably read the texts with his grandfather a decade earlier, after Kangxi gave copies of them to Li Guangdi.[62] By the winter of 1714, Yinzhi reported the completion of some chapters of the *Elements of geometry*.[63] Thus the staff of the Office of Mathematics started their compilation work with what formed the first part of the *Essence of numbers and their principles*. Although there is a close parallel between the lecture notes and chapters I.2 to 1.5 of the compendium, some changes were made in style, phrasing and terminology. Some of these changes shed light on the general project that resulted in the imperial compendium. Thus the opening item (*jie* 節) of the *Elements of geometry* (chapter I.2) reads:

In general (*fan* 凡), when discussing number and measure (*du* 度), one must begin with a point. From a point, extending it, it becomes a line. From a line, broadening it, it becomes an area. From an area, amassing it, it becomes a solid. These name the three great patterns (*dagang* 大綱). Hence what has length but no width is called line; what has length and width but no thickness is called area; what is complete with length as well as width and thickness is called solid. Only the point has no length, width, thickness or thinness. It cannot contain parts within, and is not capable of being measured by numbers. However the two extremities of a line are points, and line, area and solid all proceed from it. Although point does not enter number, it is actually the foundation of all numbers.[64]

This item groups the first three items of the Chinese version of Pardies' *Elemens de geometrie* prepared by Gerbillon and Bouvet, which read:

1st. In general, when discussing measure and number, one first begins with a point. From point, extend into line; line expands into area; area amasses into solid. These name the three measures.

2nd. In general, in measure, what has length but no breadth is called line; what has length and breadth but no thickness is called area; what has length, breadth and thickness is called solid.

3rd. In general, in measure, what has only one location, and moreover has no length, breadth or thickness, and cannot be divided is called point. In general, the two ends of a line are attached to it; they are points. Therefore although point does not enter into measure, it is actually the root of all measure.[65]

In Pardies' textbook, paragraphs are not classified as definitions, axioms, postulates and propositions as they are in many works in the

---

61 Zhongguo diyi lishi dang'anguan 1996, no. 1086.
62 See pp. 247–248.
63 Zhongguo diyi lishi dang'anguan 1996, no. 2321, 2324.
64 Guo 1993, 3: 22.
65 Taipei, NCL, Rare Books 6399, *juan* 1: 1b.

Euclidean tradition. However, the three distinct paragraphs that were preserved in Gerbillon and Bouvet's translation do address different matters, namely the generation of the three fundamental objects of geometry from a point, the characterisation of these three objects, and the special status of the point with respect to measure. When merging the three paragraphs, the editors of the imperial compendium have introduced links between them: the characterisation of line, area and solid becomes an explicit consequence of the way in which they are generated, while the exceptional status of the point is reinforced.

More importantly, two changes in the phrasing reflect a conceptual change: measure (*du* 度), which is central in the Jesuits' textbook, has been placed after number in the opening sentence; it no longer appears in the rest of the passage, and the two sentences from which it has been deleted deserve close attention. In the first one, we are told instead that line, area and solid name the 'great patterns'. The same word, *gang* 綱 ('pattern'), is used here as in the title of part I, and in the explanation of the structure of the work discussed above. Instead of representing 'only' the three fundamental objects of geometry (the particular field discussed in chapters I.2 to I.4), line, area and solid now more broadly underlie the structure of the compendium. Indeed part II is structured according to what we might call dimension (although no such concept is made explicit in the passages quoted above); the problems solved using the method of the borrowed root and powers given in the 'Final section' are also classified according to these three categories. In the second sentence from which *du* has been deleted, the notion of point, instead of being fundamental to measure, becomes fundamental to numbers, that is, not only to one of the two branches of mathematics, but to the whole of it. These two changes in wording imply that geometry is given a founding role in mathematics, even though the editors have, if one may say, demoted measure from a status on a par with number that the Jesuits had given it. That being said, there are hundreds of occurrences of *du* in the sense of 'measure' in the compendium, when dealing with geometrical objects: although the notion is denied a role in the structure of mathematics, it remains central when dealing with the 'shapes of *jihe*'.

The few changes made to the structure of the Jesuits' geometry textbook result in a stricter adherence to 'the three great patterns' as structuring categories. For example, in the manuscript ellipses are defined just after ellipsoids and in order to deal with the latter.[66] In the compendium, the items on ellipses have been moved forward so as to form part of a section devoted to plane geometric figures.[67] In the same way, the definition of similar solids, given immediately after that

---

66 Items 6.79 & 6.80, Taipei NCL, Rare Books 6399, 6: 86a–89b.
67 Item 8.12, Guo 1993, 3: 89–90.

of similar plane figures in the manuscript, has been moved into a section that deals with solids in the compendium.⁶⁸ Thus in imperial mathematics, geometry itself is organised as much as possible according to the 'great patterns' defined by its fundamental objects.

The *Elements of calculation* were reorganised to a much greater extent than the *Elements of geometry*. Although the two chapters into which the compilers divided the former work do not have titles, the first one deals with the four basic operations and the simplification and homogenisation of fractions, whereas the second one deals with ratios. The two chapters therefore give foundations respectively for the 'Initial section' and the 'Line section' of part II; just like the *Elements of geometry*, the *Elements of calculation* have been restructured according to the 'great patterns' that underlie imperial mathematics. And as in the *Elements of geometry*, small changes in phrasing compared with the manuscript textbook can reflect quite important differences. Thus the latter opens as follows:

One is the root of numbers. A multiplicity of ones mutually combined is called a number.⁶⁹

This is a definition; the imperial compendium, on the other hand, reads:

One is the origin of numbers. A multiplicity of ones mutually combined makes the numbers proliferate.⁷⁰

From the standpoint of Euclidean 'orthodoxy', this second version is not a definition. However, looking at it from the more relevant standpoint of imperial orthodoxy as articulated in the first chapter of the compendium, numbers need not be defined in a mathematical context, as they are part of the cosmos that reveals their patterns to men in response to the virtue of the Sages.

14.6 Vocabulary and classification

Some features of style that unified the Jesuits' lecture notes have not been retained in the *Essence of mathematical principles*. Thus, as mentioned above, *jihe* 幾何 is used as a question marker in the problems of part II, where it alternates with *ruogan* 若干, which occurs rather less often. There seems to be no obvious criterion for using one or the other (such as the units used in the problem or

---

68 Items 6.56 to 6.58, Taipei NCL, Rare Books 6399, 6: 61a–63b; items 10.6 to 10.8 Guo 1993, 3: 100–102.
69 一者數之根也。眾一相合而謂數焉。BML, Ms. 82–90 C, 1a; see p. 195.
70 一者數之原也。眾一相合而數繁焉。Guo 1993, 3: 142.

its source).⁷¹ One possibility is that the question marker used depends on the individual who wrote down a particular problem at some stage of the compilation process. Thomas had made exclusive use of *ruogan*. But neither term would seem strange to readers in this usage: Cheng Dawei mentioned them both in his glossary, and then consistently used *ruogan*.⁷² What matters here is that in the *Essence of numbers and their principles*, *jihe* is used both as a question marker and as a concept that comes into play when defining the object of mathematics.

A feature common to the Jesuits' textbooks in their later versions is the presence in each item or problem of an explanatory passage, which opens with *heze* 何則 (why?) and closes with *ke zhi yi* 可知矣 (can be known).⁷³ This phrasing has been almost entirely eliminated in the compendium, where explanations are introduced by *ruo* 若 (if). There remain only twenty-four occurrences of *heze*, all but one in chapter I.4, that is, books 11 and 12 of the revised *Elements of geometry*.⁷⁴ This could again be the trace of a particular hand having rewritten the items in which the original term remains. The closing of an item or of an explanatory part of it with *ke zhi yi* sometimes occurs in part I of the compendium, but not always in passages opened by *heze*. Similarly, other terms that serve to articulate reasoning in geometry have been modified, although stylistic regularity does not seem to have been a major concern of the compilers. *Sheru* 設如, used by Gerbillon and Bouvet to introduce instantiation of a general statement, is mostly replaced by *ru* 如 in the revised *Elements of geometry*, but still occurs in the revised *Elements of calculation*; in part II *sheru* is used systematically and exclusively to introduce problems. These now all have the same status, in contrast with Thomas' choice of stating each type of problem first in general terms (opening with *fan* 凡), then with abstract numbers (opening with *jiaru* 假如), and then finally in one or more 'concrete situations' (opening with *sheru* 設如 and then *you sheru* 又設如).⁷⁵ On the whole, the unified status of problems is the most systematic and significant change made to Thomas' textbook as regards the organisation of material in the text. There are two occurrences of *sheru* in the notes in small characters, where the term refers to the whole of a problem and its solution:⁷⁶ a word that indicates supposition is thus

---

71 *Ruogan* is entirely absent from problems of the Area section and Solid section. Taking some chapters as samples, one finds that in chapter II.1 *ruogan* applies to lengths or angles just as well as *jihe*; in chapter II.6 both are used interchangeably when the amounts sought are money or goods.
72 Guo 1993, 2: 1230 ff.
73 See pp. 195, 202.
74 All counts have been made using the electronic edition of *SKQS* (accessed from the Gest Library, Princeton University, 2002 and 2003).
75 See p. 187.
76 Guo 1993, 3: 189, 993.

used as a noun to refer to the whole paragraph that follows its occurrence.

Little change has been made as regards the vocabulary of geometry and that of elementary arithmetic. The most remarkable change is the return to the way of expressing fractions of a unit common in earlier Chinese mathematical texts, including the *Mathematical Classic of Master Sun*. Among the examples of addition of fractions, Thomas gave the addition of 2/3 and 3/5, both fractions having *zhang* 丈 as unit. He expressed them respectively as *yi zhang san fen zhi er* 一丈三分之二 (lit. 'one *zhang*: out of three parts, two') and *yi zhang wu fen zhi san* 一丈五分之三 (lit. 'one *zhang*: out of five parts, three').[77] This example is taken up in chapter II.2, where fractions are expressed in the more usual way: *san fen zhang zhi er* 三分丈之二 (lit. 'out of three parts of *zhang*, two') and *wu fen zhang zhi san* 五分丈之三 (lit. 'out of five parts of *zhang*, three'). The first of these two expressions is followed by a note in small characters: 'One *zhang* is divided into three parts, and one gets two parts' (*yi zhang wei san fen, er de qi er fen* 一丈分為三分而得其二分); the second expression is followed by a similar note. Such a note occurs for each of the first four expressions of this kind found in chapter II.2. This indicates that readers were not expected to find them obvious. Since similar expressions are found, among others, in Mei Wending's and Fang Zhongtong's works, the presence of these notes suggests that the emperor and Yinzhi were better acquainted with the expression used by Thomas than with the more common one chosen for the compendium.

Ratios are given new prominence in the *Essence of numbers and their principles*. The introduction to chapter II.3 presents multiplication and division as simplified cases of the use of ratios, and goes on to present the latter as the basis of problem solving:

> As for the direct method of ratios, what it embraces is very broad. The large: computing the motion of the Seven Planets in the Heavens, measuring height, depth, breadth and distance. The small: estimating tasks and ordering affairs. Measuring the large and shifting the small are all obtained through ratios. For, taking two numbers as a ratio, from the number that one has (*jin you zhi shu* 今有之數), one can obtain the number that one does not have (*wei you zhi shu* 未有之數).[78]

Indeed the compendium contains a wealth of material dealing with ratios. Book 6 of Gerbillon and Bouvet's translation of Pardies' textbook, devoted to ratios, has been divided into five books to form books 6 to 10 of the new version of *Elements of geometry*, which constitute chapter I.3. Items 5 to 16 of the new book 6 define twelve kinds of ratios, with

---

77   BML, Ms. 82–90 A, 6a–7b.
78   Guo 1993, 3: 228.

names slightly modified from those devised by the Jesuits.[79] Further, the *Elements of calculation* define 'numbers in ratio of equal addition' (*ping jia bili* 平加比例), which correspond to numbers in arithmetical progression.[80] However, the terms defined in part I are not used in part II, although ratios play a central role there. Ten methods involving them are given in chapters II.3 to II.7. New names are coined for these methods, although most of them have equivalents in Thomas' *Outline of the essentials of calculation*. What was expressed there in terms of the 'rule of three' (*san lü qiu si lü fa* 三率求四率法, *san lü fa* 三率法), is now presented as methods involving ratios. Moreover, many problems are illustrated by the layout of four *lü* that occur in the solution, the fourth of which is the result sought, or a number that is needed to find this result (as in Box 14.3).[81]

Chapters II.8 and II.9 are devoted to methods akin to the 'rule of false position'; chapter II.8 is named after the seventh of the nine traditional branches of mathematics, 'Excess and deficit' (*Yingnü* 盈朒); the introduction to the chapter explains how the method relates to those expounded in previous chapters:

This is definitely a method of ratios. But in ratios one uses actual numbers to seek actual numbers; and in excess and deficit one uses empty numbers to seek actual numbers.[82]

The two methods expounded in chapter II.9 are given by Thomas, under the same names: 'Borrowing grades for mutual comparison' (*jie cui hu zheng* 借衰互徵) and 'Repeated borrowing for mutual comparison' (*die jie hu zheng* 疊借互徵). Here again,

[...] one only has an empty *lü* and one is obliged to borrow a graded number to form a ratio. After that one can get an actual number.[83]

Thus, with the exception of chapter II.10, devoted to Mei Wending's reworking of rectangular arrays,[84] all the 'Line section' of the work is regarded by the compilers as pertaining to ratios.

The *Calculation by borrowed root and powers* introduced by Thomas is also rephrased in terms of ratios. As the successive powers of any number or unknown form a continued proportion, the section on cossic algebra of the imperial compendium is entitled 'Ratios of borrowed root and powers' (*Jiegenfang bili* 借根方比例). However in this

---

79 These were based on the typology of proportions given in Pardies 1671, 61–63. Guo 1993, 3: 64–72; comp. Taipei NCL, Rare Books 6399, 2a–16b.
80 Item 2.31 of the *Elements of calculation* is derived from definitions 1 and 2 of Book VIII of Euclid's *Elements*; Guo 1993, 3: 171; see p. 195.
81 Guo 1993, 3: 229–385, *passim* in the 'Line section'.
82 Guo 1993, 3: 344.
83 Guo 1993, 3: 368.
84 See pp. 345–348, 352–354.

> Box 14.3 Diagram summarising the solution of the problem of the two men walking towards each other. *Essence of numbers and their principles.* (Sinica 361/34, 6: 13a; see Guo 1993, 3: 299.)
>
>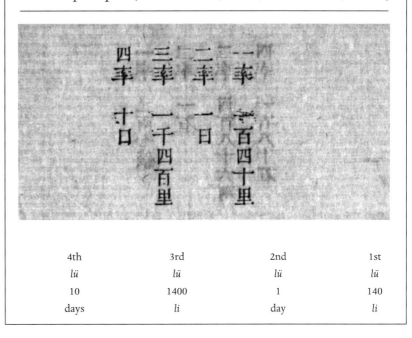
>
> | 4th | 3rd | 2nd | 1st |
> |---|---|---|---|
> | *lü* | *lü* | *lü* | *lü* |
> | 10 | 1400 | 1 | 140 |
> | days | *li* | day | *li* |

case ratios are only used to define the powers and their ranks,[85] whereas the vocabulary and notations introduced by Thomas prevail in the rest of the chapters devoted to the subject. On the whole, it seems that the term *bili* has been used to rename a number of methods in part II in order to better bring out the fact that these methods rest on notions and methods defined in part I, rather than in an attempt to unify the mathematical terminology and notation within part II. There is little connection, however, between the twelve kinds of ratios defined in the *Elements of geometry* and those used in the 'Line section'.

The main sources of the second part of the compendium are Thomas' textbooks. However, unlike the *Elements of geometry* and the *Elements of calculation*, the *Outline of the essentials of calculation* and the *Calculation by borrowed root and powers* have not been reproduced as wholes. Most of the former work seems to have been used, whereas the latter has been reduced considerably: it only takes up six chapters of the compendium.[86] Not all problems found in the manuscript textbook are taken up in the compendium, where problems from other sources

---

85 Guo 1993, 3: 940.
86 It still contains more than the *Summary of calculation by borrowed root and powers*; see pp. 209–210.

are introduced.⁸⁷ Some of the problems that are taken up are placed under a different heading. Thus the section devoted to 'Borrowing grades for mutual comparison' of Thomas' textbook contains the following problem:

Again suppose two men start walking towards each other on the same day, one from the South, the other from the North. They are 1400 *li* apart. The two men want to meet. One of them walks 80 *li* a day, the other walks 60 *li* a day. After how many days can they meet?

Now borrow one as the graded number (*cui shu* 衰數) of days. Add up the 80 *li* of the first one and the 60 *li* of the second one; one gets 140 *li* as the first *lü*; one day is the second *lü*; 1400 *li* is the third *lü*. Multiply the second and the third *lü*, divide this by the first *lü*, one gets 10 as the fourth *lü*; therefore they can meet after 10 days.⁸⁸

The derivation of the graded number is not explained in any of the problems of this section in Thomas' textbook. The fact that this number is equal to one in this particular example explains that the compilers of the *Essence of numbers and their principles* have chosen to put this problem in chapter II.6, under 'Ratios according to sums', a method that is not given by Thomas. There are minor stylistic differences between the two texts, but the significant change is that the first sentence of the solution has been deleted:⁸⁹ it is simply the sum of the distances covered by each man in one day that defines the ratio. Thus what Thomas used as a simple case of a complex method has become a standard case of a simpler method.

The Jesuits' textbooks contained cross-references that united them;⁹⁰ these also appear in the imperial compendium. The most frequent references are to the *Elements of geometry*, found both within the geometrical treatise itself and elsewhere in the compendium. The *Elements of calculation* is quoted several times in the second part; there are also cross-references within the second part itself, which suggests a generalisation of the pattern on which the *Elements of geometry* functions. There are no references to other mathematical works except in historical passages: the *Essence of numbers and their principles* is intended as a self-sufficient, comprehensive treatise. Its cross-reference system can be summarised by the diagram in Box 14.4.

---

87 For some examples see pp. 345–348, 354–356.
88 Gugong bowuyuan 2000, 403: 202; BML, Ms. 82–90 A, 1: 106b–107b; a verification follows in both versions, an elementary explanation is also added in the BML version.
89 Guo 1993, 3: 299; the diagram corresponding to this problem is reproduced in Box 14.3.
90 See p. 203.

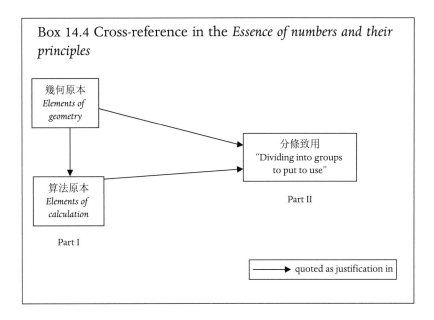

Box 14.4 Cross-reference in the *Essence of numbers and their principles*

The imperial textbook, while claiming the mathematical legacy of China's high antiquity, and borrowing freely from all sources available, rests mainly on the particular version of Western learning that Antoine Thomas and, in geometry, Bouvet and Gerbillon, had devised for Kangxi. While most of the work derives from their tutoring, its organisation and the phrasing of some passages reflect an original view of the foundations, structure and practice of mathematics. In this respect, the work fulfilled the imperial ambition from which it resulted: the reconstruction of mathematics.

CHAPTER 15

Methods and material culture in the *Essence of numbers and their principles*

The reworking of Western learning by the Office of Mathematics is anchored in Chinese learning in two ways: by a historical narrative and by the central role ascribed to numbers in the definition of the field. The dichotomy opposing Western learning to Chinese learning was certainly of some significance to scholars at the time. However, one may ask to what extent it is relevant and useful for understanding the imperial compendium, in which the dichotomy is not only passed over in silence, but is also invisible. Thus, in its second part, problems of both Chinese and Western origins appear side by side. These problems give a glimpse of the society and culture for which imperial mathematics was shaped. One purpose of the work was to make available to officials methods that would enable them to check accounts, both administrative and commercial: a significant number of problems deal with trade and taxes, in particular with the reckoning of interest. The mathematics of the compendium is that of an immense and wealthy empire where camels as well as boats are used for the transportation of goods, where pomegranates and melons as well as apples and pears can be purchased on markets, where the silk trade is thriving. Irrigation, waterworks and the management of troops are also evoked, but they take up relatively little space. As to astronomy and music, to which the two other sections of the *Origins of pitchpipes and the calendar* are devoted, planetary motion occurs in a few simple problems; music is only evoked through references to the buying and selling of instruments.

In this chapter, a few problems, their genealogy and the culture they reflect will be discussed, in an attempt to situate the emperor's mathematics both within the history of this field and within the culture of his time. The selection presented below does not exhaust all the situations in which this mathematics was to be used, nor the great variety of mathematical objects and methods brought into play. Rather, it aims at bringing out some of the classification criteria used by the compilers, and some features of the material culture that underlay their practice of mathematics. Another aim of the present chapter is to pursue to its conclusion the line of enquiry on calculation followed throughout previous chapters; therefore the problems discussed below belong to the Line Section and the Final Section.

15.1 Problems and their genealogy: Master Sun and Hieron

Some problems included by Thomas in his *Outline of the essentials of calculation* that one may think exemplify 'the Chinese' or 'the Western' mathematical traditions are taken up in the *Essence of numbers and their principles*, albeit with different numerical data. For example, the problem of the 'chickens and rabbits in a cage' (*long nei ji tu* 籠內雞兔)[1] now reads:

Suppose chickens and rabbits are in the same cage (*ji tu tong long* 雞兔同籠). One but knows that there are altogether 36 heads, and altogether 100 feet. One asks how many chickens and how many rabbits there are.

Method: multiply the total number of heads of chickens and rabbits 36 by the 2 feet of a chicken; one gets 72 feet. Subtract from the total number of feet 100; then the total feet are too many by 28. Again multiply the total number of heads of chickens and rabbits 36 by the 4 feet of a rabbit; one gets 144 feet. Subtract the total number of feet 100; then the total feet are too few by 44. Then add up the two numbers of too many and too few; one gets 72 feet, which is the first *lü* 率. The total number of heads 36 is the second *lü*. 44 feet too few is the third *lü*. One gets as fourth *lü* 22, which is the number of chickens. Subtract it from the total number of animals; the remainder 14 is the number of rabbits. If 28 feet too many is taken as the third *lü*, one gets as the forth *lü* 14, which is again the number of rabbits. In this method if one calculates taking all the 36 to be chickens, then there must be 72 feet; taking the difference with the actual total feet, then total feet are too many by 28. If one calculates taking all the 36 to be rabbits, then there must be 144 feet; taking the difference with the actual total feet, then total feet are too few by 44. Taking 36 rabbits gives 72 more feet than taking 36 chickens, that is to say taking 36 chickens gives 72 less feet than taking 36 rabbits. Thus one knows that when chickens are fewer than rabbits by 72 legs there are 36 chickens. Now chickens are fewer than rabbits by 44 legs; then there must be 22 chickens. Again when the rabbits are more than the chickens by 72 legs, there are 36 rabbits; now rabbits are more than chickens by 28 feet; then there must be 14 rabbits.[2]

Compared with Thomas' textbook, the numerical data have been changed; the solution is similar, but it has been rewritten in greater detail. This rewriting brings to the fore the meaning of 'ratios with sums of differences': the first term of the proportion is obtained by adding the differences between the given number of feet and the ones

---

1 See p. 189.
2 Guo 1993, 3: 320.

The Emperor's New Mathematics

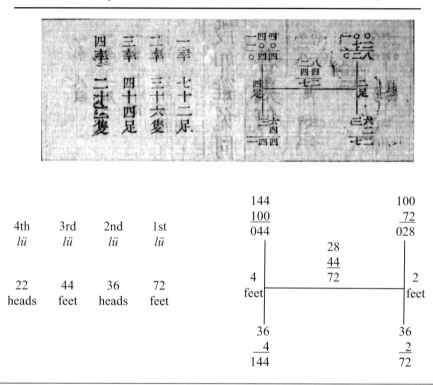

Box 15.1 First layout for the 'chickens and rabbits in the same cage' problem in the *Essence of numbers and their principles*, chapter II.7 (Bodleian Libary, Sinica 361/31, 7: 7b; see Guo 1993, 3: 320)

obtained supposing there is only one kind of animal in the cage. The use of a second proportion to obtain the number of rabbits again, after it has been obtained by a subtraction, and the lengthy explanation of the meaning of the two proportions make the text longer than would be necessary if its only purpose was to show how the solution is derived. In fact, the text not only 'sets up a method for finding numbers', but also 'clarifies the reasons why calculations are set out', as prescribed in the opening chapter of the compendium.³ Here the compendium appears to be closer to the later version of Thomas' textbook, in which solutions are followed by an explanation.⁴ A second solution follows, in which the first two terms of the proportion considered are simpler: the difference between the number of feet of rabbits and chickens being taken as the first *lü*, and 1 as the second *lü*. This lengthy treatment is not specific to this problem in the chapter.

---
3 See p. 327.
4 See pp. 187–188.

Box 15.2 Layout using borrowed root and powers for the 'chickens and rabbits in the same cage' problem in the *Essence of numbers and their principles*, chapter II.34 (Bodleian Library, Sinica 361/45, 34: 29a; see Guo 1993, 3: 1029)

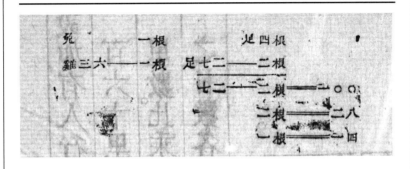

```
Rabbits      1 root         Feet  4 roots
Chickens    36 − 1 root     Feet  72 − 2 roots
                                  72 + 2 roots = 100
                                       2 roots = 28
                                       1 root  = 14
```

The problem is found again, in exactly similar terms, in chapter II.34, where it is solved using cossic notations:

Method: borrow 1 root to be the number of rabbits. Then the number of chickens is 36 − 1 root. Multiply a rabbit's 4 feet by 1 root; one gets 4 roots, as the rabbits' total number of legs. Multiply a chicken's 2 legs by 36 − 1 root; one gets 72 − 2 roots, as the chickens' total number of legs. Add them up; one gets 72 + 2 roots equal to 100. Subtract 72 on each [side]; then the remainders 2 roots and 28 are equal. Given that 2 roots are equal to 28, then 1 root must be equal to 14, which is the number of rabbits. From the total 36 subtract 14 rabbits. The remainder is 22, which is the number of chickens. Multiply the 14 rabbits by 4 legs, one gets 56 legs, which is the total number of rabbit legs. Multiply the 22 chickens by 2 legs, one gets 44; it is the total number of legs of chickens. Add them up, one gets 100, which matches the given number. (This [comes under] the method of ratios with sums of differences.)[5]

The solution given here is not specific to the type of problem identified in the final note in small characters, and does not make use of ratios.

---

5 Guo 1993, 3: 1029.

Table 15.3 The 'chickens and rabbits in the same cage' problem in various mathematical works

Work	heads	feet	chickens	rabbits
Mathematical classic of Master Sun (*Sunzi suanjing* 孫子算經)[1]	35	94	23	12
Instructions for calculation in common script (*Tongwen suanzhi* 同文算指)[2]	96	308	38	58
Number and magnitude expanded (*Shuduyan* 數度衍)[3]	96	308	38	58
Outline of the essentials of calculation (*Suanfa zuanyao zonggang* 算法纂要總綱)[4]	80	200	60	20
Essence of numbers and their principles II-7[5]	36	100	22	14
Essence of numbers and their principles II-34[6]	36	100	22	14

1 Guo 1993, 1: 244 (problem 3–31).
2 Guo 1993, 4: 179–180.
3 Jing 1994, 2: 2831.
4 BML, Ms. 82–90 A, 1: 66a–b.
5 Guo 1993, 3: 320.
6 Guo 1993, 3: 1029.

In the *Mathematical classic of Master Sun*, where this problem first occurs, the birds involved are pheasants (*zhi* 雉) rather than chickens (*ji* 雞). The expression 'chickens and rabbits in the same cage' (*ji tu tong long* 雞兔同籠), is found in the *Instructions for calculation in common script*. In Thomas' textbook one finds instead 'chickens and rabbits inside a cage' (*long nei ji tu* 籠內雞兔). The phrase found in the *Essence of numbers and their principles* is the same as that in the *Instructions for calculation in common script*. This bears witness both to the stylistic concern of the compilers, and to their lack of concern with quoting literally from an ancient text. Examining the data given in various works where the problem is found confirms that the compilers did not take up problems as they found them. Neither did they object to applying a method taught by the Jesuits to a problem stemming from an ancient Chinese work.

Other problems taken from Thomas' textbook can be traced back more than a millennium. The famous 'problem of Hieron's crown' is stated in terms quite similar, but again with different numerical data:

Suppose there is a golden vessel, in which silver has been alloyed. Its total weight is 170 *liang* 兩 and 4 *qian* 錢. What are the weights of gold and silver?[6]

Another problem, similar to this one but involving jade and stone instead of gold and silver found in Thomas' textbook is also taken up in the compendium.[7] This one can be traced back to the *Nine chapters on mathematical procedures*,[8] but again all the numerical data are different. Thus, problems regarded as emblematic of the Chinese and the Western mathematical traditions respectively appear in the same chapter, without any comment on their historical origin or significance.[9] The solutions to both these problems run parallel to that of the problem of the 'chickens and rabbits in the same cage'. In Thomas' textbook, the weights of a cubic *cun* 寸 of gold and silver are not given in the problem, but in a table. In the imperial compendium, the table of 'fixed lü of *cun* cubes' (*cun fang ding lü* 寸方定率), is included in a series of 'Tables of fixed lü and of their logarithms' (*ding lü dui shu biao* 定率對數表) found at the end of chapter III.6;[10] the values given there function as imperial standards. Thus the *Essence of numbers and their principles* brings together 'all under heaven'; without accounting for their sources, the compilers unify and standardise all the material at their disposal.

15.2 Inkstones and brushes

Another example of this lack of engagement with earlier mathematical works is found in chapter II.10, which, as a whole, is derived from Mei Wending's *Discussion of rectangular arrays* (*Fangcheng lun* 方程論). The chapter is divided into five sections: the first four correspond to Mei's typology of problems,[11] the last one gives problems the solution of which involves a combination of the method of rectangular arrays with some other method. Mei Wending's claim that the method enables one to solve problems of various types is repeated: 'What originally does not pertain to the method of rectangular arrays can be calculated using the method of rectangular arrays'.[12] The reader's attention is brought to this fact by notes at the end of the solution of three of the problems in the chapter, which mention another method of the 'Line section' that can be used to solve the same problem.[13] The inkstones and brushes problem is given in the second section of the chapter:

---

6   Guo 1993, 3: 324; see p. 190.
7   Guo 1993, 3: 322–323; see p. 191.
8   Guo 1993, 1: 174.
9   On these two problems see Liu 2002a.
10 *SKQS* 801: 523–524.
11 See p. 95.
12 Guo 1993, 3: 397; see p. 96.
13 Guo 1993, 3: 425–427.

345

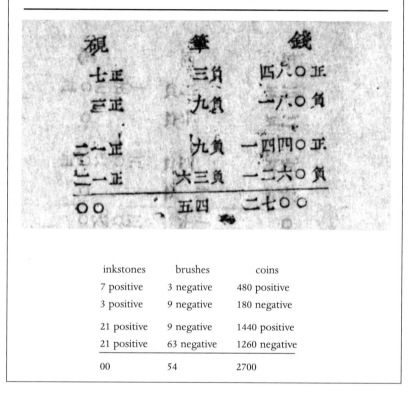

Box 15.4 Layout for the inkstones and brushes problem in the *Essence of numbers and their principles*. (Bodleian Library, Sinica 361/35, 10: 21a; see Guo 1993, 3: 405.)

inkstones	brushes	coins
7 positive	3 negative	480 positive
3 positive	9 negative	180 negative
21 positive	9 negative	1440 positive
21 positive	63 negative	1260 negative
00	54	2700

Suppose 7 inkstones cost 480 pieces (*wen* 文) more than 3 brushes, and 3 inkstones cost 180 pieces less than 9 brushes. How much do an inkstone and a brush cost each?[14]

Unlike in the problems from chapter II.7 discussed above, the data of the problem remain the same as in its earlier occurrences since the fifteenth century: the *Great classified survey of the Nine chapters on mathematical methods* (*Jiuzhang suanfa bilei daquan* 九章算法比類大全, 1450), the *Unified lineage of mathematical methods*, where the problem appears among 'Difficult problems', the *Instructions for calculations in common script*, *Number and magnitude expanded*, and the *Discussion on rectangular arrays*.[15] In the compendium as in this last work, the problem is the first example of the second category of

---

14 Guo 1993, 3: 405.
15 Guo 1993, 2: 1363–1364; 3: 422–424; 4: 194–195; Jing 1994, 2: 2852–2853.

problems, those involving 'numbers in differences' (*jiaoshu* 較數). Although its solution is expounded in detail, the problem plays no particular role. Mei's instructions for setting up the layout to do the calculations is followed, in that the numbers appear on the table in the same order as they do in the text, with the first one taken as positive. However, there are two major differences. First, the working is not inserted within the text, occupying full columns; instead it appears in the space reserved for figures and diagrams at the top of the page. The working is repeated on each of the four pages taken up by the solution. Secondly, in keeping with the choice to write numbers horizontally, consistently with their layout on the abacus rather than with the orientation of the Chinese script, the layout is transposed by 90° compared to that given by Mei Wending—and all his predecessors.

The inkstones and brushes problem was transmitted continuously and literally from at least the fourteenth century until it was included in

Table 15.5 The inkstones and brushes problem in some Ming and early Qing mathematical works

Work	Chapter	Subheading
Great classified survey of the Nine chapters on mathematical methods (1450)[1]	Rectangular arrays	Nine questions in poems (*ci shi jiuwen* 詞詩九問)
Unified lineage of mathematical methods (1592)[2]	Difficult problems	Rectangular arrays
Instructions for calculation in common script (*Tongwen suanzhi* 同文算指, 1614)[3]	Methods with miscellaneous sums, differences and multiplications	—
Number and magnitude expanded (*Shuduyan* 數度衍, 1662)[4]	Rectangular arrays	Method for establishing the positive and negative / Rectangular arrays of two kinds
Discussion of rectangular arrays (1672)[5]	Rectifying names	Numbers in differences
Essence of numbers and their principles[6]	Rectangular arrays	Numbers in differences

1 Guo 1993, 2: 266.
2 Guo 1993, 2: 1402–1403.
3 Guo 1993, 4: 187–188.
4 Jing 1994, 2: 2850.
5 Guo 1993, 4: 329–332.
6 Guo 1993, 3: 405–406.

the imperial compendium. This seems to be an exception, at least among problems pertaining to rectangular arrays.[16] Even when discussing a method that according to its author, Mei Wending, represented the acme of calculation in the Chinese tradition, the editors of the compendium appear remarkably uninterested in anchoring imperial mathematics in this tradition.

15.3 Weighing up the difficulty of problems

The choice to include a chapter of 'Difficult problems' towards the end of part II was probably inspired by the *Unified lineage of mathematical methods*. One may ask, however, how difficult the problems in chapter II.37 are, and what their difficulty consists of. The chapter opens with a brief explanation:

The study of calculation procedures is nothing but line, area and volume. Among them, methods like seeking ratios from one another, borrowing roots or borrowing powers have been put into categories above. But among problems, some have various windings and complications: how could one get their system without detailed examination and clear discussion? Here we have delved into abstruse subjects to compile this chapter of difficult problems, so that those who study think deeply and do not get confused by the ways to enter calculation. Almost all the subtleties of numbers and their principles and the skills of the human mind can be obtained from what is here, by extending it by analogy. Thus exhausting all the transformations of the universe will not be difficult either.[17]

Thus the difficulty of the problems given in the chapter does not reside in the way that they are stated, as was the case in the *Unified lineage of mathematical methods*. In fact, several criteria seem to have been used to decide where a problem belonged. This is apparent from some problems that involve weighing scales. This instrument is not associated with a scholarly approach to the mathematical sciences, nor with particular technical skills, but rather with the everyday practice of merchants. On the other hand, it had already been integrated in Western learning: scales are discussed at length in the *Illustrations and descriptions of extraordinary devices* (*Qiqi tushuo* 奇器圖說, 1627), co-authored by the Jesuit Johann Schreck (1576–1630) and Wang Zheng 王徵 (1571–1644), a Chinese convert.[18] This work, however, does not seem to have been a direct source for the problems quoted below.

---

16 Only one other such problem, found in chapter 11 of the *Unified lineage of mathematical methods* found its way into chapter II.10 of the *Essence of numbers and their principles*, albeit with a change in what we nowadays call the constant terms of its equations; it is also found in the three other works mentioned in Table 15.6; Guo 1993, 2: 1363–1364, 3: 422–424, 4: 372–374, 494–495 and Jing 1994, 2: 2852–2853.
17 Guo 1993, 3: 1110.
18 Zhang et al. 2008, 87–98.

CHAPTER 15 | Methods and material culture in the *Essence of numbers and their principles*

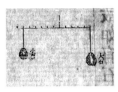

Illustration 15.6 Illustration of the copper scales problem in the *Outline of the essentials of calculation* (BML, Ms. 82–90 A, 165b) and in the *Essence of numbers and their principles* (Bodleian Library, Sinica 361/35, 9: 53b; see Guo 1993, 3: 393).

As mentioned above, weight was to be measured according to standards set by the state; the *Essence of numbers and their principles* therefore discusses the making as well as the use of scales. One problem involving both appears three times in the compendium, in chapters II.9, II.10 and II.34. In its first occurrence the problem reads:

Suppose there are two stones, a large one and a small one. Their weights are unknown. One only has a copper stick weighing 12 *liang*. If one exchanges them to weigh them, how much does each of them weigh?[19]

This problem appears in Thomas' *Outline of the essentials of calculation*. In both works the concision of the problem contrasts with the length of its solution: the one in chapter II.9 is among the longest in the compendium.[20] The illustration representing the scale has been taken up; but it is far less detailed than in the manuscript textbook.

---

19   Guo 1993, 3: 393.
20   The solution covers more than eight pages, with abundant notes in small characters; Guo 1993, 3: 393–395; BML, Ms. 82–90 A, 163a–166b.

349

The need to make a scale with the copper stick and to calibrate it, and the absence of any numerical data resulting from using this scale has not prompted the editors to regard this problem as a 'difficult' one. Interestingly, in chapters II.10 and chapter II.34, the problem itself includes the description of the weighing process and data resulting from it:

Suppose there are two stones, a large one and a small one. Their weights are unknown. One only has a copper stick weighing 12 *liang*. One parts it equally into 12 divisions. One fastens a string onto the 5th division. One end is 5 divisions [long], one end is 7 divisions [long]. One hangs up the large stone to one end of the copper stick, 5 divisions away from the fastening, and weighs it using the small stone as a sliding steelyard. When it is 6 divisions away from the fastening, [the stick] is horizontal. Again one hangs up the small stone to one end of the copper stick, 5 divisions away from the fastening, and weighs it using the large stone as a sliding steelyard. When it is 4 divisions away from the fastening, [the stick] is horizontal. One asks how much the weights of the large and small stones are.[21]

The solution involves a preliminary calibration of the scale, which makes the problem more complex than the others found in the three chapters. This explains why it is found last in chapter II.9 and then in the last section of chapter II.10, where it reads:

Method: First double 5 parts; subtract from 12 parts; the remainder is 2 parts. Halve it; one gets 1 part. Add this to 5 parts, which gives 6 parts. Then take 5 parts as the first *lü*, 6 parts as the second *lü*. Make the remainder 2 as 2 *liang*, which is the third *lü*. One gets as the fourth *lü* 2 *liang* 4 *qian*. If one adds 2 *liang* 4 *qian* to the 5 divisions side, it will be level with the 7 divisions part.[22]

This first part of the solution is identical in all three occurrences of the problems: it consists in calculating the difference between the weights of the two ends of the scale, which is taken into account in the ensuing calculations. After this the three solutions each follow the method of the corresponding chapter; those given in chapter II.10 and II.34 are about half the length of the one given in chapter II.9. The triple occurrence of one particularly complex problem exemplifies a progression among the methods discussed in these chapters, namely 'repeated borrowing for mutual comparison', 'rectangular arrays' and 'borrowed root and powers'.[23] This could be the reason why the editors repeated the problem three times, taking the method used in the second part of its solution rather than the complexity entailed by the need to combine two sets of data (those resulting from the scale and those concerning the weighing) to obtain the solution, as a classification criterion.

---

21 Guo 1993, 3: 424.
22 Guo 1993, 3; 424, comp. 3: 939, 1029.
23 See pp. 352–354.

Scales also appear in four 'difficult problems', each of which are solved in less than two pages. The first two of these also involve the construction of a scale:

Suppose there is a large stone; one does not know its weight. But one knows that a small stone weighs 4 *liang*. One seeks how much the large stone weighs.

Method: Use a wooden stick; suspend it by its middle so that the two ends are level. Hang up the large stone at one of its ends; use the small stone as a sliding weight to weigh it. If with the large stone 1 *cun* away from the hanging point and the small stone 6 *cun* away from the point of suspension, one gets evenness, then take 1 *cun* as the first *lü*, the weight of the small stone 4 *liang* as the second *lü*, 6 *cun* as the third *lü*. One obtains the fourth *lü* 24 *liang*; this is the large stone's weight. As on the figure, AB is the distance of 1 *cun* between the large stone and the hanging point. AC is the distance of 6 *cun* between the small stone and the hanging point. D is the large stone; E is the small stone. The weight of the small stone D is the graduation AB; the weight of the large stone E is the graduation AC. Therefore the ratio between AB and the small stone E is the same as the ratio between AC and the large stone.[24]

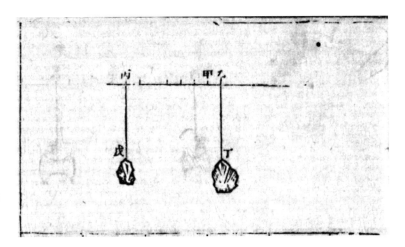

Illustration 15.7 Illustration of the wooden stick scale problem, *Essence of numbers and their principles* (Bodleian Library, Sinica 361/46, 37: 12b; see Guo 1993, 3: 1115).

The problem is solved using a simple direct proportion similar to those discussed in chapter II.3. As in the copper scale problem discussed above, the data resulting from the weighing are not given in the question but in the solution. The difficulty of the problem (that is to say, the feature that best explains why the problem is not in chapter II.3) may lie in that, or in the fact that the weight of the scale is not known, which implies the use of a weighing technique different from that of common trade scales. The explanatory part of the solution highlights the fact that the two weights are in inverse proportion to the distances between the suspension point and the two points from which they are hung. This implies that the graduations on the scale are linear, a point

---

24   Guo 1993, 3: 1114–1115.

that also underlies the instructions on how to graduate the copper scale in the other problem: this was expected to be part of readers' knowledge. The three other difficult problems on scales, which use slightly more complex ratios, could nonetheless be associated with a method of the Line Section; all involve simple calculations but non-standard weighing devices. Looking at all the problems that involve scales in the work suggests that the classification criteria applied are multilayered. The copper scale problem serves to display a hierarchy amongst the methods that applies to problems in the category of lines. The four 'difficult' scale problems, which are mathematically less complex, serve to illustrate possible ways of making weighing devices.

15.4 Algebra and the problems of the 'Line Section'

The lack of concern with what is Chinese and what is Western that is apparent in the compendium reflects Kangxi's viewpoint. In the 1660s, he already treated this dichotomy as irrelevant to imperial astronomy. He was now intent on imposing his view within imperial scholarship. The staff of the Office of Mathematics must therefore have used other criteria to operate a selection from the knowledge and methods available to them. Bearing this in mind, let us examine the choices they made as regards the methods for solving the problems of the 'Line Section'. These methods fall under three categories:

- The various techniques involving ratios presented in chapters II.3 to II.9, mostly derived from Thomas' *Outline of the essentials of calculation*; a number of them were already found in Ricci and Li Zhizao's 李之藻 *Instructions for calculation in common script*.
- The method of rectangular arrays (chapter II.10), taken up from Mei Wending's *Discussion of rectangular arrays*, where Mei severely criticised the methods given in the *Instructions for calculations in common script*, and claimed the superiority of the Chinese tradition in the field of calculation. He claimed that the method of rectangular arrays as he reconstructed it was part of the legacy of Chinese antiquity, and that it subsumed a number of other calculation methods.[25]
- Cossic algebra as introduced by Thomas in his *Calculation by borrowed root and powers*; it is quite possible that this introduction was a response to Mei's severe criticism of Western calculation.[26]

Some problems in Thomas' work used a symbolic notation to solve systems of linear equations in several unknowns; this notation, which

---

25 See p. 93.
26 See p. 217.

CHAPTER 15 | Methods and material culture in the *Essence of numbers and their principles*

effectively provides an alternative to Mei's method of rectangular arrays, is similar to the one Foucquet tried to teach Kangxi—except that the French Jesuit proposed to use literal instead of numerical coefficients.

The choices made by the compilers reflect the imperial assessment of these methods:[27] as one might have expected, no trace of Foucquet's enterprise is found in the compendium. Neither is Thomas' symbolic notation taken up. On the other hand both his cossic algebra and Mei's rectangular arrays are incorporated as wholes. The position of the latter at the end of the 'Line Section' echoes Mei's claim that it is 'the acme of calculation'. As mentioned above, annotations at the end of each of the three last problems of chapter II.10 state that these problems pertain to earlier chapters of the section, again echoing Mei's claim that rectangular arrays could be used to solve problems of different categories. In the imperial compendium rectangular arrays are the acme of calculation, at least within the category of lines.

The claims concerning the 'borrowed root and powers' are stronger. At the opening of chapter II.31, one is told that 'this method can manage all the sections of line, area and solid'.[28] Not only is it included in the 'Final Section', but the chapters devoted to it are structured according to headings and subheadings of some of the four previous sections. Chapter II.31 is divided into five parts: methods for fixing ranks, addition, subtraction, multiplication and division; so that the structure of the chapter is the same as that of chapter II.1. This brings out the analogy between 'borrowed root and powers' and numbers. Most of the sixty-three problems that constitute chapter II.34 are closed by a note stating that the problem can also be solved using one or two of the methods of the 'Line section'; when two methods are mentioned, the second one proposed is mostly that of rectangular arrays. The repetition of the problems of the 'chickens and rabbits in the same cage' and of the copper scale illustrates the links mentioned in these notes—both among the methods of the 'Line section' and between these and the 'borrowed root and powers'.

The compendium does not mention the existence of the methods that the compilers chose to exclude, let alone explain why they have been excluded; but one can only infer from the criterion given at the opening of the work[29] that Thomas' symbolic notation for linear problems in several unknowns was not deemed to be 'excellent'

---

27 See p. 330.
28 Guo 1993, 3: 940.
29 See p. 330.

353

enough to deserve inclusion. Two facts explain this. The notation was introduced in a rather hasty way, and used without much explanation. Moreover, Mei Wending's method of rectangular arrays, which included thorough explanations, could solve exactly the same problems. Nowhere does one find a claim of superiority of one method over another. However the order in which the methods are expounded integrates the recognition that some are more general, in the precise sense that they can be used, among other purposes, to solve types of problems that fall under one or several of the categories presented earlier. Thus cross-references serve to bring out the fact that while rectangular arrays subsumes a number of methods in the category of lines, borrowed root and powers encompasses both the former method and those it subsumes. However in this elegant construction the compilers pass over one fact in silence: whereas the rectangular arrays method can solve problems with any number of unknowns (se 色),[30] the borrowed root and powers is used to solve problems that have only up to three unknowns. In this respect, there is an element of generality in the method of rectangular arrays that is not matched by cossic algebra.

15.5 The remainder problem

Not all the 'difficult problems' entail know-how related to instruments or the supplement of insufficient data. Some present calculation techniques that do not fit elsewhere in the work. Typically, only one problem is given for each of these techniques. For example, the first problem of the chapter reads:

Suppose three men, A, B and C, are on duty, A once every 3 days, B once every 4 days, C every 5 days. What days will the three men be on duty together?
  Method: Multiply 3 days by 4 days, one obtains 12 days; again multiply by 5 days, one gets 60 days; these are the days when the three men are on duty together. This method relies on the fact that 60 is a number which 3, 4 and 5 can all measure exhaustively (*dujin* 度盡).[31]

A more detailed explanation follows. The notion of 'measuring exhaustively' defined and used extensively in the *Elements of calculation*,[32] is hardly used in part II of the compendium: it appears only in the simplification of fractions. This problem is immediately followed by the remainder problem found in the *Mathematical classic of Master Sun*, one of the most famous in the history of Chinese mathematics, and the earliest evidence of interest in matters relevant to what is known

---

30 This fact is mentioned at the beginning of chapter II.10, which gives examples in up to four unknowns (Guo 1993, 3: 396).
31 Guo 1993, 3: 1110.
32 See p. 198.

nowadays as the 'Chinese remainder theorem'. The only change compared with the original is that instead of an unknown number of things one is now dealing with an unknown sum of money:

Suppose there is some money, of which one does not know the total amount. Counting it by threes the remainder is 2 pieces (*wen* 文); counting it by fives the remainder is 3 pieces; counting it by sevens the remainder is 2 pieces. One asks how much the total amount of money is.[33]

The general algorithm constructed by Qin Jiushao 秦九韶 in the thirteenth century for solving problems of this type was not known at the time of the compilation of the compendium.[34] This is a striking example of how sparse Chinese scholars' knowledge of pre-Ming mathematics was at the time. Accordingly, a solution that closely follows that of the *Mathematical classic of Master Sun* is proposed:[35]

Method: first fix the *lü* of counting by threes to be 70; fix the *lü* of counting by fives to be 21; fix the *lü* of counting by sevens to be 15. Then multiply the *lü* 70 and the remainder 2 of counting by threes; one gets 140. Multiply the *lü* 21 and the remainder 3 of counting by fives; one gets 63. Multiply the *lü* 15 and the remainder 2 of counting by sevens; one gets 30. Add up these three numbers; one gets 233. Again, multiply 3, 5 and 7 successively; one gets 105. Subtract it twice from 233; the remainder is 23; it is the total amount of money.[36]

The main difference with the *Mathematical Classic of Master Sun* is the use in the compendium of the word *lü* to refer to the partial products. The word is mainly used to name the terms of ratios; its occurrence in this problem evokes the fact that the use of *lü* as a mathematical term long predates the introduction of ratios in the context of Euclidean geometry. Its earlier meaning was broader: *lü* more generally refers to numbers given in relation to one another (such as exchange rates for various kinds of cereals).[37] Evidently the integration of *lü* into Euclidean terminology did not preclude its use in this broader sense.

The explanation given after this solution is much longer than the one found in the *Mathematical Classic of Master Sun*; the choice of the first *lü* is accounted for in the following way:

In this method the *lü* for counting by threes is fixed as 70: counting by fives and by sevens both exhaust it; only when counting by threes is there a remainder of 1. Now multiplying it by the remainder 2, one gets 140; then counting by fives and by sevens both exhaust it; only when counting by threes is there a remainder of 2.[38]

---

33 Guo 1993, 3: 1110.
34 On this algorithm, see Libbrecht 1973, 328–357.
35 Guo 1993, 1: 243; see Lam & Ang 2004, 138–140.
36 Guo 1993, 3: 1110.
37 See Chemla & Guo 2004, 956–959.
38 Guo 1993, 3: 1110.

Parallel explanations are given for the two other *lü*. The reasoning goes on to point out that the sum of the three numbers thus obtained will meet the conditions of the problem, and that all the numbers obtained by repeatedly adding or subtracting 105 (the product of the three numbers used to count up the money) once or more from this result will also meet the conditions; this explicit statement of how to obtain all solutions of the problem is absent from the *Mathematical Classic of Master Sun*.[39] Finally, the way to proceed if one counted by twos, threes and fives instead of threes, fives and sevens is briefly sketched.

As is well known, the two problems above are relevant to calendar-making, where various astronomical cycles need to be combined; however there is no hint of this link in the compendium. A historical reading of these two problems locates them at the confluence of two traditions: Chinese mathematics as shaped by its uses in astronomy, and elementary number theory in the Euclidean style. However, the two problems are found in the *Essence of numbers and their principles* not because of the prestige conferred to them by their genealogy (as it was known at the time), but because, in the eyes of the compilers, the method used to solve them deserved to be incorporated into imperial mathematics.

15.6 The construction and use of instruments

The calculation techniques available in China at the time included the four listed by Fang Zhongtong in his *Number and magnitude expanded*: beads (the abacus), brush (written calculation), rods (Napier's rods) and ruler (the proportional compass) (*zhu bi chou chi* 珠筆籌尺);[40] we know that all of these and more were present at court. The emperor himself was familiar with abacus calculation. Like him and like Chen Houyao, the readers for whom the *Essence of numbers and their principles* was intended would have been familiar with abacus calculation and were expected to learn written calculation from the imperial compendium.[41] The absence of Napier's rods from it suggests that they were not deemed indispensable aids to written calculation. Neither are the multiplication tables that underlie the practice of written multiplication, division and root extraction given in the compendium; to my knowledge these tables were not included in any of the mathematical works that the Jesuits wrote in Chinese. On the other hand, operation tables associated with the use of the abacus are found at the beginning of Cheng Dawei's *United lineage of mathematical methods*:[42] it is therefore reasonable to suppose that anyone who could perform the four basic

---

39 Only positive numbers are considered.
40 See p. 45.
41 Jami 2002b, 32–33; see p. 258.
42 Guo 1993, 2: 1239–1242.

operations of arithmetic on an abacus was familiar with them. This means that the imperial compendium is not intended for complete beginners, but for readers who are already numerate in the Chinese tradition. We do not know whether these readers actually adopted written calculation or whether they continued to use the abacus. In any case the detailed description of place value notation and of the layout of the four elementary operations of arithmetic served several purposes. First, they formed the bases of a structured corpus of problems and methods for solving these, with calculations laid out in diagrams. Secondly, they provided foundations for 'calculation by the borrowed root and powers'; the ranks (*wei* 位) of successive powers and the four basic operations for expressions written in the cossic notation are defined by analogy and in parallel with those of arithmetic. Last but not least, they assert a rupture with the Chinese tradition of calculation, situating imperial mathematics among the disciplines practised using the 'four treasures of the study' (*wen fang si bao* 文房四寶), namely brush, inkstone, ink stick and paper. Thus Mei Wending's view that mathematics should be practised by scholars was taken up, even though his vertical layout of operations was not.[43]

Instruments, however, are not entirely absent from the compendium. They are first discussed in items 19 to 22 of book 12 of the *Elements of geometry* (chapter I.4), where the construction of the four main lines of the proportional compass is explained.[44] Chapters II.39 and II.40, entitled 'Explanation of the proportional compass' (*Bili gui jie* 比例規解) are mainly devoted to solving problems with the help of this instrument. There is no mention of the use of the instrument for military purposes; it is merely presented as a substitute for calculation.[45] The instrument is called *bili chi* 比例尺 (lit. 'proportional ruler') rather than *bili gui* 比例規 everywhere except in the title of these two chapters. This discrepancy is already found in the manuscript bearing the same title probably authored by Antoine Thomas; the title of the treatise itself is taken up from a work written by Giacomo Rho around the time of the late Ming calendar reform and included in the *Books on calendrical astronomy according to the new method.* (*Xinfa lishu* 新法曆書).[46] The construction of the lines on a sundial and on logarithmic rulers are also included in chapter II.40. The fact that the description of aids to calculation occurs in the final chapters of text of the *Essence of numbers and their principles* suggests that their use was not central to the imperial practice of mathematics.[47] Like scales,

---

43 See p. 89.
44 Guo 1993, 3: 137–141.
45 Guo 1993, 3: 1185; see Huang 1996.
46 *Xinfa suanshu* 新法算書, *SKQS* 788: 317; see pp. 193–194.
47 Sundial and logarithmic rulers; see p. 321.

 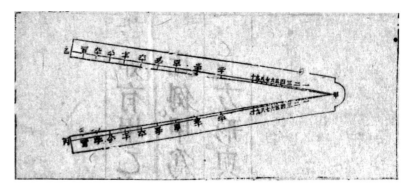

Illustration 15.8 Proportional compass in the *Outline of the essentials of calculation* (BML, Ms. 82–90 A, 188b) and in the *Essence of numbers and their principles* (Bodleian, Sinica 361/47, 39: 13b; Guo 1993, 3: 1185).

mathematical instruments are also treated as material objects on which a mathematical problem can bear: thus, in one of the 'difficult problems' of chapter II.37, one seeks the side of a square box such that a proportional compass (assumed to be of rectangular shape for the purpose) fits in it diagonally; this problem is taken up from the *Outline of the essentials of calculation*.[48]

The occurrence of a sundial evokes the links between mathematics and astronomy. However, as mentioned above, the latter subject does not appear much in the compendium.[49] In defining the border between the two disciplines, the compilers have put spherical trigonometry on the side of astronomy. On the whole, imperial mathematics is oriented towards the Earth rather than towards the heavens.

15.7 Time-keeping

Despite this orientation, time-keeping, for which the Astronomical Bureau was responsible, is mentioned several times. Time units are among those defined at the opening of part II of the compendium: they are the ones introduced by the Jesuits, which are still in use nowadays; a distinction is made between double hours (*shi* 時), of which there are twelve in a day, and hours (*xiaoshi* 小時), of which there are twenty-four—the former were more commonly used than the latter at the time. Examples then illustrate how to perform addition and subtraction of numbers given in time units.[50]

The official time-keeping device, the water clock, occurs in two problems of chapter II.9, in the part devoted to 'Borrowing grades for

---

48 Guo 1993, 3: 1124; *Outline of the essentials of calculation*. BML, Ms. 82–90 A, 2: 188a–188b.
49 Two problems in chapter II.3 deal with trajectories of planets (Guo 1993, 3: 230, 234).
50 Guo 1993, 3: 182, 184, 188, 191–192.

mutual comparison'; they deal with the time it takes for a vessel respectively to fill up and to empty. The wording of the first problem suggests that the reader is expected to be familiar with an already constructed device:[51]

Suppose there is a clepsydra (*louhu* 漏壺). At the top there is a spout (*kewu* 渴烏) for the water to pour in; in general it takes 12 double hours to fill up. At the bottom there is an aperture in the vessel to evacuate the water; in general it takes 18 double hours to empty completely. If water flows in from the top and out from the bottom, in how many double hours can one fill it up?[52]

The same chapter contains two problems presented as riddles on what the time is, in which 'small hours' are used.[53] Beside water clocks, four problems deal with the pendulum as a time-measuring device; it had first been introduced by Verbiest in the 1670s.[54] Two of these problems are in chapter II.3 and two in chapter II.37. Inverse ratios are used to determine the length of a second pendulum:

Suppose a time-checking instrument [with] plummet (*yan shi yi zhuizi* 驗時儀墜子); its string is 4 *chi* 尺, 4 *cun* 寸, 8 *fen* 分, 1 *li* 釐, 2 *hao* 毫 and 8 *si* 絲 long. In four quarters (*ke* 刻) it comes and goes (*laiwang* 來往) 3000 times. One wishes to make it come and go once a second. How long should the string be?[55]

The solution is underlain by the knowledge of the fact that the length of a pendulum is proportional to the square of the number of its beats during a given length of time. This problem is taken from the *Outline of the essentials of calculation*,[56] with a small change in the data: Thomas gave 4.4825 *chi* as the length of the original pendulum, and 3.112 *chi* as the length sought; this is slightly inaccurate, as the first figure yields a result closer to 3.113 than to 3.112. In the compendium, the former value is corrected to 4.48128 *chi*, which does yield the expected result. Obviously the problem was constructed from the length of the standard pendulum; whether this precision could be obtained in real life was irrelevant. Another problem in the same chapter depicts a possible use of the second pendulum:

Suppose [one uses] a time-checking instrument to calculate the sound of cannon. [If] from the smoke rising to hearing the sound, one counts 7 seconds, then one gets 5 *li*. Now one gets 14 seconds. How many *li* are there?[57]

---

51 On water clocks see Needham 1954–, 3: 313–329.
52 Guo 1993, 3: 371.
53 Guo 1993, 3: 372–373.
54 Verbiest discusses the pendulum in his *Compendium on the newly constructed instruments of the Observatory* (*Xinzhi lingtai yixiang zhi* 新製靈臺儀象志); Bo 1993, 7: 98–103; see p. 67.
55 Guo 1993, 3: 235.
56 BML, Ms. 82–90 A, 34b–35a.
57 Guo 1993, 3; 231.

In Thomas' textbook, the problem was again slightly different: the time measured was 12 seconds;[58] this has been changed so as to yield an integer as a result. On the other hand, the ratio of 5 *li* in 7 seconds remains the same in the compendium as in the manuscript, despite the fact that in the meantime the length of the standard *li* had changed from 250 *li* to 200 *li* for an arc of one degree of the terrestrial meridian.[59] The technique described here was potentially useful for military purposes: perhaps the lack of adjustment of the ratio according to the new standard unit, which made this ratio inaccurate, was not incidental. The same ratio is used in another problem:

Suppose one sees a village far away, and one wants to know the distance to it. One examines this using a gun and a time-checking instrument [with] plummet. How much is the distance?[60]

No information on the speed of sound is given here. This seems to be the reason why the problem is found among the 'Difficult problems' of chapter II.37. The relevant ratio is given in the solution:

Method: order a man to fire the gun near the village. As soon as smoke is seen, start using a time-checking instrument [with] plummet to watch out for it. As soon as the sound of the gun is heard, stop it. Count how many seconds there have been from seeing the smoke to hearing the sound. If one gets 3 seconds, then take 1 as the first *lü*, 128 *zhang* 丈 4 *chi* 7 *cun* as the second *lü*, 3 seconds as the third *lü*. Get the fourth *lü*: 385 *zhang* 7 *chi* 1 *cun*; this is the distance to the village. For when one has seen the smoke but not yet heard the sound, the sound has not yet arrived. Therefore the parts between seeing the smoke and hearing the sound are the parts of the distance. Having experienced that comparing the parts, if the distance is 5 *li* one gets 7 seconds, and dividing these by 7, for each second one gets 128 *zhang* 4 *chi* 7 *cun*.[61] Hearing thunder is similar: from seeing a flash of lightning to hearing the sound of thunder; one watches out for the number of seconds, and thus gets the number of *li*.[62]

The parallel between the sound of cannon and that of thunder mentioned here once more reveals Kangxi's close involvement in the work done at the Office of Mathematics: he used this parallel in one of the jottings of his *Collection of the investigation of things in leisure time* (*Jixia gewu bian* 幾暇格物編), when arguing that 'the sound of thunder does not go further than a hundred *li*'. He once discussed the speed of the sound of cannon with Li Guangdi, who appears to have been somewhat out of his depth during the conversation.[63] Thus the compendium not only incorporated what the Jesuits taught Kangxi, but also to some

---

58 BML, Ms. 82–90 A, 26b.
59 See pp. 244–245, 327. The former standard would yield 317 m/s for the speed of sound, the latter 396 m/s.
60 Guo 1993, 3: 1128.
61 One *li* contains 180 *zhang* (Guo 1993, 3: 182).
62 Guo 1993, 3: 1128.
63 Li 1993, 74; see Jami 2007b, 156–160; see p. 246.

CHAPTER 15 | Methods and material culture in the *Essence of numbers and their principles*

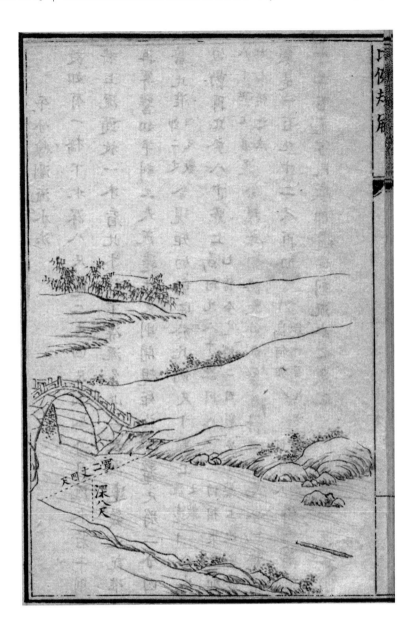

Illustration 15.9 Measuring the flow of a river, *Explanation of the proportional compass* (BML, Ms. 75-80 D, 4b)

extent his own reflections. The other occurrence of the pendulum in chapter II.37 evokes matters dear to the emperor's heart:

Suppose a river mouth has a breadth of 10 *chi* at the top, of 6 *chi* at the bottom, and a depth of 5 *chi*. One seeks how much water flows everyday.[64]

---

64 Guo 1993, 3: 1116.

361

Again the difficulty of the problem does not reside in complex calculations; this time it consists in devising a means to measure the speed of the water:

Method: place a wooden stick on the surface of the water. Watch out for it using a time-checking instrument [with] plummet. See how many *zhang* further the stick has gone in 60 seconds. [. . .]

This technique for measuring the water flow under a bridge was described in the manuscript *Explanation of the proportional compass* (*Biligui jie* 比例規解). However the device used to measure time is not mentioned there: the problem is given as an example of the use of the proportional compass for multiplication. On the other hand, the problem is nicely illustrated, like several others that dealt with surveying in this manuscript; the dimensions were given in the picture: the river is 8 *chi* deep and 2 *zhang* 4 *chi* wide; the width of the river is taken to be uniform from the surface to the bottom (see Illustration 15.9).[65]

Thus the uses of the pendulum proposed in the *Essence of numbers and their principles* all have to do with the assessment of distance; they pertain to what one might call 'outdoors mathematics', which the emperor pursued during his many trips outside the capital, often with the Jesuits' assistance.[66] The fact that these problems were published in the imperial compendium raises the issue of how widespread the use of the pendulum might have been among officials: although some of the pendulum's characteristics can be deduced from reading the problems quoted above, the work contained no precise instructions for making and using this instrument. Measuring the flow of a river could be useful to officials for purposes such as planning water conservancy works and controlling the repartition of water for irrigation; but I have found no evidence that any of them used a pendulum for that purpose.[67] The use of cannon or guns for measuring a distance is evocative of a military situation. In times of peace such as those experienced in most parts of the empire during the last four decades of the Kangxi reign, one wonders how many people, apart from the emperor, could or would order the firing of a cannon in order to assess a distance. Despite the emperor's proclaimed intention that the *Essence of numbers and their principles* should enable his officials to emulate him in using mathematics for the purpose of statecraft, when it came to the

---

65 In another copy of the *Explanation of the proportional compass*, the illustration shows neither the wooden stick nor the dimensions given in the problem (BML Ms. 34, n.p.).
66 Jami 2002b, 29, 44.
67 Incense burning, a portable time-keeping device, may well have been used for that purpose; Bedini 1994, 176–180, discusses its use for timing access to water in Japan.

latest Western techniques for time measurement, the compendium provided a display of his skills rather than the means to acquire them.

Several important features of the *Essence of numbers and their principles* emerge through the various problems discussed above. The location of problems, including the multiple occurrences of some of them, serves to bring out the classification of methods, and the links between them. While relying on the three 'great patterns' (*dagang* 大綱) provided by Euclidean geometry, the structure of the compendium entails a progression towards more general methods. Although mathematics involves the construction and use of some instruments, calculations are done mostly in writing; this turns imperial mathematics into a scholarly practice. Materials were chosen for inclusion irrespective of their origins. That these origins were disregarded and perhaps even obscured is evident from the rephrasing of problems and the changing of numerical data, as well as from the absence of reference to the sources used. Indeed the practice of quotation in scholarship did not entail the mention of the full details of the source when readers were expected to know it. But this would not apply to the Jesuits' lecture notes, which had been reserved for imperial use, nor to Chinese mathematical texts, as readers were not assumed to be familiar with them. The historicity of mathematics, it could be said, is circumscribed to the opening paragraphs of parts I and II. The implication is that history bears little relevance to its practice. Chinese and Western, ancient and modern were put on a par in imperial mathematics, thus establishing its universality.

CHAPTER 16

A new mathematical classic?

The *Essence of numbers and their principles* (*Shuli jingyun* 數理精蘊) was completed in 1721.[1] As decided back in 1713, it was to join the *Correct interpretation of [standard] pitchpipes* (*Lülü zhengyi* 律呂正義)[2] and a treatise on astronomy in a three-fold compendium entitled *Origins of pitchpipes and the calendar* (*Lüli yuanyuan* 律曆淵源).[3] In 1722, a draft of the treatise on astronomy, entitled *Thorough investigation of astronomical phenomena* (*Lixiang kaocheng* 曆象考成) was in turn completed.[4] However, Kangxi passed away in December of that year, before he could write the preface that was to assert the unity of the three treatises and to mark the conclusion of the project for which the Office of Mathematics had been founded. Instead his fourth son, Prince Yinzhen 胤禛, who succeeded him as the Yongzheng emperor, wrote the preface, which bears a symbolically important date: the first day of the tenth month of the first year of his reign (29 October 1723).[5] Each year, on that day, the following year's calendar was officially promulgated; this was the first time that the new emperor presided over such a ceremony. Starting from this preface, this final chapter gives an overview of the content of the treatises on harmonics and astronomy. The team that contributed to the imperial compendium is discussed next. Finally, the posterity of the threefold compendium is outlined, including the role played by the *Essence of numbers and their principles* in the development of mathematics during the mid-Qing period.

16.1 Yongzheng's preface

The new emperor did not share his father's interest in the mathematical sciences. Nonetheless, he dutifully paid tribute to the late monarch's enterprise:

Investigating antiquity, [one finds that] Yao 堯 employed Xi 羲 and He 和 [to make the calendar], Shun 舜 charged Kui 夔 [to be Director of music], and the

---

1 Zhao 1977, 1669.
2 Completed in December 1714; *SL* 261: 6b.
3 See pp. 272–273.
4 Zhao 1977, 1669.
5 *Da Qing lichao shilu*, Shizong 世宗 12: 2a.

Zhou enquired from Shang Gao 商高 [about numbers].⁶ And among the Histories, there is not one that does not have records about the [standard] pitchpipes and the calendar, and numbers and measures. These are used so that revering Heaven, teaching the people, regulating the spirits, harmonising the humans prevail in the country and circulate in the villages. Our Late Father, the Sagely Ancestor and Humane Emperor (*Shengzu ren huangdi* 聖祖仁皇帝),⁷ had innate knowledge, love for study, heavenly talent and multiple skills. In the leisure left to him by the myriad affairs [of the state], he applied himself to [the study of the standard] pitchpipes, the calendar and calculation methods for several decades, broadly examined their complexities, searched into their intermingled subtleties, and tied them together by one single thread. He instructed the Imperial Prince Zhuang 莊 [Yunlu 允祿] and others to lead literary officials at the School for tutoring the young in the Great Interior [Danei 大內] in the composition. Every day they presented [something that] He personally edited.⁸

The last sentence gives us a glimpse of how the staff of the Office of Mathematics must have worked: like the Jesuits had done in the 1690s, this staff submitted writings to the emperor, who edited them. This time, it was no longer imperial tutors, but disciples of Kangxi, who wrote out what was to become the standard text for all scholars and officials. The words 'imperially composed' in front of the title of the compendium and of its three parts were not merely honorific: while the emperor did not personally produce the works, he made sure they conformed to what he thought ought to be written—this was the case for most of the texts that circulated under his name.

Instead of Yunzhi 允祉, as Kangxi's Third Son was henceforth called, Yongzheng singles out Yunlu, the Sixteenth Son, as the supervisor of the editorial work. This reflects the antagonism linked to the succession to their father, which caused Yunzhi to suffer a number of vexations from the new emperor and to end his life in confinement.⁹ On the other hand, the only precedents Yongzheng mentions for his father's interest in astronomy, harmonics and mathematics are three sovereigns of Chinese antiquity. In fact there was no precedent in more recent times that he could have quoted: Kangxi was unique in Chinese history (and possibly in world history) in his study of the mathematical sciences, and in his enterprise to create an imperial body of knowledge from which officials and scholars were to study these disciplines in turn. Yongzheng mentions the three illustrious precedents in chronological order; this also corresponds to the order in which the treatises are listed:

---

6 These are reference to the Canon of Yao (*Yaodian* 堯典) in the *Book of documents* (*Shangshu* 尚書) and to the *Gnomon of Zhou* (*Zhoubi* 周髀); see Legge 1960, vol. 3: 18, 47–48; Cullen 1996, 174.
7 Kangxi's posthumous name.
8 Bo 1993, 7: 463.
9 Hummel 1943, 922–923.

The completed writings comprise a hundred chapters (*juan* 卷) and are called the *Origins of pitchpipes and the calendar*. They are ordered into three sections in all. One, called *Thorough examination of celestial phenomena*, consists of two parts. Part I is called 'Considering the heavens and scrutinizing the cycles' (*Kuitian chaji* 揆天察紀); it discusses the fundamental phenomena so as to clarify the principles. Part II is called 'Clarifying time and rectifying degrees' (*Mingshi zhengdu* 明時正度); it details the applied procedures and lays out the established tables so as to bring out the methods. One, called the *Correct interpretation of [standard] pitchpipes* (*Lülü zhengyi* 律呂正義), consists of three parts. Part I is called 'Correcting the pitchpipes and distinguishing the sounds'(*Zhenglü shenyin* 正律審音); it fixes the *chi* 尺, checks measurements and seeks the [standard] pitchpipes. Part II is called 'Harmonising notes and tuning instruments' (*Hesheng dingyue* 和聲定樂); It relies on the [standard] pitchpipes to make instruments and their eight timbres. The *Added part* (*xubian* 續編) is called 'Coordinating uniformly in order to compose music' (*Xiejun duqu* 協均度曲); it exhausts the origins of the harmony and resonance of the five tones and their two transformations. One, called *Essence of numbers and their principles*, consists of two parts. The first part is called 'Establishing the structure to clarify the substance' (*Ligang mingti* 立綱明體); it explains the Gnomon of Zhou, investigates the River [Diagram] and Luo [Inscription], explicates geometry and clarifies ratios. The second part is called 'Dividing items to convey their use' (*Fentiao zhiyong* 分條致用); it draws together the nine chapters using line, area and solid; it culminates with the borrowing of grades, the division of the circle and the seeking of solids; this is transformed with the methods of proportional compass and of proportional numbers, and the borrowed root and powers; the numbers in the tables are complete.[10]

The three treatises have similar structures: each of them consists of two main parts (*shangbian* 上編 and *xiabian* 下編, literally 'upper part' and 'lower part'), to which a third part is appended. In all cases, the first part gives some fundamentals of the subject, which are applied in the second part. In the works on mathematics and astronomy, the third part consists of tables (in respectively eight and sixteen chapters); in the works on harmonics, it consists of a separate treatise. Yongzheng goes on to discuss the origin of the knowledge found in the *Origins of pitchpipes and the calendar*:

The reputation of our vast country radiates far and long as the empire is united in Great Harmony. The countries of Europe in the Far West, which strive to cultivate their ancestors' legacy, each contributed their skills to the Palace Gate. Works and charts were brilliantly completed. This is why My Late Father made an all-embracing arbitration. Most of the harvest of the old methods had long been lost, so they were not detailed; and the mistakes and difficult words in the Western methods made them imprecise. [He] ordered all this methodically, clarifying it from beginning to end. It is accurate and appropriate, detailed and complete, although among famous specialists, none had glimpsed at one ten-thousandth of it. How could one but trust the saying that 'Only a Sage could do

---

10   Bo 1993, 7: 463–464.

this'? For numbers and principles accommodate each other and do not come apart. If one obtains numbers, then the principles are nowhere but there. These books therefore open with the *Change* and the *[Great] Plan*. They rule strings and woodwind instruments by means of lines and solids. They seek longitude and latitude by means of spherical triangles. Things of this kind are all derived from the accuracy of numerical methods and from the essentials of pitchpipes and the calendar. Therefore these three books are one another's exterior and interior altogether. Arranging the Seven Governors (*qizheng* 七政)[11] and the ordering the Five Sounds (*wuyin* 五音)[12] must go through the meaning of the Nine Chapters (*jiuzhang* 九章).[13] That which is tested by them does not mislead; [that which] uses them has efficiency.[14]

Beside comparing his father to a Sage, Yongzheng also asserts the links between harmonics and astronomy (the traditional *lüli* 律曆 category) on the one hand, and mathematics (numbers and principles being regarded as indissociable) on the other hand. His main claim here is that the *Origins of pitchpipes and the calendar* is the result of Kangxi's ordering of all knowledge, ancient and Western. The latter is described as a tribute from the Far West to the Qing empire, which suffices to account for its appropriation. The preface concludes in a similarly conventional tone:

Reverently pondering on the profundity of [Our Father's] sagely learning, how could it be easy to probe into it and to look up and gaze upon it? We have long upheld Our Father's instructions in this matter. To this book's great instructions, to its subtle meaning, to the commands it contains, We have been eagerly attentive for many years. In honour of what We have heard, We respectfully present a few words in support. Now this book surely must be the substance of the work of Our Late Father's own hand. It reveals antiquity and sets the standard for the present. In the way it assembles its great completeness, it far excels preceding ages and it bequeaths unchanging methods for a thousand myriad ages to come. In the future, those who wish to make the seasons consistent, fix proper days, bring together the standard pitchpipes and the standard units for measures and weights may seek for all that in this book. So they can set them up before Heaven and Earth without finding anything contrary to them, and they can await the rise of a sage [for a hundred ages] without misgivings.[15]

The last sentence is taken from the *Doctrine of the mean* (*Zhongyong* 中庸), one of the *Four Books*, to emphasise that, thanks to Kangxi's compilation of the *Origins of pitchpipes and the calendar*, his successors

---

11 The Sun, Moon and Five Planets.
12 The Five Notes of the pentatonic scale (*gong* 宮, *shang* 商, *jue* 角, *zhi* 徵, *yu* 羽) differ from the twelve-note system, in which the Yellow Bell (*huangzhong* 黃鐘) is the fundamental note; Needham 1954–, 4–1 160–163.
13 In an attempt at literary elegance, the three fields discussed in the compendium are referred to by expressions that contain an odd number.
14 Bo 1993, 7: 461–466.
15 Bo 1993, 7: 466. On the final sentence, see Legge 1960, 1: 426.

have the means to emulate kingly virtue as defined by Confucius himself, and thereby to ensure harmony between the human realm and the cosmos. Yongzheng also claims that the compendium supersedes all earlier enterprises, and sets definitive standards, using the received invocation of antiquity to assert the superiority of the present over less remote and eminent predecessors.[16]

16.2 Harmonics and astronomy

As indicated in the preface, the three works that form the *Origins of pitchpipes and the calendar* have parallel structures. However, the treatise on harmonics is much shorter than the two others. Its first part discusses the dimensions of the pitchpipe producing the Yellow Bell, the fundamental note of the twelve-note scale. We know that Kangxi took this matter to heart: it was directly related to that of standard length units, as the Yellow Bell pitchpipe was to be one *chi* long.[17] This definition was a prerequisite for mathematics:

Once the Yellow Bell pitchpipe's length and diameter are known, then one has measure. Only after one has measure are numbers established.[18]

Thus the order in which the compendium was compiled, namely: harmonics, then mathematics, then astronomy, corresponded to that of the constructions of disciplines : mathematics was to rest on standard units, which in turn were defined according to the length of the standard pitchpipe for the fundamental tone as it was deemed to have been defined in antiquity. The other tones were then defined by the method of augmentation and reduction (*sanfen sunyi* 三分損益, lit. three parts decreased and increased), in which the length of each pitchpipe is obtained alternatively by adding and subtracting one third of the length of the previous one:[19] obviously the four basic operations of arithmetic and ratios are used extensively for that purpose. The main reference here is Cai Yuanding's *New writing on the [standard] pitchpipes* (*Lülü xinshu* 律呂新書), the standard text on harmonics of the Cheng-Zhu philosophical school that Kangxi used to discuss with his officials.[20] However, some views of this school were tailored to suit the imperial agenda in producing a three-fold compendium. Thus, in a work that formed part of the main corpus of that school, Cai Shen 蔡沈 (1167–1230), Cai Yuanding's son, asserted: 'What has no shape is principle. What has shape is

---

16 Mei Wending used a similar argument; see p. 94.
17 See p. 246; Jami 2007b.
18 *Yuzhi lülü zhengyi* 御製律呂正義, SKQS 215: 8.
19 Needham 1954–, 4-1: 171–176.
20 See p. 230.

thing.'²¹ In the second paragraph of the treatise on harmonics, entitled 'Principles and numbers of the Yellow Bell' (*Huangzhong lishu* 黃鐘理數), this became: 'What has no shape is principle. What has shape is number. That is to say, without principles numbers have no way to exist.'²² The title of the passage and the rest of its content bring out principles and numbers as indissociable foundations of musical tones. Interestingly, this passage is included in a collection of essays by officials, where it is attributed to Yunzhi.²³ The second part of the work, which discusses fourteen different types of instruments, also opens with a paragraph attributed to Yunzhi in the same collection; it discusses the 'Eight timbres of musical instruments' (*Bayin yueqi* 八音樂器).²⁴ While this does not suffice to ascertain that the prince personally authored those particular passages, it does reflect the fact that he was as much involved in the treatise on harmonics as with the two others.

The third part of the *Correct interpretation of [standard] pitchpipes*, which formed an *Added part* (*xubian* 續編) consisting of a single chapter, is by far the most famous one.²⁵ It is a completely independent treatise, derived from the manuscript work authored by Tomás Pereira, most likely in the 1680s, entitled *Essentials of [standard] pitchpipes* (*Lülü zuanyao* 律呂纂要).²⁶ Among its sources were Gioseffo Zarlino's *Istitutione harmoniche* (1558) and Kircher's *Musurgia universalis* (1650).²⁷ Like Antoine Thomas' mathematical textbooks, however, Pereira's treatise was composed to meet the emperor's demand; it is not a translation of a specific European work. Teodorico Pedrini, the Italian Lazarist priest sent by the Pope, who succeeded Pereira as the emperor's music teacher from 1711 onwards, seems to have contributed to the version published as part of the imperial compendium, which

---

21 無形者理也。有形者物也。 *Inner chapters on the Great Plan's Supreme Rule* (*Hongfan huangji neipian* 洪範皇極內篇), SKQS 805: 705; the work is reproduced in the *Complete collection on nature and principle* (*Xingli daquan* 性理大全, 1415); SQKS 710: 521.
22 The first two sentences read: 無形者理也。有形者數也。 *Yuzhi lülü zhengyi*, SKQS 215: 5.
23 *Literary talents of the Qing dynasty* (*Huangqing wenying* 皇清文穎), SKQS 1449: 481–482; interestingly, this collection also includes six essays by different authors, all entitled *Discussion of the Yellow Bell as the foundation of all affairs* (*Huangzhong wei wanshi genben lun* 黃鐘為萬事根本論; SKQS 1449: 546–547, 549–556). This is the title of the first paragraph of the *Correct interpretation of [standard] pitchpipes*, which immediately precedes the one authored by Yunzhi; none of the six authors is known to have taken part in the compilation of the treatise on harmonics. In this treatise as in the one on mathematics, the opening matter expounds received ideas about the history and function of music.
24 These were defined according to the materials used to make the instruments: metal, stone, fired earth, skin, silk strings, wood, calabash and bamboo. SKQS 1449: 581–583.
25 For a detailed study, see Gild-Bohne 1991.
26 Wang 2002, 2003.
27 Verbiest had taken no less than twelve copies of the latter work with him when he sailed from Lisbon in 1657; Golvers 1993, 136, n. 8; Gild-Bohne 1991, 116–118. Both works are in the catalogue of the Beitang Library: Verhaeren 1949, n. 1921–23, 3542.

presents the tonal system, rhythmic and staff notation.[28] There is no evidence that it had any influence on Chinese music, ritual or otherwise.

The treatise on astronomy was completed last; it was intended as a substitute for the *Books on calendrical astronomy according to the new method* (*Xinfa lishu* 新法曆書) compiled by the Jesuits under the last Ming emperor, of which it is essentially an abridgment.[29] The first part of the treatise opens with a chapter that is a 'General discussion of astronomical principles' (*Lili zonglun* 曆理總論), which briefly introduces celestial phenomena, the Earth, the epoch (chosen as the first year of the sexagesimal cycle (*jiazi* 甲子) which had occurred during the Kangxi reign, i.e. 1685), the ecliptic, coordinates and annual difference (*suicha* 歲差, i.e. precession). After two chapters on spherical trigonometry, the astronomical principles (*lili* 曆理) of the Sun, the Moon, eclipses, the Five Planets and the fixed stars are discussed in a technical way.[30] These items are considered in the same order in the second part of the treatise, in which astronomical methods (*lifa* 曆法) related to them are presented. This structure echoes that of the *Essence of numbers and their principles*, where principles are associated with the 'why it is so' (*suoyiran* 所以然), and numbers are associated with methods.[31]

There are a few changes in astronomical constants and methods compared with the late Ming compendium. For example, the obliquity of the ecliptic is changed from 23°31′30″ to 23°29′30″; the latter value is derived from the observations done by the staff of the Office of Mathematics.[32] But, on the whole, it is the treatise's systematic organisation of knowledge that forms the main point of contrast with its predecessor, which consisted of individual books put together rather than of a single unified work produced according to a plan. The *Thorough investigation of astronomical phenomena* appears as an adequate response to the emperor's commission of an 'explanation of the principles of astronomy', 'by means of which a man who knows geometry and arithmetic could make astronomical tables'.[33] This commission, originally given to the Jesuits, was apparently fulfilled by the Office of Mathematics; to my knowledge there is no evidence as to the Jesuits' actual contribution.

---

28 Standaert 2001, 854; Lindorff 2004; Allsop & Lindorff 2007.
29 Bo 1993, 7: 460–461.
30 Bo 1993, 7: 469–470.
31 See p. 328.
32 See p. 294. Bo 1993, 7: 460–461.
33 See p. 286.

Given the few changes the *Thorough investigation of astronomical phenomena* contained compared with its late Ming predecessor, the formal adoption of the former as the basis for calculating the calendar at the beginning of the Yongzheng reign can hardly be characterised as an astronomical reform; the term does not appear anywhere in official sources. Rather, the *Thorough investigation* is presented in terms similar to its mathematical counterpart, as the outcome of Kangxi's work to make astronomy easier to learn.[34] In this respect, it seems that Yunzhi's ambition was not fulfilled, and that the emperor followed the Jesuits' advice to postpone any radical change in calendar making.[35]

Originally, the astronomical treatise was named *Book on calendrical astronomy in reverent accord* (*Qinruo lishu* 欽若曆書). This title refers to two passages of the *Book of documents* alluded to by Yongzheng in his preface, in which the mythical sovereign Yao commissions the calendar. In one of them, *qinruo* 欽若 occurs just after *shixian* 時憲 (timely modelling), the phrase that gave its name to the Qing calendar.[36] The second one famously recounts what was assumed to be the earliest commission of the calendar by a sovereign in history:

Then he [Yao] ordered Xi and He to accord reverently (*qin ruo* 欽若) with the august Heaven, and its successive phenomena (*li xiang* 歷象), with the sun, the moon and the stellar markers, and thus respectfully to bestow the seasons on the people (*jing shou ren shi* 敬授人時).[37]

This passage is quoted at the opening of the first chapter of the first part of the *Thorough investigation of astronomical phenomena*.[38] Kangxi had written a commentary on it, probably towards the end of his reign. In it he argued that observations were required in order to ensure the accuracy of the calendar, and that dividing the circle into 360 degrees (*du* 度) rather than 365.25 *du* to measure the heavens made calculations easier and did not go against the way (*dao* 道) of the calendar.[39] This would suggest that, half a century after supporting the Jesuits against Yang Guangxian on the strength of evidence provided by observation, the emperor continued to believe in observation as a means of validating astronomical knowledge, as the staff of the Office of Mathematics was now doing under Yinzhi's supervision.[40] Kangxi was by no means the first to single out this quotation in relation to calendar making. Indeed,

---

34 Zhao 1977, 1669.
35 See pp. 292–293.
36 Legge 1960, 3: 255.
37 Cullen 1996: 3; 乃命羲和欽若昊天歷象日月星辰敬授人時; comp. Legge 1960, 3: 18.
38 Bo 1993, 7: 471.
39 *Shengzu renhuangdi yuzhi wenji* 聖祖仁皇帝御製文集, SKQS 1299: 549–550; the essay is not dated. The fact that it is included in the fourth collection of essays of Kangxi suggests that it was probably written quite late (Li 2007, 2).
40 See pp. 61–62.

the name of the astronomical system used under the Yuan dynasty since 1281, the Season granting system (*Shoushi li* 授時曆) was borrowed from it.[41] In reference to the precedent set under Mongol rule, the name first chosen for the treatise that was to contain the new technical foundations of imperial astronomy was borrowed from the same sentence. It also included the phrase 'book on calendrical astronomy' (*lishu* 曆書), like the set of works hitherto used at the Astronomical Bureau, which it was intended to replace.

It is possible that the title of the work was changed after Yongzheng's accession: in a memorial in which 'He Guozong, Mei Juecheng, Wang Lansheng and others' asked Kangxi to write a preface for the *Origins of pitchpipes and the calendar*, they referred to the astronomical treatise, which was by then completed but not yet printed, as the *Books on calendrical astronomy in reverend accord*.[42] The treatise was first printed under this title; some copies of it are still extant, in which every single occurrence of the title (including those on the fold of each page) has been blocked out in black ink.[43] The new title can be regarded as more neutral in two respects. First, the phrase 'Books on calendrical astronomy' from the final title of the treatise could well have been omitted in order to avoid presenting the work as a basis for calendar reform. Secondly, unlike 'reverend accord' (and 'season granting'), the phrase 'astronomical phenomena' (*lixiang* 曆象) does not refer to imperial action, but rather directly to the heavens. The decision to change the title of the treatise may also have been linked to Yongzheng's wish to minimise its status and thereby that of his brother Yunzhi, whose contribution to it was known to all at court. The astronomical part of the *Origins of pitchpipes and the calendar* appears to have been the most controversial one. This is mainly revealed by Jesuit accounts, but, despite imperial arbitration and the fact that information on dissenting viewpoints might not have been handed down, Chinese sources also retain some traces of it—despite being blacked-out. Be that as it may, the *Thorough investigation of astronomical phenomena*, in updating the system without any claim to reform, opened the way to further similar updates.

---

41 On this astronomical system see Sivin 2009.
42 Zhongguo diyi lishi dang'an guan 1985, 1042; the document is not dated; however it mentions that the work of compilation has been going on for nine years, which allows one to date it to 1722.
43 Gugong bowuyuan tushuguan & Liaoning sheng tushuguan 1995, 331; I have consulted a copy of the main text at the library of Tsinghua University, Beijing, and a copy of the tables at the Library of Institute for Research in Humanities, Kyoto University. A quick overview of these has revealed no difference between their content and that of the work printed under the title *Thorough investigation of astronomical phenomena*.

16.3 The contributors

Yongzheng's preface is followed by a list, dated to 7 July 1724,[44] of forty-seven contributors to the imperial compendium; as was usual with imperial editorial projects, the rank of each of them as of that date is mentioned.[45] They have served in various capacities: there are two Editors by Edict (*zhi zuanxiu* 旨纂修), two Compilers (*huibian* 彙編), three Proofreaders (*fenjiao* 分校), ten Observers (*kaoce* 考測), fifteen Calculators (*jiaosuan* 校算), and fifteen Copyeditors (*jiaolu* 校錄).

Yunlu's name appears before that of Yunzhi on this list. This is in accordance with Yongzheng's account of the compilation of the *Origins of pitchpipes and the calendar*, but not with the evidence presented in previous chapters of the present work as to the princes' respective roles in the compilation. He Guozong and Mei Juecheng were the two Compilers; Wei Tingzhen, Wang Lansheng and Fang Bao were the three Proofreaders. These five scholars, all of which held the title of Metropolitan Graduates, belonged to the team who had worked on other imperial compilations under the supervision of Li Guangdi. The expertise of He and Mei in the mathematical sciences, which the three other scholars do not seem to have shared, is reflected in their role in

Table 16.1 The contributors to the *Origins of pitchpipes and the calendar*

Function	Total number	Metropolitan Graduates	Present in 1713	Bannermen	Li Guangdi's protégés	Official Astronomers and families
Editors by Edict (*zhizuanxiu* 旨纂修)	2		2	2 (imperial princes)		
Compilers (*huibian* 彙編)	2	2**	2		1	1
Proofreaders (*fenjiao* 分校)	3	3 (incl. 1**)			3	
Observers (*kaoce* 考測)	10		5*	8*		3*
Calculators (*jiaosuan* 校算)	15		1*	9*		5*
Copyeditors (*jiaolu* 校錄)	15	2 (incl. 1**)	3*	6*	1	
Total	47	7 (incl. 4**)	13*	25*	5	9*

** Metropolitan degree granted by the emperor (*qinci* 欽賜).

* Lower estimate; we do not have a complete list of those who were working on the project since 1713; see p. 268. Evidence as to the origin of some contributors has been found; some others' names, like that of the Calculator Sige 四格, are clearly Chinese transcriptions of Manchu or Mongol names. But for others it is not the case; moreover there were Chinese Bannermen. Therefore the number of Bannermen given is a lower estimate. There are twelve men on whom I have located no other information than that mentioned on the list.

---

44   Yongzheng 2/5/17.
45   See Illustration 16.2.

雍正二年五月十七日奉

旨開載纂修編校諸臣職名

旨纂修

承

　和碩莊親王臣允祿

　和碩誠親王臣允祉

彙編

　日講官起居注詹事府詹事兼翰林院侍講學士加一級臣何國宗

　翰林院編修臣梅瑴成

分校

　原任湖南巡撫都察院右副都御史臣魏廷珍

　翰林院編修臣王蘭生

原

考測

　進士臣方苞

會考府郎中臣成德

參

　原任吏部員外郎臣顧琮

　原任吏部員外郎加一級臣照海

　工部員外郎加一級臣阿齊圖

　兵部主事加一級臣平安

　福建汀州府知府臣何國棟

　江西袁州府知府臣李英

　翰林院筆帖式加一級臣那海

　候補筆帖式臣豐盛額

校算

　兵部郎中兼管欽天監左監副事加二級臣何國柱

　刑部員外郎臣倫大理

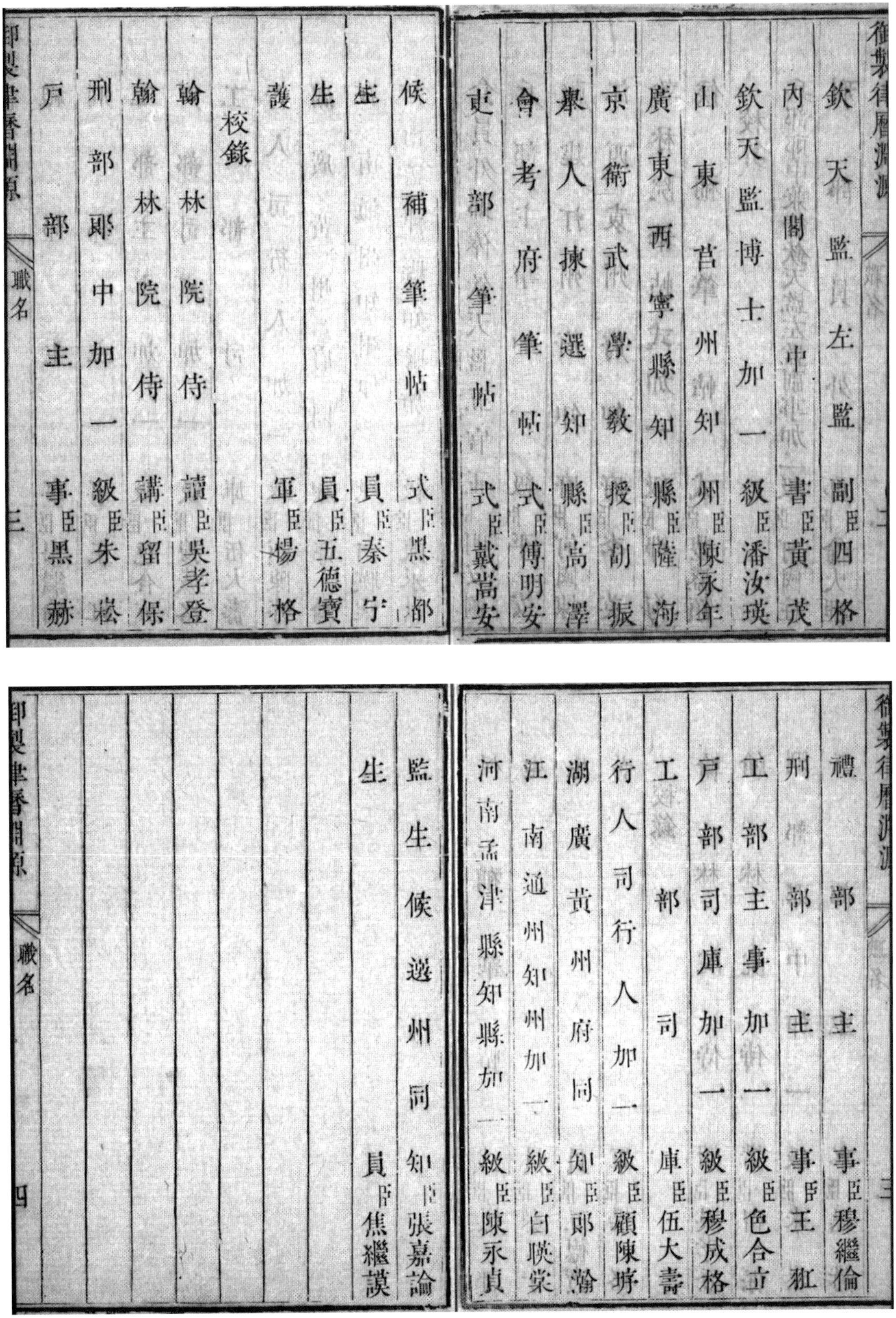

御製律曆淵源

欽天監副臣四格
欽天監員外郎臣黃茂
內閣撰文舍人供事儒士臣潘如瑛
欽天監博士加一級臣薩海
山東莒州知州臣陳永年
廣東西寧縣知縣臣郝振
京衛武學教授臣高澤
舉人揀選知縣臣傅明安
會考府筆帖式臣戴嵩
吏部筆帖式臣黑都
生員臣秦寧
候補筆帖式臣吳孝登
護軍臣楊格
校錄
生員臣伍德寶
生員臣吳孝登
翰林院侍讀臣留保
翰林院侍講臣朱崟
刑部郎中加一級臣朱崟
戶部主事臣黑赫

御製律曆淵源

禮部主事臣穆繼倫
刑部主事臣王狅
工部主事加一級臣色合立
戶部司庫加一級臣穆成格
工部司庫臣伍大壽
行人司行人加一級臣顧陳玠
行人司行人加一級臣郎瀚
湖廣黃州府同知臣白暎棠
江南通州知州加一級臣陳永貞
河南孟津縣知縣加一級臣張嘉謨
監生候選州同知臣焦繼謨
生員

Illustration 16.2 List of contributors to the *Origins of pitchpipes and the calendar* (Bodleian Library, Sinica 361/1, 1a–4a; see Bo 1993, 7: 467–469).

this particular compilation. The two other Metropolitan Graduates among the contributors were the leading Copyeditors: Wu Xiaodeng had worked on the editorial projects supervised by Li Guangdi[46] Liubao 留保, a Manchu bondservant, was granted the Metropolitan degree in 1721, at the same time as Wang Lansheng.[47] Thus of the seven Metropolitan Graduates involved in the compendium, four were granted the title by Kangxi in reward for their contribution to imperial editorial projects; only two (Wu Xiaodeng and Liubao) were Bannermen. Altogether, more than half of the contributors can be identified as Bannermen. If one examines the two groups of contributors who applied their technical expertise to the project, namely Observers and Calculators, one finds that a majority of them were Bannermen (at least 80% and 60% respectively). The more 'literary' tasks, on the other hand, were mainly fulfilled by Han Chinese. The number of Bannermen involved in this work is significantly larger than for other editorial projects of the same period. This reflects Kangxi's concern that Bannermen should be well versed in the mathematical sciences.[48]

If one compares the list of contributors to that of the staff of the Office of Mathematics as partially reconstructed above,[49] it appears that there was a degree of continuity. Five of the 'six disciples' of the emperor who were on his retinue in the summer of 1712 appear on the list. The sixth one, Chen Houyao, seems to have retired in 1719; he died in 1722.[50] At least half of the observers had started working on the project around the time of the emperor's sixtieth birthday.[51] Gu Chenxu, who had ranked first in the 1713 special examination in mathematics but had failed the second test later that year, appears among the Copyeditors; he had continued on the team, albeit in a literary rather than technical capacity.

Let us now consider more generally the division of labour among contributors. There is evidence that most if not all of the Observers were actually involved in observations during the decade that the Office operated, whether for astronomical or cartographical purposes;[52] this confirms that the categories corresponded to different fields of expertise. These do not necessarily overlap with the three disciplines covered in the *Origins of pitchpipes and the calendar*: thus

---

46 See p. 269.
47 *SL* 291: 24a–25a; *Cilin diangu* 詞林典故, *SKQS* 599: 651.
48 Jami 1994a, 237.
49 See p. 268.
50 Ruan 1982, 510.
51 These included Cengde, Gucong, Zhaohai, Minggantu and He Guodong, three of whom had been Kangxi's 'disciples'.
52 Bo 1978, Qu 1997, *SL* 261: 4b–5a; Gucong is the only one among them for whom I have found no such evidence.

Calculators must have been needed both for the astronomical and the mathematical treatises, in particular for the tables appended to both. One could also consider the likely contribution of individuals. Wang Lansheng, for example, was well versed in phonology and music; he probably played a more important role in the compilation of the *Correct interpretation of [standard] pitchpipes* than in that of the astronomical treatise.[53] On the whole, neither evidence internal to the treatises nor other materials located so far permit us to understand exactly how work was divided between the contributors.

As was usual with imperial compilations, the position that each contributor occupied when the list was drawn is mentioned on it. Several of them seem to have taken up a new position once the project was completed. Thus, Sahai 薩海, a Calculator, was appointed a District Magistrate in Xining District, Guangdong (*Guangdong Xining xian zhixian* 廣東西寧縣知縣) in 1722.[54] The following year, Bai Yingtang 白暎棠, a Copyeditor, was appointed a Department Magistrate in Tongzhou, Jiangnan (*Jiangnan Tongzhou zhizhou* 江南通州知州).[55] Most of the contributors, however, held positions in the capital, with fifteen positions in one of the Six Ministries (*liu bu* 六部). Only four contributors are listed as working for the Astronomical Bureau;[56] however, many more were involved in this Bureau's activities in the course of their careers.

The case of Chen Houyao illustrates the fact that not all of those who had been involved in the Office of Mathematics are mentioned on the list. Neither is it known to what extent expertise was sought outside the Office. Mei Wending was sent a copy of the *Correct interpretation of the [standard] pitchpipes* in 1714, as soon as the work was completed, and was asked to correct it; there is no evidence, however, as to whether he proposed corrections or merely responded by formulaic praise. In 1721, Mei Juecheng was granted permission to return home; it is possible that he took some of the writings on mathematics and astronomy with him. In any case, as his grandfather passed away a month later, at the age of eighty-eight, he would not have had the time to read them in great detail.[57] It is unknown whether any other scholar had the privilege of receiving a copy of the imperial treatises before their publication.

The Yongzheng emperor does not seem to have made direct use of the expertise of the staff of the Office of Mathematics. In a sense, this was quite natural: once the project was completed, the Office was

---

53 Ruan 1982, 719; Zhao 1977, 10272.
54 Bo 1993, 7: 468; *Wanshou shengdian chuji* 萬壽盛典初集, SKQS 563: 282.
55 Bo 1993, 7: 468; *Jiangnan tongzhi* 江南通志, SKQS 510: 207.
56 Minggantu, He Guozhu, Sige and Pan Ruying 潘汝瑛; Bo 1978, Qu 1997.
57 Fang 1983, 1: 335. Li 2006, 51–52, 54.

closed down and members of the staff continued their careers. The new emperor's lack of interest in the mathematical sciences suffices to explain that they were no longer needed at court; there is no evidence that they suffered from the enmity he felt towards his brother Yunzhi. On the contrary, He Guozong and Mei Juecheng in particular seem to have subsequently held rather high posts.[58] However, when He put forward a proposal that Mei should replace the Jesuit then in charge of the Astronomical Bureau, the emperor rejected it.[59] In 1725, Ignatius Kögler (1680–1746), who has succeeded Stumpf in 1720 as Administrator of the Calendar, was given the title of Director of the Bureau (*jianzheng* 監正).[60] In this respect Yongzheng followed the example of his father, whose attitude he characterised as follows:

The Sagely Ancestor and Humane Emperor had great leniency. There was nothing in the world that he did not tolerate. The lamas and Westerners as well as monks, Daoists and the like that he supported were many.

The Sagely Ancestor and Humane Emperor only regarded them as craftsmen of various skills.[61]

Yongzheng, while continuing to make use of the Jesuits' 'craftsmanship', did not entirely approve of his father's tolerance, which, in his view, could lead to disorder: in 1724, Christianity was prohibited as a heterodox sect.[62]

16.4 Supplements to the *Origins of pitchpipes and the calendar*

Despite the closing down of the Office of Mathematics and the limitation of the Jesuits' activities to Beijing, both groups patronised by Kangxi remained active as experts during the two following reigns. In particular some of them took part in the compilation of sequels to two of the three treatises that made up the *Origins of pitchpipes and the calendar*.

Within a few years of its publication, the *Thorough investigation of astronomical phenomena* was criticised by imperial astronomers. In 1730, after a prediction for a solar eclipse turned out to be inaccurate, Mingtu, the Manchu Director of Astronomical Bureau, memorialised to ask that the work should be revised.[63] Stemming from Kögler and

---

58 Hummel 1943, 285 & 569.
59 Rodrigues 1990, 33.
60 Standaert 2001, 721; on Kögler, see Sun 2006.
61 *Shizong xianhuangdi shangyu neige* 世宗憲皇帝上諭內閣, *SKQS* 415: 166; dated to 9 January 1729. I am grateful to Prof. Nicolas Standaert for drawing my attention to this edict.
62 Standaert 2001, 499.
63 Bo 1993, 7: 965.

Andre Pereira (1689–1743), in their respective capacities of Director and Vice-Director of the Bureau, this memorial was a counter attack after He Guozong's attempt to displace them. The Jesuits proposed to substitute the tables of the *Thorough investigation of astronomical phenomena* with Newtonian tables that they had calculated themselves, which they presented to the throne in 1731.[64] Shortly after ascending the throne, the Qianlong emperor commissioned a *Later part of the thorough investigation of astronomical phenomena* (*Lixiang kaocheng houbian* 曆象考成後編), in response to a proposal made by Gucong.[65] The work was intended to incorporate these new tables into the treatise that was compiled under his grandfather. Astronomical methods were thus changed without formally dismissing the work done under Kangxi, which would have been a breach of filial piety. Completed in 1742, the *Later part* was structured like its predecessor, separating principles from methods. It deals with solar and lunar motion and eclipses. Chapters 1 to 3 discuss 'numbers and principles' (*shuli* 數理); chapters 4 to 6 deal with 'pacing methods' (*bufa* 步法, i.e. calculation of the position of celestial bodies in their orbits); the tables form the last four chapters. Just like its predecessor, the *Later part* provided the basis on which the new tables appended to it had been calculated.

The work paralleled its predecessor not only as regards its structure but also as regards its contributors. It was supervised by two imperial princes: Yunlu and Hongzhou 弘晝 (1712–1770, the fifth son of Yongzheng); the latter was then in charge of the Imperial Press. There were eight compilers altogether. The first four, Gucong, Zhang Zhao 張照 (1691–1745), He Guozong and Mei Juecheng, were not affiliated to the Astronomical Bureau. The four others, namely Jin'ai 進愛,[66] Kögler, Pereira and Minggantu, all belong to it. This list reflects the complexity of the situation: while the *Later part* can be said to have resulted from a Jesuit initiative against some of the staff of the former Office of Mathematics, both groups appear to have produced it together. At the time, the four former 'disciples' of Kangxi were not all in the same position. Minggantu, the only one among them who remained at the Astronomical Bureau throughout his career, had worked on the tables with the two Jesuits since the early 1730s. On the other hand, Gucong's memorial recommending He and Mei as editors for the project reflected the eagerness of the former Office staff to be involved in the official sanctioning of astronomical methods that were already in use.[67] That is not to say, of course, that He and Mei did not effectively contribute to the work: the process of compilation of the *Later part* remains to be investigated. In one respect the contrast with the

---

64 Shi & Xing 2006; see also Hashimoto 1971, Han 2001b.
65 Bo 1993, 7: 967.
66 The Manchu Director of the Bureau in 1739–1743 and 1744–1745.
67 Bo 1993, 7: 967.

previous project was sharp: this time most of the contributors belonged to the Astronomical Bureau.⁶⁸

Two years after the completion of the *Later part*, another project was set in motion, the outcomes of which were a star map and a star catalogue, with a description of the instruments and methods used to observe them and determine their position. It was published in 1752 under the title *Thorough investigation of instruments and phenomena imperially commissioned* (*Yuding yixiang kaocheng* 御定儀象考成). Here again Yunlu, Zhang Zhao and He Guozong endorsed the work carried out by the staff of the Astronomical Bureau. Mei Juecheng's name does not appear among the contributors.⁶⁹

Both He and Mei were involved in updating another imperial publication of the Kangxi period that dealt with a related topic, namely divination. Like the astronomical treatise, the *Investigation of the origins of planetary ephemerides* (*Xingli kaoyuan* 星曆考原, 1713), compiled under the supervision of Li Guangdi, was deemed to be in need of revision.⁷⁰ A much longer work, the *Book of harmonising the times and distinguishing the directions imperially commissioned* (*Qinding xieji bianfang shu* 欽定協紀辨方書), was commissioned for this purpose.⁷¹ Completed a few months before the *Later part*, it was supervised by Yunlu and Hongzhou, and edited by four Director-generals (*zongcai* 總裁): Zhang Zhao, Mei Juecheng, He Guozong and Jin'ai. Again, apart from the staff of the Imperial Press, all collaborators belonged to the Astronomical Bureau; but this time neither Minggantu nor the two Jesuit astronomers were involved. The creation of an Office for the revision of the Books on Calculation of the Timely Modelling [System] (*Zengxiu shixian suanshu guan* 增修時憲算書館), supervised by Yunlu, seems to have been decided in relation to this compilation as well as for that of the *Later part*.⁷² Qianlong's interests and motivations may have differed from those of his grandfather, but the Office of Mathematics created by Kangxi provided him with an institutional model and a group of experts for the compilation of more imperial works related to the mathematical sciences.

Officials of the same group were also involved in producing a *Later part of the Correct interpretation of the [standard] pitchpipes* (*Lülü zhengyi houbian* 律呂正義後編, 1746): the treatise on harmonics compiled as part of the *Origins of pitchpipes and the calendar* was deemed to be in need of supplementing rather than revision. In 1729, the Music Office

---

68 Shi 2008b provides a detailed account of the process that led to the compilation of the *Later part of the thorough investigation of astronomical phenomena* and an analysis of its main features.
69 Bo 1993, 7: 1360–1361.
70 *Yuding xingli kaoyuan* 御定星曆考原, SKQS 811: 110.
71 On this work and its role in Qing official cosmology, see Smith 1991; 51–91.
72 SKQS 811: 111.

(*Jiaofang si* 教坊司), an agency of the Ministry of Rites, had been transformed into an independent Ministry of Music (*Yuebu* 樂部).⁷³ An Office for the Correct interpretation of the [standard] pitchpipes (*Lülü zhengyi guan* 律呂正義館) was created under that Ministry in 1741, with Yunlu and Zhang Zhao in charge. The *Later part* that it produced included 120 chapters, and was thus considerably longer than the original work.⁷⁴ Its scope was also much broader; it aimed to codify and unify the music and dances to be performed during official rituals throughout the empire.⁷⁵ Obviously, collaboration on the part of Jesuits was no more necessary or even relevant for this than for divination: in harmonics, Western learning seems to have had no posterity at all. Zhang Zhao and He Guozong were among the compilers; but three more prominent officials took precedence over them in the list of contributors; these included Zhang Tingyu 張廷玉 (1672–1755, the son of Zhang Ying), under whose supervision the *Ming History* project undertaken in 1679 had finally been completed in 1739.⁷⁶

During the Qianlong reign, He Guozong's official career was once more shaped by his expertise in the mathematical sciences. In 1745 he was appointed 'simultaneously Director in charge of the general affairs' (*jianguan jianzheng shiwu* 兼管監正事物) of the Astronomical Bureau on top of his main position. The function, for which there was, as usual, a Manchu parallel, appears to have been created for him: the post was never filled again after he retired from it in 1757, when he became a Minister of Rites. His Manchu counterpart, on the other hand, remained in post, and had successors to the end of the dynasty.⁷⁷ It is not entirely clear what He Guozong's role at the Astronomical Bureau was during the twelve years that he held the post: there continued to be both a Manchu and a Jesuit Director. Another title he cumulated with his principal post was that of Grand Minister of the Ministry of Music (*Yuebu dachen* 樂部大臣), which was also granted to Zhang Zhao.⁷⁸ Thus rather than limiting factors in a career, a high official's expertise in astronomy and music could bring about recognition in the form of titles that reflected the contribution he made in those fields. This does not seem to have been the case for Mei Juecheng and Gucong; neither of them took part in all of the four projects discussed above.

Qianlong took up the pattern set up by Kangxi for these editorial projects. The contents of two of the four works thus produced in the mid-eighteenth century do not pertain to what counts as mathematical

---

73 Hucker 1985, no. 8269.
74 A further supplement of eight chapters was added in 1789; Hummel 1943, 24.
75 Standaert 2005, 134–138.
76 *Yuzhi lülü zhengyi houbian* 御製律呂正義後編, *SKQS* 215: 229–231; on Zhang Tingyu see Hummel 1943, 54–56.
77 Bo 1978, 90–101.
78 *SKQS* 215: 229–230; Bo 1993, 7: 1360.

or scientific knowledge in the twenty-first century, but might rather be regarded as pertaining to the realm of religion (divination and ritual). This highlights the fact that fields of learning were then configured very differently from the way that they are now. In fact, the close connection between mathematics, astronomy and harmonics on the one hand, and divination and ritual on the other hand is one of the reasons that made it possible for Kangxi to commission a work that focused on the former subjects in a technical manner.

16.5 The study of mathematics in mid-Qing China

Of the three parts of the *Origins of pitchpipes and the calendar*, only the *Essence of numbers and their principles* was handed down without further addition. This might mean either that the work was uncontroversial and authoritative in mathematics, or that Kangxi's successors did not regard this field as worthy of their patronage. None of them is known to have spent time learning mathematics. But this does not imply that the training of officials in the mathematical sciences emphasised by their 'Sagely Ancestor' was entirely neglected.

Beside the Astronomical Bureau, the Banner Schools were the main institutions for which we have evidence that mathematics was part of the curriculum. In 1734 Yunli 允禮 (1697–1738), Kangxi's seventeenth son, one of the few whom Yongzheng highly favoured, but who is not known to have been versed in mathematics, memorialised to ask that sixteen teachers be appointed to teach this discipline to Bannermen, and that at least thirty bright students be made to work on the subject for two hours every day. Four years later, however, this teaching was discontinued on the grounds that it was too difficult for youngsters who mainly applied themselves to reading and translation.[79] But in 1739 the teaching of mathematics was transferred to the Imperial University (*Guozijian* 國子監), and thus became available to Chinese as well to as Bannermen. It was supervised by Cengde, then a Vice-Minister of the Court of the Imperial Stud (*Taipu si shaoqing* 太僕寺少卿); one of the 'six disciples' of Kangxi in mathematics in 1712, he had worked at the Office of Mathematics and had carried out some astronomical observations during the compilation of the *Origins of pitchpipes and the calendar*.[80] He Guodong, one of He Guozong's brothers and also a contributor to that work, was Instructor in mathematics at the Imperial University (*Guozijian suanxue zhujiao* 國子監算學助教) in the 1740s.[81] The syllabus was the *Essence of numbers and their principles*, with one year devoted to each of the three sections of the work on lines,

---

79 *Qinding baqi tongzhi* 欽定八旗通志, *SKQS* 665: 744–745.
80 See pp. 263–264.
81 Bo 1993 7, 972; *SKQS* 215: 230; 811: 127.

areas and solids, followed by two years devoted to astronomy; it is unclear whether algebra was studied. The studies were sanctioned by examinations; they seem to have been limited to students who were good in these subjects, who also took the regular examinations.[82] Thus the teaching in mathematics ceased to be reserved for Bannermen and trainee astronomers; the entry of the discipline into a major imperial educational institution suggests a further enhancement of its status. At that juncture, both teachers and textbooks were available as a result of Kangxi's promotion of mathematics. This follow up to the completion of the *Origins of pitchpipes and the calendar* was the institutional outcome of his uncommon taste for what had hitherto been a technical subject rather than a scholarly pursuit.

Kangxi's mathematical textbook was also influential beyond the capital. During the Qianlong (1636–1795) and Jiaqing (1796–1820) reigns, evidential scholarship (*kaozhengxue* 考證學), which occupied the centre of the scholarly scene, incorporated mathematics and astronomy both as techniques and as objects of learning.[83] On the one hand, it was necessary to master them in order to deal with chronology and the dating of ancient texts. On the other hand, ancient texts, such as that of the *Nine chapters on mathematical procedures* (*Jiuzhang suanshu* 九章算術) reconstructed by the famous scholar Dai Zhen 戴震 (1724–1777), were gradually retrieved and became an object of study for scholars versed in mathematics.[84] For most of them, the content of the *Essence of numbers and their principles* was the background culture from which they approached these texts. One of the earliest and most famous examples is Mei Juecheng's rediscovery of the procedure of the celestial element (*tianyuan shu* 天元術), the place-value notation for polynomials and equations developed during the Song dynasty: he interpreted it in terms of cossic algebra as taught to Kangxi by Antoine Thomas.[85] The apogee of evidential scholarship is mostly regarded as a period almost devoid of innovation in the mathematical sciences.[86] However, recent studies have brought to light the richness and complexity of investigations and debates that went on at the time, such as those on the respective potential of the celestial element procedure and cossic algebra.[87] This general orientation may have been closer to the taste and interests of Mei Wending than to those of Kangxi. Although the emperor would probably have regarded most of mid-Qing scholars' engagement with the mathematical

---

82 *Qinding Guozijian zhi* 欽定國子監志, *SKQS* 600: 445.
83 Elman 1984, 180–184.
84 Chemla & Guo 2004, 74–79.
85 *Meishi congshu jiyao* 1761, *juan* 61: 8b–11b; see p. 279.
86 See e.g. Horng 1993, 180; Engelfriet 1998, 446.
87 Tian 1999; Tian 2005, 134–181.

sciences as 'empty discussions' typical of 'cultivated talents', their approach would have seemed far from alien to him: his own investigation of the Old Man Star used astronomical observation as evidence to assess historical material.[88] In any case, his great mathematical enterprise provided the basis for later scholars to cultivate the discipline as they saw fit, rather than along the lines he had advocated.

The *Essence of numbers and their principles* continued to be used in imperial institutions until the end of the nineteenth century. In 1882, a revised edition was published under the supervision of some officials of the Ministry of War, including Zeng Guoquan 曾國荃 (1824–1890), a younger brother of Zeng Guofan 曾國藩 (1811–1872), and a general whose military successes played an important role in the suppression of the Taiping rebellion.[89] More than a century and a half after its completion, Kangxi's mathematical treatise continued to be regarded as useful by those who advocated the study of the discipline. Mathematics was then promoted as a key to military affairs rather than to calendrical and cosmological issues. It was in this connection that it once more appeared as a crucial tool that could and should be put in the service of the Qing, whose future was at stake. The dynasty had, after all, accomplished the unprecedented task of bringing together all mathematics under Heaven.

---

88 See pp. 133–135.
89 A copy of this edition, printed in Canton, is kept at the Library of the Centre for Research in the Humanities of Kyoto University; on Zeng Guoquan, see Hummel 1943, 749–751.

Conclusion

Understood in the light of cultural politics, and in particular of Kangxi's relationship with his Chinese officials, his engagement with the mathematical sciences appears as a significant element in his rulership. This was crucial to the consolidation of the Manchu dynasty. As the first Qing emperor to reign over China for a prolonged period as an adult, Kangxi had no fatherly role model to emulate: he had to invent more than any of his successors, and he was also freer to do so. The Chinese facet of his multi-cultural persona shows him enthroned at the centre of a millenary cosmological order, and drawing on all resources available within and without the territory under his control to master both military (*wu* 武) and civil/cultural (*wen* 文) affairs. As regards the latter, he turned himself into a learned patron of scholarship, and conscientiously strove to personify the Sagely Emperor. His unusual interpretation of this role added to the obligatory imperial study of the Classics and Histories that of the mathematical sciences, which had hitherto been regarded as mere technical skills, with little claim to the dignity of learning.

Kangxi initially encountered astronomy when he was only fourteen and still subject to a regency (although already a father), in the context of his first attempt to arbitrate an apparently minor affair of state. The reinstatement of the Jesuits at the Astronomical Bureau, coinciding as it did with the young emperor's successful assumption of personal rule, appears as seminal for his decision-making strategy. Thereafter his displayed mastery of even the most technical matters underlay his claim to supreme arbitration. His pursuit of the mathematical sciences was ostensibly motivated by the fact that they provided a representation of his control of Heaven and Earth as based on a thorough understanding of them. This understanding enabled him to fulfil his role of intermediary between the cosmos and human society, and also to verify the competence and the loyalty of his subjects. Thus his mastery extended to the Three Powers (*san cai* 三才, Heaven, Earth and Man) essential to Chinese cosmology, in more than a purely symbolic fashion. In that sense, mathematics provided the Manchu emperor with a means to beat Chinese scholars at their own game. His undoubted taste for this discipline, which was evident to his Jesuit tutors, was however never publicly mentioned as a justification for studying it. In sum, the staging of the emperor's engagement with mathematics was no less crucial to his construction of rulership than this engagement itself.

The continued emphasis on technical knowledge as a tool for statecraft is one of the features that situate Kangxi's rulership in the early modern world, in many parts of which the idea attributed to Francis Bacon that 'knowledge is power' was increasingly applied in government. This being said, another rationale, certainly more familiar to Chinese scholars of the late imperial period albeit unknown to twenty-first century historians of science, may also account for his imperial persona. Its sources lay in diverse currents of thought in the late Ming dynasty. One of its aspects was articulated by the Ming philosopher Wang Yangming 王陽明 (1472–1528), who claimed that 'knowledge and action are one' (zhixing heyi 知行合一). Admittedly Kangxi patronised the Cheng-Zhu school, which opposed the intellectual heirs of Wang Yangming. Nonetheless a number of Qing thinkers continued the emphasis placed on the value of practical learning (shixue 實學) by Xu Guangqi and other late Ming scholars and statesmen. The emperor's engagement with the mathematical sciences seems to actualise the ideal Confucian monarch according to another philosopher, who lived in his own time, namely Yan Yuan 顏元 (1635–1704):

For the Sage, studying, teaching and ruling are one and the same thing.[1]

Yan Yuan was little known in his lifetime, so he certainly had no personal influence on the monarch: rather, the latter was well in tune with the intellectual trends of his time. His choice not to limit his study and teaching of his subjects to what was thought to be the Confucian tradition was closer to the viewpoint of Yan Yuan than to the example set by high officials.[2]

Indeed for several decades Kangxi complained that mathematics was beyond the competence of the literati from whom his high officials were drawn: to study it he relied on a group of specialists who were foreign to East Asian civilisation, the Jesuits. Although the latter had already circulated some of their mathematical knowledge in the form of printed books in Chinese, the emperor had them compose a new body of knowledge, using Manchu as well as classical Chinese as vehicles. After monopolising this corpus for a long time, he finally had it rewritten by literati so as to bestow it on his subjects, present and to come, as the Sages had bestowed the Classics in high Antiquity. This was a fitting conclusion to a reign throughout which he simultaneously established himself as a teacher on the Confucian model and asserted

---

1　聖人學教治皆一致也。Yan 1987, 39; Cheng 1997, 557.
2　Cheng 1997, 554–557.

mathematics as a branch of learning. The knowledge that allegedly founded his control of Heaven and Earth (and Man) was turned into a branch of imperial—and thereby universal—learning that he circulated throughout the empire.

Producing texts in order to spread knowledge was a literally 'bookish' approach to the sciences, which were thus turned into a branch of Chinese learning. This was the aspect of imperial mathematics most widely taken up by scholars in the following century. However, 'science as discourse', as one might call it, represents only one side of the coin. On the other side, which one might call 'science as action', objects of a very different nature were produced, which effectively served as tools of imperial control. Beside cannon, these tools included the calendar issued every year, a general map of the empire, and the instruments needed for producing these. Kangxi learned the practice of such instruments, and repeatedly staged his mastery of their use during scenes that mostly took place in the open air, and often outside the capital. On these occasions he skilfully combined display of 'science as action'—astronomical observation, surveying, firing cannon...—with 'science as discourse'—lecturing. The former functioned as evidence for what was asserted in the latter. That imperial utterances were accompanied by such manner of proof is another element of Kangxi's rulership. It took more than the rhetoric of Chinese political cosmology to assert Manchu authority: evidence had to be drawn from all under heaven (*tianxia* 天下). The Chinese notion of imperial space encompassed in this phrase was thereby recast in a universalistic manner characteristic of the Qing empire—and perhaps of empire in general.

The symbolic relationship between Kangxi and his officials expressed by such a mode of imperial discourse, that of teacher to students, was also shaped by Chinese cultural imperatives. A similar relationship was at play, albeit with very different modalities, between the emperor and his Jesuit tutors, so long as the missionaries continued to be accepted as teachers. This was first and foremost the case for Ferdinand Verbiest, who was a generation older than his student, and was perceived as one of his early supporters. It still functioned, it seems, with Verbiest's direct successors such as Antoine Thomas, who also tutored the emperor's sons. During the last decade of the reign, by contrast, the court Jesuits were relative newcomers and younger than the emperor. Moreover his attitude towards the Catholic enterprise had undergone significant change: he came to treat the missionaries who served him as subjects whose loyalty and competence could not be taken for granted. The effective obliteration of the Jesuits' seminal contribution to imperial mathematics in official written accounts was certainly a consequence of this. But it was also a prerequisite for the universality of the mathematics that he had codified. It enhanced the emperor's image as a Sage: like Confucius, he would go down in history as having studied without a teacher. His knowledge thereby acquired a transcendental dimension. In the process, he appropriated not only the Jesuits' knowledge, but also the role they had striven to play within Chinese society. Matteo Ricci and his successors combined the two main targets that the Society of Jesus had set itself since its foundations, namely mission and education. They aptly translated the role of professors of mathematics at a Jesuit college into that of masters in the Confucian tradition, thereby creating a niche for themselves within elite circles. By first monopolising them as his tutors and then denying them the credit of the work accomplished in that capacity, Kangxi contributed to severing the links they had long cultivated with literati.

Imperial sagehood, however, did not solely proceed from the appropriation of Jesuit knowledge. It was also made possible by the fact that the emperor stood at the centre of the Qing state's sophisticated network of communication; he was therefore in a unique position to check the elements of information he received against one another. Thus he used astronomy as practised by the Jesuits to refute literati interpretations of celestial phenomena as portents. But he also checked astronomers' reports against his own observations. Similarly, he commanded abundant cartographic data, including some that the Jesuits had no part in gathering. In checking information, he made use of each of the groups that provided it to control the others. This is an example of another feature of his rulership, often regarded as negative: he played the various factions of his court against each other, which was a way of neutralising them, but also maintained their existence.[3] In the sciences, several groups had mutually exclusive claims to possessing valid and relevant knowledge. The Office of Mathematics may well have served a faction in the context of the rivalry between Kangxi's sons for his succession. At all events, it functioned as a place of arbitration between exclusive claims: imperial learning was the outcome of this arbitration. In order that this learning be the object of a consensus, the Office was staffed with representatives of several groups, excluding the Jesuits, who were denied participation in 'science as discourse'. Within the Office, Chinese scholars were represented by Li Guangdi's protégés, while official astronomers were mainly represented by the He family; bannermen formed the majority of the staff. In short, Kangxi's mathematical sagehood was constructed through conciliation, albeit temporary, between the main groups who were to outlast his reign. The audience given to Mei Wending in 1705 can be regarded as the moment when this conciliation took shape: to imperial recognition of Chinese literati's competence in the mathematical sciences responded scholarly legitimisation of official astronomy.

---

3 Spence 2002, 160–169.

However, this transaction did not bring the imperial monopoly on Western learning to an end. Significant elements of it remained the monarch's privilege. Neither the mathematics nor the technology specifically related to military and cartographic know-how, which pertained to 'science as action', were circulated. The *Essence of numbers and their principles* only gives a glimpse of the imperial command of techniques relevant to these fields, but does not provide its readers with the means to acquire a similar command. Here, 'science as discourse' served to display the continued imperial monopoly of 'science as action'. This monopoly vis-à-vis scholars and officials entailed the Jesuits' continued service to the dynasty as experts in astronomy, artillery and cartography into the Qianlong reign. Perhaps no other group of experts who served the Qing were sufficiently cut off from the literati milieu to preserve this monopoly safely.

Kangxi's appropriation shaped mathematics in a way that forces us to reconsider the dominant narrative of Western learning. This narrative presents us with a body of knowledge, allegedly consistent and radically foreign to Chinese elites, and recounts the understandings and misunderstandings, distortions and rejections to which it was subjected in China. In other words, one considers the ways in which a 'Western entity' exerts an action that arouses a 'Chinese reaction'.[4] This narrative has been typically applied to Euclidean geometry. However, considering the evolution of the teaching of this branch of mathematics in China from Ricci's translation of Euclid's *Elements* to the French Jesuits' translation of Pardies' textbook, and in particular the conditions under which the latter was carried out, reveals the amount of negotiation and adaptation to the explicit demands of the intended readership that shaped the successive geometrical works published in Chinese. There is a close parallel between this evolution and the one that occurred in education in Europe during the same period. So the 'action–reaction' model may well be a hindrance rather than a help for understanding the history of Euclidean geometry in the late Ming and early Qing period. Complex *interactions* were evidently at work in the construction of mathematics first for the emperor, then for the empire.[5]

The prominent role of calculation in the lecture notes produced by the Jesuits for the emperor allows us to further question the 'action–reaction' narrative. The Jesuit texts in Chinese and in Manchu on this subject themselves show signs of hybridisation, and a variety of sources for them have been identified. On the other hand, the recurrence of particular problems in a number of late Ming and early Qing works reveals the measure of continuity that guaranteed the identity of the

---

4 See e.g. Martzloff 1980.
5 'Action–reaction' and 'interaction' are two of the models discussed in Standaert 2002.

discipline. Tracing problems from one work to another and studying their environment in each of these shows that it is impossible to delineate a meaningful opposition between 'Chinese' and 'Western' elements in mathematics. This opposition, central in much of the discourse on mathematics, turns out to be less relevant within mathematical discourse itself. These problems served to situate the author of a work vis-à-vis his predecessors, and to show the superiority or the greater generality of a method compared with others, within a given work or compared with earlier ones. In particular, the continued use of such problems as illustrative examples allowed Western learning to be integrated within the field represented by Cheng Dawei's *Unified lineage of mathematical methods*, and therefore paved the way for the Jesuit assertion that Euclidean geometry also belonged to this field. This in turn led to reconsideration of the ways in which such knowledge could be set out in written form. The judgement given in the *Essence of numbers and their principles* was that both problems and statements were relevant forms, the former for establishing (*li* 立) and the latter for clarifying (*ming* 明): two distinctive modes of mathematical discourse were jointly legitimised.

Considering calculation on a par with geometry also sheds light on the successive proposals for structuring mathematics and for reconfiguring its various branches. The duality between geometry and arithmetic proposed by Ricci was rephrased several times. This finally resulted in articulating the foundations of the discipline in terms of numbers and principles. European scholastic concepts gave way to Neo-Confucian ones in the structure of mathematics, and concomitantly in relating it to other forms of knowledge. This integration of mathematics among the objects of scholarship 'from the inside' was an essential element in Kangxi's appropriation of Western learning as part of 'science as discourse'.

The present book's title is borrowed from the famous Andersen tale *The Emperor's new clothes*: two swindlers passing off as weavers promise an emperor a new suit of clothes invisible to those unfit for their positions or who are incompetent; when the emperor parades before his subjects in his new clothes, a child cries out: 'But he hasn't got anything on!'[6] Before closing, let me address the question that this tale hints at: for all his talent for staging and self-display, was Kangxi naked? If this question is taken to mean 'did he really understand mathematics, or are all the accounts that suggest this merely courtly flattery?' then one can reply that ample evidence has been provided here that he did not parade naked. To be more specific, it is quite clear from his words, actions and writings that he understood, or had understood at some

---

6 http://www.andersen.sdu.dk/vaerk/hersholt/TheEmperorsNewClothes_e.html

stage of his study, the contents of the *Essence of numbers and their principles*, which he intended to define the mathematical culture of officials. If, on the other hand, one further uses the tale's metaphor to account for the viewpoints of the actors, one might say that most Chinese literati who served the emperor considered the fabric of which Kangxi's new clothes were made as unworthy of imperial robes, and even of literati garb; they also regarded the Jesuits as swindlers rather than trustworthy interlocutors that could safely be integrated within their community. This was the main predicament of the appropriation of Western learning during the Kangxi reign; it was best summarised by Li Guangdi's words to the emperor:

> [The Westerners'] methods (*fa* 法) are in effect completely accurate, but it seems that their learning (*xue* 學) is completely absurd.[7]

Since learning, in the sense of the rational systematisation of thought, was the touchstone of belonging to the literati class and of trustworthiness, how much confidence could one place in the Jesuits' methods? Kangxi's response to this predicament was first to study these methods; secondly, to adhere to Mei Wending's claim that they were ultimately of Chinese origin and, finally, to appropriate these method as imperial scholarship and bestow on his officials the means to understand them. But this was as far as he could go. For all the praise and veneration that his work received from his officials and successors as emperor, the study of mathematics in eighteenth century China remained restricted to a few specialised scholars. Kangxi's solution to the predicament posed by Western learning did not become a dynastic policy. Sixty years after his death, the dominant attitude was a denial that the Jesuits had in anyway contributed to imperial scholarship:

> Our dynasty, which limits itself to using [the Jesuits'] technical skills (*jineng* 技能) while forbidding the spread of their learning (*xueshu* 學術), possesses deep insight.[8]

Thus, in accordance with Li Guangdi's view and despite Kangxi's great enterprise, the Jesuits' mathematics continued to be confined to the rank of techniques. It took an independent appropriation of the subject by scholars to relate it to their own preoccupations: the most visible feature of the mathematical sciences as practised by Chinese literati in the eighteenth century was a textual approach, largely motivated by the search for the knowledge that their ancestors had once possessed. Thus the scholars' response to Kangxi's claim that in constructing imperial mathematics, which was universal, he was effectively occupying what ought to be their intellectual territory, was to propose a

---

7 Li 1995, 2: 487–488.
8 *SKQS* 3: 709; see Gernet 1982, 85.

different construction of mathematics, which was deeply intertwined with evidential scholarship, and that emphasised the Chinese roots of the discipline. Thus one might say that they declined to follow the mathematical fashion that the emperor had sought to launch.

The choices made in the present study have been to bring together all available sources, to rely on the contemporary actors' categories rather than on those of twenty-first century observers, and to carry out the analysis at a local level, centring the narrative in China. They were not made in an attempt to deny or to ignore the need for studies on a larger scale that entail a comparative dimension. Beyond the Jesuits' apologetic accounts, it makes historical sense to confront the ways in which power and knowledge—action and discourse— interacted in Kangxi's empire and Louis XIV's France, and also, perhaps even more interestingly in Peter I's Russia. Indeed the fact that individuals, trade goods and knowledge circulated between these three countries (and many others) at the time makes such comparisons crucial to historical enquiry, if we aim to understand the early modern world in its complexity. But for a comparison to be undertaken on equal terms presupposes enquiries of similar depth into both sides of the comparison. In that sense the present study is offered as a prerequisite to rather than a substitute for comparative and global history. There is no doubt that further research is needed to fill up the discrepancy between our understanding of how states in East Asia and in Europe related to the sciences at a time when the connections between these parts of the world are abundantly documented.

Main Units

1. Length units

li 里	360 bu = 1800 chi
zhang 丈	10 chi
bu 步	5 chi
chi 尺[1]	10 cun
cun 寸[2]	10 fen
fen 分	10^{-1} cun
li 釐	10^{-2} cun
hao 毫	10^{-3} cun
si 絲	10^{-4} cun
hu 忽	10^{-5} cun

Value of the *li* to be used in surveys:
- Until 1702: 1/250 length of degree of meridian: 444 m (1 *cun* = 2.47 cm)
- From 1702 on: 1/200 length of degree of meridian: 556 m (1 *cun* = 3.09 cm)

Length of standard *cun* according to some instruments made at the Imperial Workshops during the Kangxi reign:

Jade ruler: 1 *cun* = 3.17 cm

Bamboo ruler: 1 *cun* = 3.2 cm[3]

Other units mentioned[4]

yingzao chi 營造尺	builder's chi	31.793 cm
caiyi chi 裁衣尺	tailor's chi	34 cm
liangdi chi 量地尺	surveyor's chi	32.64 cm

---

[1] Often rendered as 'foot'.
[2] Often rendered as 'inch'.
[3] As measured by 20th century curators: Liu 1999, nr. 51 & 52; see Qiu *et al.* 2001, 423; various *chi* continued to be used throughout the dynasty; Luo 1957, 52–57.
[4] See p. 246; Qiu *et al.* 2001, 406–409.

2. Time units

- Before 1644

    1 day (*ri* 日) = 12 'double hours' (*shi* 時) = 100 *ke* 刻

- After 1644

    1 day (*ri* 日) = 12 'double hours' (*shi* 時) = 24 hours (*xiaoshi* 小時)
    1 hour = four quarters (*ke* 刻) = 60 minutes (*fen* 分)
    1 minute = 60 seconds (*miao* 秒)

3. Weight units

1 *liang* 兩 (ounce) = 10 *qian* 錢
1 *qian* = 10 *fen* 分

Units below the *fen* of weight are named in the same way as the submultiples of the length unit *fen* given above.

The first three units were also used for copper currency, although the only one represented by an actual coin was the *qian*. It should be noted that in different periods of the Kangxi reign the official weight of a copper coin varied between 0.1 and 0.14 ounce.[5] This is reflected in the fact that while the ratio 1 *liang* = 10 *qian* is used in some problems in the *Essence of numbers and their principles*, it is not specified in the passage where official units are defined.[6]

In most problems translated in the present book, the money unit used is *wen* 文, which is a measure word for money.

Weight of standard *liang* according to some weights kept at the Palace Museum in Beijing: 1 *liang* = 37.3 g.[7]

---

[5] Morse 1921, 143.
[6] Guo 1993, 3: 182.
[7] Qiu *et al.* 2001, 430.

BIBLIOGRAPHY

ABBREVIATIONS FOR LIBRARIES AND ARCHIVES

AdS: Archives de l'Académie des sciences, Paris
AMEP: Archives des missions étrangères de Paris
APF: Archives of the Congregatio Sacra de Propaganda Fide, Rome
ARSI: Archivum Romanum Societatis Iesu
BAV: Vatican Library
Beitu: National Library, Beijing
BL: British Library
BML: Bibliothèque municipale de Lyon
BnF: Bibliothèque nationale de France, Paris
Gugong: Library of the Palace Museum, Beijing
IHNS: Library of the Institute for the history of natural sciences, Chinese Academy of Sciences, Beijing
IMARL: Inner Mongolia Autonomous Region Library, Hohhot
St Petersburg: St Petersburg Branch of the Institute of Oriental Studies, Russian Academy of Sciences
Taipei NCL: National Central Library, Taipei
Taipei NPML: Library of the National Palace Museum, Taipei
Tohoku: Library of Tohoku University, Sendai
Toyo Bunko, Tokyo

1. Mathematical and astronomical manuscripts from the Kangxi court

Library, Reference	Title
BAV, Borg. Cin. 318 (3)	五緯曆指
BAV, Borg. Cin. 319 (1), (2) & (3)	交食曆指
BAV, Borg. Cin. 319 (4)	阿爾熱巴拉新法
	數表問答
BAV, Borg. Cin. 518 (4)	對數廣運
Beitu, 善本 2725	借根方算法節要
Beitu, 善本 2726	測量儀器用法
	比例規解
	八線表根
Beitu, 善本 2727	勾股相求之法
Beitu, 善本 6523	算法纂要總綱
Beitu, 善本 11739	幾何原本

Library, Reference	Title
Beitu, 普通古籍 科 130/8036	借根方算法
BL, OC Add 16634	曆法問答
BML, Ms. 75–80 C	數表
BML, Ms. 39–43	借根方算法
BML, Ms. 74	比例規解
BML, Ms. 75–80 A	算法纂要總綱
BML, Ms. 75–80 B	數表用法
BML, Ms. 75–80 D	測量儀器用法
	比例規解
BML, Ms. 75–80 E	地平線離地球圓面表
BML, Ms. 81	推算日影法
BML, Ms. 82–90 A	算法原本
BML, Ms. 82–90 B	算法纂要總綱
BML, Ms. 82–90 C	測量儀器用法
	比例規解
	八線表根
BML, Ms. 82–90 D	借根方算法節要
BML, Ms. 82–90 E	比例表用法
	數表用法
BML, Ms. 82–90 F	蒙求各法細草
BML, Ms. 82–90 G	勾股相求之法
BML, Ms. 92	數表
BML, Ms. 92–93	數表
BML, Ms. 94	度數表
BnF, Mandchou 191	*Bodoro arga-i oyonggo be araha uheri hešen-i bithe*
Gugong	*Gi ho yuwan ben bithe*
Gugong 律八三五 29	幾何原本 (7 chapters)
Gugong	幾何原本 (12 chapters)
Gugong	借根方算法節要°
Gugong	借根方算法°
Gugong	數表問答°
Gugong, Ms. 13274–75	算法纂要總綱
Gugong, Ms. 13276–81	算法纂要總綱°
IMARL, Manchu 210	*Gi ho yuwan ben bithe*
IHNS, 子 530.7.224 [1]	算法纂要總綱
IHNS, 子 530.7.224 [2]	算法纂要總綱
St Petersburg, Blockprint C 291	*Gi ho yuwan ben bithe* *
Taipei NCL, 善本 6398	幾何原本
Taipei NCL, 善本 6399	幾何原本
Tohoku, Fuji 4215	數表
Tohoku, Fuji 4216	度數表
Tohoku, Fuji 4217	算法纂要總綱
Tohoku, Fuji 4218	借根方算法節要
Toyo Bunko, Manchu 502	*Suwan yuwan ben bithe* *

* indicates manuscripts that I have not seen.
° published in Gugong bowuyuan 2000.

2. Editions used for main other Chinese works on mathematics, astronomy and harmonics

(references are to works given below in 'Other Sources')

Biligui jie 比例規解, 1630. *SKQS* 788: 613–662

Bisuan 筆算, 1706.

Cuozong fayi 錯綜法義, in Guo 1993, 4: 685–688.

Fangcheng lun 方程論, in Guo 1993, 4: 323–408; in *SKQS* 795: 64–208.

Geyuan milü jiefa 割圜密率捷法, in Guo 1993, 4: 865–943.

Jihe yuanben 幾何原本 (1607, 1859 edition), in Guo 1993, 5: 1151–1301.

Jiuzhang suanfa bilei daquan 九章算法比類大全, in Guo 1993, 2: 5–333.

Jiuzhang suanshu 九章算術: Chemla & Guo 1984.

Lisuan quanshu 曆算全書. *SKQS* 794 & 795: 1–818.

Meishi congshu jiyao 梅氏叢書輯要, 1761.

Ouluoba xijinglu 歐羅巴西鏡錄, in Guo 1993, 4: 281–302.

Shixue 視學, in Guo 1993, 4; 711–853.

Shudu yan 數度衍, in Jing 1994, 2: 2555–2872 ; *SKQS* 802: 233–592.

Suanfa tongzong 算法統宗, in Guo 1993, 2: 1217–1420.

Sunzi suanjing 孫子算經, in Guo 1993, 1: 7–77.

Tongwen suanzhi 同文算指, in Guo 1993, 4: 77–278.

Xinfa suanshu 新法算書. *SKQS* 788 & 789: 1–814.

Xinzhi lingtai yixiangzhi 新製靈臺儀象志, in Bo 1993, 7: 7–457.

Yuzhi lixiang kaocheng 御製曆象考成, in Bo 1993, 7: 463–957; Bodleian Library, Oxford, Sinica 361.

Yuzhi lixiang kaocheng houbian 御製曆象考成後編, in Bo 1993, 7: 965–1338.

Yuzhi lülü zhengyi 御製律呂正義. *SKQS* 215: 1–221.

Yuzhi lülü zhengyi houbian 御製律呂正義後編. *SKQS* 215 (223–367) & 216–218.

Yuzhi shuli jingyun 御製數理精蘊, in Guo 1993: 3; Bodleian Library, Oxford, Sinica 361.

Zhoubi suanjing 周髀算經, in Guo 1993, 1: 7–77.

3. Other sources

Arnauld, Antoine. *Nouveaux elemens de geometrie*. Paris: Ch. Savreux, 1667.

Beijing tianwentai 北京天文台. *Zhongguo gudai tianxiang jilu zongji* 中國古代天象記錄總集. Nanjing: Jiangsu kexue chubanshe, 1988.

Bo Shuren 薄樹人 ed. *Zhongguo kexue jishu dianji tonghui. Tianwen juan* 中國科學技術典籍通彙. 天文卷. 8 vols. Zhengzhou: Henan jiaoyu chubanshe, 1993.

Bouvet, Joachim. *Portrait historique de l'empereur de la Chine présenté au Roy*. Paris: Estienne Michallet, 1697.

Charas, Moyse. *Pharmacopée royale galénique et chymique*. Paris: l'auteur, 1676.

Chen Houyao 陳厚耀. *Gougu fayi* 勾股法義. Manuscript, n.d. IHNS Library, Beijing, 善子 530 284.

Chen Menglei 陳夢雷. *Songhe shan fang wenji* 松鶴山房文集. *Xuxiu Siku quanshu* 續修四庫全書, vol. 1416. Shanghai: Shanghai guji chubanshe, 1995.

Chen Shou 陳壽. *Sanguo zhi* 三國志. 5 vols. Beijing: Zhonghua shuju, 1982.

Choisy, Abbé de. *Journal du voyage de Siam fait en 1685 & 1686*. Edited by Dirk Van der Cruysse. Paris, Fayard, 1995.

Clavius, Christophorus. *Epitome arithmeticæ practicæ*. Rome, 1583.

——. *Algebra*. In *Opera mathematica*, vol. 2. Mainz: A. Hierat, 1612. http://mathematics.library.nd.edu/clavius.

D'Anville, Jean-Baptiste Bourguignon. *Nouvel atlas de la Chine, de la Tartarie chinoise et du Thibet: contenant les cartes générales & particulieres de ces pays, ainsi que la carte du royaume de Corée; la plupart levées sur les lieux par ordre de l'empereur Cang-Hi avec toute l'exactitude imaginable, soit par les PP. Jésuites missionnaires à la Chine, soit par les mêmes peres: rédigées par M. d'Anville. Précedé d'une description de la Boucharie par un officier suedois qui a fait quelque sejour dans ce pays*. La Haye, Scheurleer, 1737.

Da Qing huidian 大清會典. Shangwu yinshu guan, 1908.

Da Qing huidian shili 大清會典事例. Shangwu yinshuguan, 1908, 30 vols.

Da Qing lichao shilu 大清歷朝實錄. Tokyo: Daizō shuppan (reprint), 1937, 1200 vols.

Dechales, Claude-François Milliet. *Les elemens d'Euclide expliquez d'une maniere nouvelle & tres-facile: avec l'usage de chaque proposition pour toutes les parties des mathematiques*. Paris: Estienne Michallet, 1677.

Dechales, Claudius Franciscus Milliet. *Claudii Francisci Milliet Dechales. Cursus seu mundus mathematicus*. Lugduni: apud Anissonios [etc.], 1690.

Dionis, Pierre, *L'Anatomie de l'homme*. Paris: Laurent d'Houry, 1691.

Du Halde, Jean-Baptiste. *Description geographique, historique, chronologique, politique et physique de l'empire de la Chine et de la Tartarie chinoise*. 4 vols. (1st edn: Paris, 1735). The Hague: H. Scheuleer, 1736.

Du Hamel, Jean-Baptiste. *Philosophia vetus et nova ad usum scholæ accomodata, in regia Burgundia olim pertractata*. 1st edn. 4 vols. Paris: Etienne Michallet, 1678.

Euclid. *The Thirteen books of the Elements*. Translated by Thomas L. Heath. Reprint of the 2nd edn. 3 vols. New York: Dover, 1956.

Fang Bao 方苞. *Fang Bao ji* 方苞集. Edited by Liu Jigao 劉季高. Shanghai: Shanghai guji chubanshe, 1983, 2 vols.

Fontenelle, Bernard le Bovier de. *Histoire du renouvellement de l'Académie royale des sciences en MDCXCIX; et les éloges historiques de tous les académiciens morts depuis ce renouvellement...* Amsterdam: Pierre de Coup, 1709.

Foucquet, Jean-François. *Relation exacte de ce qui s'est passé à Péking par raport à l'astronomie européane depuis juin 1711 jusqu'au commencement de novembre 1716*. 1716. Manuscript, ARSI, Jap. Sin. II 154.

Gao Shiqi 高士奇. 'Pengshan miji 蓬山密記.' In *Congshu jicheng xubian* 叢書集成續編, 40: 31–33. Shanghai: Shanghai shudian, 1994.

Gatty, Geneviève, ed. *Voiage de Siam du Père Bouvet: précédé d'une introduction avec une biographie et une bibliographie de son auteur*. Brill: Leiden, 1963.

Gaubil, Antoine. *Correspondance de Pékin*. Edited by Renée Simon. Genève: Droz, 1970.

Gaubil, Father. 'A description of the plan of Peking, the capital of China; sent to the Royal Society by Father Gaubil, e Societate Jesu: Translated from the French.' *Philosophical Transactions* 50 (1757–1758): 704–726.

Golvers, Noël. *The Astronomia Europaea of Ferdinand Verbiest, S.J. (Dillingen, 1687): Text, translation, notes and commentaries, Monumenta Serica monograph series*, 28. Nettetal: Steyler Verlag, 1993.

Goüye, Thomas. *Observations physiques et mathématiques pour servir à l'histoire naturelle et à la perfection de l'astronomie et de la géographie: envoyées de Siam à l'Academie Royale des Sciences à Paris, par les Peres Jesuites François qui vont à la Chine en qualité de Mathematiciens du Roy: avec les reflexions de Messieurs de l'Academie et quelques Notes du P. Goüye, de la Compagnie de Jesus*. Paris: La veuve d'Edme Martin, Jean Boudot, & Estienne Martin, 1688.

——. *Observations physiques et mathématiques, pour servir à l'histoire naturelle & à la perfection de l'astronomie & de la géographie: envoyées des Indes et de la Chine à l'Academie royale des sciences à Paris, par les Peres jesuites. Avec les reflexions de Mrs de l'Academie, & les notes du P. Goüye, de la Compagnie de Jesus*. Paris: Imprimerie royale, 1692.

Greslon, Adrien. *Suite de l'histoire de la Chine*. Paris: Jean Henault, 1671–1672.

Gugong bowuyuan 故宮博物院 ed. *Gugong zhenben congkan* 故宮珍本叢刊. Haikou: Hainan chubanshe, 2000, 731 vols.

Guo Shuchun 郭書春, ed. *Zhongguo kexue jishu dianji tonghui. Shuxue juan* 中國科學技術典籍通彙。數學卷. Zhengzhou: Henan jiaoyu chubanshe, 1993.

Guo Shuchun 郭書春, and Liu Dun 劉鈍, eds. *Suanjing shi shu* 算經十書. Taipei: Jiuzhang chubanshe, 2001.

Han Qi 韓琦, and Wu Min 吳旻, eds. *Xichao chongzheng ji Xichao ding'an (wai san zhong)* 熙朝崇正集 熙朝定案（外三種）. Beijing: Zhonghua shuju, 2006.

Hong, Chŏng-Ha 洪正夏. *Kuilchip* 九一集. Edited by Han'guk Kwahaksa Hakhoe 韓國科學史學會. Seoul: Yŏja Taehakkyo Chulpanbu, 1983.

Hu Jing 胡敬. 'Guochao yuanhua lu 國朝院畫錄.' In *Huashi congshu* 畫史叢書, ed. Yu Anlan 于安瀾. Shanghai: Shanghai renmin meishu chubanshe, (1816) 1963.

Huangchao wenying 皇朝文穎. SKQS 1449–1450.

Huang Chongjun 黃鍾駿. *Chouren zhuan sibian* 疇人傳四編. Shanghai: Shangwu yinshu guan, 1955; reprint in Ruan 1982.

Ides, Evert Ysbrants. *Three years travels from Moscow over-land to China: thro' Great Ustiga, Siriania, Permia, Sibiria, Saour, Great Tartary &c. to Peking [...] printed in Dutch. and now faithfully done into English*. London: printed for W. Freeman, J. Walthoe, T. Newborough, J. Nicholson and R. Parker, 1706.

Jiang Liangqi 蔣良騏. *Donghua lu* 東華錄. Beijing: Zhonghua shuju, 1980.

Jiao Xun 焦循. 'Litang daoting lu 里堂道聽錄.' In *Beijing tushuguan guji zhenben congkan* 北京圖書館古籍珍本叢刊, vol. 69. Beijing: Shumu wenxian chubanshe, 1998.

Jing Yushu 靖玉樹. *Zhongguo lidai suanxue jicheng* 中國歷代算學集成. 3 vols. Jinan: Shandong renmin chubanshe, 1994.

Josson, Henry and Léopold Willaert. *Correspondance de Ferdinand Verbiest de la Compagnie de Jésus (1623–1688), Directeur de l'Observatoire de Pékin*. Bruxelles: Palais des Académies, 1938.

Kangxi 康熙. *Yuzhi sanjiaoxing tuisuanfa lun* 御製三角形推算法論. Bilingual (Manchu and Chinese) text, 1704. IMARL, no. 2335.

Kircher, Athanasius. *Musurgia universalis, sive Ars magna Consoni et Dissoni*. Rome, 1650.

Klopp, Onno. *Die Werke von Leibniz gemäss seinem handschiftlichen Nachlasse in der Königlichen Bibliothek zu Hannover*. 11 vols. Hannover: Klindworth, 1864–84.

La Hire, Philippe de. *Tabulae astronomicae Ludovici magni jussu et munificentia exaratae et in lucem editae: adjecta svnt descriptio, constructio & vsus instrumentorum astronomiae; ad meridianum observatorii Regii Parisiensis in quo habitae sunt observationes ab ipso autore Philippo de la Hire*. Paris: J. Boudot, 1702.

Le Blanc, Marcel. 'Relation du voyage et de l'entrée de cinq P.P. Jesuites Missionnaires Apostoliques et Mathematiciens du Roy dans l'Empire de Chine.' Manuscript. ARSI, Jap. Sin. 127, ff. 127–170. Rome, n.d.

Le Comte, Louis. *Nouveaux mémoires sur l'état présent de la Chine*. 2 vols. Paris: Jean Anisson, 1696.

Le Gobien, Charles. *Histoire de l'edit de l'empereur de la Chine en faveur de la religion chrestienne: avec un éclaircissement sur les honneurs que les Chinois rendent à Confucius & aux Morts*. Paris: Jean Anisson, 1698.

Lecomte (Le Comte), Louis. *Un jésuite à Pékin: Nouveaux mémoires sur l'état présent de la Chine 1687–1692*. Edited by Frédérique Touboul-Bayeure. Paris: Phébus, 1990.

LEC. *Lettres édifiantes et curieuses, écrites des Missions étrangères, par quelques missionnaires de la Compagnie de Jésus*. 34 vols. Paris: Nicolas Le Clerc, 1703–1776.

Legge, James. *The Chinese classics; with a translation, critical and exegetical notes, prolegomena, and copious indexes*. 5 vols. Hong Kong: Hong Kong University Press, 1960.

Li Di 李迪 ed. *Kangxi Jixia gewu bian yizhu* 康熙幾暇格物編譯注. Shanghai: Shanghai guji chubanshe, 1993 (2nd edition 2007).

Li Guangdi 李光地. *Rongcun ji* 榕村集. SKQS 1324: 525–1074.

——. *Rongcun yulu; Rongcun xu yulu* 榕村語錄 榕村續語錄. 2 vols. Edited by Chen Zuwu 陳祖武. Beijing: Zhonghua shuju, 1995.

Li Qingfu 李清馥. *Rongcun pulu hekao* 榕村譜錄合考. 1826. Facsimile reprint in Beijing tushuguan ed., *Beijing tushuguan cang zhenben nianpu congkan* 北京圖書館藏珍本年譜叢刊. Beijing: Beijing tushuguan chubanshe, 1999, vol. 85: 403–688.

Li Qingzhi 李清植. *Wenzhen gong nianpu* 文貞公年譜. 2 vols, 1825.

Mei Wending 梅文鼎. *Jixuetang shiwenchao* 績學堂詩文鈔. Edited by He Jingheng 何靜恆 and Zhang Jinghe 張靜河. Hefei: Huangshan shushe, 1995.

——. *Wu'an lisuan shuji* 勿安曆算書記. SKQS 795: 961–992.

——. *Wu'an lisuan shumu* 勿安曆算書目. Shanghai: Shangwu yinshuguan, 1939.

Mémoires concernant l'histoire, les sciences, les arts, les moeurs, les usages, &c. des Chinois par les missionnaires de Pe-kin. 16 vols. Paris: Nyon aîné, 1776–1814.

Midian zhulin Shiqu baoji hebian 秘殿珠林石渠寶笈合編. 12 vols. Shanghai: Shanghai shuju, 1988.

Milon, Alfred. *Mémoires de la Congrégation de la Mission: La Congrégation de la Mission en Chine*. 3 vols. Paris/1911–1912.

Ozanam, Jacques. *Tables des sinus, tangentes et sécantes, et des logarithmes des sinus et des tangentes; avec un traité de trigonométrie par de nouvelles démonstrations*. Paris: E. Michallet, 1685.

Pardies, Ignace Gaston S.J. *Elémens de géométrie*. Paris: Sébastien Mabre-Cramoisy, 1671; 2nd edition 1673.

Pralon-Julia, Dolorès et al. *Ratio studiorum: Plan raisonné et institution des études dans la Compagnie de Jésus*. Paris: Belin, 1997.

Prestet, Jean. *Nouveaux elemens des mathematiques, ou principes generaux de toutes les sciences qui ont les grandeurs pour objet*. 2 vols. Paris: André Pralard, 1689.

Qinding huangchao wenxian tongkao 欽定皇朝文獻通考. SKQS 632–638.

Qinding xieji bianfang shu 欽定協紀辨方書. SKQS 811: 109–1022.

Renaldini, Carlo. *Opus mathematicum in quo utraque algebra . . . pertractata novis praeceptis, novisque demonstratibus illustratur . . . Pars prior numerosam algebram complectens*. Bologna: H. H. Duci, 1657.

Ricci, Matthieu, and Nicolas Trigault. *Histoire de l'expédition chrétienne au royaume de la Chine 1582–1610*. Edited by Georges Bessières and Joseph Dehergne. Paris: Desclée de Brouwer, 1978 (1617).

Ripa, Matteo. *Memoirs of Father Ripa during thirteen years' residence at the court of Peking at the service of the emperor of China*. Edited and translated by Fortunato Prandi. London: John Murray, 1844.

——. *Giornale (1705–1724)*. Edited by Michele Fatica. 2 vols. Naples: Istituto Universitario Orientale, 1996.

Ruan Yuan 阮元. *Chouren zhuan* 疇人傳. 2 vols. Taipei: Xuesheng shuju, 1982.

Shen Gua 沈括. *Mengxi bitan* 夢溪筆談. Edited by Hu Daojing 胡道靜. Beijing, Zhonghua shuju, 1975.

Shengzu renhuangdi shengxun 聖祖仁皇帝聖訓. SKQS 411.

Shengzu renhuangdi tingxun geyan 聖祖仁皇帝庭訓格言. SKQS 717: 613–662.

Shengzu renhuangdi yuzhi wenji 聖祖仁皇帝御製文集. SKQS 1298–1299.

Sima Qian 司馬遷. *Shiji* 史記. 10 vols. Beijing: Zhonghua shuju, 1959.

SKQS. *Yingyin Wenyuange Siku quanshu* 景印文淵閣四庫全書. 1500 vols. Taipei: Taibei shangwu yinshuguan, 1986.

SL. *Shengzu renhuangdi shilu* 聖祖仁皇帝實錄. Tokyo: Daizō shuppan (reprint), 1937.

Tachard, Guy S.J. *Voyage de Siam des pères Jésuites, envoyéz par le Roy, aux Indes et à la Chine; Avec leurs observations astronomiques, et leurs remarques de physique, de géographie, d'hydrographie, et d'histoire.* Paris: Arnould Seneuze & Daniel Hortemels, 1686.

Thomas, Antoine. *Synopsis mathematica, complectens varios tractatus quos hujus scientiae tyronibus et missionis sinicae candidatis breviter et clare concinnavit P. Antonius Thomas e Societate Iesu.* 2 vols. Douai: Mairesse, 1685. 2nd edn Douai: Derbaix, 1729.

Tuotuo 脫脫 et al. *Liaoshi* 遼史. 5 vols. (1343). Beijing: Zhonghua shuju, 1974.

Vissière, Isabelle and Jean-Louis Vissière ed. *Lettres édifiantes et curieuses de Chine par des missionnaires jésuites 1702–1776.* Paris: Garnier-Flammarion, 1979.

Vlacq, Adriaan. *Arithmetica logarithmica, sive logarithmorum, chiliades centvm, pro numeris naturali serie crescentibus ab vnitate ad 100000... Gouda. Excudebat Petrus Rammasenius*, 1628.

——. *Tabulae sinuum, tangentium et secantium et logarithmi sinuum, tangentium et numerorum ab unitate ad 10000 cum methodo facillima illarum ope resolvendi omnia triangula rectilinea et sphaerica et plurimas quaestiones astronomicas, ab A. Vlacq. Editio ultima.* Lyon: J. Thioly, 1670.

Von Collani, Claudia, ed. *Joachim Bouvet S.J.: Journal des voyages.* Vol. 95, *Variétés sinologiques new series.* Taipei: Ricci Institute, 2005.

Wanshou shengdian chuji 萬壽盛典初集. SKQS 653–654.

Wang Lansheng 王蘭生. *Jiaohe ji* 交河集, 1836.

Wang Shizhen 王士禛. *Juyi lu* 居易錄. SKQS 869: 309–754.

Wang Zhongmin 王重民, ed. *Xu Guangqi ji* 徐光啓集. 2 vols. Shanghai: Shanghai guji chubanshe, 1984.

Wu Xiangxiang 吳相湘, ed. *Tianzhujiao dongchuan wenxian xubian* 天主教東傳文獻續編. 3 vols. Taipei: Xuesheng shuju, 1966.

Xi Yufu 席裕福, ed. *Huangchao zhengdian leizuan* 皇朝政典類纂. 30 vols. Taibei: Chengwen chubanshe, 1969.

Yan Yuan 顏元. *Yan Yuan ji* 顏元集. 2 vols. Beijing, Zhonghua shuju, 1987.

Yang Guangxian 楊光先. *Budeyi fu er zhong* 不得已附二種. Edited by Chen Zhanshan 陳占山. Hefei: Huangshan shushe, 2000.

Xichao ding'an 熙朝定案. BnF, Chinois 1330.

Yuding xingli kaoyuan 御定星曆考原. SKQS 811: 1–107.

Yuding yinyun chanwei 御定音韻闡微. SKQS 240: 1–357.

Yuzuan Zhouyi zhezhong 御纂周易折中. SKQS 38: 1–564.

Zarlino, Gioseffo. *Le Istitutioni harmoniche di M. Gioseffo Zarlino da Chioggia, nelle quali, oltra le materie appartenenti alla musica, si trovano dichiarati molti luoghi di poeti, d'historici et di filosofi.* Venice, 1558.

Zhang Ying 張英. *Nanxun hucong jilue* 南巡扈從紀略. *Congshu jicheng xubian* 叢書集成續編. Shanghai: Shanghai shudian, 1994, 180 vols, vol. 65: 209–229.

Zhang Qin 章梫. *Kangxi zhengyao* 康熙政要. Edited by Chu Jiawei 褚家偉 et al. Beijing: Zhonggong zhongyang dangxiao chubanshe, 1994.

Zhang Yushu 張玉書. *Zhang Wenzhen ji* 張文貞集. SKQS 1322: 389–687.

Zhao Erxun 趙爾巽, ed. *Qingshi gao* 清史稿. 28 vols. Beijing: Zhonghua shuju, 1977.

Zhongguo diyi lishi dang'an guan 中國第一歷史檔案館. *Kangxi qiju zhu* 康熙起居注. 3 vols. Beijing: Zhonghua shuju, 1984.

——. *Kangxi chao hanwen zhupi zouzhe* 康熙朝漢文硃批奏摺. 8 vols. Beijing: Dang'an chubanshe, 1985.

——. *Kangxi chao manwen zhupi zouzhe quanyi* 康熙朝滿文硃批奏摺全譯. Beijing: Zhongguo shehui kexue chubanshe, 1996.

——. *Qing zhongqianqi xiyang tianzhujiao zai Hua huodong dang'an shiliao* 清中前期西洋天主教在華活動檔案史料. 4 vols. Beijing: Zhonghua shuju, 2003.

Zhongguo dizhen ju 中國地震局, and Zhongguo diyi lishi dang'anguan 中國第一歷史檔案館, eds. *Ming-Qing gongcang dizhen dang'an* 明清宮藏地震檔案. 2 vols. Beijing: Dizhen chubanshe, 2005.

4. Secondary literature

Ahn Daeok 安大玉. *Minmatsu seiyō kagaku tōdenshi: Tengaku shohen no kenkyū* 明末西洋科学東伝史：天学初函の研究. Tokyo: Chisen, 2007.

Aiton, E. J. *Leibniz: a biography.* Bristol: Adam Hilger Ltd, 1985.

Alden, Dauril. *The making of an enterprise: the Society of Jesus in Portugal, its empire, and beyond.* Stanford: Stanford University Press, 1996.

Allsop, Peter and Joyce Lindorff. 'Teodorico Pedrini: the music and letters of an 18th-century missionary in China.' *Vincentian heritage* 27, no. 2 (2007): 43–59.

Ames, Glenn. *Colbert, mercantilism and the French quest of Asian trade.* DeKalb Illinois: Northern Illinois University Press, 1996.

Ang, Tianse. 'Zu Chongzhi.' In *Encyclopedia of the history of science, technology and medicine in non-western cultures*, ed. Helaine Selin, 1060–1061. Dordrecht: Kluwer Academic Publishers, 1997.

Arai, Shinji. 'Astronomical studies by Zhao Youqin.' *Taiwanese journal for philosophy and history of science* 8, no. 1 (1996): 59–102.

Arrault, Alain. *Shao Yong (1012–1077), poète et cosmologue*. Paris: Collège de France, Institut des Hautes Études Chinoises, 2002.

Bai, Limin. 'Mathematical study and intellectual transition in the early and mid-Qing.' *Late Imperial China* 16, no. 2 (1995): 23–61.

Bai Shangshu 白尚恕 and Li Di 李迪. 'Liuhe yanshiyi 六合驗時儀.' *Kejishi wenji* 科技史文集 12 (1984): 153–156.

Baldini, Ugo. *Legem impone subactis. Studi su filosofia e scienza dei Gesuiti in Italia, 1540–1632*. Rome: Bulzoni Editore, 1992.

———. 'The Portuguese Assistancy of the Society and scientific activities in its Asian mission until 1640.' In *História das ciências matematicas: Portugal e o Oriente*, 49–104. Lisbon: Fundação Oriente, 2000.

———. 'The Academy of Mathematics at the Collegio Romano from 1553 to 1612.' In *Jesuit science and the Republic of Letters*, ed. Mordechai Feingold, 47–98. Cambridge, Mass.: MIT Press, 2002.

———, ed. *Christoph Clavius e l'attività scientifica dei gesuiti nell'età di Galileo*. Rome: Bulzoni, 1995.

Bartlett, Beatrice S. *Monarchs and ministers: the Grand Council in mid-Ch'ing China, 1723–1820*. Berkeley/Los Angeles/Oxford: University of California Press, 1991.

Bedini, Silvio. *The trail of time: time measurement with incense in East Asia*. Cambridge: Cambridge University Press, 1994.

Bennett, J. A. 'Geometry and surveying in early-seventeenth-century England.' *Annals of science*, 48 (1991): 345–354.

Bernard-Maître, Henri. 'Le voyage du Père de Fontaney.' *Bulletin de l'Université Aurore* sér. 3, 3, no. 2 (1942): 227–280.

———. 'Les adaptations chinoises d'ouvrages européens: bibliographie chronologique depuis la venue des Portugais à Canton jusqu'à la Mission française de Pékin 1514–1688.' *Monumenta Serica* 10 (1945): 1–57, 309–388.

———. 'De la question des termes à la querelle des rites de Chine: le dossier Foucquet de 1711.' *Neue Zeitschrift für Missionswissenschaft* 15 (1958): 178–195, 267–275.

———. 'Les adaptations chinoises d'ouvrages européens: bibliographie chronologique. Deuxième partie: depuis la fondation de la mission française de Pékin jusqu'à la mort de l'empereur K'ien-long, 1689–1799.' *Monumenta Serica* (1960): 349–383.

Biagioli, Mario. 'Le Prince et les savants. La civilité scientifique au XVIIe siècle.' *Annales HSS* 50, no. 6 (1995): 1417–1453.

Bo, Shuren 薄樹人. 'Qing Qintianjian renshi nianbiao 清欽天監人事年表.' *Kejishi wenji* 科技史文集 1 (1978): 86–101.

Bosmans, Henri. 'Ferdinand Verbiest, directeur de l'Observatoire de Peking.' *Revue des questions scientifiques* 3rd series, 21 (1912): 195–173 & 375–474.

———. 'L'œuvre scientifique d'Antoine Thomas de Namur, S.J. (1644–1709).' *Annales de la société scientifique de Bruxelles* 44, 46 (1924, 1926): 169–208, 154–181.

Bray, Francesca. 'Agricultural illustrations: blueprint or icon?' In *Graphics and text in the production of technical knowledge in China: the warp and the weft*, eds Francesca Bray, Vera Dorofeeva-Lichtmann and Georges Métailié, 521–567. Leiden: Brill, 2007.

Bray, Francesca and Georges Métailié. 'Who was the author of the *Nongzheng quanshu*?' In *Statecraft and intellectual renewal: The cross-cultural synthesis of Xu Guangqi (1562–1633)*, ed. Catherine Jami, Peter M. Engelfriet and Gregory Blue, 322-359. Leiden: Brill, 2001.

Bréard, Andrea. 'Knowledge and practice of mathematics in late Ming daily life encyclopedias.' In *Looking at it from Asia: the processes that shaped the sources of history of science*, ed. Florence Bretelle-Establet Dordrecht/Heidelberg/London/New York: Springer, 2010, 305–331.

Brian, Eric, Demeulenaere-Douyère, Christiane. *Histoire et mémoire de l'Académie des Sciences: guide de recherches*. Paris: Lavoisier, 1996.

Brinker, Helmut and Albert Lutz. *Chinese cloisonné: the Pierre Uldry collection*. Translated by Susanna Swoboda. New York: Asia Society Galleries in association with Bamboo Publishing, 1989.

Brockey, Liam. 'The harvest of the vine: the Jesuit missionary enterprise in China, 1579–1710.' PhD dissertation, Brown University, 2002.

———. *Journey to the East: The Jesuit Mission to China, 1579–1724*. Cambridge, Mass.: Harvard University Press, 2007.

Ch'en, Chieh-hsien. 'The decline of the Manchu language in China during the Ch'ing period.' In *Altaica Collecta*, ed. Walter Heissig, 137–154. Wiesbaden: Otto Harrassowitz, 1976.

Chabbert, Pierre. 'Jacques Borelly, membre de l'Académie royale des Sciences.' *Revue d'histoire des sciences et de leurs applications* 27, no. 3 (1970): 203–227.

Chan, Albert. 'Johann Adam Schall in the *Pei-yu lu* of T'an Ch'ien and in the eyes of his contemporaries.' In *Western learning and Christianity in China: the contribution and impact of Johann Adam Schall von Bell, S.J. (1592–1666)*, ed. Roman Malek, 273–362. Netteral: Steyler Verlag, 1998.

Chan, Wing-tsit. 'The *Hsing-li ching-i* and the Ch'eng-Chu school of the seventeenth century.' In *The unfolding of neo-confucianism*, ed. Wm.Theodore De Bary, 543–579. New York: Columbia University Press, 1975.

Chang, Chia-feng. 'Disease and its impact on politics, diplomacy and the military: The case of smallpox and the Manchus (1613–1795).' *Journal of the history of medicine and allied sciences* 57, no. 2 (2002): 177–197.

Chang, Michael G. *A court on horseback: imperial touring and the construction of Qing rule, 1680–1785.* Cambridge, Mass.: Harvard University Asia Center, 2007.

Chapman, Allan. 'Tycho Brahe in China: The Jesuit mission to Peking and the iconography of European instrument-making processes.' *Annals of science* 41 (1984): 417–443.

Chemla, Karine, and Guo Shuchun. *Les Neuf chapitres: le classique mathématique de la Chine ancienne et ses commentaires.* Paris: Dunod, 2004.

Chen, Hui-hung. 'Encounters in peoples, religions, and sciences: Jesuit visual culture in seventeenth century China.' PhD dissertation, Brown University, 2004.

Chen, Jiang-Ping Jeff. 'The evolution of transformation media in spherical trigonometry in 17th- and 18th-century China, and its relation to "Western learning".' *Historia Mathematica* 37, no. 1 (2010) 62–109.

Chen Yinke 陳寅恪. 'Jihe yuanben manwen yiben ba 幾何原本滿文譯本跋.' *Zhongyang yanjiu yuan lishi yuyan yanjiu jikan* 中央研究院歷史語言研究集刊 2, no. 3 (1931): 281–282.

Cheng, Anne. *Histoire de la pensée chinoise.* Paris: Editions du Seuil, 1997.

Chengzhi 承志 (Kicengge). 'Manwen "Wula dengchu difang tu" kao 滿文《烏喇等處地方圖》考'. *Gugong xueshu jikan* 故宮學術季刊 26, no. 4 (2009) 1–74.

Cheung, Anita. *Drawing boundaries: architectural images in Qing China.* Honolulu: Hawaii University Press, 2004.

Chiu, Wai Yee Lulu. 'The function of Western music in the eighteenth century Chinese court.' PhD Dissertation, Chinese University of Hong Kong, 2007.

Chu, Pingyi 祝平一. 'Technical knowledge, cultural practices and social boudaries: Wan-nan scholars and the recasting of Jesuit astronomy, 1600–1800.' PhD dissertation, UCLA, 1994.

——. 'Scientific dispute in the imperial court: the 1664 calendar case.' *Chinese Science* 14 (1997): 7–34.

——. 'Trust, instruments, and cross-cultural scientific exchanges: Chinese debate over the shape of the earth 1600–1800.' *Science in Context* 12, no. 3 (1999): 385–411.

——. 'Remembering our grand tradition: the historical memory of the scientific exchanges between China and Europe, 1600–1800.' *History of science* 41 (2003): 193–215.

——. 'Fudu Shengcai – "Lixue yiwen bu" yu "Sanjiaoxing tuisuanfa lun" 伏讀聖裁 — 《曆學疑問》與《三角推算法論》.' *Xin shixue* 新史學 16, no. 1 (2005): 51–84.

——. 'Archiving knowledge: a life history of the *Calendrical treatises of the Chongzhen reign* (*Chongzhen lishu*).' In *Qu'était-ce qu'écrire une encyclopédie en Chine?* eds. Florence Bretelle-Establet and Karine Chemla, 159–184. Saint-Denis: Presses Universitaires de Vincennes, 2007a.

——. 'Problems with the brain: European anatomical knowledge in China and Japan.' Paper presented at the Kyujanggak International Workshop 'Comparative Perspectives on the Introduction of Western Science into East Asian Countries in the late Chosŏn Period', 16–18 October 2007, Seoul, 2007b.

——. 'Shouer daxue Kuizhangge cang *Chongzhen lishu* ji qi xiangguan shiliao yanjiu 首爾大學奎章閣藏《崇禎曆書》及其相關史料研究.' *Kyujanggak* 奎章閣 34 (2009): 250–262.

Cordier, Henri. 'Cinq lettres inédites du Père Gerbillon S.J.' *T'oung-pao* II-7, no. 4 (1906): 437–468.

Corsi, Elisabetta. 'Late baroque painting in China prior to the arrival of Matteo Ripa: Giovanni Gherardini and the perspective painting called *Xianfa*.' In *La missione cattolica in Cina tra il secoli XVII–XIX: Matteo Ripa e il Collegio dei Cinesi*, eds Michele Fatica and Francesco D'Arelli, 103–122. Napoli: Isistuto Universitario Orientale, 1999.

——. 'Nian Xiyao 年希堯 (1671–1738)'s rendering of Western perspective in the prologues to "The Science of Vision".' In *A life journey to the East: Sinological studies in memory of Giuliano Bertuccioli (1923–2001)*, eds Antonino Forte and Federico Masini, 201–243. Kyoto: Italian School for East Asian Studies, 2002.

——. *La Fabrica de las illusiones: Los Jesuitas y la difusion de la perspectiva lineal en China, 1698–1766*. Mexico CIty: El Colegio de México, 2004.

Crossley, Pamela Kyle. 'Manchu education.' In *Education and society in late imperial China, 1600–1900*, eds Benjamin A. Elman and Alexander Woodside, 340–378. Berkeley/Los Angeles/London: Univeristy of California Press, 1994.

——. *A translucent mirror: history and identity in Qing imperial ideology*. Berkeley: University of California Press, 1999.

Crossley, Pamela Kyle and Evelyn S. Rawski. 'A profile of the Manchu language in Ch'ing history.' *Harvard Journal of Asiatic Studies* 53, no. 1 (1993): 63–102.

Cui Zhenhua 崔振華 and Zhang Shucai 張書才. *Qingdai tianwen dang'an shiliao huibian* 清代天文檔案史料彙編. Zhengzhou: Daxiang chubanshe, 1997.

Cullen, Christopher. 'An eighth century Chinese table of tangents.' *Chinese Science* 5 (1982): 1–33.

——. *Astronomy and mathematics in ancient China: the Zhoubi suanjing*. Cambridge: Cambridge University Press, 1996.

——. *The Suan shu shu 'Writings on reckoning': a draft translation of a Chinese mathematical collection of the second century BC, with explanatory commentary*. Cambridge: Needham Research Institute, 2004.

——. 'The *Suan shu shu* 筭數書, 'Writings on reckoning': Rewriting the history of early Chinese mathematics in the light of an excavated manuscript.' *Historia Mathematica* 34, no. 1 (2007), 10–43.

Curtis, Emily Byrne. 'A plan of the emperor's glassworks.' *Arts asiatiques* 56 (2001): 81–90.

——. 'Complete plan of the glass workshop.' In *Pure brightness shines everywhere: the glass of China*, ed. Emily Byrne Curtis, 49–58. Aldershot: Ashgate, 2004.

——. *Glass exchange between Europe and China, 1550–1800*. Farnham: Ashgate, 2009.

D'Elia, Pasquale. *Galileo in Cina: relazioni attraverso il Collegio Romano tra Galileo e i gesuiti scienziati missionari in Cina (1610–1640)*. Rome: Gregorian University, 1947.

Dai, Yingcong. 'To nourish a strong military: Kangxi's preferential treatment of his military officials.' *War and Society* 18 (2000): 71–91.

De Dainville, François. 'L'enseignement des mathématiques dans les collèges jésuites de France du XVIe au XVIIIe siècle.' *Revue d'histoire des sciences* 7 (1954): 6–21, 109–123.

De Ribou, Marie-Hélène. 'L'histoire de l'empereur de la Chine.' In *Kangxi empereur de Chine 1662–1722: la Cité interdite à Versailles*, 220–225. Paris: Réunion des musées nationaux, 2004.

De Thomaz de Bossierre, Mme Yves. *Un Belge mandarin à la cour de Chine aux XVIIe et XVIIIe siècles: Antoine Thomas, 1644–1709, Ngan To P'ing-che*. Paris: Les Belles Lettres, 1977.

——. *Jean-François Gerbillon, S.J. (1654–1707): un des cinq mathématiciens envoyés en Chine par Louis XIV*. Leuven: Ferdinand Verbiest Foundation, 1994.

Deane, Thatcher. 'The Chinese imperial astronomical Bureau: form and function of the Ming dynasty "Qintianjian" from 1365 to 1627.' PhD dissertation, University of Washington, 1989.

Debergh, Minako. 'Ecrits géographiques et cartes du monde illustrées du P. Ferdinand Verbiest: transformations de l'image du monde.' In *L'Europe en Chine: interactions scientifiques, culturelles et religieuses aux XVIIe et XVIIIe siècles*, eds Catherine Jami and Hubert Delahaye, 205–216. Paris: Collège de France, Institut des Hautes Études Chinoises, 1993.

Dehergne, Joseph. *Répertoire des jésuites de Chine de 1552 à 1800*. Rome/Paris: IHSI/Letouzey & Ané, 1973.

Dennerline, Jerry. 'The Shunzhi reign.' In *The Cambridge history of China*, Vol. 9, Part 1, The Ch'ing empire to 1800, ed. Willard Peterson, 73–119. Cambridge: Cambridge University Press, 2002.

Di Cosmo, Nicola. 'Manchu shamanic ceremonies at the Qing court.' In *State and court ritual in China*, ed. Joseph P. McDermott, 352–398. Cambridge: Cambridge University Press, 1999.

——. 'European technology and Manchu power: reflections on the 'military revolution' in seventeenth-century China.' In *Making sense of global history*, ed. Sølvi Sogner, 119–139. Oslo: Unversitetsforlaget, 2001.

Dijksterhuis, Eduard Jan. *Archimedes*. Princeton: Princeton University Press, 1987.

Dong Shaoxin 董少新. 'Xiyang chuanjiaoshi zai Hua zaoqi xingyi shiji kaoshu 西洋傳教士在華早期行醫事跡考述.' PhD dissertation, Zhongshan daxue, 2004.

Douglas, Robert Kennaway. *Catalogue of Chinese printed books, manuscripts and drawings in the Library of the British Museum*. London: Longmans & Co., 1877.

Dray-Novey, Alison. 'Spatial order and police in imperial Beijing.' *Journal of Asian Studies* 52, no. 4 (1993): 885–922.

Du Shiran 杜石然. *Zhongguo gudai kexuejia zhuanji* 中國古代科學家傳集. 2 vols. Beijing: Kexue chubanshe, 1993.

Dudink, Ad. 'The rediscovery of a seventeenth century collection of Chinese Christian texts: the manuscript

Tianxue jijie.' *Sino-Western cultural relations journal* 15 (1993): 1–26.

———. 'Opposition to the introduction of Western science and the Nanjing persecution (1616–1617).' In *Statecraft and intellectual renewal: The cross-cultural synthesis of Xu Guangqi (1562–1633)*, eds Catherine Jami, Peter M. Engelfriet and Gregory Blue, 191–224. Leiden: Brill, 2001.

Dudink, Ad and Nicolas Standaert. 'Ferdinand Verbiest's *Qionglixue* (1683).' In *The Christian mission in China in the Verbiest era: Some aspects of the missionary approach*, ed. Noël Golvers, 11–31. Leuven: Leuven University Press, 1999.

———. 'Chinese Christian Texts Database' (CCT-Database) (http://www.arts.kuleuven.be/sinology/cct)

Dunne, George H. *Generation of giants: the story of the Jesuits in China during the last decades of the Ming dynasty*. Notre Dame (Indiana): University of Notre Dame Press, 1962.

Durand, Pierre-Henri. *Lettrés et pouvoirs: un procès littéraire dans la Chine impériale*. Paris: EHESS, 1992.

Elliott, Mark C. 'The limits of Tartary: Manchuria in imperial and national geographies.' *Journal of Asian Studies* 59, no. 3 (2000): 603–646.

———. *The Manchu way: the Eight Banners and ethnic identity in late imperial China*. Stanford: Stanford University Press, 2001.

Elman, Benjamin A. *From philosophy to philology: intellectual and social aspects of change in late imperial China*. Cambridge, Mass.: Council on East Asian Studies Harvard University, 1984.

———. *Classicism, politics, and kinship: the Ch'ang-chou school of new text Confucianism in late imperial China*. Berkeley: University of California Press, 1990.

———. *A cultural history of civil examinations in late imperial China*. Berkeley: University of California Press, 2000.

———. *On their own terms: science in China 1550–1900*. Cambridge, Mass.: Harvard University Press, 2005.

Elman, Benjamin A. and Alexander Woodside, eds. *Education and society in late imperial China, 1600–1900*. Berkeley: University of California Press, 1994.

Elvin, Mark. 'Who was responsible for the weather? Moral meteorology in late imperial China.' *Osiris* 13 (1998): 213–237.

———. *The retreat of the elephants: an environmental history of China*. New Haven/London: Yale University Press, 2004.

Engelfriet, Peter M. *Euclid in China: The genesis of the first Chinese translation of Euclid's Elements, books I–VI (Jihe yuanben, Beijing, 1607) and its reception up to 1723*. Leiden; Boston: Brill, 1998.

———. 'The transmission of Western scientific knowledge in Nanjing during the Shunzhi reign.' Paper presented at the 'Europe in China III' Conference, Berlin 1998a.

Engelfriet, Peter M. and Siu Man-Keung. 'Xu Guangqi's attempts to integrate Western and Chinese mathematics.' In *Statecraft and intellectual renewal: The cross-cultural synthesis of Xu Guangqi*, eds Catherine Jami, Peter M. Engelfriet and Gregory Blue, 279–310. Leiden: Brill, 2001.

Fan Jingzhong 范景中. 'Tonghuozi taoyin ben "Yuzhi shuli jingyun" 銅活字套印本《御製數理精蘊》.' *Gugong bowuyuan yuankan* 故宮博物院院刊 84 (1999): 88–91.

Feingold, Mordechai, ed. *Jesuit science and the Republic of Letters*. Cambridge, Mass.: MIT Press, 2002.

Feldhay, Rivka. 'The cultural field of Jesuit science.' In *The Jesuits: Cultures, sciences and the arts 1540–1773*, eds John W. O'Malley, Gauvin Alexander Bailey, Steven J. Harris and T. Frank Kennedy, 107–130. Toronto: University of Toronto Press, 1999.

Forêt, Philippe. *Mapping Chengde: the Qing landscape enterprise*. Honolulu: University of Hawai'i Press, 2000.

Foss, Theodore N. 'A Western interpretation of China: Jesuit cartography.' In *East meets West: the Jesuits in China, 1582–1773*, eds Charles E. Ronan and Bonnie B.C. Oh, 209–251. Chicago: Loyola University Press, 1988.

———. 'The European sojourn of Philippe Couplet and Michael Shen Fuzong, 1683–1692.' In *Philippe Couplet, S.J. (1623–1693): The man who brought China to Europe*, ed. Jerome Heyndrickx. Nettetal: Steyler Verlag, 1990.

Frémontier, Camille. 'Un objet dans les collections de l'Académie des sciences: la machine de Roemer.' In *Histoire et mémoire de l'Académie des sciences: guide de recherche*, eds Eric Brian and Christiane Demeulenaere-Douyère, 319–324. Paris: Lavoisier Tec & Doc, 1996.

Fu, Lo-shu. 'The two Portuguese embassies to China during the K'ang-hsi period.' *T'oung-pao* 43 (1955): 75–94.

———. *A documentary chronicle of sino-western relations (1644–1820)*. 2 vols. Tucson: University of Arizona Press, 1966.

Fuchs, Walter. *Der Jesuiten-Atlas der Kanghsi-Zeit: seine Entstehungsgeschichte nebst Namensindices für die Karten der Mandjurei, Mongolei, Ostturkestan und Tibet, mit Wiedergabe der Jesuiten-Karten in Original Grösse*. Beijing: Fu-jen Universität, 1943.

Funakoshi, Akio 船越昭生. *Sakoku Nihon ni kita 'Kōki zu' no chirigaku shiteki kenkyū* 鎖国日本にきた「康熙図」の地理学史的研究. Tokyo: Hōsei University Press, 1986.

Fung Kam-Wing 馮錦榮. 'Fang Zhongtong ji qi "Shudu yan" 方中通及其《數度衍》.' *Lunheng* 論衡 2, no. 1 (1995): 128–209.

——. 'Minmatsu Shinsho ni okeru Kô Hyakuka (Huang Baijia) no seigai to chosaku 明末清初における黄百家の生涯と著作.' *Chūgoku shisōshi kenkyū* 中国思想史研究 20 (1997): 61–92.

——. 'Qingchu xiyang celiangxue de fazhan 清初西洋測量學的發展.' In *Dongxi fang wenhua chengchuan yu chuangxin* 東西方文化承傳與創新, eds Sin Chou Yiu 單周堯 Lee Cheuk Yin 李焯然 and Wong Yoon Wah 王潤華, 136–160. Singapore: Bafang wenhua chuangzuo shi, 2004.

Gardner, Daniel K. 'Modes of thinking and modes of discourse in the Sung: some thoughts on the Yü-lu ('Recorded conversations') texts.' *Journal of Asian Studies* 50, no. 3 (1991): 574–603.

——. *The Four Books: the basic teachings of the later Confucian tradition*. Translations, with introduction and commentary. Indianapolis: Hackett Publishing Company Inc., 2007.

Ge Rongjin 葛榮晉. *Zhongguo shixue sixiang shi* 中國實學思想史. 3 vols. Beijing: Shoudu shifan daxue chubanshe, 1994.

Gernet, Jacques. *Le Monde chinois*. Paris: Armand Colin, 1972.

——. *Chine et christianisme: action et réaction*. Paris: Gallimard, 1982.

Gild-Bohne, Gerlinde. *Das Lülü zhengyi xubian: ein Jesuiten traktat über die europaïsche Notation in China von 1713*. Göttingen: Edition Re, 1991.

——. 'The introduction of European musical theory during the early Qing dynasty: The achievements of Thomas Pereira and Theodorico Pedrini.' In *Western learning and Christianity in China: the contribution and impact of Johann Adam Schall von Bell, S.J. (1592–1666)*, ed. Roman Malek, 1189–1200. Nettetal: Steyler Verlag, 1998.

Gillispie, Charles Coulston. *Dictionary of scientific biography*. 16 vols. New York: Charles Scribner, 1970–1980.

Golvers, Noël. 'La mission des jésuites en Chine en 1678: un cri d'alarme!' *Courrier Verbiest* 5 & 6 (1993–94): 2–5 & 4–9.

——. 'A Chinese imitation of a Flemish allegorical picture representing the Muses of European sciences', *T'oung-pao* 2nd Series, 81, no. 4/5 (1995), 303–314.

——. 'Verbiest's introduction of *Aristoteles latinus* (Coimbra) in China: New Western evidence.' In *The Christian mission in China in the Verbiest era: Some aspects of the missionary approach*, ed. Noël Golvers, 33–53. Leuven: Leuven University Press, 1999.

——. 'An unnoticed letter of F. Verbiest, S.J., on his geodesic operations in Tartary (1683/1684).' *Archives internationales d'histoire des sciences* 50 (2000): 86–102.

——. *Ferdinand Verbiest, S.J. (1623–1688) and the Chinese Heaven*. Vol. XII, *Leuven Chinese Studies*: Leuven University Press, 2003.

——. 'Ferdinand Verbiest à Pékin: un enseignement dans une perspective missionnaire.' *Courrier Verbiest* 16 (2004a): 3–5.

——. 'F. Verbiest's mathematical formation: some observations on post-Clavian Jesuit mathematics in mid-17th century Europe.' *Archives internationales d'histoire des sciences* 54, no. 153 (2004b): 29–47.

——. 'Foreign Jesuits *Indipetae*. Mathematical teachings and mathematical books at the Colégio des Artes in Coimbra in the 2nd half of the 17th century.' *Bulletin of Portuguese–Japanese studies* 14 (2007), 21–42.

——. 'F. Verbiest, G. Magalhães, T. Pereyra and the others. The Jesuit Xitang College in Peking (1670–1688) as an extra-ordinary professional milieu.' In *Tomás Pereira, S.J. (1646–1708); life, work and world*, ed. Luis Filipe Barreto, 277–298. Lisbon: Centro Científico e Cultural de Macau, 2010.

Gorman, Michael John. 'From "the eyes of all" to "useful Quarries in philosophy and good literature": consuming Jesuit science 1600–1665.' In *The Jesuits: Cultures, sciences and the arts 1540–1773*, eds John W. O'Malley, Gauvin Alexander Bailey, Steven J. Harris and T. Frank Kennedy, 170–189. Toronto: University of Toronto Press, 1999.

Gotō, Sueo. 'Le goût scientifique de K'ang-hi, empereur de Chine.' *Bulletin de la Maison Franco-Japonaise* 4 (1933): 117–132.

Gouzévitch, Irina and Dmitri Gouzévitch. 'Introducing mathematics, building and empire: Russia and Peter I.' In *The Oxford Handbook of the history of mathematics*, eds Eleanor Robson and Jacqueline Stedall, 353–373. Oxford: Oxford University Press, 2008.

Gugong bowuyuan tushuguan 故宮博物院圖書館 & Liaoning sheng tushuguan 遼寧省圖書館 eds. *Qingdai Neiwufu ke shu mulu jieti* 清代內務府刻書目錄解題. Beijing: Zijincheng chubanshe, 1995.

Guo Fuxiang 郭福祥. 'Kangxi shiqi de Yangxindian 康熙時期的養心殿.' *Gugong bowuyuan yuankan* 故宮博物院院刊 108 (2003): 30–34.

Guo Shirong 郭世榮. *Suanfa tongzong daodu* 算法統宗導讀. Wuhan, Hubei jiaoyu chubanshe, 2000.

Guo Shirong 郭世榮 and Li Di 李迪. 'He Guozhu yu Choaxian Hong Shengxia taolun shuxue wenti de youlai 何國柱與朝鮮洪正夏討論數學問題的由來.' *Nei Menggu shifan daxue bao (ziran kexue)* 內蒙古師範大學報（自然科學） 33, no. 2 (2004): 209–212.

Guy, R. Kent. *The emperor's four treasuries: scholars and the state in the late Ch'ien-lung era*. Cambridge, Mass.: Council on East Asian Studies Harvard University, 1987.

——. 'Who were the Manchus? A review essay.' *Journal of Asian Studies* 61, no. 1 (2002): 151–164.

Hahn, Roger. *The anatomy of a French scientific institution: the Paris Academy of Sciences (1666–1803)*. Berkeley & Los Angeles: University of California Press, 1971.

Halsberghe, Nicole. 'The resemblances and differences of the construction of Ferdinand Verbiest's astronomical instruments, as compared with those of Tycho Brahe.' In *Ferdinand Verbiest, S.J. (1623–1688), Jesuit missionary, scientist, engineer and diplomat*, ed. John W. Witek, 85–92. Nettetal: Steyler Verlag, 1994.

——. 'Sources and interpretation of chapters one to four in Ferdinand Verbiest's *Xin zhi lingtai yixiang zhi* (Discourse on the newly-built astronomical instruments in the observatory) Beijing, 1674.' *Review of culture* 2nd ser., no. 21 (1994b): 213–234.

Han Qi 韓琦. 'Kangxi shidai chuanru de xifang shuxue ji qi dui Zhongguo shuxue de yingxiang 康熙時代傳入的西方數學及其對中國數學的影響.' PhD dissertation, Beijing, IHNS, 1991.

——. '*Shuli jingyun* duishu zao biao fa yu Dai Xu de erxiang zhankaishi de yanjiu 數理精蘊對數造表法與戴煦的二項展開式的研究.' *Ziran kexue shi yanjiu* 自然科學史研究 11, no. 2 (1992): 109–119.

——. 'Junzhu he buyi zhijian: Li Guangdi zai Kangxi shidai de huodong ji qi dui kexue de yingxiang 君主和布衣之間：李光地在康熙時代的活動及其對科學的影響.' *Qinghua xuebao* 清華學報 26, no. 4 (1996): 421–445.

——. 'Patronage scientifique et carrière politique: Li Guangdi entre Kangxi et Mei Wending.' *Etudes chinoises* 16, no. 2 (1997): 7–37.

——. 'Bai Jin de "Yijing" yanjiu he Kangxi shidai de "Xixue zhongyuan" shuo 白晉的《易經》研究和康熙時代的「西學中源」說.' *Hanxue yanjiu* 漢學研究 16, no. 1 (1998): 185–201.

——. *Zhongguo kexue jishu de xichuan ji qi yingxiang* 中國科學技術的西傳及其影響. Shijiazhuang: Hebei renmin chubanshe, 1999a.

——. '"Gewu qiongli yuan" yu Mengyangzhai – 17, 18 shiji zhi Zhong-Fa kexue jiaoliu "格物窮理院" 與蒙養齋 – 17、18 世紀中法科學交流.' *Faguo Hanxue* 法國漢學 4 (1999b): 302–324.

——. 'Astronomy, Chinese and Western: the influence of Xu Guangqi's views in the early and mid-Qing.' In *Statecraft and intellectual renewal in late Ming China: The cross-cultural synthesis of Xu Guangqi*, eds Catherine Jami, Peter M. Engelfriet and Gregory Blue, 360–379. Leiden: Brill, 2001a.

——. 'The compilation of the *Lixiang kaocheng houbian*, its origin, sources and social context.' In *Scientific practices and the Portuguese expansion in Asia (1498–1759)*, ed. Luis Saraiva, 147–152. Lisbon: EMAF-UF, 2001b.

——. '"Zili" jingshen yu lisuan huodong – Kang-Qian zhiji wenren dui xixue taidu zhi gaibian ji qi beijing "自立" 精神與曆算活動 — 康乾之際文人對西學態度之改變及其背景.' *Ziran kexue shi yanjiu* 自然科學史研究 21, no. 3 (2002): 210–221.

——. 'Antoine Thomas, SJ, and his mathematical activities in China: a preliminary research through Chinese sources.' In *The history of relations between the Low Countries and China in the Qing era (1644–1911)*, eds Willy Vande Walle and Noël Golvers, 105–114. Leuven: Leuven University Press Ferdinand & Verbiest Foundation, K.U. Leuven, 2003a.

——. 'L'enseignement des sciences mathématiques sous le règne de Kangxi (1662–1722) et son contexte social.' In *Education et Instruction en Chine. II. Les formations spécialisées*, eds Christine Nguyen-Tri and Catherine Despeux, 69–88. Paris/Louvain: Centre d'études chinoises/Peeters, 2003b.

——. 'Chen Houyao "Zhaodui jiyan" shizheng 陳厚耀《召對紀言》釋證.' In *Wenshi xinlan: Zhejiang guji chubanshe jianshe ershi zhounian jinian lunwen ji* 文史新瀾：浙江古籍出版社建社二十周年紀念論文集, 458–475. Hangzhou: Zhejiang guji chubanshe, 2003c.

——. 'Kangxi shidai de shuxue jiaoyu ji qi shehui beijing 康熙時代的數學教育及其社會背景.' *Faguo hanxue* 法國漢學 8 (2003d): 434–448.

——. 'Zailun Bai Jin de "Yijing" yanjiu – cong Fantikang tushuguan suo cang shougao fenxi qi yanjiu beijing, mudi ji fanxiang 再論白晉的《易經》研究—從梵蒂岡圖書館所藏手稿分析其研究背景、目的及反響.' In *Zhongwai guanxi shi: xin cailiao yu xin wenti* 中外關係史：新材料與新問題, eds Rong Xinjiang 榮新江 and Li Xiaocong 李孝聰, 315–323. Beijing: Kexue chubanshe, 2004.

——. 'Fengjiao tianwenxuejia yu "Liyi zhi zheng" (1700–1702) 奉教天文學家與 '禮儀之爭' (1700－1702).' In *Encounters and dialogues: changing perspectives on Chinese–Western exchanges from the sixteenth to the eighteenth centuries*, ed. Xiaoxin Wu, 197–209. Nettetal: Steyler Verlag, 2005a.

——. 'Shanshan laichi de 'Xiyang xiaoxi' – 1709 nian jiaohuang zhi Kangxi xin daoda gongting shimo 姍姍來遲的《西洋消息》—1709 年教皇致康熙信到達宮廷始末.' *Wenhua zazhi* 文化雜誌 15 (2005b): 1–14.

——. 'Qingchu lisuan yu jingxue guanxi jianlun 清初曆算與經學關係簡論.' In *Qingdai jingxue yu wenhua* 清代經學與文化, ed. Peng Lin 彭林, 409–418. Beijing: Beijing daxue chubanshe, 2005c.

——. 'Kangxi shidai de lisuan huodong: jiyu dang'an ziliao de xin yanjiu 康熙時代的曆算活動：基於檔案資料的新研究.' In *Shiliao yu shijie — zhongwen wenxian yu Zhongguo jidujiao shi yanjiu* 史料與視界—中文文獻與中國基督教史研究, ed. Zhang Xianqing 張先清, 40–60. Shanghai: Shanghai remin chubanshe, 2007a.

——. 'Ming-Qing zhiji "Lishi qiuye" lun zhi yuan yu liu 明清之際 "禮失求野" 論之源與流.' *Ziran kexue shi yanjiu* 自然科學史研究 26, no. 3 (2007b): 303–311.

Han Qi 韓琦, and Catherine Jami 詹嘉玲. 'Kangxi shidai xifang shuxue zai gongting de chuanbo – yi An Duo he "Suanfa zuanyao zonggang" de bianzuan wei li 康熙時代西方數學在宮廷的傳播 — 以安多和《算法纂要總綱》的編纂為例.' *Ziran kexue shi yanjiu* 自然科學史研究 22, no. 2 (2003): 145–156.

Hanson, Marta. 'The significance of Manchu medical sources in the Qing.' In *Proceedings of the First North American Conference on Manchu Studies (Portland, OR, May 9–10, 2003): Volume 1: Studies in Manchu Literature and History*, eds Stephen Wadley, Carsten Naeher and Keith Dede, 131–175. Wiesbaden: Harrassowitz Verlag, 2006.

Hart, Roger. 'Proof, propaganda and patronage: A cultural history of the dissemination of Western studies in seventeenth-century China.' PhD dissertation. UCLA, 1997.

Hashimoto, Keizō 橋本敬造. 'Rekishō kōsei no seiritsu 曆象考成の成立.' In *Min-Shin jidai no kagaku gijutsu shi* 明清時代の科學技術史, eds Yabuuti Kiyosi 薮内清 and Yoshida Mitsukuni 吉田光邦, 49–92. Kyoto: Kyōto daigaku Jinbun kagaku kenkyūjo, 1970a.

——. 'Bai Buntei no rekisan gaku – Kōki nenkan no tenmon rekisangaku 梅文鼎の曆算学ー康熙年間の天文曆算学.' *Tōhō gakuhō* 東方学報 41 (1970b): 491–518.

——. 'Daen hōno tenkai – "Rekishō kōsei gohen" no naiyō ni tsuite 楕円法の展開ー曆象考成後編の內容について.' *Tōhō gakuhō* 東洋学報 42 (1971): 245–272.

——. 'Bai Buntei no sūgaku kenkyū 梅文鼎の数学研究.' *Tōhō gakuhō* 東方学報 44 (1973): 233–279.

——. *Hsü Kuang-ch'i and astronomical reform: The process of the Chinese acceptance of Western astronomy, 1629–1635*. Osaka: Kansai University Press, 1988.

——. 'Jean-François Foucquet no kagaku chishiki to "Rekihō montō" ni okeru Kepler hōsoku no ichi ジャン－フランスワ＝フーケの科学知識と『曆法問答』におけるケプラー法則の位置.' In *Chūgoku gijutsushi no kenkyū* 中国技術史の研究, ed. Tanaka Tan 田中淡. Kyoto: Kyōto daigaku jinbunkagaku kenkyūsho, 1998.

——. 'Seihō hihan no-naka no tengaku: Kōki shonen no "rekigoku" wo chûshin ni shite 西法批判のなかの天学: 康熙初年の曆獄を中心にして.' *Kansai daigaku Tōzai gakujutsu kenkyūjo kiyō* 関西大学東西学術研究所紀要 40 (2007): 2138.

Hashimoto, Keizo and Catherine Jami. 'Kepler in China: a missing link? Jean-François Foucquet's *Lifa wenda*.' *Historia Scientiarum* 6, no. 3 (1997): 171–185.

——. 'From the Elements to calendar reform: Xu Guangqi's shaping of mathematics and astronomy.' In *Statecraft and intellectual renewal in late Ming China: The cross-cultural synthesis of Xu Guangqi*, eds Catherine Jami, Peter M. Engelfriet and Gregory Blue, 263–279. Leiden: Brill, 2001.

Hay, Jonathan. 'Ming palace and tomb in early Qing Jiangning: dynastic memory and the openness of history.' *Late Imperial China* 20, no. 1 (1999): 1–48.

——. 'The Kangxi emperor's brush-traces: Calligraphy, writing and the art of imperial authority.' In *Body and face in Chinese visual culture*, eds Wu Hung and Katherine R. Tsiang, 311–334. Cambridge, Mass.: Harvard University Asia Center, 2005.

Henderson, John B. *The development and decline of Chinese cosmology*. New York: Columbia University Press, 1984.

——. 'Ch'ing scholars' views of Western astronomy.' *Harvard Journal of Asiatic Studies* 6, no. 3 (1986): 171–185.

Héraud, Bénédicte. 'Les fonds chinois de la Bibliothèque du roi 1719–1742.' DEA Sciences de l'information et de la communication, ENSSIB, Université de Lyon 2, Université de Lyon 3, 1993.

Heyndrickx, Jerome, ed. *Philippe Couplet, S.J. (1623–1693), the man who brought China to Europe*. Nettetal: Steyler Verlag, 1990.

Hibbert, Eloise Talcott. *K'ang Hsi emperor of China*. London: Kegan Paul, Trench, Trubner & Co, 1940.

Hilairet, Jacques. *Dictionnaire historique des rues de Paris*. 2 vols. Paris: Editions de Minuit, 1963.

Ho, Peng Yoke. 'Natural phenomena recorded in the Đai-Việt Su'-ky Toan-Thu': an early Annamese historical source.' *Journal of the American Oriental Society* 84, no. 2 (1964): 127–149.

——. *Chinese mathematical astrology: reaching out to the stars*. London/New York: RoutledgeCurzon, 2003.

Horng, Wann-Sheng. 'Nineteenth century Chinese mathematics.' In *Philosophy and conceptual history of science in Taiwan*, eds Cheng-Hung Lin and Daiwie Fu, 167–208. Dodrecht: Kluwers, 1993.

Horng Wann-Sheng 洪萬生. 'Shiba shiji dongsuan yu zhongsuan de yi duan duihua: Hong Zhengxia vs. He Guozhu 十八世紀東算與中算的一段對話：洪正夏vs. 何國柱.' *Hanxue yanjiu* 漢學研究 20, no. 2 (2002): 57–80.

Hostetler, Laura. *Qing colonial enterprise: Ethnography and cartography in early modern China*. Chicago / London: The University of Chicago Press, 2001.

——. 'Global or local? Exploring connections between Chinese and European geographical knowledge during the early modern period.' *East Asian Science, Technology and Medicine* 26 (2007): 117–135.

Hou Gang 侯鋼; 'Liang Song yishu ji qi yu shuxue zhi guanxi chulun 兩宋易數及其與數學致關係初論.' PhD dissertation. Beijing, IHNS, 2006.

Hsia, Florence C. 'French Jesuits and the mission to China: science, religion, history.' PhD dissertation. The University of Chicago, 1999.

——. *Sojourners in a strange land: Jesuits and their scientific missions in late imperial China*. Chicago/London: The University of Chicago Press, 2009.

Hsu, Kuang-tai 徐光台. 'Xifang jidu shenxue dui Donglin renshi Xiong Mingyu de chongji ji qi fanying 西方基督神學對東林人士熊明遇的衝激及其反應.' *Hanxue yanjiu* 漢學研究 26, no. 3 (2008): 191–224.

——. 'Shiqi shiji chuan Hua xixue dui fenye shuo de chongji 十七世紀傳華西學對分野說的衝激.' *Jiuzhou xuelin* 九州學林 7, no. 2 (2009): 2–42.

Hu, Minghui. 'Provenance in Contest: Searching for the Origins of Jesuit Astronomy in Late Imperial China.' *The International History Review* 24, no. 1 (2002): 1–36.

——. 'Cosmopolitan Confucianism: China's road to modern science.' PhD dissertation. UCLA, 2004.

Huang, Chin-shing. *Philosophy, philology, and politics in eighteenth century China: Li Fu and the Lu-Wang school under the Ch'ing*. Cambridge: Cambridge University Press, 1995.

Huang Runhua 黃潤華, and Qu Liusheng 屈六生. *Quanguo manwen tushu ziliao lianhe mulu* 全國滿文圖書資料聯合目錄. Beijing: Shumu wenxian chubanshe, 1991.

Huang, Xiang. 'The trading zone communication of scientific knowledge: an examination of Jesuit science in China (1582–1773).' *Science in Context* 13, no. 3 (2005): 393–427.

Huang Yi-Long 黃一農. 'Qingchu Qintianjian zhong ge minzu tianwenjia de quanli qifu 清初欽天監中各民族天文家的權力起伏.' *Xin shixue* 新史學 2, no. 2 (1991a): 75–108.

——. 'Zeri zhi zheng yu "Kangxi liyu" 擇日之爭與 "康熙曆獄".' *Qinghua xuebao* 清華學報 21, no. 2 (1991b): 247–280.

——. 'Court divination and Christianity in the K'ang-hsi era.' *Chinese Science* 10 (1991c): 1–20.

——. 'Qing chuqi dui Zui, Shen liang xiu xianhou cixu de zhengzhi: shehui tianwenxue shi zhi yi ge an yanjiu 清初期對嘴、參兩宿先後次須序之爭執：社會天文學史之一個案研究.' In *Jindai Zhongguo kejishi lunji* 近代中國科技史論集, eds Yang Cuihua 楊翠華 and Huang Yi-Long 黃一農, 71–93. Taipei/Xinzhu: Zhongyang yanjiu yuan jindaishi yanjiusuo/Guoli Qinghua daxue lishi yanjiusuo, 1991d.

——. 'Wu Mingxuan yu Wu Mingxuan — Qing chu yu xi fa kangzheng de yi dui huihui tianwenxue jia xiongdi? 吳明炫與吳明烜——清初與西法相抗爭的一對回回天文家兄弟?' *Dalu zazhi* 大陸雜誌 84, no. 4 (1992): 145–149.

——. 'L'attitude des missionnaires jésuites face à l'astrologie et à la divination chinoises.' In *L'Europe en Chine: interactions scientifiques, religieuses et culturelles aux XVIIe et XVIIIe siècles*, eds Catherine Jami and Hubert Delahaye, 87–108. Paris: Collège de France, Institut des Hautes Études Chinoises, 1993a.

——. 'Qingchu tianzhujiao yu huijiao tianwenjia de douzheng 清初天主教與回教天文家的鬥爭.' *Jiuzhou xuekan* 九州學刊 5, no. 3 (1993b): 47–69.

——. 'Biligui zai huopaoxue shang de yingyong 比例規在火砲學上的應用.' *Kexueshi tongxun* 科學史通訊 15 (1996) 4–11.

——. 'Sun Yuanhua (1581–1632): A Christian convert who put Xu Guangqi's military reform policy into practice.' In *Statecraft and intellectual renewal in late Ming China: The cross-cultural synthesis of Xu Guangqi (1562–1633)*, eds

Catherine Jami, Peter M. Engelfriet and Gregory Blue, 235–259. Leiden: Brill, 2001.

———. *Liang tou she: Mingmo Qingchu de diyi dai tianzhujiao tu* 兩頭蛇: 明末清初的第一代天主教徒. Hsinchu: Guoli Qinghua daxue chubanshe, 2005.

Huang Yi-Long 黃一農 and Zhang Zhicheng 張志誠. 'Zhongguo chuantong qihou shuo de yanjin yu shuaitui 中國傳統氣候說的演進與衰退.' *Qinghua xuebao* 清華學報 23, no. 2 (1993): 125–147.

Hucker, Charles O. *A dictionary of official titles in imperial China*. Stanford: Stanford University Press, 1985.

Hummel, Arthur W. *Eminent Chinese of the Ch'ing period, 1644–1912*. 2 vols. Washington, D.C.: U.S. Government Printing Office, 1943.

Iannaccone, Isaia. *Johann Schreck Terrentius: Le scienze rinascimentali e lo spirito dell'Accademia dei Lincei nella Cina dei Ming*. Naples: Istituto Universitario Orientale, 1998.

Jami, Catherine. 'Jean-François Foucquet et la modernisation de la science en Chine: la "Nouvelle méthode d'algèbre".' Mémoire de maîtrise, Université de Paris 7, 1986.

———. 'Western influence and Chinese tradition in an eighteenth century mathematical work.' *Historia Mathematica* 15 (1988): 311–331.

———. *Les méthodes rapides pour la trigonométrie et le rapport précis du cercle (1774): tradition chinoise et apport occidental en mathématiques*. Paris: Collège de France, Institut des hautes études chinoises, 1990.

———. 'Rencontre entre arithmétiques chinoise et occidentale au XVIIe siècle.' In *Histoire de fractions, fractions d'histoire*, eds Paul Benoit, Karine Chemla and Jim Ritter, 351–373. Basel: Birkhaüser, 1992.

———. 'Learning the mathematical sciences in early and mid-Ch'ing China.' In *Education and society in late imperial China*, eds Benjamin A. Elman and Alexander Woodside, 223–256. Berkeley: University of California Press, 1994a.

———. 'The French mission and Verbiest's scientific legacy.' In *Ferdinand Verbiest, S.J. (1623–1688), Jesuit missionary, scientist, engineer and diplomat*, ed. John W. Witek, 531–542. Nettetal: Steyler, 1994b.

———. 'History of mathematics in Mei Wending's (1633–1721) work.' *Historia Scientiarum* 53 (1994c): 157–172.

———. 'L'empereur Kangxi (1662–1722) et la diffusion des sciences occidentales en Chine.' In *Nombres, astres, plantes et viscères: sept essais sur l'histoire des sciences en Asie orientale*, eds Pierre-Etienne Will and Isabelle Ang, 193–209. Paris: Collège de France, Institut des Hautes Études Chinoises, 1994d; 493–499.

———. 'From Clavius to Pardies: the geometry transmitted to China by Jesuits (1607–1723).' In *Western humanistic culture presented to China by Jesuit missionaries (17th–18th centuries)*, ed. Federico Masini, 175–199. Rome: Institutum Historicum S.I., 1996.

———. 'Giulio Aleni's contribution to geometry in China: the *Jihe yaofa*.' In *'Scholar from the West': Giulio Aleni S.J. and the dialogue between Christianity and China*, eds Tiziana Lippiello and Roman Malek, 553–572. Nettetal: Steyler Verlag, 1997.

———. 'Mathematical knowledge in the *Chongzhen lishu*.' In *Western learning and christianity in China: The contribution of Johann Adam Schall von Bell, S.J. (1592–1666)*, ed. Roman Malek, 661–674. Sankt Augustin: China-Zentrum & Monumenta Serica Institute, 1998.

———. 'Teachers of mathematics in China: The Jesuits and their textbooks (1580–1723).' *Archives internationales d'histoire des sciences* 52 (2002a): 159–175.

———. 'Western learning and imperial control: the Kangxi emperor's (r. 1662–1722) performance.' *Late Imperial China* 23, no. 1 (2002b): 28–49.

———. 'Légitimité dynastique et reconstruction des sciences en chine au XVIIe siècle: Mei Wending (1633–1721).' *Annales* 59, no. 4 (2004): 701–727.

———. 'L'empereur Kangxi et les sciences: réflexion sur l'histoire comparée.' *Etudes chinoises* 25 (2006a): 13–40.

———. 'Review of Benjamin A. Elman, *On their own terms. Science in China 1550–1900* (Cambridge, Mass, 2005).' *Journal of Asian Studies* 65, no. 2 (2006b) 405–407.

———. 'A discreet mathematician: Antoine Thomas (1644–1709) and his textbooks.' In *A lifelong dedication to the China mission: Essays presented in honor of Father Jeroom Heyndrickx, CICM, on the occasion of his 75th birthday and the 25th anniversary of the F. Verbiest institut K.U. Leuven*, eds Noël Golvers and Sara Lievens, 447–468. Leuven: Ferdinand Verbiest Institute, 2007a.

———. 'Western learning and imperial scholarship: the Kangxi emperor's study.' *East Asian Science, Technology and Medicine* 27 (2007b): 146–172.

———. 'Pékin au début de la dynastie Qing: capitale des savoirs impériaux et relais de l'Académie royale des sciences de Paris.' *Revue d'histoire moderne et contemporaine* 55, no. 2 (2008a): 43–70.

———. 'Tomé Pereira (1645–1708), clockmaker, musician and interpreter at the Kangxi court: Portuguese interests and the transmission of science.' In *The Jesuits, the Padroado and East Asian science (1552–1773)*, eds Luis Saraiva and

Catherine Jami, 187–204. Singapore: World Scientific Publishing, 2008b.

———. 'Science in Manchu in early Qing China: does it matter?' In *Looking at it from Asia: the processes that shaped the sources of history of science*, ed. Florence Bretelle-Establet, 371–391. Dordrecht/Heidelberg/London/New York: Springer, 2010.

Jami, Catherine and Han Qi. 'The reconstruction of imperial mathematics in China during the Kangxi reign (1662–1722).' *Early science and medicine* 8, no. 2 (2003): 88–110.

Jiang Qingbo 江慶柏. *Qingdai renwu shengzu nianbiao* 清代人物生卒年表. Beijing: Renmin wenxue chubanshe, 2005.

Jiang Xiaoyuan 江曉原. 'Shilun Qingdai xixue zhongyuan shuo 試論清代西學中源說.' *Ziran kexue shi yanjiu* 自然科學史研究 7 (1988): 101–108.

Jin Fu 金福. 'Qingchu gaili douzheng yu Kangxi di tiansuan xueshu 清初改曆鬥爭與康熙帝天算學術.' *Nei Menggu shifan daxue bao (ziran kexue)* 内蒙古師範大學報（自然科學）1 (1989): 16–23.

Jun, Yong Hoon. 'Mathematics in context: a case in early nineteenth-century Korea.' *Science in context* 19, no. 4 (2006): 475–512.

Kangxi and Jiao Bingzhen 焦秉貞. *Le Gengzhitu: le livre du riz et de la soie. Poèmes de l'empereur Kangxi; peintures sur soie de Jiao Bingzhen*. Translated by Bernard Fuhrer. Paris: Jean-Claude Lattès, 2003.

Karpinski, L. C., and F. W. Kokomoor. 'The teaching of elementary geometry in the seventeenth century.' *Isis* 10, no. 1 (1928): 21–32.

Kawahara Hideki 川原秀城. 'Ritsureki engen to Kato Rakusho 律曆淵源と河図洛書.' *Chūgoku kenkyū shūkan* 中国研究週刊 16 (1995): 1319–1410.

Kessler, Lawrence D. 'Chinese scholars and the early Manchu state.' *Harvard Journal of Asiatic Studies* 31 (1971): 179–200.

———. *K'ang-hsi and the consolidation of Ch'ing rule, 1661–1684*. Chicago: The University of Chicago Press, 1976.

Kim, Seonmin. 'Ginseng and border trespassing between Qing China and Chosŏn Korea.' *Late imperial China* 28, no. 1 (2007) 33–61.

Kim, Yung Sik. *The natural philosophy of Chu Hsi (1130–1200)*. Philadelphia: American Philosopical Society, 2000.

Kobayashi, Hiromitsu. 'Suzhou print and Western perspective: the painting techniques of Jesuit artists at the Qing court, and dissemination of the contemporary court style of painting to mid-eighteenth-century Chinese society through blockprints.' In *The Jesuits II: cultures, sciences and the arts, 1540–1773*, eds John W. O'Malley, Gauvin Alexander Bailey, Steven J. Harris and T. Frank Kennedy, 262–286. Toronto: Toronto University Press, 2006.

Kokomoor, F. W. 'The teaching of elementary geometry in the seventeenth century.' *Isis* 11, no. 1 (1928): 85–110.

Kolmaš, Josef. 'Father Karel (Carolus) Slavíček, S.J. (1678–1735): the first Bohemian sinologist.' *Monumenta Serica* 54 (2006): 243–251.

Kurtz, Joachim. 'Anatomy of a textual monstrosity: dissecting the *Minglitan* (*De Logica*, 1931).' In *Linguistic exchanges between Europe, China and Japan*, ed. Federica Casalin. Rome: Tiellemedia, 2008, 35–57.

Lam, Lay Yong. *A critical study of the Yang Hui suan fa: a thirteenth-century Chinese mathematical treatise*. Singapore: Singapore University Press, 1977.

Lam, Lay Yong and Tian Se Ang. *Fleeting footsteps: tracing the conception of arithmetic and algebra in ancient China*. River Edge, NJ: World Scientific, 2004.

Landry-Deron, Isabelle. 'Les leçons de sciences occidentales de l'empereur de Chine Kangxi (1662–1722): textes des journaux des Pères Bouvet et Gerbillon.' 2 vols. Diplôme de l'EHESS, Paris, 1995.

———. 'Les mathématiciens envoyés en Chine par Louis XIV en 1685.' *Archive for the history of exact science* 55 (2001): 423–463.

———. *La preuve par la Chine: la 'Description' de J.-B. Du Halde, jésuite, 1735*. Paris: Editions de l'Ecole des Hautes Etudes en Sciences Sociales, 2002.

Lau, D.C. *Mencius*. Harmondsworth: Penguin Books, 1970.

———. *Confucius. The Analects*. Harmondsworth: Penguin Books, 1979.

Ledyard, Gari. 'Cartography in Korea.' In *The history of cartography, volume 2, book 2: Cartography in the traditional East and Southeast Asia societies*, eds J.B. Harley and David Woodward. Chicago: The University of Chicago Press, 1994, 235–345.

Leibundgut, Brice. *La rhubarbe et la pivoine: Dominique Parrenin, 1665–1741, missionnaire jésuite à a cour des empereurs mandchous*. Morteau: Comtois illustres, 2007.

Leitão, Henrique. 'Jesuit mathematical practice in Portugal, 1540–1759.' In *The new science and Jesuit science: seventeenth century perspectives*, ed. Mordechai Feingold, 229–247. Dordrecht; Boston; London: Kluwer Academic Publishers, 2002.

——. 'A periphery between two centres? Portugal on the scientific route from Europe to China (sixteenth and seventeenth centuries).' In *Travels of learning: a geography of science in Europe*, eds Ana Simões, Ana Carneiro and Maria Paula Diogo, 19–46. Dordrecht: Kluwer academic publishers, 2003.

——. 'The content and context of Manuel Dias' *Tianwenlüe* 天問略.' In *The Jesuits, the Padroado and East Asian science (1552–1773)*, eds Luis Saraiva and Catherine Jami, 99–121. Singapore: World Scientific, 2008.

Leung, Beatrice. *Sino-Vatican relations: problems in conflicting authorities 1976–1986*. Cambridge: Cambridge University Press, 1992.

Li Bozhong 李伯重. 'Qingdai Zhongguo dushuren de shuxue zhishi 清代中國讀書人的數學知識.' *Dushu* 讀書, no. 9 (2006a): 81–89.

Li Di 李迪. *Zhongguo shuxue shi daxi, fujuan dier juan: Zhongguo suanxue shumu huibian* 中國數學史大系副卷第二卷：中國算學書目彙編. Beijing: Beijing shifan daxue chubanshe, 2000a.

——. 'Kangxi di yu shuxue 康熙帝與數學.' *Kexue jishu yu bianzhengfa* 科學技術與辯證法 17, no. 2 (2000b): 28–31.

——. *Zhongguo shuxue tongshi: Ming-Qing juan* 中國數學通史明清卷. Nanjing: Jiangsu jiaoyu chubanshe, 2004.

——. *Mei Wending ping zhuan* 梅文鼎評傳. Nanjing: Nanjing daxue chubanshe, 2006.

Li Di 李迪 and Cha Yongping 查永平. *Zhongguo lidai keji renwu shengzu nianbiao* 中國科技人物生卒年表. Beijing: Kexue chubanshe, 2002.

Li Di 李迪 and Guo Shirong 郭世榮. *Qingdai zhuming tianwen shuxue jia Mei Wending* 清代著名天文數學家梅文鼎. Shanghai: Shanghai kexue jishu wenxian chubanshe, 1988.

Li Huan 李歡. 'Qinggong jiucang manwen "Xiyang yaoshu" 清宮舊藏滿文《西洋藥書》.' *Zijincheng* 紫禁城 105 (1999): 30.

Li Shenwen. *Stratégies missionnaires des jésuites français en Nouvelle-France et en Chine au XVIIe siècle*. Paris: L'Harmattan, 2001.

Li Yan 李嚴. 'Mei Wending nianpu 梅文鼎年譜.' In *Li Yan & Qiang Baocong* 1998, vol. 7, 515–545. Shenyang: Liaoning jiaoyu chubanshe, 1998.

Li Yan and Du Shiran. *Chinese mathematics: a concise history*. Translated by John N. Crossley and Anthony W.-C. Lun. Oxford: Clarendon Press, 1987.

Li Yan 李嚴 and Qian Baocong 錢寶琮. *Li Yan Qian Baocong kexueshi quanji* 李嚴錢寶琮科學史全集. 10 vols. Shenyang: Liaoning jiaoyu chubanshe, 1998.

Li Zhaohua 李兆華. 'Jihe yuanben manwen chaoben de laiyuan 幾何原本滿文鈔本的來源.' *Gugong bowuyuan yuankan* 故宮博物院院刊 24 (1984): 67–69.

——. 'Guanyu "Shuli jingyun" de ruogan wenti 關於《數理精蘊》的若干問題.' In *Gusuan jinlun* 古算今論, 1–25. Tianjin: Tianjin kexue jishu chubanshe, 2000.

Li Zhaohua 李兆華 and Mei Rongzhao 梅榮照. *Suanfa tongzong jiaoshi* 算法統宗較釋. Hefei: Anhui jiaoyu chubanshe, 1990.

Libbrecht, Ulrich. *Chinese mathematics in the thirteenth century: the Shu-shu chiu-chang of Ch'in Chiu-shao*. Cambridge, Mass.: MIT Press, 1973.

——. 'What kind of science did the Jesuits bring to China?' In *Western humanistic culture presented to China by Jesuit missionaries (17th–18th centuries)*, ed. Federico Masini, 221–234. Rome: Institutum Historicum S.I., 1996.

Lim Jongtae. 'The introduction of Western science and the rationalization of traditional astronomy: reevaluating Yi Ik's "On field-allocation".' *Seoul Journal of Korean Studies* 17 (2004): 45–65.

Lindorff, Joyce. 'The harpsichord and clavichord in China during the Ming and Qing dynasties.' *Early keyboard studies* 8, no. 4 (1994): 1–8.

——. 'Missionaries, keyboards and musical exchange in the Ming and Qing courts.' *Early music* 32, no. 2 (2004): 403–414.

Lippiello, Tiziana. 'Astronomy and astrology: Johann Adam Schall von Bell.' In *Western learning in China: The contribution and impact of Johann Adam Schall von Bell, S.J. (1592–1666)*, ed. Roman Malek, 402–430. Nettetal: Steyler Verlag, 1998.

Liu Baojian 劉寶建. 'Gaishu Kangxi chao dadi celiang 概述康熙朝大地測量.' In *Qingdai gongshi luncong* 清代宮史論叢, ed. Qingdai gongshi yanjiuhui 清代宮史研究會, 244–256. Beijing: Zijingcheng chubanshe, 2001.

Liu Dun 劉鈍. 'Qingchu lisuan dashi Mei Wending 清初曆算大師梅文鼎.' *Ziran bianzhengfa tongxun* 自然辯證法通訊 8, no. 1 (1986): 53–64.

——. '"Shuli jingyun" zhong "Jihe yuanben" de diben wenti 《數理精蘊》中《幾何原本》的本問題.' *Zhongguo keji shiliao* 中國科技史料 12, no. 3 (1991): 88–96.

——. 'Fang Tai suo jian shuxue zhenji 訪台所見數學珍籍.' *Zhongguo keji shiliao* 中國科技史料 16, no. 4 (1995): 8–21.

———. 'A homecoming stranger: transmission of the method of double false position and the story of Hiero's crown.' In *From China to Paris: 2000 years of transmission of mathematical ideas*, eds Yvonne Dold-Samplonius, Joseph W. Dauben, Menso Folkerts and Benno Van Dalen, 157–166. Stuttgart: Franz Steiner Verlag, 2002a.

———. 'Cong "Laozi hua Hu" dao "xixue zhongyuan": "Yi Xia zhi bian" beijing xia wailai wenhua zai Zhongguo de qite jingli 從"老子化胡"到"西學中源":"夷夏之變"背景下外來文化在中國的奇特經歷.' *Faguo Hanxue* 法國漢學 6 (2002b): 538–564.

———. '"Liangfa" yu "jihe" — cong Qing ren de jihe guankan xueshu, zhengzhi yu wenhua de jiaohu yingxiang 量法與幾何－從清人的幾何觀看學術、政治與文化的交互影響.' Paper presented at *Li Madou yu Xu Guangxi heyi 'Jihe yuanben' sibai zhounian jinian yantaohui* 利瑪竇與徐光啟合譯《幾何原本》四百週年紀念研討會 (*Conference to commemorate the 400th anniversary of Ricci and Xu Guangqi's translation of Euclid's Elements*). Institute of Mathematics, Academia Sinica, Taibei, 2007.

Liu Lu 劉潞, ed. *Qinggong xiyang yiqi* 清宮西洋儀器. Hong Kong: Shangwu yinshuguan, 1998.

———. 'Faguo kexuejia zai Kangxi gongting 法國科學家在康熙宮廷.' *Zijincheng* 紫禁城 no. 2 (2004): 14–19.

Loewe, Michael, ed. *Early Chinese texts: a bibliographical guide*. Berkeley: Society for the Study of Early China & Institute of East Asian Studies, University of California, 1993.

Luo Fuyi 羅福頤. *Chuanshi lidai guchi tulu* 傳世歷代古尺圖錄. Beijing: Wenwu chubanshe, 1957.

Löwendahl, Björn. *Sino-Western relations, conceptions of China, cultural influences and the development of sinology disclosed in Western printed books 1477–1877: the catalogue of the Löwendahl-von der Burg collection*. 2 vols. Hua Hin: Elephant Press, 2008.

Lü Lingfeng 呂凌峰. 'Eclipses and the victory of European astronomy in China.' *East Asian Science, Technology and Medicine* 27 (2007) : 127–145.

Madrolle, Claudius. *Les premiers voyages français à la Chine: la Compagnie de la Chine*. Paris: A. Chamallel, 1901.

Magone, Rui. 'Once every three years: people and papers at the Metropolitan examination of 1685.' PhD dissertation, Free University of Berlin, 2001.

———. 'The textual tradition of Manuel Dias' *Tianwenlüe* 天問略.' In *The Jesuits, the Padroado and East Asian science (1552–1773)*, eds Luis Saraiva and Catherine Jami, 123–138. Singapore: World Scientific, 2008.

Manders, Kenneth. 'Algebra in Roth, Faulhaber, and Descartes.' *Historia Mathematica* 33 (2006): 184–209.

Martzloff, Jean-Claude. 'La compréhension chinoise des méthodes démonstratives euclidiennes au XVIIe siècle et au début du XVIIIe.' In *Actes du IIe Colloque international de sinologie: les rapports entre l'Europe et la Chine au temps des Lumières. Chantilly, 16–18 septembre 1977*. Paris: Les Belles Lettres, 1980.

———. *Recherches sur l'œuvre mathématique de Mei Wending (1633–1721)*. Paris: Collège de France, Institut des Hautes Études Chinoises, 1981a.

———. 'La géométrie euclidienne selon Mei Wending.' *Historia Scientiarum* 21 (1981b): 27–42.

———. 'The Manchu mathematical manuscript Bodoro arga i oyonggongge be araha uheri hesen i bithe of the Bibliothèque nationale, Paris: preliminary investigations.' Paper presented at the 3rd International Conference on the History of Chinese Science, Beijing 1984.

———. 'La science astronomique européenne au service de la diffusion du catholicisme en Chine: l'œuvre astronomique de Jean-François Foucquet (1665–1741).' *Mélanges de l'Ecole Française de Rome: Italie et Méditerranée* 101, no. 2 (1989): 973–989.

———. 'Eléments de réflexion sur les réactions chinoises à la géométrie euclidienne à la fin du XVIIe siècle—Le *Jihe lunyue* de Du Zhigeng vu principalement à partir de la préface de l'auteur et deux notices bibliographiques rédigées par des lettrés illustres.' *Historia Mathematica* 20, no. 2 (1993a): 160–179.

———. 'Note sur les traductions chinoises et mandchoues des *Éléments d'Euclide* effectuées entre 1690 et 1723.' In *Actes du Ve Colloque international de sinologie -Chantilly 1986*, eds Edward J. Malatesta and Yves Raguin, 201–212. Paris/Taipei: Institut Ricci/Ricci Institute, 1993b.

———. 'A glimpse of the post-Verbiet period: Jean-François Foucquet's *Lifa wenda* (Dialogues on calendrical techniques) and the modernization of Chinese astronomy or Urania's feet unbound.' In *Ferdinand Verbiest (1623–1688) Jesuit missionary, scientist, engineer and diplomat*, ed. John W. Witek, 519–529. Nettetal: Steyler Verlag, 1994.

———. 'Clavius traduit en chinois.' In *Les jésuites à la Renaissance: Système éducatif et production du savoir*, ed. Luce Giard, 309–332. Paris: PUF, 1995.

———. *A history of Chinese mathematics*. Berlin: Springer, 1997.

Masini, Federico. 'Bio-bibliographical notes on Claudio Filippo Grimaldi S.J, missionary in China (1638–1712).'

In *A life journey to the East: Sinological studies in memory of Giuliano Bertuccioli (1923–2001)*, eds Antonino Forte and Federico Masini, 185–200. Kyoto: Italian School of East Asian Studies, 2002.

McCune, Shannon. 'Jean-Baptiste Régis, S.J. an extraordinary cartographer.' In *Actes du IVe Colloque international de sinologie de Chantilly (8–11 septembre 1983). Chine et Europe: évolution et particularité des rapports Est-Ouest du XVIe au XXe siècle*, 237–248. Paris: Institut Ricci-Centre d'études chinoises, 1991.

McDermott, Joseph P. 'Chinese lenses and Chinese art.' *Kaikodo Journal* (2001): 9–29.

———. *A social history of the Chinese book: books and literati culture in late imperial China*. Hong Kong: Hong Kong University Press, 2006.

McGuire, J. E. and Martin Tamny. 'Newton's Astronomical Apprenticeship: Notes of 1664/5.' *Isis* 76, no. 3 (1985): 349–365.

Menegon, Eugenio. 'Yang Guangxian's opposition to Johann Adam Schall: Christianity and Western science in his work *Budeyi*.' In *Western learning and Christianity in China: The contribution and impact of Johann Adam Schall von Bell, S.J. (1592–1666)*, ed. John W. Witek, 311–338. Nettetal: Steyler Verlag, 1998.

Miyazaki, Kentarō. 'Roman Catholic mission in pre-modern Japan.' In *Handbook of Christianity in Japan*, ed. Mark R. Mullins, 1–18. Leiden: Brill, 2003.

Moortgat, Grete. 'Verbiest et la sphéricité de la Terre.' In *L'Europe en Chine: interactions scientifiques, religieuses et culturelles aux XVIIe et XVIIIe siècles*, eds Catherine Jami and Hubert Delahaye. Paris: Collège de France, Institut des Hautes Études Chinoises, 1993, 171–204.

———. 'Substance versus function (*ti* vs. *yong*): the humanistic relevance of Yang Guangxian's objection to Western astronomy.' In *Western humanistic culture presented to China by Jesuit missionaries (17th–18th centuries)*, ed. Federico Masini, 259–277. Rome: Institutum Historicum S.I., 1996.

Morse, Hosea Ballou. *The trade and administration of China*. London/New York: Longmans, Green, 1921.

Mungello, David E. ed. *The Chinese Rites Controversy: its history and meaning*. Nettetal: Steyler Verlag, 1994.

Naquin, Susan. *Peking: Temples and city life 1400–1900*. Berkeley/London: University of California Press, 2000.

Needham, Joseph. *Science and civilisation in China*. 23 vols. published. Cambridge: Cambridge University Press, 1954–.

Ng, On-cho. *Cheng-Zhu Confucianism in the early Qing: Li Guangdi (1642–1718) and Qing learning*. Albany: SUNY Press, 2001.

Nie Chongzheng 聂崇正, ed. *Qingdai gongting huihua* 清代宮廷繪畫. Hong Kong: Shangwu yinshuguan, 1996.

Nielsen, Bent. *A companion to Yi jing numerology and cosmology: Chinese studies of images and numbers from Han* 漢 *(202 BCE–220 CE) to Song* 宋 *(960–1279 CE)*. London: Routledge Curzon, 2003.

O'Malley, John W., Gauvin Alexander Bailey, Steven J. Harris and T. Frank Kennedy, eds *The Jesuits: cultures, sciences, and the arts, 1540–1773*. Toronto: University of Toronto Press, 1999.

———. *The Jesuits II: cultures, sciences, and the arts, 1540–1773*. Toronto: University of Toronto Press, 2006.

Okada, Hidehiro. 'Jesuit influence in emperor K'ang-hsi's Manchu letters.' In *Proceedings of the XXVIII permanent international altaistic conference: Venice 8–14 July 1985*, ed. Giovanni Stary, 165–171. Wiesbaden: Otto Harrassowitz, 1989.

Olivova, Lucie. 'Tobacco smoking in Qing China.' *Asia Major* 18, no. 1 (2005): 225–260.

Oxnam, Robert B. *Ruling from horseback: Manchu politics in the Oboi regency, 1661–1669*. Chicago: The University of Chicago Press, 1975.

Oyunbilig, Borjigidai. *Zur Überlieferungsgeschichte des Berichts über den persönlichen Feldzug des Kangxi Kaisers gegen Galdan (1696–1697)*. Wiesbaden: Otto Harrassowitz, 1999.

Pagani, Catherine. *Eastern magnificence and European ingenuity: clocks of late imperial China*. Ann Arbor: The University of Michigan Press, 2004.

Pan Yining 潘亦寧. 'Zhongxi shuxue huitong de changshi: yi "Tongwen suanzhi" (1614 nian) de bianzuan weili 中西數學會通的嘗試：以"同文算指"（1614年）的編纂為例.' *Ziran kexue shi yanjiu* 自然科學史研究 25, no. 3 (2006): 215–226.

Pang, Tatiana A. and Giovanni Stary. 'On the discovery of a printed manchu text based on Euclid's "Elements".' *Manuscripta orientalia* 6, no. 4 (2000): 49–56.

Pang, Tatjana A. *Descriptive catalogue of Manchu manuscripts and blockprints in the St Petersburg branch of the Institute of Oriental Studies, Russian Academy of Sciences*. Wiesbaden: Harrassowitz Verlag, 2001.

Pelliot, Paul. *Inventaire sommaire des manuscripts et imprimés chinois de la Bibliothèque vaticane*. ed. Takata Tokio. Kyoto: Istituto Italiano di Cultura / Scuola di Studi sull'Asia Orientale, 1995.

Peng, Rita Hsiao-fu. 'The K'ang-hsi emperor's absorption in Western mathematics and his extensive application of scientific knowledge.' *Shida lishi xuebao* 師大歷史學報 *(Bulletin of historical research, Graduate Institute of History, National Taiwan Normal University)*, no. 3 (1975): 349–442.

Perdue, Peter C. *China marches West: the Qing conquest of Central Eurasia.* Cambridge, Mass.: The Belknap Press of Harvard University Press, 2005.

Peterson, Willard. 'Calendar reform prior to the arrival of missionaries at the Ming court.' *Ming Studies* 21 (1968): 45–61.

——. 'Western natural philosophy published in late Ming China.' *Proceedings of the American Philosophical Society* 117, no. 4 (1973): 295–322.

——. 'From interest to indifference: Fang I-chih and Western learning.' *Ch'ing-shih wen-t'i* 3, no. 5 (1976): 72–85.

——. 'Learning from Heaven: the introduction of Christianity and other Western ideas into late Ming China.' In *Cambridge History of China: The Ming dynasty 1368–1644*, eds Denis Twitchett and Frederick W. Mote, 708–788. Cambridge: Cambridge University Press, 1998.

——. 'Changing literati attitudes toward new learning in astronomy and mathematics in early Qing.' *Monumenta Serica* 50 (2002): 375–390.

——. ed. *The Cambridge history of China.* Vol. 9 part 1: The Ch'ing empire to 1800. Cambridge: Cambridge University Press, 2002.

Pfister, Louis. *Notices biographiques et bibliographiques sur les jésuites de l'ancienne mission de Chine.* 2 vols. Shanghai: Imprimerie de la mission catholique, 1932–34.

Pih, Irène. *Le Père Gabriel de Magalhães: Un jésuite portugais en chine au XVIIe siècle.* Paris: Fundação Calouste Gulbenkian / Centro Cultural Português, 1979.

Pinot, Virgile. *Documents inédits relatifs à la connaissance de la Chine en France de 1685 à 1740.* Paris: Paul Geuthner, 1932a.

——. *La Chine et la formation de l'esprit philosophique en France (1640–1740).* Paris: Paul Geuthner, 1932b.

Pirazzoli-t'Serstevens, Michèle. *Giuseppe Castiglione, 1688–1766, peintre et architecte à la cour de Chine.* Paris: Thalia Edition, 2007.

Poppe, Nikolai Nikolaevitch, Leon Hurvitz and Hidehiro Okada. *Catalogue of the Manchu-Mongol section of the Toyo Bunko.* Tokyo / Seattle: Toyo Bunko / University of Washington Press, 1964.

Puente Ballesteros, Beatriz. '¿Quinquina o 金吉那 jinjina?: La misión jesuita francesa entre la estrategia, la fe y la medicina.' In *Proceedings of the First National Conference of Spanish Researchers in Asia–Pacific*, ed. Pedro San Ginés Aguilar. Granada: Editorial Universidad de Granada, 2007, 993–1006.

——. 'De París a Pekín, de Pekín a París: la misión francesa como interlocutor medico en la China de la era Kangxi (r. 1662–1722).' PhD dissertation, Universidad Complutense de Madrid, 2009.

Puyraimond, Marie-Jeanne. *Catalogue du fonds mandchou.* Paris: Bibliothèque nationale, 1979.

Qian Baocong 錢寶琮. 'Mei Wu'an xiansheng nianpu 梅勿安先生年譜.' In *Li Yan, Qian Baocong kexueshi quanji* 李儼錢寶琮科學史全集, 9: 107–142. Shenyang: Liaoning chubanshe, 1998.

Qiu Guangming 丘光明, Qiu Long 邱隆, and Yang Ping 楊平. *Zhongguo kexue jishu shi: Duliangheng juan* 中國科學技術史：度量衡卷. Beijing: Kexue chubanshe, 2001.

Qu Chunhai 屈春海. 'Qingdai Qintianjian ji Shixianke zhiguan nianbiao 清代欽天監暨時憲科職官年表.' *Zhongguo keji shiliao* 中國科技史料 18, no. 3 (1997): 45–71.

Rawski, Evelyn S. 'Presidential address: Reenvisioning the Qing: the significance of the Qing period in Chinese history.' *Journal of Asian Studies* 55, no. 4 (1996): 829–850.

——. *The last emperors: A social history of Qing imperial institutions.* Berkeley: University of California Press, 1998.

——. 'Qing publishing in non-Han languages.' In *Printing and book culture in late imperial China*, eds Cynthia J. Brokaw and Kai-wing Chow, 304–331. Berkeley / Los Angeles / London: University of California Press, 2005.

Rawski, Evelyn S. and Jessica Rawson, eds. *China: the three emperors 1662–1796.* London: Royal Academy of Arts, 2005.

Reich, Karen. 'The "Coss" tradition in algebra.' In *Companion encyclopedia of the history and philosophy of the mathematical sciences*, ed. Ivor Grattan-Guinness, 192–199. London: Routledge, 1994.

Reil, Sebald. *Kilian Stumpf, 1655–1720: ein Würzburger Jesuit am Kaiserhof zu Peking.* Münster: Aschendorff, 1978.

'Ricci 21st century roundtable database.' Ricci Institute, University of San Francisco: http://ricci.rt.usfca.edu/

Rider, Robin E. *A bibliography of early modern algebra.* Berkeley: Office for History of Science and Technology, University of California, Berkeley, 1982.

Rodrigues, Francisco. *Jesuitas portugueses astronómos na China 1583–1805.* Macao: Instituto cultural de Macau, 1990.

Rogaski, Ruth. 'In search of Mount Changbai: creating imperial knowledge of a Manchu homeland in the early Qing.' Unpublished paper, 2006.

Romano, Antonella. *La Contre-réforme mathématique: constitution et diffusion d'une culture mathématique jésuite à la Renaissance (1560–1640)*. Rome: Ecole Française de Rome, 1999.

———. 'Arpenter la "vigne du Seigneur"? Note sur l'activité scientifique des jésuites dans les provinces extra-européennes (XVIe–XVIIe siècles).' *Archives Internationales d'Histoire des Sciences* 52, no. 148 (2002): 73–101.

———. 'Observer, vénérer, servir: une polémique jésuite autour du Tribunal des mathématiques de Pékin.' *Annales* 59, no. 4 (2004): 729–756.

———. 'Teaching mathematics in Jesuit schools: programs, course content, and classroom practices.' In *The Jesuits II: cultures, sciences, and the arts, 1540–1773*, eds John W. O'Malley, Gauvin Alexander Bailey, Steven J. Harris and T. Frank Kennedy, 355–370. Toronto; Buffalo; London: University of Toronto Press, 2006.

Romanovsky, Wolfgang. *Die Kriege des Qing-Kaisers Kangxi gegen den Oiratenfürsten Galdan: eine Darstellung der Ereignisse und ihrer Ursachen anhand der Dokumentensammlung 'Qing Shilu' vorgelegt von Wolfgang Romanovsky*. Vienna: Verlag der Österreichischen Akademie der Wissenschaften, 1998.

Rommevaux, Sabine. *Clavius: une clé pour Euclide au XVIe siècle*. Paris: Vrin, 2005.

Roth Li, Gertraude. 'State building before 1644.' In *The Cambridge history of China*. Vol. 9, Part 1, The Ch'ing empire to 1800, ed. Willard Peterson, 9–72. Cambridge: Cambridge University Press, 2002.

Rouleau, Francis A. 'Maillard de Tournon, Papal legate at the court of Peking: the first imperial audience (31 December 1705).' *Archivum historicum Societatis Iesu* 31 (1962): 264–323.

Rouleau, Francis A. and Edward J. Malatesta. 'The "excommunication" of Ferdinand Verbiest.' In *Ferdinand Verbiest, S.J. (1623–1688): Jesuit missionary, scientist, engineer and diplomat*, ed. John W. Witek, 485–494. Nettetal: Steyler Verlag, 1994.

Rule, Paul. 'The *Acta Pekinensia* project.' *Sino-Western cultural relations journal* 30 (2008): 17–29.

Saunders, John B. and Francis R. Lee. *The Manchu anatomy and its historical origin*. Taipei: Li Ming Cultural Enterprise Co, 1981.

Sebes, Joseph. *The Jesuits and the Sino-Russian treaty of Nerchinsk (1689)*. Rome: Institutum Historicum Societatis Iesu, 1961.

Selin, Helaine, ed. *Encyclopedia of the history of science, technology and medicine in non-western cultures*. Dordrecht: Kluwer Academic publishers, 1997.

Sharma, Virendra Nath. *Sawai Jai Singh and his astronomy*. Delhi: Motilal Banarsidass Publ., 1995.

Shi Yumin 史玉民. 'Qintianjian tianwenke zhiguan nianbiao 欽天監天文科職官年表.' *Zhongguo keji shiliao* 中國科技史料 21, no. 1 (2000): 34–47.

———. 'Qing Qintianjian yashu weizhi ji xieyu guimo kao 清欽天監衙署位置及廨宇規模考.' *Zhongguo keji shiliao* 中國科技史料 24, no. 1 (2003): 58–68.

Shi Yunli 石云里. '*Lixiang kaocheng* tiyao《曆象考成》提要.' In *Zhongguo kexue jishu dianji tonghui: tianwen juan* 中国科学技术典籍通汇.天文卷, ed. Bo Shuren 薄樹人, 459–462. Zhengzhou: Henan jiaoyu chubanshe, 1995.

———. '*Tianbu zhenyuan* yu Gebaini tianwenxue zai Zhongguo de zaoqi chuanbo《天步真原》與哥白尼在中國的早期傳播.' *Zhongguo keji shiliao* 21, no. 1 (2000): 83–91.

———. 'Nikolaus Smogulecki and Xue Fengzuo's *Tianbu zhenyuan* 天步真原: its production, publication and reception.' *East Asian Science, Technology and Medicine* 27 (2007): 63–126.

———. 'The *Yuzhi lixiang kaocheng houbian* 曆象考成後編 in Korea.' In *The Jesuits, the Padroado and East Asian science*, eds Luis Saraiva and Catherine Jami, 205–229. Singapore: World Scientific, 2008a.

———. 'Reforming astronomy and compiling imperial science in the post-Kangxi era: the social dimension of the *Yuzhi lixiang kaocheng houbian* 曆象考成後編.' *East Asian Science, Technology and Medicine* 28 (2008b), 36–81.

Shi Yunli and Xing Gang. 'The first Chinese version of the Newtonian tables of the Sun and Moon.' In *Proceedings of the Fifth International Conference on Oriental Astronomy*, eds K-Y. Chen, W. Orchiston, B. Soonthornthum and R. Strom, 91–96. Chiang-Mai: University of Chiang-Mai Press, 2006.

Shu, Liguang. 'Ferdinand Verbiest and the casting of cannons in the Qing dynasty.' In *Ferdinand Verbiest, S.J. (1623–1688): Jesuit missionary, scientist, engineer and diplomat*, ed. John W. Witek, 227–244. Nettetal: Steyler Verlag, 1994.

Siu Man-Keung, and Alexeï Volkov. 'Official curriculum in traditional Chinese mathematics: how did candidates pass the examinations?' *Historia Scientiarum* 9, no. 1 (1999): 85–99.

Sivin, Nathan. 'Copernicus in China.' *Studia Copernicana* 6 (1973): 63–122.

——. 'Wang Hsi-shan (1628–1682).' In *Dictionary of Scientific Biography*, ed. Charles C. Gillispie. New York: Charles Scribner's Sons, 1976, 14: 159–168.

——. Why the scientific revolution did not take place in China–Or didn't it?' *Chinese Science* 5 (1982): 45–66.

——. *Science in ancient China: researches and reflections*. Aldershot: Variorum, 1995.

——. *Granting the seasons: the Chinese astronomical reform of 1280, with a study of its many dimensions and an annotated translation of its records*. New York: Springer, 2009.

Smith, David E. *History of mathematics*. 2 vols. Boston: Ginn and Company, 1925.

Smith, Jay M. *The culture of merit: nobility, royal service, and the making of absolute monarchy in France, 1600–1789*. Ann Arbor: Michigan University Press, 1996.

Smith, Kidder Jr., Peter K. Bol, Joseph A. Adler and Don J. Wyatt. *Sung dynasty uses of the I Ching*. Princeton: Princeton University Press, 1990.

Smith, Richard J. *Fortune tellers and philosophers: divination in traditional Chinese society*. Boulder/San Francisco/Oxford: Westview Press, 1991.

——. 'Mapping China's world: cultural cartography in late imperial times.' In *Landscape, culture, and power in Chinese society*, ed. Wen-hsin Yeh, 52–105. Berkeley: Institute of East Asian Studies, University of California Berkeley, 1998.

——. 'Jesuit interpretations of the *Yijing* (Classic of changes) in historical and comparative perspective.' In *Matteo Ricci and after: Four centuries of cultural interactions between China and the West*. Hong Kong, 2001.

Sommervogel, Carlos. *Bibliothèque de la Compagnie de Jésus*. 10 vols. Bruxelles / Paris: O. Schepens / A. Picard, 1890–1912.

Spence, Jonathan. *Ts'ao Yin and the K'ang-hsi Emperor: Bondservant and master*. New Haven: Yale University Press, 1966.

——. 'The seven ages of K'ang-hsi (1654–1722).' *Journal of Asian Studies* 26, no. 2 (1967): 205–211.

——. *Emperor of China: Self-portrait of K'ang-hsi*. New York: Vintage Books, (1974) 1975.

——. *The memory palace of Matteo Ricci*. New York: Vintage, 1984.

——. *The question of Hu*. New York: A. Knopf, 1988.

——. 'Claims and counter-claims: the Kangxi emperor and the Europeans (1661–1722).' In *The Chinese rites controversy: its history and meaning*, ed. David E. Mungello, 15–28. Nettetal: Steyler Verlag, 1994.

——. 'The K'ang-hsi reign.' In *The Cambridge history of China*. Vol. 9, Part 1, The Ch'ing empire to 1800, ed. Willard Peterson, 120–182. Cambridge: Cambridge University Press, 2002.

Standaert, Nicolas. 'The investigation of things and fathoming of principles (*gewu qiongli*) in the seventeenth-century contact between Jesuits and Chinese scholars.' In *Ferdinand Verbiest (1623–1688), Jesuit missionary, scientist, engineer and diplomat*, ed. John W. Witek, 395–420. Nettetal: Steyler Verlag, 1994.

——. 'Jesuit corporate culture as shaped by the Chinese.' In *The Jesuits: Cultures, sciences and the arts 1540–1773*, eds John W. O'Malley, Gauvin Alexander Bailey, Steven J. Harris and T. Frank Kennedy, 352–363. Toronto: University of Toronto Press, 1999a.

——. 'A Chinese translation of Ambroise Paré's Anatomy.' *Sino-western cultural relations journal*, 21 (1999b), 9–33.

——, ed. *Handbook of Christianity in China*. Leiden; Boston: Brill, 2001.

——. 'European astrology in early Qing China: Xue Fengzuo's and Smogulecki's translation of Cardano's Commentaries on Ptolemy's Tetrabiblos.' *Sino-Western cultural relations journal* 23 (2001a): 50–79.

——. *Methodology in view of contact between cultures: the China case in the seventeenth century*. CSRCS Occasional papers No. 11. Hong Kong: Centre for the study of religion and Chinese society, Chung Chi College, The Chinese University of Hong Kong, 2002.

——. 'Ritual dances and their visual representations in Ming and Qing.' *East Asian Library Journal* 12, no. 1 (2005): 68–181.

——. *The interweaving of rituals: funerals in the cultural exchange between China and Europe*. Seattle: University of Washington Press, 2008.

Stary, Giovanni. *Manchu studies: an international bibliography*. 3 vols. Wiesbaden: Otto Harrassowitz, 1990.

——. 'The "Manchu cannons" cast by Ferdand verbiest and the hitherto unknown title of his instructions.' In *Ferdinand Verbiest, S.J. (1623–1688): Jesuit missionary, scientist, engineer and diplomat*, ed. John W. Witek, 215–225. Nettetal: Steyler Verlag, 1994.

——. *A Dictionary of Manchu names*. Wiesbaden: Harrassowitz Verlag, 2000a.

——. 'Review: *Zur Überlieferungsgeschichte des Berichts über den persönlichen Feldzug des Kangxi Kaisers gegen Galdan (1696–1697)* by Borjigidai Oyunbilig.' *Journal of Asian Studies* 59, no. 3 (2000b): 732–734.

——. 'The Kangxi emperor's linguistic corrections to Dominique Parrenin's translation of the "Manchu Anatomy".' *Altaic Hakpo* 13 (2003a): 41–60.

——. *Manchu studies: an international bibliography; vol. 4: 1988–2002.* Wiesbaden: Otto Harrassowitz, 2003b.

Strassberg, Richard E. *The World of K'ung Shang-jen: A man of letters in early Ch'ing China.* New York, 1983.

——. 'War and peace: four intellectual landscapes.' In *China on paper: European and Chinese works from the late sixteenth to the early nineteenth century*, eds Marcia Reed and Paola Dematté. Los Angeles: Getty Research Institute, 2007, 89–137.

Stroup, Alice. *Royal Funding of the Parisian Académie Royale des Sciences during the 1690s.* Philadelphia: The American Philosophical Society, 1987.

——. *A company of scientists: botany, patronage, and community at the seventeenth-century Parisian Royal Academy of Sciences.* Berkeley/Los Angeles/Oxford: University of California Press, 1990.

Struve, Lynn A. 'The Hsü brothers and semiofficial patronage of scholars in the K'ang-hsi period.' *Harvard Journal of Asiatic Studies* 42, no. 1 (1982): 231–266.

——. *The Southern Ming, 1644–1662.* New Haven: Yale University Press, 1984.

——. 'Ruling from sedan chair: Wei Yijie (1616–1686) and the examination reform of the "Oboi" regency.' *Late Imperial China* 25, no. 2 (2004): 1–32.

Sun Chengsheng 孫承晟. 'Ming Qing zhiji shiren dui xifang ziran zhexue de fanying: yi Jie Xuan "Haoshu" he "Xuanji yishu" wei zhongxin 明清之際士人對西方自然哲學的反應—以揭暄《昊書》和《璇璣遺述》為中心.' PhD dissertation, Zhongguo kexueyuan yanjiusheng yuan, 2005.

Sun, Xi. *Bedeutung und Rolle des Jesuitenmissionars Ignaz Kögler (1680–1746) in China: aus chinesischer Sicht.* Frankfurt: Peter Lang, 2006.

Subrahmanyam, Sanjay. 'Par-delà l'incommensurabilité: pour une histoire connectée des empires aux temps modernes.' *Revue d'histoire moderne et contemporaine* 54, no.5 (2007), 34–53.

Szczesniak, Boleslaw. 'Diplomatic relations between emperor K'ang Hsi and King John III of Poland.' *Journal of the American Oriental Society* 89, no. 1 (1969): 157–161.

Takeda Kusuo 武田楠雄. '*Dōbun sanshi no seiritsu* 同文算指の成立.' *Kagaku shi kenkyū* 科学史研究 30 (1954): 7–14.

Tian Miao 田淼. "Jiegenfang, tianyuan, and daishu: algebra in Qing China", in *Historia Scientiarum* 9, no. 1 (1999): 101–119.

——. *Zhongguo shuxue de xihua licheng* 中國數學的西化歷程. Jinan: Shandong jiaoyu chubanshe, 2005.

——. 'The transmission of European mathematics in the Kangxi reign (1662–1722) – Looking at the international role China could play from an historical perspective.' In *Chinas' new role in the international community: challenges and expectations for the 21st century*, eds Heinz-Dieter Assmann and Karin Moser v. Filseck, 217–235. Frankfurt: Peter Lang, 2005.

Tian Miao 田淼 and Zhang Baichun 張柏春. 'Mei Wending "Yuanxi qiqi tushuo luzui" zhu yanjiu 梅文鼎《遠西奇器圖說錄最》注研究.' *Zhongguo keji shi zazhi* 中國科技史杂志, no. 4 (2006): 330–339.

——. 'Qiqi tushuo zai Zhongguo de yingxiang《奇奇圖說》在中國的影響'. in *Chuanbo yu huitong: 'Qiqi tushuo' yanjiu yu jiaozhu* 傳播與會通:《奇器圖說》研究與會通, ed. Zhang Baichun 張柏春 et al. Nanjing: Fenghuang chuban chuanmo jituan & Jiangsu kexue jishu chubanshe, 2008, 2 vols., vol 1: 226–298.

Tits-Dieuaide, Marie-Jeanne. 'Une institution sans statuts: l'Académie royale des sciences de 1666 à 1699.' In *Histoire et mémoire de l'Académie des sciences: guide de recherches*, eds Eric Brian and Christiane Demeulenaere-Douyère. Paris: Lavoisier. Tec & Doc, 1996.

Tong Qingjun 童慶鈞 and Feng Lisheng 馮立昇. 'Mei Wending "Zhongxi suanxue tong" tanyuan 梅文鼎《中西算學通》探原.' In *Nei Menggu shifan daxue xuebao (ziran kexue hanwen ban)* 內蒙古師範大學學報（自然科學漢文版）36, no. 6 (2007) 717–721.

Treutlein, Theodore E. 'Jesuit missions in China in the last years of K'ang Hsi.' *The Pacific historical review* 10, no. 4 (1941): 435–446.

Udias, Agustin. *Searching the Heavens and the Earth: The history of Jesuit Observatories.* Dordrecht: Kluwer, 2003.

Van der Cruysse, Dirk. *Louis XIV et le Siam.* Paris: Fayard, 1991.

Väth, Alfons. *Johann Adam Schall von Bell SJ: Missionar in China, kaiserlicher Astronom und Ratgeber am Hofe von Peking, 1592–1666.* Nettetal: Steyler Verlag, (1933) 1991.

Verhaeren, Hubert. *Catalogue de la Bibliothèque du Pé-Tang.* Beijing: Imprimerie des Lazaristes, 1949.

Von Collani, Claudia, ed. *Eine wissenschaftliche Akademie für China: Briefe des Chinamissionars Joachim Bouvet S.J. an Gottfried Wilhelm Leibniz und Jean-Paul Bignon über die Erforschung der chinesischen Kultur, Sprache und Geschichte.* Stuttgart: Franz Steiner, 1989.

Wadley, Stephen A. 'Altaic influences on Beijing dialect: the Manchu case.' *Journal of the American Oriental Society* 116, no. 1 (1996): 99–104.

Wakeman, Frederic Jr. *The Great Enterprise: The Manchu reconstruction of imperial order in seventeenth-century China.* 2 vols. Berkeley: University of California Press, 1985.

Waley-Cohen, Joanna. 'China and Western technology in the late eighteenth century.' *American Historical Review* 98, no. 5 (1993): 1525–1544.

——. *The sextants of Beijing: Global currents in Chinese history.* New York / London: W.W. Norton & Company, 1999.

——. 'The new Qing history', *Radical history review* 88 (2004), 193–206.

Walravens, Hartmut. 'Medical knowledge of the Manchus and the Manchu Anatomy.' *Études mongoles et sibériennes* 27 (1996): 359–374.

——. 'Mandjurische Medizin – eine Bibliographie der originalsprachigen Quellen.' in *Zentralasiatische Studien des Seminars für Sprach- und Kulturwissenschaft Zentralasiens der Universität Bonn* 30 (2000): 91–102.

Wang Bing 王冰. '"Lülü zuanyao" zhi yanjiu 《律呂纂要》之研究.' *Gugong bowuyuan yuankan* 故宮博物院院刊 102 (2002): 68–81.

——. 'Xu Risheng he xifang yinyue zhishi zai Zhongguo de chuanbo 徐日昇和西方音樂在中國的傳播.' *Wenhua zazhi* 文化雜誌 47 (2003): 71–90.

Wang Hui 王慧. 'Zhongwai keji jiaoliu de wuzheng 中外科技交流的物證.' *Zijincheng* 紫禁城, no. 2 (2004): 20–23.

Wang Miao 王淼. 'Xing Yunlu yu Mingmo chuantong lifa gaige 邢雲路與明末傳統曆法改革.' *Ziran bianzhengfa tongxun* 自然辯證法通訊 24, no. 4 (2004): 79–85.

Wang Ping 王萍. *Xifang lisuan zhi shuru* 西方曆算之輸入. Taipei: Zhongyuan yanjiuyuan Jindai shi yanjiusuo, 1966.

Wang Qianjin 汪前進. '"Huangyu quanlantu" cehui 《皇輿全覽圖》測繪.' In *Zhongguo jinxiandai kexue jishu shi* 中國近現代科學技術史, ed. Dong Guangbi 董光璧, 131–174. Changsha: Hunan jiaoyu chubanshe, 1995.

Wang Yangzong 王揚宗. 'Mingmo Qingchu "Xixue zhongyuan" shuo xinkao 明末清初 "西學中源" 說新考.' In *Keshi xinzhuan* 科史薪傳, eds Liu Dun 劉鈍 and Han Qi 韓琦, 71–83. Shenyang: Liaoning jiaoyu chubanshe, 1997.

——. 'Kangxi "Sanjiaoxing tuisuanfa lun" jianlun 康熙《三角形推算法論》簡論.' *Huowen* 或問 12 (2006a): 117–123.

——. 'Kangxi, Mei Wending yu "Xixue zhongyuan" shuo zai shangque 康熙, 梅文鼎與 "西學中源" 說再商榷.' *Zhonghua kejishi xuehui huikan* 中華科技史學會會刊 10 (2006b): 59–63.

Wardy, Robert. *Aristotle in China: Language, categories, and translation.* Cambridge: Cambridge University Press, 2000.

Watanabe Junsei 度辺純成. 'Manshūgo shiryō kara mita "kika" no gogen ni tsuite 満州語のオクリッド　東洋文庫所蔵の滿文算法原本について.' *Sūgaku shi kenkyū* 数学史研究 (2004a): 34–42.

——. 'Manshūgo no Euclid – Tōyō Bunko shozō no manbun Sanpō genpon ni tsuite 満州語のオクリッド-東洋文庫所蔵の滿文算法原本について.' *Manzokushi kenkyū* 満族史研究 3 (2004b): 40–90.

——. 'A Manchu Manuscript on arithmetic owned by Tōyō Bunko: 'Suwan fa yuwan ben bithe.' *Sciamus* 6 (2005a): 177–264.

——. 'Manshūgo igakusho *Kakutai zenroku* ni tsuite 満洲語医学書格體全録について.' *Manzokushi kenkyū* 満族史研究 5 (2005b): 22–113.

Watson, E. C. 'The early days of the Académie des sciences as portrayed in the engravings of Sébastien Le Clerc.' *Osiris* 7 (1939): 556–587.

Widmaier, Rita, ed. *Leibniz korrespondiert mit China.* Frankfurt: Vittorio Klostermann, 1990.

Wilhelm, Hellmut. 'The *Po-hsüeh hung-ju* examination of 1679.' *Journal of the American Oriental Society* 71 (1951): 60–76.

Will, Pierre-Étienne. 'Developing forensic medicine through cases in the Qing dynasty.' In *Thinking with cases: specialist knowledge in Chinese cultural history*, eds Charlotte Furth, Judith T. Zeitlin and Ping-chen Hsiong, 62–100. Honolulu: University of Hawai'i Press, 2007.

——. 'Views of the realm in crisis: testimonies on imperial audiences in the nineteenth century.' *Late Imperial China* 29, no. 1 (2008): 125–159.

——. *Official handbooks and anthologies of imperial China: a descriptive and critical bibliography*, forthcoming (10 June 2008 version).

Willard, David P. 'Chen Mao's 1717 memorial to the Kangxi emperor: perspectives on the prohibition of Catholicism in the early Qing period.' Junior Thesis, revised version, Princeton University, 2006.

Wills Jr., John E. *Embassies and illusions: Dutch and Portuguese envoys to K'ang-hsi, 1666–1687.* Cambridge, Mass.: Council on East Asian Studies, Harvard University, 1984.

——. 'Some Dutch sources on the Jesuit China mission, 1662–1687.' *Archivum historicum Societatis Iesu*, no. 54 (1985): 267–294.

——. *1688: A global history*. New York: Norton, 2001.

Witek, John W. 'An eighteenth-century Frenchman at the court of hte K'ang-hsi emperor: a study of the early life of Jean-François Foucquet.' PhD dissertation, Georgetown University, 1973.

——. *Controversial ideas in China and Europe: A biography of Jean-François Foucquet S.J. (1665–1741)*. Rome: Institutum Historicum Societatis Iesu, 1982.

——. 'Understanding the Chinese: a comparision of Matteo Ricci and the French Jesuit mathematicians sent by Louis XIV.' In *East meets West: the Jesuits in China, 1582–1773*, eds Charles E. Ronan and Bonnie B.C. Oh, 62–102. Chicago: Loyola University Press, 1988.

——. 'Johann Adam Schall von Bell and the transition from the Ming to the Ch'ing dynasty.' In *Western learning and Christianity in China: the contribution and impact of Johann Adam Schall von Bell, S.J. (1592–1666)*, ed. Roman Malek, 109–124. Nettetal: Steyler Verlag, 1998.

——. 'The role of Antoine Thomas, SJ, (1644–1709) in determining the terrestrial meridian line in eighteenth-century China.' In *The history of the relations between the Low Countries and China in the Qing era (1644–1911)*, eds Willy Vande Walle and Noël Golvers. Leuven: Leuven University Press / Ferdinand Verbiest Foundation, 2003.

Wong, R. Bin. *China transformed: historical change and the limits of European experience*. Ithaca/London: Cornell University Press, 1997.

Wu Boya 吳伯婭. 'Kangxi yu "Lüli yuanyuan" de bianzuan 康熙與《律曆淵源》的編纂.' *Gugong bowuyuan yuankan* 故宮博物院院刊 102 (2002): 62–81.

Wu Guangyou 吳光酉 et al. *Lu Longqi nianpu* 陸隴其年譜. Beijing: Zhonghua shuju, 1993.

Wu, Silas H. L. *Communication and imperial control in China: Evolution of the palace memorial system 1693–1735.* Cambridge, Mass.: Harvard University Press, 1970.

——. *Passage to power: K'ang-hsi and his Heir Apparent, 1661–1722.* Cambridge, Mass.: Harvard University Press, 1979.

Wu Wenjun 吳文俊 ed. *Zhongguo shuxue shi daxi: Ming mo dao Qing zhongqi* 中國數學史大系：明末到清中期. (Vol. 7) Beijing: Beijing shifan daxue chubanshe, 2000.

Wu Xiuliang (Silas) 吳秀良. 'Nanshufang zhi jianzhi ji qi qianqi zhi fazhan 南書房之建置及其前期之發展.' *Si yu yan* 思與言 5, no. 6 (1968): 1428–1434.

Wu Zhuo 吳焯. 'Lai Hua yesuhuishi yu Qingting Neiwufu Zaobanchu 來華耶穌會士與清廷內務府造辦處.' *Jiuzhou xuelin* 九州學林 2, no. 2 (2004): 65–102.

Xi Zezong 席澤宗. 'Lun Kangxi kexue zhengce de shiwu 論康熙科學政策的失誤.' *Ziran kexue shi yanjiu* 自然科學史研究 19, no. 1 (2000): 18–29.

Xu Haisong 徐海松. 'The reaction of scholars to the works of Ferdinand Verbiest, S.J. during the Kangxi-Qianlong reign.' In *The Christian mission in China in the Verbiest era: some aspects of the missionary approach*, ed. Noël Golvers. Leuven: Leuven University Press, 1999.

——. *Qingchu shiren yu xixue* 清初士人與西學. Beijing: Dongfang chubanshe, 2000.

——. 'Xixue dongjian yu Qing dai Zhedong xuepai 西學東漸與清代浙東學派.' In *Encounters and dialogues: changing perspectives on Chinese–Western echanges from the sixteenth to the eighteenth centuries*, ed. Xiaoxin Wu, 141–160. Sankt Augustin / San Francisco: Monumenta Serica Insitute / Ricci Institute, University of San Francisco, 2005.

Xu Yipu 徐藝圃. 'Shilun Kangxi yumen tingzheng 識論康熙御門聽政.' *Gugong bowuyuan yuankan* 故宮博物院院刊, no. 1 (1983): 3–19.

Xu Zongze 徐宗澤. *Ming-Qing Yesu huishi yizhu tiyao* 明清耶穌會士譯著提要. Taipei: Zhonghua shuju, 1958.

Xue Zhongsan 薛仲三 and Ouyang Yi 歐陽頤. *Liangqian nian zhongxi li duizhao biao* 兩千年中西曆對照錶. Hong Kong: Commercial Press, 1956.

Yabuuti Kiyosi 薮内清. *Chūgoku no tenmon rekihō* 中国の天文曆法. Tokyo: Heibonsha, 1969.

——. *Une histoire des mathématiques chinoises*. Translated by Catherine Jami and Kaoru Baba. Paris: Belin, 2000.

Yabuuti Kiyosi 薮内清 and Nakayama Shigeru 中山茂. *Jūjireki – yakuchu to kenkyū* 授時曆－訳注と研究. Kawasaki: Ai Kei Corporation, 2006.

Yang Xiaoming 楊小明. 'Huang Zongxi de kexue yanjiu 黃宗羲的科學研究.' *Zhongguo keji shiliao* 中國科技史料 18, no. 4 (1997), 20–27.

——. *Qingdai Zhedong xuepai yu kexue* 清代浙東學派與科學. Beijing: Zhongguo wenlian chubanshe, 2001.

——. 'Huang Baijia yu ri yue wuxing zuo, you xuan zhi zheng 黃百家與日月五星左右旋之爭.' *Ziran kexue shi yanjiu* 自然科學史研究 21, no. 3 (2002): 222–231.

Yang Zhen 楊珍. *Kangxi huangdi yijia* 康熙皇帝一家. Beijing: Xueyuan chubanshe, 1994.

Yee, Cordell D.K. 'Traditional Chinese cartography and the myth of westernization.' In *Cartography in the traditional East and Southeast Asian societies*, eds J.B. Harley and David Woodward, 170–402. Chicago: The University of Chicago Press, 1994.

Zhang Baichun 張柏春 et al. *Chuanbo yu huitong: 'Qiqi tushuo' yanjiu yu jiaozhu* 傳播與會通：《奇器圖說》研究與校注. Nanjing: Fenghuang chuban chuanmo jituan & Jiangsu kexue jishu chubanshe, 2008, 2 vols.

Zhang Naiwei 張乃煒 and Wang Airen 王藹人. *Qing gong shuwen chu, xubian hebian ben* 清宮述聞初、續編合編本. Beijing: Zijincheng chubanshe, 1990.

Zhang Peiyu 張培瑜. *Sanqian wubai nian liri tianxiang* 三千五百年曆日天象. Zhengzhou: Henan jiaoyu chubanshe, 1990.

Zhang Xiping 張西平. '"Yijing" yanjiu: Kangxi he Faguo chuanjiaoshi Bai Jin de wenhua duihua《易經》研究：康熙和法國傳教士白晉的文化對話.' *Wenhua zazhi* 文化雜誌 14 (2005): 83–93.

Zhang Yongtang 張永堂. *Mingmo Fangshi xuepai yanjiu chubian* 明末方氏學派研究初編. Taibei: Wenjing wenhua shiye, 1987.

——. *Mingmo Qingchu lixue yu kexue guanxi zailun* 明末清初理學與科學關係再論. Taipei: Taiwan xuesheng shuju, 1994.

Zhu Zuyan 朱祖延 and Guo Kangsong 郭康松. *Qing shilu leizuan. Kexue jishu juan* 清實錄類鑽。科學技術卷. Wuhan: Wuhan chubanshe, 2005.

Ziggelaar, August. *Le physicien Ignace Gaston Pardies, S.J. (1636–1673)*. Odense: Odense Universitetsforlag, 1971.

Zürcher, Erik. 'In the Yellow Tiger's den: Buglio and Magalhães at the court of Zhang Xianzhong, 1644–1647.' *Monumenta Serica* 50 (2002): 355–374.

——. *Kouduo richao – Li Jiubiao's Diary of Oral Admonitions: a late Ming Christian journal*. 2 vols. Sankt Augustin: Institut Monumenta Serica, 2007.

Zürcher, Erik, Nicolas Standaert and Adrianus Dudink. *Bibliography of the Jesuit mission in China ca.1580 – ca.1680*. Leiden: Centre of Non-Western Studies, Leiden University, 1991.

Zurndorfer, Harriet T. '"One Adam having driven us out of paradise, another has driven us out of China": Yang Kuang-hsien's challenge of Adam Schall von Bell', in *Conflict and accommodation in early modern China: essays in honour of Erik Zürcher*, eds Leonard Blussé and Harriet T. Zurndorfer. Leiden: Brill, 1993, 141–168.

INDEX

abacus 14, 356–7
  and popular mathematics 17–18
  and speed of 194
  as universal counting device 15
Académie royale des sciences (Royal Academy of Sciences, Paris):
  and foundation of 102
  and Jesuits 103
  and plans mission to China 103–4
  and questions about China 106–7
  and training of Jesuits 103
  and travels and observations of mission to China 116–19
  and work on navigation 102–3
Aerrebala xinfa 阿爾熱巴拉新法, **see** *New method of algebra*
Alantai 阿蘭泰 232
Aleni, Giulio:
  and *Areas outside the concern of the Imperial Geographer* (*Zhifang waiji* 職方外紀) 29, 150
  and *Main methods of geometry* (*Jihe yaofa* 幾何要法) 34
  and *Outline of Western learning* (*Xixue fan* 西學凡) 29
algebra:
  and *Calculation by borrowed root and powers* (*Jiegenfang suanfa* 借根方算法):
    abridgment of 209–10
    ambiguity of Chinese terms 206
    composition of 200–1
    contents 204–5
    foreword to 202–6
    problem solving 208–9
    structure of 201–2
    symbolic linear algebra 210–12
    symbols 207
    table of power numbers 206–7
  and celestial element (*tianyuan* 天元) place-value algebra 15, 279, 383
  and cossic algebra 24, 201
  and Foucquet's writings on 294–6, 299
    symbols in 300–5
  and Kangxi's study of 200, 201, 294–6, 298–9, 302, 304
Algebra (Clavius) 201, 206, 208
anatomy:
  and Chinese forensic medicine 308
  and Kangxi's commission of writings on 307–8
  and Kangxi's study of 146–8
Andersen, Hans Christian 390

annual difference (*suicha* 歲差) (precession) 123, 370
Archimedes 190
Areas outside the concern of the Imperial Geographer (*Zhifang waiji* 職方外紀) 29, 150
Aristotelian philosophy:
  and *Conimbricenses* 23, 30, 31, 78
  of less relevance for evangelising mission 145
  and the structure of mathematics 22
  and Verbiest's *Study of the fathoming of principles* (*Qiongli xue* 窮理學) 78
Arnauld, Antoine 163, 164
astrology, and Schall 40
  see also divination
Astronomical Bureau (*Qintianjian* 欽天監):
  as agency of Ministry of Rites 38
  and arbitration over Foucquet's innovations 290–3
  and autonomy of 38, 65
  and calculations checked by Kangxi 260–1
  and *Great concordance* (*Datong* 大統) system 36, 59, 67
  and Grimaldi appointed Administrator of the Calendar (*zhili lifa* 治理曆法) 116
  and hostility to Jesuit presence in 40, 53–4
    death sentences 52–3
    death sentences rescinded 53
    house arrests 54
    Yang Guangxian 49–50, 51–2
  and increase in students 66–7
  and internal structure of 38, 39, 65
  and Kangxi's intervention 1
  and Muslim Section (*Huihui ke* 回回科) 40–1, 57, 65
  and observations by Section of heavenly signs (*Tianwen ke* 天文科) 66
  and rehabilitation of Jesuit astronomy 58–65
    dispute over Wu Mingxuan's calendar 59–64
  and Schall:
    appointed head of 37
    arrested 52
    conflict with Chinese subordinates 40
    criticism of 38–40
    death sentence 52–3
    death sentence rescinded 53
    as imperial astronomer 38–41

and solar eclipses:
  divinatory significance 225, 228–9
  prediction of 223
and Verbiest:
  appointed Administrator of the Calendar (*zhili lifa* 治理曆法) 64
  increase in student numbers 66–7
  responsibilities of 65–6
  teaching duties 66
  updating instruments 67–8
and Wu Mingxuan:
  appointed Vice-Director 54
  first calendar submitted by 59–60
  and Yang Guangxian appointed Director 54
astronomy 49–54
  and autonomy of imperial astronomy 293–4
  and calendar reform 32–4, 36–7
  and court concern over reliance on Europeans 286
  and *Dialogue on astronomical methods* (*Lifa wenda* 曆法問答) (Foucquet):
    aims to secure long-term need for Jesuit astronomers 288–9
    arbitration on innovations of 290–3
    dialogic structure 287–8
    heliocentrism 288
    innovations 288, 289
    Kangxi's response to 289
    shortcomings of Tycho Brahe's astronomy 288
    treatises included in 289
  and *Doubts concerning the study of astronomy* (*Lixue yiwen* 曆學疑問) 218–22, 247, 248
  and *Great concordance* (*Datong* 大統) system 36, 59, 67, 214–15
  and increase in students 57, 66–7
  and Kangxi:
    challenges sinocentric view 226–7
    commissions textbook from Jesuits 261
    contempt for superstition surrounding 225–6
    demonstration of skills 233
    Discussion of triangles and computation (*Sanjiaoxing tuisuanfa lun* 三角形推算法論) 248–51
    rehabilitation of Jesuit astronomy 58–65
    role in Chinese cosmology 227–9

421

Index

astronomy (cont.)
  and *Later part of the thorough investigation of astronomical phenomena* (*Lixiang kaocheng houbian* 曆象考成後編) 379–80
  and reform of 32–4
  and religion 70–2
  and *Season granting* (*Shoushi* 授時) system 214–15
  and solar eclipses 222–9
    divinatory significance 223–4, 225, 228–9
    prediction of 222–3
  and summer solstice of 1711 260–1
  and *Thorough investigation of astronomical phenomena* (*Lixiang kaocheng* 曆象考成) 364, 370–2
    criticism of 378–9
  and twenty-eight lodges 122
  see also Astronomical Bureau; Nanjing Observatory, Kangxi's visit to
Aveiro, Duchess of (Maria-Guadalupe de Lencastre) 182, 183

Bacon, Francis 386
Bai Yingtang 白暎棠 377
Bannermen 7n12, 66, 67, 82, 143, 151, 157, 167, 229, 264, 373, 376, 382, 383, 388
Bartholinus, Thomas 308n109
base and altitude (*gougu* 句股) 45, 46, 47, 87, 90, 92, 202, 245, 251, 306, 316, 320, 322, 330
Basics of the eight [trigonometric] lines (*Baxian biao gen* 八線表根) (Thomas) 191
Baudino, Giuseppe 240
Baxian biao gen 八線表根, *see Basics of the eight [trigonometric] lines*
Beijing:
  and French mission to China 112–16
  and Jesuits arrested 52
  and Jesuits' residences 240–1, 242
  and Kangxi's 60th birthday ceremony 265–6
  and Le Comte on 113–14
  and Manchu entry into 36
Beitang 北堂, *see* Northern Church
Belleville, Charles de 241
Bencao gangmu 本草綱目, *see Systematic materia medica*
bili 比例, *see* ratios and proportions
Biligui jie 比例規解, *see Explanation of the proportional compass*
Boerhe 博爾和 268
book market, and broadening of 15
Book of change (*Yijing* 易經) 16, 17, 18, 27, 42, 54, 73n93, 83, 95–6, 126, 128, 269, 281, 282, 286, 295, 303, 324

Book of Documents (*Shangshu* 尚書) 16, 18, 36, 123n17, 124n20, 247n45, 252, 323n27, 365n6
Books on calendrical astronomy according to the new method (*Xinfa lishu* 新法曆書) 68–9, 75, 84, 93, 288, 289, 292, 294, 321, 357, 370
  see also Books on calendrical astronomy according to the new Western Method; *Books on calendrical astronomy of the Chongzhen reign*
Books on calendrical astronomy according to the new Western method (*Xiyang xinfa lishu* 西洋新法曆書) 37, 321
Books on calendrical astronomy of the Chongzhen reign (*Chongzhen lishu* 崇禎曆書) 33, 34, 37, 42
Borelly, Jacques 107
Bouvet, Joachim 107, 115, 118–19, 265, 270
  and *Book of change* (*Yijing* 易經), interest in 286
  and *Elements of geometry* (*Jihe yuanben* 幾何原本) (Bouvet and Gerbillon) 167–8, 180, 247, 275
    Kangxi's role in composition 176–9
    lü 率 (term of a proportion) 172–3
    practical geometry 173–6
    proportions 169–71
    structure and style of 168–9
  and Figurist interpretation of Chinese classics 286
  and *Historical Portrait of the Emperor of China* 140, 144, 147
  and Manchu language 142, 156, 157–8
  and *Mirror of ancient and modern revering of Heaven* (*Gujin jingtian jian* 古今敬天鑒) 286
  and *Original meaning of heavenly studies* (*Tianxue benyi* 天學本義) 286
  and solar eclipses 224
  and survey of the empire 256
  as tutor to Kangxi 139, 141, 142, 143
    algebra 200
    Euclidean geometry 160–2
    instruments 155
    location of tutoring 151–2
    medicine and anatomy 146–8, 156
    philosophy 144–5, 146, 148
    plan for 144
    typical lesson 148–50
Boym, Michael 37
Brahe, Tycho 67, 115
  see also Tychonic system
Brocard, Jacques 285, 305
Brush calculation (*Bisuan* 筆算) (Mei Wending) 86–90, 91, 93, 99, 326
Budeyi 不得已, *see I cannot do otherwise*

Budeyi bian 不得已辨, *see Refutation of 'I cannot do otherwise'*
Buglio, Lodovico 37, 52, 54, 57–8, 241
  and *Refutation of 'I cannot do otherwise'* (*Budeyi bian* 不得已辨) 51n71, 68

Cai Rui 蔡霷 85
Cai Shen 蔡沈 368–9
Cai Yuanding 蔡元定 230, 232, 368
Calculating rods (*Chousuan* 籌算) (Mei Wending) 85, 86, 216
calculation:
  and *Brush calculation* (*Bisuan* 筆算) 86
    layout of basic operations 89
    unity of mathematics 87–8
    written calculation 88–90
  and *Calculation by borrowed root and powers* (*Jiegenfang suanfa* 借根方算法) 184, 217, 300, 302, 304, 336, 337, 352
    abridgment of 209–10
    ambiguity of Chinese terms 206
    composition of 200–1
    contents 204–5
    foreword to 202–6
    problem solving 208–9
    structure of 201–2
    symbolic linear algebra 210–12
    symbols 207
    table of power numbers 206–7
  and chickens and rabbits in a cage problems 189, 341–4
  and *Discussion of rectangular arrays* (*Fangcheng lun* 方程論), primacy of calculation 92, 96
  and *Elements of calculation* (*Suanfa yuanben* 算法原本) 188, 195, 202, 203, 247, 264, 278, 322, 331, 333, 334, 336, 337, 338, 339, 354
    authorship 197–8
    comparison of Chinese and Manchu versions 198
    comparison with Euclid's *Elements* 195–6
    Kangxi as possible author of preface 198–200
    and number theory 195
    organisation of 195
  and *Essence of numbers and their principles* (*Shuli jingyun* 數理精蘊) 348–52
    chickens and rabbits in a cage problem 341–4
    Hieron's crown problem 344–5
    inkstones and brushes problem 345–8
    methods for solving 'line section' problems 352–4
    pendulums 359–62
    remainder problem 354–6

Index

calculation
  and *Essence of numbers and their principles* (cont.)
    time-keeping 358–62
    water clocks 358–9
    weighing scale problems 348–52
    and Hieron's crown problem 190–1, 344–5
    and inkstones and brushes problem 20–2, 27–9, 47–8, 97–101, 345–8
    and *Instructions for calculation in common script* (*Tongwen suanzhi* 同文算指) 26, 27–9, 184, 189, 190, 191
    and *Integration of Chinese and Western mathematics* (*Zhongxi suanxue tong* 中西算學通), written calculation 88–90
    and *Outline of the essentials of calculation* (*Suanfa zuanyao zonggang* 算法纂要總綱) 184–91
  and techniques 356
Calculation by borrowed root and powers (*Jiegenfang suanfa* 借根方算法) 184, 217, 300, 302, 304, 336, 337, 352
  abridgment of 209–10
  ambiguity of Chinese terms 206
  composition of 200–1
  contents 204–5
  foreword to 202–6
  problem solving 208–9
  structure of 201–2
  symbolic linear algebra 210–12
  symbols 207
  table of power numbers 206–7
calendar:
  and astronomical reform 32–4
  and first calendar submitted by Wu Mingxuan 59–60
  and intercalary months 60–1
  and political and symbolic importance in China 32
  and reform of 36–7
Calendar Department (*Liju* 曆局) 33
calligraphy 73, 233n106, 253
Canon of Yao (*Yaodian* 堯典) 123n17, 365n6
cartography:
  and *Gazetteer of the Great Qing unification* (*Da Qing yitong zhi* 大清一統志) 256
  and survey of the empire 255–7, 277–8, 279–80
Cassini, Gian-Domenico 102, 103–4, 107
Castiglione, Giuseppe 322
celestial element (*tianyuan* 天元) place-value algebra 15, 279, 383
Celiang fayi 測量法義, see *Meaning of measurement methods*
Celiang gaoyuan yiqi yongfa 測量高遠儀器用法, see *Practice of instruments for measuring heights and distances*

Celiang quanyi 測量全義, see *Complete meaning of measurement*
Cengde 成德 264, 268, 376n51, 382
Chaize, François de la 104, 106
Changchunyuan 暢春園, see *Garden of pervading spring*
Changning 常寧 217, 218
Charas, Moyse 148
Chen Houyao 陳厚耀 263, 268, 306, 321, 376, 377
  and *Exploring the mystery of mathematics* (*Suanyi tan'ao* 算義探奥) 321–2
  and interviewed by Kangxi 257–9
  and *Mathematical writings of Chen Houyao* (*Chen Houyao suanshu* 陳厚耀算書) 322
  and *Meaning of base and altitude* (*Gougu fayi* 句股法義) 306
  and *Significance of an intricate method* (*Cuozong fayi* 錯綜法義) 322
Chen Houyao suanshu 陳厚耀算書, see *Mathematical writings of Chen Houyao*
Chen Menglei 陳夢雷 271, 274
Chen Shiming 陳世明 268, 271
Chen Tingjing 陳廷敬 232
Chen Wance 陳萬策 247, 269
Cheng Dawei 程大位 17, 258
  and *Unified lineage of mathematical methods* (*Suanfa tongzong* 算法統宗) 15, 17–20, 27, 95, 96, 191, 258, 264, 295, 315n3, 321, 348, 390
  inkstones and brushes problem 20–2, 48, 97, 346, 347
  miscellaneous methods (*zafa* 雜法) 22
Cheng Hao 程顥 222
Cheng Shisui 程世綏 264, 267
Cheng Yi 程頤 16
Cheng-Zhu 程朱 school 15–16
chickens and rabbits in a cage problem 189, 341–4
  in mathematical works 344
Chishui yizhen 赤水遺珍, see *Lost Treasures from the Red Waters*
Chongzhen 崇禎 Emperor 36
Chongzhen lishu 崇禎曆書, see *Books on calendrical astronomy of the Chongzhen reign*
Chousuan 籌算, see *Calculating rods*
Christianity:
  and Edict of Toleration (1692) 142, 239, 253
  as marginal religion in China 13
  and number of Christians in China 13n4, 253
  and prohibition of 378
  and Rites Controversy 254
churches of Beijing:

Eastern Church (*Dongtang* 東堂) 57, 58, 240, 287
Northern Church (*Beitang* 北堂) 156n87, 240, 241, 287
Southern Church (*Nantang* 南堂) 240, 266n30
Cicero, Alessandro 200
cinchona 156, 200
circulation, and historiography of science 4
Clarification of the subtleties of phonetics (*Yinyun chanwei* 音韻闡微) 269, 276
Classified conversations of Master Zhu (*Zhuzi yulei* 朱子語類) 287n15
Clavius, Christoph:
  and *Algebra* 201, 206, 208
  and *Epitome arithmeticæ* 27, 184, 187
  and establishes mathematics in Jesuit curriculum 23
  and Gregorian Calendar Reform 23
  and influence of 34
  and Ptolemaic system 24
  and role in European science 24
  and structure of mathematics 23
  and translations of 25–6
  and writings of 23
clocks:
  and imperial collections 152
  and Kangxi's interest in 154–5
Coimbra, Jesuit College of 23
  and Thomas 181
Colbert, Jean-Baptiste 102, 103, 105, 106
Collection of the investigation of things in leisure time (*Jixia gewu bian* 幾暇格物編) (Kangxi) 133, 360
Collected poems and prose from the Hall of Erudite Learning (*Jixuetang shiwenchao* 繢學堂詩文鈔) (Mei Wending) 253
College of Mathematics (*Suanxue* 算學), and syllabus 15
comets 58n7, 225
  and 1664 comet 52, 53
Compagnie française des Indes orientales (French East India Company) 102
compass, see proportional compass
Compendium on the newly constructed instruments of the Observatory (*Xinzhi lingtai yixiang zhi* 新制靈臺儀象志) (Verbiest) 67–8, 77, 359n54
Complete collection on nature and principle (*Xingli daquan* 性理大全) 229, 230, 369n21
Complete library of the four treasuries (*Siku quanshu* 四庫全書) 82
Complete meaning of measurement (*Celiang quanyi* 測量全義) (Rho) 34, 278
Complete treatise on agricultural administration (*Nongzheng quanshu* 農政全書) (Xu Guangqi) 31

423

Index

Complete works of Master Zhu (*Zhuzi quanshu* 朱子全書) 255n85, 269, 270
Complete writings on astronomy and mathematics (*Lisuan quanshu* 曆算全書) (Mei Wending) 82, 216n22, 283n103
Comprehensive gazetteer of the Eight Banners imperially commissioned (*Qinding baqi tongzhi* 欽定八旗通志) 66n58
Confucius/Confucianism 324, 368, 387
    and Cheng-Zhu school's interpretation of 16
    and core of teachings 15
    and heavenly learning 31
    and Jesuit interpretation of 254
    and Kangxi's demonstration of ruler's virtues 122, 125
    and Kangxi's education 73
    and sage ruler 74
conic sections 289, 293
Conimbricenses 23, 30
Contancin, Cyr 285
converts to Christianity:
    and translations by 25, 31
    and understanding of heavenly learning 31
Copernican 43, 287, 288
Correct interpretation of [standard] pitchpipes (*Lülü zhengyi* 律呂正義) 281n93, 364, 366, 368–70, 377
cossic algebra 24
    and *Calculation by borrowed root and powers* (*Jiegenfang suanfa* 借根方算法) 201
        abridgment of 209–10
        ambiguity of Chinese terms 206
        contents 204–5
        foreword to 202–6
        problem solving 208–9
        structure of 201–2
        symbolic linear algebra 210–12
        symbols 207
        table of power numbers 206–7
    and *Essence of numbers and their principles* (*Shuli jingyun* 數理精蘊) 352–4
    and terminology of 208
counting devices, *see* abacus; counting rods
counting rods 4, 14, 15, 22, 278–9, 325
Couplet, Philippe 106
craftsmanship, and Jesuits 240–4
cross-cultural contacts, and analytical frameworks 4
cuozong 錯綜 322
Cuozong fayi 錯綜法義, *see* *Significance of an intricate method*

Da Qing yitong zhi 大清一統志, *see* *Gazetteer of the Great Qing unification*
Dai Mingshi 戴名世 216, 271

Dai Zhen 戴震 383
Dalai Lama 41
Datong 大統, *see* Great concordance
Daxue 大學, *see* Great learning
Dechales, Claude-François Milliet 299
    and arithmetic in Euclid 196–7
    and *Elements of Geometry* 162
Deliberations on hemerology (*Xuanze yi* 選擇議) (Yang Guangxian) 49–50
derivation of principles (*tuili* 推理) 79, 145
Descartes, René 164
Dialogue on astronomical methods (*Lifa wenda* 曆法問答) (Foucquet)
    and aims to secure long-term need for Jesuit astronomers 288–9
    and arbitration on innovations of 290–3
    and dialogic structure 287–8
    and heliocentrism 288
    and innovations 288, 289
    and Kangxi's response to 289
    and shortcomings of Tycho Brahe's astronomy 288
    and treatises included in 289
Dialogue on numerical tables (*Shubiao wenda* 數表問答) (Foucquet) 297–8, 302
Dias Jr, Manuel 29–30
difficult problems (*nanti* 難題) 20, 295, 316
Dionis, Pierre 308
Dipingxian li diqiumian biao 地平線離地球面表, *see* *Tables of distance between the horizon and the terrestrial sphere*
Discussion of rectangular arrays (*Fangcheng lun* 方程論) (Mei Wending) 83, 86, 217, 221, 304, 321, 345, 347, 352
    and ambition of work 90
    and assessment of Western methods 92–3
    and branches of mathematics 90–1, 92
    and classification of problems 94, 95
    and inkstones and brushes problem 97–101
    and preface to 216
    and primacy of calculation 92, 96
    and restoring one of the 'nine reckonings' (*jiushu* 九數) 90–3
    and status of mathematics 91–2
    and structure of 93–6
    and style of 101
Discussion of triangles and computation imperially composed (*Yuzhi sanjiaoxing tuisuanfa lun* 御製三角形推算法論) (Kangxi) 60n17, 248–51
divination, and celestial phenomena 16, 17, 52, 58–9, 66, 83, 124, 125, 135, 225, 227–9, 382
Dodart, Denis 107
Dominicans 102
Dongtai 董泰 265, 268
Dongtang 東堂, *see* Eastern Church

Dorgon 多爾袞 35, 36, 37
double false position 28, 186n5, 189
Doubts concerning the study of astronomy (*Lixue yiwen* 曆學疑問) (Mei Wending) 218–22
    and ancients and moderns 219–20
    and compares Chinese and Western systems 220–1
    and dialogic structure 219
    and Kangxi's opinion of 247, 248
    and Muslim astronomy 221
    and roundness of the Earth 221–2
Draft for the Astronomical Records of the Ming History (*Mingshi lizhi nigao* 明史曆志擬稿) (Mei Wending) 215
du 度 5, 26, 27, 91, 328, 331, 332
Du Hamel, Jean-Baptiste 144–5, 146
Duverney, Joseph Guichard 147–8

Eastern Church (Dongtang 東堂) 57, 58, 240, 287
Ebilun 遏必隆 58, 64
Eight Banners (*Baqi* 八旗) 7n12, 35, 73
Eight poems on the Southern Tour (*Nanxun shi ba shou* 南巡詩八首) (Zhang Ying) 131n48
Elements of calculation (*Suanfa yuanben* 算法原本) 188, 195, 202, 203, 247, 264, 278, 322, 331, 333, 336, 337, 338, 339, 354
    and authorship of 197–8
    and comparison of Chinese and Manchu versions 198
    and comparison with Euclid's *Elements* 195–6
    and Kangxi as possible author of preface 198–200
    and number theory 195
    and organisation of 195
Elements of geometry (*Jihe yuanben* 幾何原本) 6, 278
    and Bouvet and Gerbillon's translation 167–9, 172–9, 180, 247, 275
    and Manchu translation by Verbiest 76, 160
    and Ricci and Xu Guangqi's translation 25, 26–7, 77–8, 160, 172
Epitome arithmeticæ (Clavius) 27, 184, 187
Essence of numbers and their principles imperially composed (*Yuzhi shuli jingyun* 御製數理精蘊) 8, 366, 370, 389, 390, 391
    and anchorage in Chinese learning 340
    and calculation techniques 356–7
        rupture from Chinese tradition 357
    and completion of 364
    and context of 340

Essence of numbers and their principles (cont.)
   and cossic algebra 336–7, 352–4
   and cross-references 338–9
   and features of 363
   and fractions 335
   and geometry 335
   and historical narrative of
      mathematics 323–7
   and influence of 382–4
   and instruments 357–8
      proportional compass 357
   and intended readership 356, 357
   and longevity of 384
   and material culture 340–1
   and nineteenth-century revision of 384
   and outline of 315–16
      contents 317–19
   as part of *Origins of pitchpipes and the
      calendar* (*Lüli yuanyuan* 律曆淵源)
      364
   and problems in 341
      chickens and rabbits in a cage 341–4
      difficult problems 348–52
      Hieron's crown problem 344–5
      inkstones and brushes problem 345–8
      methods for solving 'line section'
         problems 352–4
      pendulums 359–62
      remainder problem 354–6
      time-keeping 358–62
      use of instruments 357–8
      water clocks 358–9
      weighing scale problems 348–52
   and ratios 335–7
   and rectangular arrays 352–4
   and rephrasing of Jesuit textbooks 330–3
   and size of 315
   and sources for 320–3
   and statecraft 362
   and structure of imperial
      mathematics 327–30
      jihe 幾何 327–9
   and universality of imperial
      mathematics 363
   and vocabulary and classification 333–9
Essential details of numerical tables (*Shubiao
   jingxiang* 數表精詳) 309n115
Essentials of [standard] pitchpipes (*Lülü
   zuanyao* 律呂纂要) (Pereira) 72, 369
Essentials of Trigonometry (*Sanjiaofa juyao* 三
   角法舉要) (Mei Wending) 251–2
Euclid, and translations and editions of:
   and Dechales' *Elements of Geometry* 162
   and *Elements of geometry* (*Jihe yuanben* 幾
      何原本) (Bouvet and
      Gerbillon) 167–8
      Kangxi's role in composition 176–9
      lü 率 (term of a proportion) 172–3

      practical geometry 173–6
      proportions 169–71
      structure and style of 168–9
   Elements of geometry (*Jihe yuanben* 幾何原本)
      (Manchu translation by
      Verbiest) 76, 160
   Elements of geometry (*Jihe yuanben* 幾何原
      本) (Ricci and Xu Guangqi) 25,
      26–7, 160
   and Pardies' *Elements of Geometry* 161,
      162–6
      approach of 162
      contents 165
      dedication of 163
      practical geometry 173–4
      structure and style of 165–6
   and seventeenth-century rewritings of 162
   Euclidean geometry:
      and *Number and magnitude expanded*
         (*Shudu yan* 數度衍) 48–9
      and Kangxi's study of 75–6, 143, 164, 173
         Pardies' textbook 161
         translations used 160–2
         use of models 171
      and Nicole's disapproval of 164–5
      and proportions 169–73
         lü 率 (term of a proportion) 171–2
      and religion 164
Europe, and transformation of intellectual
   cultures 3
European philosophy, and Kangxi's study
   of 143, 144–5, 146
evidential scholarship 134, 222, 383, 392
examinations:
   and literati culture in China *c.*1600 17
   and mathematical sciences 280–3
excess and deficit (*ying buzu* 盈不足, *yingnü*
   盈朒) 92–3, 96, 189, 191, 336
Expanded meaning of the Classic of filial piety
   (*Xiaojing yanyi* 孝經衍義) 74n96
Explanation of the examination of qi (*Yanqi shuo*
   驗氣說) (Verbiest) 77
Explanation of the proportional compass (*Biligui
   jie* 比例規解) (Schall and Rho) 193
Explanation of the proportional compass (*Biligui
   jie* 比例規解) (Thomas) 193–4, 362
Exploitation of the works of nature (*Tiangong
   kaiwu* 天工開物) (Song Yingxing) 15
Exploring the mystery of mathematics (*Suanyi
   tan'ao* 算義探奧) (Chen Houyao)
   321–2

Fabre-Bonjour, Guillaume 284
false position 91, 93, 186n5, 189, 205,
   326, 336
Fang Bao 方苞 271–2, 373
Fang Kongzhao 方孔炤 42
Fang Yizhi 方以智 42, 43

Fang Zhongtong 方中通 42, 43, 85
   and appropriation of Western learning 49
   and meeting with Schall 43–4
   and *Number and magnitude expanded*
      (*Shudu yan* 數度衍) 44–9, 84, 99,
      344, 346, 347, 356
      Euclidean geometry 48–9
      inkstones and brushes problem 47–8
      purpose of 46
      reorganisation of mathematical
         knowledge 47
      structure of 47
fangcheng 方程, **see** rectangular arrays
Fangcheng lun 方程論, **see** *Discussion of
   rectangular arrays*
fathoming the principles (*qiongli* 窮理) 16
Feng Rui 馮霦 268
figures and numbers (*xiangshu* 象數) 17,
   83, 324, 326
Figurism 286, 291, 305, 310
Filippucci, Francesco Saviero 117, 157
fine arts, and Jesuits 241
   Chinese apprentices 241–2
First collection of heavenly learning (*Tianxue
   chuhan* 天學初函) (Li Zhizao)
   29–30, 31, 42, 79
First collection of the imperial birthday ceremony
   (*Wanshou shengdian chuji* 萬壽盛典
   初集) 265, 268n1–2
Five Classics (*Wujing* 五經) 15, 16, 17, 36, 74
Five Phases (*wuxing* 五行) 47, 69n78,
   134n68
Fontaney, Jean de 24, 104, 105
   on Chinese instruments 115
   and French mission to China 113
      chosen to lead 103–4, 107
      programme of work 111
      travels and observations in China
         116–17
   on Kangxi's study of Western science 140
   and Kangxi's visit to Nanjing
      Observatory 131–3
Foucquet, Jean-François 260–1, 262, 273,
   274–5
   and arrives in Beijing 286
   and astronomy:
      imperial request to write astronomy
         text 286–7
      report to Kangxi 284–5
   and criticism of 289–90
   and *Dialogue on astronomical methods* (*Lifa
      wenda* 曆法問答):
      aims to secure long-term need for Jesuit
         astronomers 288–9
      arbitration on innovations of 290–3
      dialogic structure 287–8
      heliocentrism 288
      innovations 288, 289

425

Index

Foucquet, Jean-François (*cont.*)
    Kangxi's response to 289
    shortcomings of Tycho Brahe's astronomy 288
    treatises included in 289
    and *Dialogue on numerical tables* (*Shubiao wenda* 數表問答) 297–8, 302
    and failure to influence imperial mathematics 310–11
    and *New method of algebra* (*Aerrebala xinfa* 阿爾熱巴拉新法) 294–6, 298, 299
    symbols in 300–5
    and tables 309–10
    and writings on mathematics 294–9
        algebra 294–6, 299
        logarithms 296, 297–8
Four Books (*Sishu* 四書) 15, 16, 17, 74, 158, 367
fractions 335
France:
    and *Académie royale des sciences* 102–3
        questions about China 106–7
    and image of China 103
    and interest in Asia 102, 106
    and Jesuit mission to China 103–4, 107–8
        in Beijing 112–16
        instruments as aid to evangelisation 105–6
        journey to China 108–11
        motivations for 104–5
        programme of work 111
        religious and scientific motivations 105
        travels and observations in China 116–19
Franciscans 102
French East India Company 102
Fundamentals of astronomical phenomena (*Lixiang benyao* 曆象本要) 219n33
Fuquan 福全 216, 218
Furtado, Francisco 78
Fuxi 伏羲 324

Gabiani, Giandomenico 131, 133, 152–3, 154–5
Galdan 201
Galileo 30, 193
Gao Shiqi 高士奇 121, 243–4, 253n79
Garden of pervading spring (*Changchunyuan* 暢春園) 149, 192n65, 243, 257, 265, 266, 271, 279, 281, 293, 305
Gazetteer of the Great Qing unification (*Da Qing yitong zhi* 大清一統志) 256
Gemei 戈枚 225
General study of astronomical methods (*Lifa tongkao* 曆法通考) (Mei Wending) 218–19
Gengzhitu 耕織圖, see *Pictures of tilling and weaving*

Gerbillon, Jean-François 107, 115, 116, 119
    and death of 265
    and *Elements of geometry* (*Jihe yuanben* 幾何原本) (Bouvet and Gerbillon) 167–8, 180, 247, 275
        Kangxi's role in composition 176–9
        lü 率 (term of a proportion) 172–3
        practical geometry 173–6
        proportions 169–71
        structure and style of 168–9
    on Imperial Workshops 151–2
    and Manchu language 142, 156, 157–8
    and Qing delegation to Nerchinsk 141–2
    and solar eclipses 224–5
    as tutor to Kangxi 139, 141, 142, 143
        Euclidean geometry 160–2
        follows Moderns' style 163–4
        instruments 155
        location of tutoring 151–2
        medicine and anatomy 146–8, 156
        philosophy 144–5, 146, 148
        plan for 144
        typical lesson 148–50
gewu 格物, see investigation of things
Gexiang xinshu 革象新書, see *New writing on the image of [the hexagram] alteration*
Gherardini, Giovanni 241
glassmaking 240, 243
Gnomon of Zhou (*Zhoubi* 周髀) 45, 46–7, 317, 320, 325, 366
Gonzaga, Luigi 270, 297
gougu 句股, see base and altitude
Gougu fayi 句股法義, see *Meaning of base and altitude*
Goüye, Thomas 117–18, 132
Great classified survey of the Nine chapters on mathematical methods (*Jiuzhang suanfa bilei daquan* 九章算法比類大全) 17, 22, 191, 346, 347
Great concordance (*Datong* 大統) system 33, 36, 59, 67, 71, 214–15
Great learning (*Daxue* 大學) 16, 199n87
Gregory XV, Pope 284
Grienberger, Christoph 33
Grimaldi, Carlo Filippo 72, 183, 200
    and appointed Administrator of the Calendar (*zhili lifa* 治理曆法) 116
    and death of 265
Gu Chenxu 顧陳垿 263, 268, 271, 376
Gu Yanwu 顧炎武 215
Gubadai 顧八代 262
Gucong 顧琮 262–3, 376n51, 379, 381
Gujin jingtian jian 古今敬天鑒, see *Mirror of ancient and modern revering of Heaven*
Guo Shoujing 郭守敬 123, 124, 126, 214, 250, 326, 327
Guozi jian 國子監, see Imperial University

Hall of Military Glory (Wuyingdian 武英殿) 264, 309
Hall of the Nourishment of the Mind (Yangxindian 養心殿) 149, 151–2, 153, 156, 159, 184, 194, 200, 243, 265
Han Chinese 1n1
harmonics:
    and *Correct interpretation of [standard] pitchpipes* (*Lülü zhengyi* 律呂正義) 364, 366, 368–70
        Later part (*houbian* 後編) supplement 380–1
    and Kangxi's pronouncement on 229–33
        officials' response to 233–6
    see also music
He Guodong 何國棟 266, 268, 279, 376n51, 382
He Guozhu 何國柱 263, 268, 377n56
    and discussions with Korean scholars 278–9
    and survey of the empire 277–8
He Guozong 何國宗 263, 268, 305, 373, 378, 379, 380, 381
    and imperial editorial projects 269, 277
He Junxi 何君錫 263, 266
heavenly learning (*tianxue* 天學):
    and Chinese translators' understanding of 31
    and Confucianism 31
    and foreignness an obstacle to adoption 32
    and term supplanted by Western learning (*xixue* 西學) 30
    and Xu Guangqi's rationale for adoption of 32–3
heliocentrism 288
Hieron's crown problem 190–1, 344–5
historiography:
    and circulation of knowledge 4
    and cross-cultural contacts 3, 4
    and Europe 3
    and new understanding of Qing period 2–3
Hong Chŏng-Ha 洪正夏 278–9
Hong Taiji 皇太極 35
Hongzhou 弘晝 379, 380
Hu Zhenyue 胡振鉞 63n37
Huang Baijia 黃百家 42
Huang Yuji 黃虞稷 84, 85
Huang Zongxi 黃宗羲 42, 215
Huangdi neijing 黃帝內徑, see *Inner Canon of the Yellow Emperor*
Hungai tongxian tushuo 渾蓋通憲圖說, see *Illustrated explanation of the sphere and the astrolabe*

Index

I cannot do otherwise (*Budeyi* 不得已) (Yang Guangxian) 51–2
Illustrated explanation of the sphere and the astrolabe (*Hungai tongxian tushuo* 渾蓋通憲圖說) (Ricci and Li Zhizao) 25
Imperial Diary (*Qiju zhu* 起居注) 129, 130, 229, 233, 234
    and Kangxi's reign 121
    and Kangxi's visit to Nanjing Observatory 121–6
Imperial Printing Office, **see** Hall of Military Glory
Imperial University (*Guozi jian* 國子監) 382
Imperial Workshops 151–2, 243
inkstones and brushes problem:
    and *Unified lineage of mathematical methods* (*Suanfa tongzong* 算法統宗) 20–2, 48, 97, 346, 347
    and *Essence of numbers and their principles* (*Shuli jingyun* 數理精蘊) 345–8
    and *Number and magnitude expanded* (*Shudu yan* 數度衍) 47–8
    and *Instructions for calculation in common script* (*Tongwen suanzhi* 同文算指) 27–9
    in mathematical works 347
    and *Discussion of rectangular arrays* (*Fangcheng lun* 方程論) 97–101
Inner Canon of the Yellow Emperor (*Huangdi neijing* 黃帝內徑) 222
Inner Mongolia 35, 244
Instructions for calculation in common script (*Tongwen suanzhi* 同文算指) (Ricci and Li Zhizao) 26, 27–9, 31, 45, 47, 48, 49, 89, 92, 93, 96, 99, 184, 185, 187, 189, 190, 191, 321, 326, 344, 346, 347, 352
instruments:
    as aid to evangelisation 105–6
    and Astronomical Bureau 67–8
    and *Essence of numbers and their principles* (*Shuli jingyun* 數理精蘊) 357–8
    and de Fontaney on Chinese instruments 115
    and Gerbillon on Chinese instruments 151–2
    as gifts 152–3
    and Imperial Workshops 151–2
    and Kangxi's study of 152, 154–6, 193–4, 387
    and Le Comte on Chinese instruments 114–15
    and Thomas' *Explanation of the proportional compass* (*Biligui jie* 比例規解) 193–4, 357, 361, 362
    and Thomas' *Practice of instruments for measuring heights and distances* (*Celiang gaoyuan yiqi yongfa* 測量高遠儀器用法) 194, 251, 258, 267, 283
intercalary months, and calendar 60–1
Integration of Chinese and Western mathematics (*Zhongxi suanxue tong* 中西算學通) (Mei Wending) 85, 86–90, 91
Integration of learning in calendrical astronomy (*Lixue huitong* 曆學會通) (Xue Fengzuo) 43, 84
Intorcetta, Prospero 85, 111
Investigation into the principles of names (*Mingli tan* 名理探) (Furtado and Li Zhizao) 78
Investigation of the origins of planetary ephemerides (*Xingli kaoyuan* 星曆考原) 269, 380
investigation of things (*gewu* 格物) 16, 150

Jai Singh 1n2
Japan, and Jesuit mission 24–5
Jartoux, Pierre 256, 285, 290, 291, 304–5, 307
Jehol (Rehe 熱河) 260n2
    and Jesuits in 261, 270, 287, 294
    and mathematicians in 263, 264, 269–70, 271
Jesuits:
    and Académie royale des sciences 103
    and Aristotelian philosophy 30
    and attacks on:
        Beijing Jesuits arrested 52
        death sentences 52–3
        death sentences rescinded 53
        Yang Guangxian 49–50, 51–2
    and calendar reform 32–4, 36–7
    and changes in engagement with science 70–2
    and changing situation of mission 239
    and China mission 24–5
        teaching 25, 30
        translations 25–9
    as craftsmen and artists 240–4
    and decline in prestige and authority of 24, 253–4
        impact of papal legation 254–5
    and developments in education 163
    and dissolution of 240
    and divisions within 285–6, 310–11
        Foucquet's innovations 291–3
    and dynastic transition 36–7, 41
    and entry into China 13
    and evangelisation 25
        science as tool of 13
    and favour of imperial princes 217–18
    and French mission to China 103–4, 107–8
        in Beijing 112–16
        instruments as aid to evangelisation 105–6
        journey to China 108–11
        motivations for 104–5
        programme of work 111
        religious and scientific motivations 105
        travels and observations in China 116–19
    and Kangxi commissions astronomy textbook 261
    and Kangxi's 60th birthday ceremony 265, 266
    and Kangxi's distancing from 239, 253–4
    and literati's view of 391
    and mathematics 22–4
        education in 22–4
        importance for 13
    and missionaries banished to Canton 54
    and Office of Mathematics 305–9
    and philosophy 78
    and recent research on 3
    and Rites Controversy 254
        papal legation to China 254–5
    and rival missionaries 284
    and science and religion 70–2
    and solar eclipses 224–5
    and surveying 244–5
        meridian expedition 245
        survey of the empire 255–7
    and tutoring of Kangxi 2, 184
        agenda 140
        algebra 200, 201
        chronology of lessons 142–3
        diverse interests of 150–1
        ending of 244
        Euclidean geometry 75–6, 143, 160–2, 164
        French science 144–8
        instruments 152, 154–6, 194
        location of tutoring 151–2
        medicine and anatomy 146–8, 156
        motives of 140, 149, 154
        philosophy 143, 144–5, 146, 148
        practical geometry 143, 173
        proportional compass 193–4
        as route to conversion 144
        series of tutorials 139
        sources for 139–40
        tables 192, 193
        treatment of tutors 143
        tutors 139
        typical lesson 148–51
        use of Manchu language 156–8
Jiangnan 江南, and resistance to Manchu conquest 41
    see also Nanjing
Jiangning 江寧, **see** Nanjing
Jiao Bingzhen 焦秉貞 241–2, 243, 244
Jiegenfang suanfa 借根方算法, **see** *Calculation by borrowed root and powers*

427

Index

Jiegenfang suanfa jieyao 借根方算法節要, see *Summary of calculation by borrowed root and powers*
jihe 幾何 26–7, 327–9, 333, 334
Jihe yaofa 幾何要法, see *Main methods of geometry*
Jihe yuanben 幾何原本, see *Elements of geometry*
Jin'ai 進愛 379, 380
jiushu 九數, see nine reckonings
jiuzhang 九章, see nine chapters
Jiuzhang suanfa bilei daquan 九章算法比類大全, see *Great classified survey of the Nine chapters on mathematical methods*
Jiuzhang suanshu 九章算術, see *Nine chapters on mathematical procedures*
Jixia gewu bian 幾暇格物編, see *Collection of the investigation of things in leisure time*
Jixuetang shiwenchao 績學堂詩文鈔, see *Collected poems and prose from the Hall of Erudite Learning*

Kangxi 康熙 Emperor:
   and appropriation of Western learning 270, 391
   and Astronomical Bureau:
      checks calculations of 260–1
      dispute over Wu Mingxuan's calendar 59–64
      Grimaldi's appointment to 116
      increase in students 66–7
      rehabilitation of Jesuit astronomy 58–65
   and astronomy 385
      arbitration on Foucquet's innovations 290–1, 293
      challenges sinocentric view of 226–7
      commissions textbook from Jesuits 261, 286–7
      contempt for superstition surrounding 225–6
      court concern over reliance on Europeans 286
      demonstration of skills 233
      Discussion of triangles and computation (*Sanjiaoxing tuisuanfa lun* 三角形推算法論) 248–51
      eclipse prediction 222–3
      Foucquet's report 284–5
      investigation of the Old Man Star (*Laorenxing* 老人星) 133–5
      opinion of Mei Wending's work 247, 248
      role in Chinese cosmology 227–9
      study of 74–5
      summer solstice of 1711 260–1
   and birthday (60th) ceremony 265–6
   and campaign against Oirats 201
   at centre of network of communication 388
   and *Collection of the investigation of things in leisure time* (*Jixia gewu bian* 幾暇格物編) 133, 360
   and Confucian virtues 122
   and consolidation of Manchu dynasty 385
   and death of 364
   and *Discussion of triangles and computation* (*Sanjiaoxing tuisuanfa lun* 三角形推算法論) 60n17, 248–51
   and distancing from Jesuit mission 239, 253–4
      impact of papal legation 254–5
   and Edict of Toleration (1692) 142, 239
   and education of sons 275, 276
   and harmonics 229–31, 368
   and identity as ruler 73
   and imperial discourse 387
   and imperial editorial projects:
      Clarification of the subtleties of phonetics (*Yinyun chanwei* 音韻闡微) 269, 276
      Complete works of Master Zhu (*Zhuzi quanshu* 朱子全書) 255n85, 269, 270
      editing of mathematical writings 270
      Essential significance of nature and principle (*Xingli jingyi* 性理精義) 269, 282
      Investigation of the origins of planetary ephemerides (*Xingli kaoyuan* 星曆考原) 269, 380
      Kangxi Character Dictionary (*Kangxi zidian* 康熙字典) 267
      Middle way to the Change of Zhou (*Zhouyi zhezhong* 周易折中) 269, 281, 324
      Quintessence of the Masters and Histories (*Zishi jinghua* 子史精華) 265, 267
      Rhyme Repository of the Adorned Literature Studio (*Peiwen yunfu* 佩文韻府) 267
      Summary of the Monthly ordinances (*Yueling jiyao* 月令輯要) 269
      see also *Essence of numbers and their principles* (*Shuli jingyun* 數理精蘊); *Ming History* (*Mingshi* 明史); Office of Mathematics; *Origins of pitchpipes and the calendar* (*Lüli yuanyuan* 律曆淵源)
   and increased interest in Chinese mathematical scholarship 239
   and information sources 133, 388
   and instruments 152, 154–6, 193–4, 387
   and language:
      attitude towards 156–7, 158
      preference for Manchu 159
   and mathematics:
      acknowledges value of Chinese mathematics 264
      algebra 200, 201, 294–6, 298–9, 302, 304
      assessment of knowledge of 297, 390–1
      attitude towards 198–200
      demonstration of skills 233
      examinations 280–3
      imperial pronouncement on 229–33
      interviews Chen Houyao 257–9
      logarithms 296, 298
      low opinion of Chinese scholars' skills in mathematics 246–7
      military motivations for study of 77–8
      officials' response to pronouncement on 233–6
      reprimands Chinese officials for ignorance of 230–2
      significance of engagement with 385–6
      study of 74–7
      tables 309, 310
      unique role in 365
   and *Maxims of fatherly advice* (*Tingxun geyan* 庭訓格言) 276
   and measurement:
      standard units 246
      surveying techniques 245
   and medicine, commissions writings on 307–8
   and Mei Wending:
      grants audience to 251–2
      opinion of work of 247, 248
      praise of 253
   and military prowess 1
   and monopoly on Western learning 7, 305, 311, 389
   and music 229–31, 243
      approach to 277
   and Nanjing Observatory, visit to 120
      account in *Sagely instructions* (*Shengxun* 聖訓) 126
      commends Jesuits' astronomical system 123
      emphasis of different accounts 131
      Jesuits' role 131–3
      Li Guangdi's account of 127–31
      limits of literati's authority on astronomy 125
      questions Li Guangdi 122–4, 128–30
      record of *Imperial Diary* (*Qijuzhu* 起居注) 121–6
      role as teacher 125
   and Office of Mathematics:
      foundation of 269–73
      instructions for organisation of 271
      recruitment of scholars 262–4
   and official *Ming History* 74
   and official rule begins 58
   as personification of Sagely Emperor 385
   and playing factions against each other 388

Kangxi 康熙 Emperor (cont.)
  and recruitment of Western experts 182–3
  and regency 50
    and condemnation of Schall and colleagues 49–54
  and rejection of Western philosophy 80–1
  and relationship with officials 387
  and *Sagely instructions* (*Shengxun* 聖訓) 229, 271, 282
  and Southern inspection tours 120
    purpose of 120–1
    retinue 121
  and studies of:
    Chinese 73
    Classics 73–4
    daily tutoring 73
    Western learning 139, 184
    algebra 200, 201
    attitude as student 171, 194
    chronology of lessons 142–3
    diversity of interests 150–1
    ends Jesuit tutoring 244
    Euclidean geometry 75–6, 143, 160–2, 164, 171
    European philosophy 143, 144–5, 146, 148
    French science 144–8
    instruments 152, 154–6, 194
    Jesuits' agenda 140
    limits circulation of 158
    location of tutoring 151–2
    mathematics and astronomy 74–7
    medicine and anatomy 146–8, 156
    military motivations for 77–8
    motives for 140, 149, 154
    possible preface to *Elements of calculation* (*Suanfa yuanben* 算法原本) 198–200
    practical geometry 143, 173
    proportional compass 193–4
    publication of tutors' lecture notes 73–4, 139
    relationship with Jesuit tutors 387–8
    role in composition of *Outline of the essentials of calculation* (*Suanfa zuanyao zonggang* 算法纂要總綱) 188–9
    role in composition of the new *Elements* 176–9
    series of tutorials 139
    sources for 139–40
    tables 192, 193
    treatment of tutors 143
    tutors 2, 139
    typical lesson 148–51
    use of Manchu language 156–8
    use of models 171
  and succession crises 239, 253–4
    strife amongst sons 273–4, 275–6, 365
  and surveying:
    methods 245
    survey of the empire 255–7, 277–8, 279–80
  and takeover of power 58
  and tests Mei Wending 231
  and Tibetan Buddhism 223
  and tolerance 378
  and value of practical learning 386
  and Western curios in private apartments 243–4
  and Western learning as tool of statecraft 4, 15, 311, 386
Kepler, Johannes, 287–288
Kircher, Athanasius 369
Koffler, Andreas Xavier 37
Kögler, Ignatius 378–9
Korea 35, 228, 277–9
Kuixu 揆敘 257

La Hire, Philippe de 104
  and machine 305
  and tables 309
Lansberge, Philippe van 43
Laorenxing 老人星, see Old Man Star
Later part of the *Correct interpretation of [standard] pitchpipes* (*Lülü zhengyi houbian* 律呂正義後編) 380–1
Later part of the *Thorough investigation of astronomical phenomena* (*Lixiang kaocheng houbian* 曆象考成後編) 379–80
Le Comte, Louis 107, 113–14, 114–15
Leibniz, Gottfried Wilhelm 105, 106
li 理, see principle
Li Chunfeng 李淳風 27
Li Guangdi 李光地 121, 229, 245, 391
  and career of 127
  and examinations in mathematics 281–2
  and exchange of books with Kangxi 247
  and imperial editorial projects 269
  and Kangxi's visit to Nanjing Observatory:
    account of 127–31
    questioned by Kangxi 122–4, 128–30
  and Mei Wending 247–8
    commissions work on astronomy from 218–19
    patronage of 215–16
    presents work of to Kangxi 247
    promotion of 251
  and spectacles 129
  and visits Verbiest 69–70
Li Guangxian 李光顯 63n37
Li Shizhen 李時珍 15, 330n60
Li Yu 李煦 257
Li Zhizao 李之藻 25, 78

  and astronomical reform 32
  and 'concrete things' 42
  and *First collection of heavenly learning* (*Tianxue chuhan* 天學初函) 29–30, 31, 42, 79
  and *Illustrated explanation of the sphere and the astrolabe* (*Hungai tongxian tushuo* 渾蓋通憲圖說) 25
  and *Instructions for calculation in common script* (*Tongwen suanzhi* 同文算指) 26, 27–9, 31, 45, 47, 48, 89, 92, 93, 96, 99, 184, 185, 187, 189, 190, 191, 321, 326, 344, 346, 347, 352
  and *Investigation into the principles of names* (*Mingli tan* 名理探) 78
  and *Meaning of compared figures* (*Yuanrong jiaoyi* 圜容較義) 26
  and *Meaning of Heavenly and Earthly forms* (*Qiankun tiyi* 乾坤體義) 25–6, 29
  and *Memorial on ritual and music at local schools* (*Pangong liyue shu* 頖宮禮樂疏) 31
  and understanding of heavenly learning 31
  and wide interests of 31
Li Zhonglun 李鍾倫 247
Li Zhongqiao 李鍾僑 269
Li Zicheng 李自成 35–6
Li Zubai 李祖白 51, 53
liangfa 量法, see measurement
Lifa tongkao 曆法通考, see *General study of astronomical methods*
Lifa wenda 曆法問答, see *Dialogue on astronomical methods*
Lisuan quanshu 曆算全書, see *Complete writings on astronomy and mathematics*
Liu Hui 劉徽 171–2
Liubao 留保 376
Lixiang benyao 曆象本要, see *Fundamentals of astronomical phenomena*
Lixiang kaocheng 曆象考成, see *Thorough investigation of astronomical phenomena*
Lixiang kaocheng houbian 曆象考成後編, see *Later part of the thorough investigation of astronomical phenomena*
lixue 理學, see principle, study of
Lixue huitong 曆學會通, see *Integration of learning in calendrical astronomy*
Lixue pianzhi 曆學駢枝, see *Superfluous learning on calendrical astronomy*
Lixue yiwen 曆學疑問, see *Doubts concerning the study of astronomy*
Lixue yiwen bu 曆學疑問補, see *Sequel to the Doubts concerning the study of astronomy*
lodges (*xiu* 宿) 122, 224

429

Index

logarithms, and Foucquet's writings on 294, 296, 297–8
Lost Treasures from the Red Waters (*Chishui yizhen* 赤水遺珍) (Mei Juecheng) 306–7
Lou Shu 樓璹 242–3, 244
Louis XIV 102, 104, 106, 107, 140, 145
Louvois, Marquis de 103
Lu Jiuyuan 陸九淵 32
Lu Longqi 陸隴其 68–9, 216
Lundali 倫達禮 268
lü 率 (term of a proportion) 92n44, 101, 171–3, 189, 190, 307, 336–8, 341, 355–6, 360
Lüli yuanyuan 律曆淵源, see *Origins of pitchpipes and the calendar*
lülü 律呂, see pitchpipe [standard]
Lülü xinshu 律呂新書, see *New writing on the [standard] pitchpipes*
Lülü zhengyi 律呂正義, see *Correct interpretation of [standard] pitchpipes*
Lülü zhengyi houbian 律呂正義後編, see *Later part of the Correct interpretation of [standard] pitchpipes*
Lülü zuanyao 律呂纂要, see *Essentials of [standard] pitchpipes*
Lüxue xinshuo 律學新說, see *New explanation of the study of [standard] pitchpipes*

Ma Decheng 馬德稱 85
Macao 24
Magalhães, Gabriel de 37, 65, 72
    and arrested 52
    and criticism of Schall 38–40
    and house arrest 54, 57–8
magic squares 20, 22, 324
Mahu 馬祜 57, 62, 63, 67n61
Maigrot, Charles 254
Main methods of geometry (*Jihe yaofa* 幾何要法) (Aleni) 34
Manchu dynasty, see Qing dynasty
Manchu language:
    and creation of literary corpus in 74, 76
    and Kangxi's preference for 159
    and mathematical terminology 167
    and use by Jesuit tutors of Kangxi 156–8
    and used to restrict access to scientific writings 158
Mao Qiling 毛奇齡 234–5
maps:
    and *Gazetteer of the Great Qing unification* (*Da Qing yitong zhi* 大清一統志) 256
    and survey of the empire 255–7, 277–8, 279–80
Mathematical classic of Master Sun (*Sunzi suanjing* 孫子算經) 189, 320, 327, 335, 344, 354, 355, 356

Mathematical procedures of Master Sun (*Sunzi suanshu* 孫子算術) 320n10
Mathematical writings of Chen Houyao (*Chen Houyao suanshu* 陳厚耀算書) 322
mathematics:
    and Clavius' classification of 23
    and evidential scholarship 383
    and examinations in 280–3
    and Foucquet's writings on 294–9
        algebra 294–6
    and historical narrative of 323–7
    and Jesuits:
        education system 22–4
        teaching 25, 30
        translations 25–9
    and Kangxi:
        acknowledges value of Chinese mathematics 264
        demonstration of skills 233
        knowledge of 297, 390–1
        low opinion of Chinese scholars' skills 246–7
        officials' response to pronouncement of 233–6
        pronouncement on 229–33
        reprimands Chinese officials for ignorance of 230–2
        significance of engagement with 385–6
    and literati culture in China c.1600 14–22
        Cheng Dawei 17–21
        Cheng-Zhu school 15–16
        decline in 14–15
        examinations 17
        numbers 16–17
        recognition as member of literati class 17
        renaissance in 15
    and religion 164
    and structure of 390
    and teaching of 382–3
    see also *Essence of numbers and their principles* (*Shuli jingyun* 數理精蘊); Office of Mathematics; *Origins of pitchpipes and the calendar* (*Lüli yuanyuan* 律曆淵源)
Maxims of fatherly advice (*Tingxun geyan* 庭訓格言) (Kangxi) 276
Meaning of base and altitude (*Gougu fayi* 句股法義) (Chen Houyao) 306
Meaning of compared figures (*Yuanrong jiaoyi* 圓容較義) (Ricci & Li Zhizao) 26
Meaning of Heavenly and Earthly forms (*Qiankun tiyi* 乾坤體義) (Ricci & Li Zhizao) 25–6, 29
Meaning of measurement methods (*Celiang fayi* 測量法義) (Ricci & Xu Guangqi) 26

measurement (*liangfa* 量法) 91, 328
    and Kangxi on techniques 245
    and standardisation of units 245–6
medicine:
    and Chinese forensic medicine 308
    and Kangxi's commission of writings on 307–8
    and Kangxi's study of 146–8, 156
Mei Geng 梅庚 84
Mei Juecheng 梅瑴成 85, 247, 264, 266, 268, 270, 279, 331, 373, 377, 378, 379, 380
    and imperial editorial projects 269, 277
    and *Lost Treasures from the Red Waters* (*Chishui yizhen* 赤水遺珍) 306–7
    and rediscovery of the procedure of the celestial element (*tianyuan* 天元) 383
    and *Selected essentials of Mr Mei's collection* (*Meishi congshu jiyao* 梅氏叢書輯要) 82, 96n57, 306
Mei Wending 梅文鼎 42, 377
    and audience with Kangxi 251–2
    and Beijing:
        arrival in 86, 120, 214
        network of contacts 215–17
    and *Brush calculation* (*Bisuan* 筆算) 86, 91, 93, 99, 326
        layout of basic operations 89
        unity of mathematics 87–8
        written calculation 88–90
    and *Calculating rods* (*Chousuan* 籌算) 85, 86, 216
    and Chinese origin of Western learning 87, 250, 252–3
    and *Collected poems and prose from the Hall of Erudite Learning* (*Jixuetang shiwenchao* 績學堂詩文鈔) 253
    and *Complete writings on astronomy and mathematics* (*Lisuan quanshu* 曆算全書) 82, 216n22, 283n103
    and counting rods 279
    and *Discussion of rectangular arrays* (*Fangcheng lun* 方程論) 83, 86, 217, 221, 304, 321, 345, 347, 352
        ambition of work 90
        assessment of Western methods 92–3
        branches of mathematics 90–1, 92
        classification of problems 94, 95
        inkstones and brushes problem 97–101
        preface to 216
        primacy of calculation 92, 96
        restoring one of the 'nine reckonings' 90–3
        status of mathematics 91–2
        structure of 93–6
        style of 101
    and *Doubts concerning the study of astronomy* (*Lixue yiwen* 曆學疑問) 218–22

Mei Wending 梅文鼎
  and *Doubts concerning the study of astronomy* (cont.)
    ancients and moderns 219–20
    compares Chinese and Western systems 220–1
    dialogic structure 219
    Kangxi's opinion of 247, 248
    Muslim astronomy 221
    roundness of the Earth 221–2
  and *Draft for the Astronomical Records of the Ming History* (*Mingshi lizhi nigao* 明史曆志擬稿) 215
  and early career of 83–6
  and *Essentials of Trigonometry* (*Sanjiaofa juyao* 三角法舉要) 251–2
  and criticised by Kangxi 231
  and *General study of astronomical methods* (*Lifa tongkao* 曆法通考) 218–19
  and *Integration of Chinese and Western mathematics* (*Zhongxi suanxue tong* 中西算學通) 85, 86–90, 91
    coherent system 88
    as first collection of works 86–7
    reasons for subjects chosen 87
    titles for inclusion in 86
    unity of mathematics 87–8
    unity of numbers and principles 88
    written calculation 88–90
  and Li Guangdi 247–8
    as patron 215–16
    promotion by 251
    work on astronomy commissioned by 218–19
  and *Ming History* (*Mingshi* 明史):
    asked to work on 214
    Draft for the Astronomical Records (*Mingshi lizhi nigao* 明史曆志擬稿) 215
  and Muslim scholarship 85
  and praised by Kangxi 253
  and prolific writings of 82
  and *Selected essentials of Mr Mei's collection* (*Meishi congshu jiyao* 梅氏叢書輯要) 82, 306
  and *Sequel to the Doubts concerning the study of astronomy* (*Lixue yiwen bu* 曆學疑問補) 252–3
  and *Superfluous learning on calendrical astronomy* (*Lixue pianzhi* 曆學駢枝) 83
  and Thomas 216–17
  and Western learning 85
Mei Xuan 梅瑴 214
Meishi congshu jiyao 梅氏叢書輯要, see *Selected essentials of Mr Mei's collection*
Memorial for the ordering of musical tones and of mathematics (*Qing bianci yuelü suanshu shushu* 請編次樂律算數書疏) 233–4
Memorial on ritual and music at local schools (*Pangong liyue shu* 頖宮禮樂疏) (Li Zhizao) 31
Mengyangzhai 蒙養齋, see School for tutoring the young
Metonic cycle 60n22
Middle way to the Change of Zhou (*Zhouyi zhezhong* 周易折中) 269, 281, 324
Min Am 閔黯 228
Ming History (*Mingshi* 明史) 74, 121n2, 216, 232n101, 280
  and completion of 381
  and contributors to 84, 215, 234, 266
  and Mei Wending's work on 120, 214, 215, 221
  and winning over of Chinese elites by Manchu rulers 215
Minggantu 明安圖 264, 268, 307, 376n51, 377n56, 379
Mingju 明珠 80, 182, 257n99
Mingli tan 名理探, see *Investigation into the principles of names*
Mingshi 明史, see *Ming History*
Mingshi lizhi nigao 明史曆志擬稿, see *Draft for the Astronomical Records of the Ming History*
Mingtu 明圖 245, 265–6, 378–9
Ministry of Rites (*Libu* 禮部) 38, 65
  and French mission to China 109–10
  and Yang Guangxian's accusations against Jesuits 52
Mirror of ancient and modern revering of Heaven (*Gujin jingtian jian* 古今敬天鑒) (Bouvet) 286
missionaries:
  in China 13n2, 102
  and *Propaganda Fide* [Sacred Congregation for the] Propagation of the Faith 284
  see also Jesuits
Mucengge 穆成格 (d. 1689) 131
Mucengge 穆成格 (member of the 'staff serving in mathematics' in 1713) 268
Mushitai 穆世泰 268
music:
  and education of Kangxi's sons 276
  and Jesuits 241
  and Kangxi 229–31, 243
    approach of 277
  and Office of Mathematics 273–4
  see also harmonics

Nanjing 南京:
  as centre of Western learning 35
  as intellectual centre 41–2
  and renamed Jiangning 江寧 120
  and scholars of Western learning 42–4
Nanjing Observatory, Kangxi's visit to 120
  and account in *Sagely instructions* (*Shengxun* 聖訓) 126
  and commends Jesuits' astronomical system 123
  and emphasis of different accounts 131
  and *Imperial Diary* (*Qiju zhu* 起居注) record of 121–6
  and Jesuits' role 131–3
  and Li Guangdi's account of 127–31
  and limits of literati's authority on astronomy 125
  and questions Li Guangdi 122–4, 128–30
  and role as teacher 125
Nanjing persecution 32
Nantang 南堂, see Southern Church
nanti 難題, see difficult problems
Nanxun shi ba shou 南巡詩八首, see *Eight poems on the Southern Tour*
Needham, Joseph 3, 13
Neo-Confucianism 16
Nerchinsk, and Qing delegation to 141–2
networks, and Western learning 4
New explanation of the study of [standard] pitchpipes (*Lüxue xinshuo* 律學新說) (Zhu Zaiyu) 15, 230
New method of algebra (*Aerrebala xinfa* 阿爾熱巴拉新法) (Foucquet) 294–6, 298, 299, 300n76
  symbols in 300–5
New writing on the image of [the hexagram] alteration (*Gexiang xinshu* 革象新書) (Zhao Youqin) 218
New writing on the [standard] pitchpipes (*Lülü xinshu* 律呂新書) (Cai Yuanding) 230, 232, 368
Newton, Isaac 52n77, 289n22
  and tables 379
Ni Guanhu 倪觀湖 83
Nian Xiyao 年希堯 243, 322–3
Nicole, Pierre 164–5
nine chapters (*jiuzhang* 九章), and classification of mathematics 18, 19, 20, 45, 46, 47, 87, 90, 91, 92–3, 96, 258, 320, 325, 330, 366, see also nine reckonings
Nine chapters on mathematical procedures (*Jiuzhang suanshu* 九章算術) 14–15, 17, 21, 45, 85, 97, 191, 171–2, 320, 342n33, 345, 383
nine reckonings (*jiushu* 九數) 86, 87, 90–1, see also nine chapters
Nongzheng quanshu 農政全書, see *Complete treatise on agricultural administration*
Northern Church (Beitang 北堂) 156n87, 240, 241, 287
numbers 16–17
  and cosmology 16–17

Index

numbers (cont.)
  and definition 196
  and duality with magnitude 5, 26, 91, 328
  and Kangxi on 198–200
  and origins of 18, 323–4
  and prime numbers 196
  and relationship with principles 81, 88, 91, 198, 328, 367, 369
  and representation by counting rods 14
  and representation on abacus 14
  see also Essence of numbers and their principles (Shuli jingyun 數理精蘊)
Number and magnitude expanded (Shudu yan 數度衍) (Fang Zhongtong) 44–9, 84, 99, 344, 346, 347, 356
Numerical tables (Shubiao 數表) (Thomas) 192
Nurhaci 努爾哈赤 35

Oboi 鰲拜 1, 58, 64
Office for Ming History (Mingshi guan 明史館) 214
Office for the Complete View of the Imperial Territory (Huangyu quanlan guan 皇輿全覽館) 256
Office for the revision of the Books on Calculation of the Timely Modelling [System] (Zengxiu shixian suanshu guan 增修時憲算書館) 380
Office of Mathematics (Suanxue guan 算學館):
  and appropriation of Jesuit sciences 270
  and astronomical observations 279
  and autonomy of imperial astronomy 293–4
  and closure of 377–8
  and exclusion of Jesuit scholars 270
  and foundation of 269–73
  and function of 273
  and Jesuits 305–9
  and Kangxi sets work for 269–70
  and Kangxi's instructions for organisation of 271
  and method of working 365
  and music 273–4
  and recruitment of scholars 262–4
    characteristics of 264
  and staff of 265–7, 268, 388
    conquest elites 267
    hierarchy among 266
  and succession struggle 273–4
  and survey of the empire 277–8, 279–80
  and tables 309–11
  and Yinzhi:
    appointed head of 269–70, 274
    qualifications for post 274–5
    reports to Kangxi 276
  see also Essence of numbers and their principles (Shuli jingyun 數理精蘊); Origins of pitchpipes and the calendar (Lüli yuanyuan 製律曆淵源)
Old Man Star (Laorenxing 老人星):
  and de Fontaney's explanation 132–3
  and Kangxi's investigation of 133–5
  and Kangxi's visit to Nanjing Observatory 125, 129, 131–3
Original meaning of heavenly studies (Tianxue benyi 天學本義) (Bouvet) 286
Origins of pitchpipes and the calendar (Lüli yuanyuan 律曆淵源) 1, 5, 8, 139, 273, 279, 283, 315n2, 340
  and components of 364
  and contributors to 373–8
  and Correct interpretation of [standard] pitchpipes (Lülü zhengyi 律呂正義) 364, 366, 368–70
    Later part supplement 380–1
  and harmonics treatise 368–9
  and influence of 382–4
  and Later part of the thorough investigation of astronomical phenomena (Lixiang kaocheng houbian 曆象考成後編) 379–80
  and Office of Mathematics 273
  and order of composition 368
  and structure of 366, 368
  and supplements to 378–82
  and Thorough investigation of astronomical phenomena (Lixiang kaocheng 曆象考成) 364, 370–2
    criticism of 378–9
  and Thorough investigation of instruments and phenomena (Yixiang kaocheng 儀象考成) 380
  and Yongzheng's preface 364–8
    origin of knowledge in 366–7
  structure of work 366
Ouluoba xijing lu 歐羅巴西鏡錄, see Western mirror of Europe
Outline of the essentials of calculation (Suanfa zuanyao zonggang 算法纂要總綱) (Thomas) 184–91, 201, 202, 206, 209, 246, 322, 336, 337, 341, 344, 349, 352, 358, 359
Outline of Western learning (Xixue fan 西學凡) (Aleni) 29

painting, and Jesuits 58, 72, 153–4, 241–2
Pan Lei 潘耒 84, 215–16
Pan Ruying 潘汝瑛 377n56
Pan Yunhong 潘蘊洪 268
Pangong liyue shu 頖宮禮樂疏, see Memorial on ritual and music at local schools
papacy:
  and papal legation to China 254–5
  and Propaganda Fide [Sacred Congregation for the] Propagation of the Faith 284
Pardies, Ignace Gaston 103
  and Elements of Geometry 161, 162–6
    approach of 162
    contents 165
    dedication of 163
    practical geometry 173–4
    structure and style of 165–6
    use of models 171
  and mathematics and religion 164
Parrenin, Dominique 265, 307–8
Pedrini, Teodorico 270, 276, 284, 369–70
pendulum 246, 359–62
Pereira, Andre 378–9
Pereira, Tomás 72, 270
  and death of 265
  and Essentials of [standard] pitchpipes (Lülü zuanyao 律呂纂要) 72, 369
  and French mission to China 112–13, 115–16
  and instruments 155–6, 241
  and Qing delegation to Nerchinsk 141–2
  as tutor to Kangxi 139, 141, 142–3, 180
Pernon, Louis de 241
perspective 58, 241, 242, 243
Peter the Great, Tsar 1n2, 284
Pictures of tilling and weaving (Gengzhitu 耕織圖) (Lou Shu) 242–3, 244
pitchpipe [standard] (lülü 律呂) 230, 365, 366, 367, 368, 381
popular mathematics 17–18
Portugal, and declining power of 104
Pozzo, Andrea 322
practical geometry:
  and Kangxi's study of 143, 173
  in translations of Elements of geometry 173–6
practical learning (shixue 實學) 31, 215
  and renewed interest in 13–14
  and value of 15, 386
Practice of instruments for measuring heights and distances (Celiang gaoyuan yiqi yongfa 測量高遠儀器用法) (Thomas) 194, 245, 251–2, 258, 267
Practice of numerical tables (Shubiao yongfa 數表用法) (Thomas) 192–3
Precious mirror of mathematics (Suanxue baojian 算學寶鑑) 18n26
principle:
  and notion of (li 理) 16, 51, 81, 88, 124, 173, 177, 198, 199
  and study of (lixue 理學) 15, 80
  and Verbiest's Study of the fathoming of principles (Qiongli xue 窮理學) 78–80

432

Index

printing, and decline in costs of 15
Proclus 23
proportional compass 357
    and Thomas' *Explanation of the proportional compass* (*Biligui jie* 比例規解) 193–4, 362
proportions, **see** ratios and proportions

qi 器 ('concrete things') 29, 42
Qiankun tiyi 乾坤體義, **see** *Meaning of Heavenly and Earthly forms*
Qianlong 乾隆 Emperor 126, 264, 274
    and *Complete library of the four treasuries* (*Siku quanshu* 四庫全書) 82
    and editorial projects 378–82
Qiju zhu 起居注, **see** *Imperial Diary*
Qin Jiushao 秦九韶 355
Qing dynasty:
    and appropriation of Western learning 54, 270
    and calendar reform 36–7
    and consolidation of 385
    and formation of 35–6
    and historical understanding of 2–3
    and longevity of 37
Qing bianci yuelü suanshu shushu 請編次樂律算數書疏, **see** *Memorial for the ordering of musical tones and of mathematics*
Qintianjian 欽天監, **see** Astronomical Bureau
qiongli 窮理, **see** fathoming the principles
Qiongli xue 窮理學, **see** *Study of the fathoming of principles*
Qiu Qiyuan 丘起元 265
Qu Rukui 瞿汝夔 25

Ramus, Petrus 162–3
ratios and proportions:
    and Euclidean geometry 169–73
    and *lü* 率 (term of a proportion) 171–2
    Rebellion of the Three Feudatories 76, 141
rectangular arrays (*fangcheng* 方程) 27–8, 29
    and *Essence of numbers and their principles* (*Shuli jingyun* 數理精蘊) 352–4
    and inkstones and brushes problem 21, 47–8
    and Mei Wending's *Discussion of rectangular arrays* (*Fangcheng lun* 方程論) 83, 86
    ambition of work 90
    assessment of Western methods 92–3
    branches of mathematics 90–1, 92
    classification of problems 94, 95
    inkstones and brushes problem 97–101
    primacy of calculation 92, 96
    restoring one of the 'nine reckonings' 90–3
    status of mathematics 91–2
    structure of 93–6
    style of 101
Refutation of 'I cannot do otherwise' (*Budeyi bian* 不得已辨) (Buglio & Verbiest) 51n71, 68
Régis, Jean-Baptiste 240, 256, 285
religion:
    and mathematics 164
    and science 70–2, 105
remainder problem 354–6
Renaldini, Carlo 206
Rho, Giacomo 357
    and astronomical reform 33
    and *Complete meaning of measurement* (*Celiang quanyi* 測量全義) 34, 278
    and *Explanation of the proportional compass* (*Biligui jie* 比例規解) 193
Rhodes, Bernard 240, 285
Ricci, Matteo 24, 164, 388
    and Chinese version of world map 25
    and *Elements of geometry* (*Jihe yuanben* 幾何原本) 25, 26–7, 160
    lü 率 (term of a proportion) 172
    preface to 77–8
    and *Illustrated explanation of the sphere and the astrolabe* (*Hungai tongxian tushuo* 渾蓋通憲圖說) 25
    and *Instructions for calculation in common script* (*Tongwen suanzhi* 同文算指) 26, 27–9, 31, 45, 47, 48, 89, 92, 93, 96, 99, 184, 185, 187, 189, 190, 191, 321, 326, 344, 346, 347, 352
    and Jesuit strategy in China 25
    and *Meaning of compared figures* (*Yuanrong jiaoyi* 圓容較義) 26
    and *Meaning of Heavenly and Earthly forms* (*Qiankun tiyi* 乾坤體義) 25–6, 29
    and *Meaning of measurement methods* (*Celiang fayi* 測量法義) 26
    and science and religion 70
    and structure of mathematics 26–7
    and teaching 25
Ripa, Matteo 284, 297
Rites Controversy 254
    and papal legation to China 254–5
Roemer's machines 105, 108, 154, 155, 305
Roman College 22–3
    and Jesuit missionaries to China 26, 29, 33
Rong 榮, Prince 49, 51, 53
Royal Academy of Sciences, Paris, **see** Académie royale des sciences
Ruggieri, Michele 25
Russia, and Orthodox mission to China 284

Sagely instructions (*Shengxun* 聖訓) (Kangxi) 126, 229, 271, 282
Sahai 薩海 377
Sambiasi, Francesco 37
San yuan 三垣, **see** Three Enclosures
Sanjiaofa juyao 三角法舉要, **see** *Essentials of Trigonometry*
Sanjiaoxing tuisuanfa lun 三角形推算法論, **see** *Discussion of triangles and computation*
Schall von Bell, Johann Adam:
    and arrested 52
    and astrology 40
    and Astronomical Bureau:
        conflict with Chinese subordinates 40
        as imperial astronomer 38–41
        put in charge of 37
    and calendar reform 33, 36–7
    and criticism of 38–40
        Yang Guangxian 50
    and death of 57
    and death sentence 52–3
    and death sentence rescinded 53
    and dynastic transition 36–7
    and *Explanation of the proportional compass* (*Biligui jie* 比例規解) 193
    and funeral of 64
    and house arrest 54, 57
School for tutoring the young (Mengyangzhai 蒙養齋) 266, 271, 272, 365
Schreck, Johann (Terrenz) 33, 348
science:
    and circulation of 3, 4
    and religion 70–2, 105
se 色 (kinds) 47, 95, 354
Season granting (*Shoushi* 授時) system 214–15, 250n60, 372
Selected essentials of Mr Mei's collection (*Meishi congshu jiyao* 梅氏叢書輯要) 82, 96n57, 306
Sequel to the Doubts concerning the study of astronomy (*Lixue yiwen bu* 曆學疑問補) (Mei Wending) 252–3
Shang Gao 商高 326
Shangshu 尚書, **see** *Book of Documents*
Shao Yong 邵雍 44, 258
    and numbers 16–17
Shen Chenglie 沈盛烈 268
Shen Gua 沈括 17
Shengxun 聖訓, **see** *Sagely instructions*
Shi Runzhang 施閏章 83, 214
Shilu 實錄, **see** *Veritable records*
shixue 實學, **see** practical learning
Shixue 視學, **see** *Study of Vision*
Shoushi 授時, **see** Season granting
Shubiao 數表, **see** *Numerical tables*

433

Index

Shubiao jingxiang 數表精詳, **see** *Essential details of numerical tables*
Shubiao wenda 數表問答, **see** *Dialogue on numerical tables*
Shubiao yongfa 數表用法, **see** *Practice of numerical tables*
Shudu yan 數度衍, **see** *Number and magnitude expanded*
Shuli jingyun 數理精蘊, **see** *Essence of numbers and their principles*
Shunzhi 順治 Emperor 35
   and assumption of personal rule 41
   and death of 50
   and education 73
   and *Expanded meaning of the Classic of filial piety* (*Xiaojing yanyi* 孝經衍義) 74n96
Sige 四格 373n*, 377n56
Significance of an intricate method (*Cuozong fayi* 錯綜法義) (Chen Houyao) 322
Siku quanshu 四庫全書, **see** *Complete library of the four treasuries*
Singde 性德 257n99
Sishu 四書, **see** *Four Books*
Slaviček, Karel 307
smallpox 226
Smogulecki, Nikolaus 43, 45, 84
   and *True source of the pacing of heavens* (*Tianbu zhenyuan* 天步真原) 43
Societé des missions étrangères de Paris 102
solar eclipses 222–9
   and divinatory significance 223–4, 228
      Kangxi's contempt for 225–6
      Kangxi's role in Chinese cosmology 227–9
   and prediction of 222–3
Song Yingxing 宋應星 15
Songgotu 索額圖 142, 253
Soni 索尼 58
Southern Church (Nantang 南堂) 240, 266n30
statecraft, and Western learning as tool of 4, 6, 15, 45, 311, 386
Study of the fathoming of principles (*Qiongli xue* 窮理學) (Verbiest) 78–81, 145, 159
Study of Vision (*Shixue* 視學) (Nian Xiyao) 243, 322
Stumpf, Kilian 240, 265–6, 287, 293, 299, 326
suan 算 4–5, 8, 18, 91, 92
suanfa 算法 27, 196, 329
Suanfa tongzong 算法統宗, **see** *Unified lineage of mathematical methods*
Suanfa yuanben 算法原本, **see** *Elements of calculation*
Suanfa zuanyao zonggang 算法纂要總綱, **see** *Outline of the essentials of calculation*
Suanjing shishu 算經十書, **see** *Ten mathematical classics*

Suanxue baojian 算學寶鑑, **see** *Precious mirror of mathematics*
Suanxue guan 算學館, **see** *Office of Mathematics*
Suanyi tan'ao 算義探奧, **see** *Exploring the mystery of mathematics*
suicha 歲差, **see** annual difference 123, 370
Sukjong 肅宗 (Yi Sun 李焞), King of Korea 228n77, 277
Suksaha 蘇克薩哈 57, 58
Summary of calculation by borrowed root and powers (*Jiegenfang suanfa jieyao* 借根方算法節要) 209–10, 337n86
Sunzi suanjing 孫子算經, **see** *Mathematical classic of Master Sun*
Sunzi suanshu 孫子算術, **see** *Mathematical procedures of Master Sun*
suodangran 所當然 (what should be so) 16, 328
suoyiran 所以然 (why it is so) 16, 310, 328, 370
Superfluous learning on calendrical astronomy (*Lixue pianzhi* 曆學駢枝) (Mei Wending) 83
surveying 244–5
   and Kangxi on techniques 245
   and meridian expedition 245
   and standardisation of units 245–6
   and survey of the empire 255–7
Synopsis mathematica (Thomas) 181, 183, 186n5, 187, 188, 190, 202
   and contents 181–2
   and purpose of 182
   and structure of 184–5
Systematic materia medica (*Bencao gangmu* 本草綱目) (Li Shizhen) 15, 330n60

tables:
   and standardisation of mathematical sciences 309–11
   and Foucquet's *Dialogue on numerical tables* (*Shubiao wenda* 數表問答) 297–8, 302
   and La Hire's 309
   and Thomas' *Numerical tables* (*Shubiao* 數表) 192
   and Thomas' *Practice of numerical tables* (*Shubiao yongfa* 數表用法) 192–3
   and Thomas' *Tables of distance between the horizon and the terrestrial sphere* (*Dipingxian li diqiumian biao* 地平線離地球面表) 193–4
   and Vlacq's 192, 296, 309
Tables of distance between the horizon and the terrestrial sphere (*Dipingxian li diqiumian biao* 地平線離地球面表) (Thomas) 193–4

Tachard, Guy 107, 109, 117
Tartre, Pierre Vincent de 285
technical learning, and revival of interest in 15
Ten mathematical classics (*Suanjing shishu* 算經十書) 5, 14, 22, 27, 46, 320
   and *Gnomon of Zhou* 45, 46–7, 317, 320, 325, 366
   and *Mathematical classic of Master Sun* (*Sunzi suanjing* 孫子算經) 189, 320, 327, 335, 344, 354, 355, 356
   as syllabus for College of Mathematics (*Suanxue* 算學) 15
theriac 150, 156
Thillisch, Franz 287, 299, 306
Thomas, Antoine 72, 116, 118, 132, 240, 387
   and arrives in Beijing 183
   and attempts to enter Japan 182
   and *Basics of the eight [trigonometric] lines* (*Baxian biao gen* 八線表根) 191
   and *Calculation by borrowed root and powers* (*Jiegenfang suanfa* 借根方算法) 184, 217, 300, 302, 304, 336, 337, 352
      abridgment of 209–10
      ambiguity of Chinese terms 206
      composition of 200–1
      contents 204–5
      foreword to 202–6
      problem solving 208–9
      structure of 201–2
      symbolic linear algebra 210–12
      symbols 207
      table of power numbers 206–7
   and career of 181
   and death of 265
   and *Elements of calculation* (*Suanfa yuanben* 算法原本), possible author of 198
   and *Explanation of the proportional compass* (*Biligui jie* 比例規解) 193–4, 362
   and in Coimbra 181
   and Kangxi's recruitment of 182–3
   and Mei Wending 216–17
   and *Numerical tables* (*Shubiao* 數表) 192
   and *Outline of the essentials of calculation* (*Suanfa zuanyao zonggang* 算法纂要總綱) 184–91, 201, 202, 206, 209, 246, 322, 336, 337, 341, 344, 349, 352, 358, 359
      contents 186
      evolution of text 188
      extant copies of 187
      Kangxi's role in composition 188–9
      problem of chickens and rabbits 189
      problem of Hieron's crown 190–1
      scope of 187
      structure of 185–7
      versions of text 187–8
   and patronage 182, 183

Thomas, Antoine (cont.)
  and *Practice of instruments for measuring heights and distances* (*Celiang gaoyuan yiqi yongfa* 測量高遠儀器用法) 194, 245, 251–2, 258, 267
  and *Practice of numerical tables* (*Shubiao yongfa* 數表用法) 192–3
  and *Synopsis mathematica* 181, 183, 186n5, 188, 190, 201
    contents 181–2
    purpose of 182
    structure of 184–5
  and *Tables of distance between the horizon and the terrestrial sphere* (*Dipingxian li diqumian biao* 地平線離地球面表) 193–4
  and translations of Euclid 168
  as tutor to Kangxi 139, 142, 180
  subjects covered 184
Thorough investigation of astronomical phenomena (*Lixiang kaocheng* 曆象考成) 364, 370–2, 378–9
Thorough investigation of instruments and phenomena (*Yixiang kaocheng* 儀象考成) 380
Three Enclosures (*San yuan* 三垣) 125
Thuret, Isaac II 155
Tianbu zhenyuan 天步真原, see *True source of the pacing of heavens*
Tiangong kaiwu 天工開物, see *Exploitation of the works of nature*
Tianxue benyi 天學本義, see *Original meaning of heavenly studies*
Tianxue chuhan 天學初函, see *First collection of heavenly learning*
tianyuan 天元, see celestial element
Tibetan Buddhism 41, 223
time-keeping 358–9
Tingxun geyan 庭訓格言, see *Maxims of fatherly advice*
Toleration, Edict of (1692) 142, 239, 253
Tong Guogang 佟國綱 142, 218, 224
Tongwen suanzhi 同文算指, see *Instructions for calculation in common script*
Tordesillas, Treaty of (1494) 24
Tournon, Charles-Thomas Maillard de 254–5, 286
trigonometry 43, 181, 191, 245, 247, 250, 251, 252, 358
True source of the pacing of heavens (*Tianbu zhenyuan* 天步真原) (Smogulecki & Xue Fengzuo) 43
tuili 推理, see derivation of principles
Tuna 圖納 121, 125
Tychonic system 24, 33
  and Foucquet's criticism of 288–9, 293–4

Unified lineage of mathematical methods (*Suanfa tongzong* 算法統宗) (Cheng Dawei) 15, 17–20, 27, 95, 96, 191, 258, 264, 295, 315n3, 321, 348, 390
  inkstones and brushes problem 20–2, 48, 97, 346, 347
  miscellaneous methods (*zafa* 雜法) 22, 96
Ursis, Sabatino de 29
Verbiest, Ferdinand 387
  and arrested 52
  and *Astronomia Europæa* 64–5
  and Astronomical Bureau:
    appointed Administrator of the Calendar (*zhili lifa* 治理曆法) 64
    increase in student numbers 66–7
    responsibilities in 65–6
    teaching duties 66
    updating instruments 67–8
  and calls for more missionaries 104
  and casting of cannon 68
  and *Compendium on the newly constructed instruments of the Observatory* (*Xinzhi lingtai yixiang zhi* 新製靈臺儀象志) 67–8, 77, 359n54
  and death of 112
  and dispute over Wu Mingxuan's calendar 60, 61, 62–3, 64
  and *Explanation of the examination of qi* (*Yanqi shuo* 驗氣說) 77
  and French mission to China 110
  and house arrest 54, 57–8
  and network of connections with Chinese officials 68–70
  and *Refutation of 'I cannot do otherwise'* (*Budeyi bian* 不得已辨) 51n71, 68
  and rehabilitation of Jesuit astronomy 64–5
  and science and religion 71–2
  and *Study of the fathoming of principles* (*Qiongli xue* 窮理學) 78–9, 145, 159
    Kangxi's rejection of 80–1
    motives for writing 79–80
  as tutor to Kangxi 74, 75–7
  religious aims in 78
Veritable records (*Shilu* 實錄) 60, 74, 124, 126, 149, 229, 251n65, 273, 282
  and 1664 comet 52
  and death sentence on Schall 52–3
  and Kangxi's reign 121
Viète, François 201
Visdelou, Claude de 107, 192
Vlacq, Adriaan:
  and tables 192, 296, 309

waifa 外法 (external method) 45, 47
Wang Chong 王翀 268

Wang Hongxu 王鴻緒 266
Wang Lansheng 王蘭生 247, 270, 276, 373, 376, 377
  and imperial editorial projects 269
  and Office of Mathematics 271
Wang Shizhen 王士禎 86, 87
Wang Xichan 王錫闡 42, 86
Wang Yangming 王陽明 386
Wang Yuanzheng 王元正 268
Wang Zheng 王徵 348
Wanshou shengdian chuji 萬壽盛典初集, see *First collection of the imperial birthday ceremony*
water clocks 358–9
Wei Tingzhen 魏廷珍 247, 269, 271, 272, 373
Western learning (*xixue* 西學) 3, 30
  and actor networks 4
  and appropriation of 54, 270, 391
  and Chinese origin of 87, 250, 252–3, 259
  and dominant narrative of 389–90
  and double rhetoric about 222
  and imperial monopoly of 7, 305, 311, 389
  and integration of 390
  and Jesuits' entry into China 13
  and Nanjing scholars 42–4
  as tool of statecraft 4, 6, 15, 45, 311, 386
Western mirror of Europe (*Ouluoba xijing lu* 歐羅巴西鏡錄) 190
Wu Cheng 吳澄 (Caolu 草廬) 69n77
Wu Han 吳涵 255n85
Wu Mingxuan 吳明炫 40, 41, 50, 51
  and appointed Vice-Director of Astronomical Bureau 54
  and faults with Great concordance (*Datong* 大統) system 59
  and first calendar submitted by 59–60
  dispute over 59–64
Wu Xiaodeng 吳孝登 269, 376
Wu Yongxi 吳用錫 247
Wujing 五經, see *Five Classics*
wuxing 五行, see Five Phases
Wuyingdian 武英殿, see Hall of Military Glory

xiangshu 象數, see figures and numbers
Xiao Siwen 蕭思溫 134
Xiaojing yanyi 孝經衍義, see *Expanded meaning of the Classic of filial piety*
Xiaozhuang 孝莊, Grand Empress Dowager 53, 58, 73, 112
Xichao ding'an 熙朝定案 110n39–41 132n58
Xinfa lishu 新法曆書, see *Books on calendrical astronomy according to the new method*
Xingli kaoyuan 星曆考原, see *Investigation of the origins of planetary ephemerides*

Index

Xinzhi lingtai yixiang zhi 新制靈臺儀象志, **see** *Compendium on the newly constructed instruments of the Observatory*
Xiong Cilü 熊賜履 127, 128, 230, 231–2, 255
Xiong Mingyu 熊明遇 42
xiu 宿, **see** lodges
Xixue fan 西學凡, **see** *Outline of Western learning*
Xiyang xinfa lishu 西洋新法曆書, **see** *Books on calendrical astronomy according to the new Western method*
Xu Guangqi 徐光啟 25, 29, 386
    and army reform 32
    and astronomical reform 32–3
    and base and altitude (*gougu* 句股) 47
    and *Complete treatise on agricultural administration* (*Nongzheng quanshu* 農政全書) 31
    and death of 34
    and *Elements of geometry* (*Jihe yuanben* 幾何原本) (Ricci and Xu Guangqi) 25, 26–7
        lü 率 (term of a proportion) 172
    and *Meaning of measurement methods* (*Celiang fayi* 測量法義) 26
    and wide interests of 31
Xu Juemin 徐覺民 268
Xu Qianxue 徐乾學 127, 128
Xu Zhijian 許之漸 51
Xuanze yi 選擇議, **see** *Deliberations on hemerology*
Xue Fengzuo 薛鳳祚 42, 43, 84–5
    and *Integration of learning in calendrical astronomy* (*Lixue huitong* 曆學會通) 43, 84
    and reputation of 86
    and *True source of the pacing of heavens* (*Tianbu zhenyuan* 天步真原) 43

Yan Yuan 顏元 386
Yang Guangxian 楊光先 49, 263
    and appointed Director of Astronomical Bureau 54
    and attacks Jesuits 49–50, 51–2
    and *Deliberations on hemerology* (*Xuanze yi* 選擇議) 49–50
    and dispute over Wu Mingxuan's calendar 64
    and *I cannot do otherwise* (*Budeyi* 不得已) 51–2
    and sentenced to death and pardoned 64
Yang Hui 楊輝 18
Yangxindian 養心殿, **see** *Hall of the Nourishment of the Mind*

Yanqi shuo 驗氣說, **see** *Explanation of the examination of qi*
Yaodian 堯典, **see** *Canon of Yao*
Ye Changyang 葉長揚 268
Yijing 易經, **see** *Book of change*
ying buzu 盈不足, **see** excess and deficit
yingnü 盈朒, **see** excess and deficit
Yinlu 胤祿 (Yunlu 允祿) 273, 274, 275, 365, 373, 379, 380
Yinreng 胤礽 217–18, 239, 253, 274, 286
Yinsi 胤禩 274
Yintao 胤祹 276
Yinti 胤禔 121, 217, 243, 274
Yinwu 胤禑 274, 276
Yinyun chanwei 音韻闡微, **see** *Clarification of the subtleties of phonetics*
Yinzhi 胤祉 (Yunzhi 允祉) 244–5, 273–6, 365, 373
    and algebra 299
    and astronomical reform 293
    and astronomical tables 309–10
    and music 369
    and Office of Mathematics:
        appointed head of 269–70
        foundation of 271, 273
        qualifications for leading 274–5
        reports to Kangxi 276
    and requests Kangxi's arbitration over Foucquet's innovations 290–1
    and Yongzheng Emperor 273–4
Yitala 宜塔喇 63n37, 67n61
Yixiang kaocheng 儀象考成, **see** *Thorough investigation of instruments and phenomena*
Yixing 一行 44, 250
Yongzheng 雍正 Emperor 126, 264, 274
    and astronomy 275
    and brothers 273–4
    and lack of interest in mathematics 378
    and *Maxims of fatherly advice* (*Tingxun geyan* 庭訓格言) 276
    and preface to *Origins of pitchpipes and the calendar* 364–8
Yu Chenglong 于成龍 232
Yu 禹 the Great 323n27, 324
Yuanrong jiaoyi 圜容較義, **see** *Meaning of compared figures*
Yuding yixiang kaocheng 御定儀象考成, **see** *Thorough investigation of instruments and phenomena*
Yulun 余掄 268
Yunli 允禮 382
Yunlu 允祿, **see** Yinlu
Yunzhi 允祉, **see** Yinzhi

Yuzhi lixiang kaocheng 御製曆象考成, **see** *Thorough investigation of astronomical phenomena*
Yuzhi lüli yuanyuan 御製律曆淵源, **see** *Origins of pitchpipes and the calendar*
Yuzhi lülü zhengyi 御製律呂正義, **see** *Correct interpretation of [standard] pitchpipes*
Yuzhi sanjiaoxing tuisuanfa 御製三角形推算法論, **see** *Discussion of triangles and computation*
Yuzhi shubiao jingxiang 御製數表精詳, **see** *Essential details of numerical tables*
Yuzhi shuli jingyun 御製數理精蘊, **see** *Essence of numbers and their principles*

Zarlino, Gioseffo 369
Zeng Guofan 曾國藩 384
Zeng Guoquan 曾國荃 384
Zhang Heng 張衡 44, 69n75, 325
Zhang Shoujie 張守節 134, 135
Zhang Tingyu 張廷玉 381
Zhang Wei 張位 265
Zhang Xianzhong 張獻忠 37
Zhang Ying 張英 121, 131, 232, 246, 381
    and *Eight poems on the Southern Tour* (*Nanxun shi ba shou* 南巡詩八首) 131n48
Zhang Yushu 張玉書 121, 125, 130, 229, 231, 232, 245, 250
    and 'Memorial for the ordering of musical tones and of mathematics' (*Qing bianci yuelü suanshu shushu* 請編次樂律算數書疏) 233–4
Zhang Zhao 張照 379, 380, 381
Zhao Youqin 趙友欽 218
Zhaochang 趙昌 132, 133, 153, 158, 160, 194
Zhaohai 照海 268, 271, 376n51
Zhifang waiji 職方外紀, **see** *Areas outside the concern of the Imperial Geographer*
Zhongxi suanxue tong 中西算學通, **see** *Integration of Chinese and Western mathematics*
Zhou 周, Duke of 326
Zhoubi 周髀, **see** *Gnomon of Zhou*
Zhouyi zhezhong 周易折中, **see** *Middle way to the Change of Zhou*
Zhu Song 朱崧 268
Zhu Xi 朱熹 15–16, 288
Zhu Yizun 朱彝尊 216
Zhu Zaiyu 朱載堉 15, 230, 231, 232
Zhuzi quanshu 朱子全書, **see** *Complete works of Master Zhu*
Zhuzi yulei 朱子語類, **see** *Classified conversations of Master Zhu*
Zu Chongzhi 祖沖之 259, 325, 326, 327

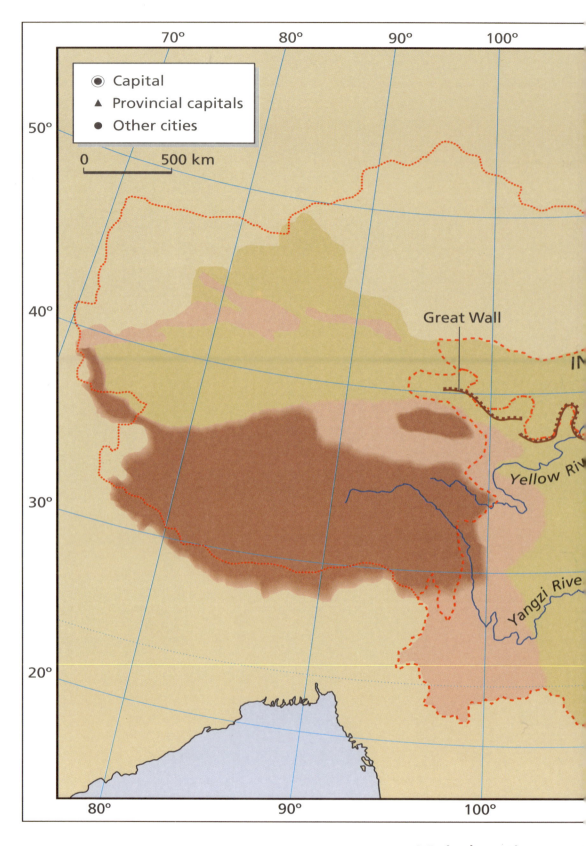